THE AGE OF DINOSAURS IN SOUTH AMERICA

Life of the Past

James O. Farlow, editor

THE AGE OF DINOSAURS IN SOUTH AMERICA

Fernando E. Novas

Indiana University Press
Bloomington & Indianapolis

Gray-scale illustrations facing front and back matter openers are details from color plates 14, 15, 16 © by Maurilio Olveira and color plates 17, 18 © by Gabriel Lio.

This book is a publication of

Indiana University Press
601 North Morton Street
Bloomington, IN 47404-3797 USA

www.iupress.indiana.edu

Telephone orders 800-842-6796
Fax orders 812-855-7931
Orders by e-mail iuporder@indiana.edu

© 2009 by Fernando Novas
All rights reserved

No part of this book may be reproduced or utilized in any form or by any means, electronic or mechanical, including photocopying and recording, or by any information storage and retrieval system, without permission in writing from the publisher. The Association of American University Presses' Resolution on Permissions constitutes the only exception to this prohibition.

⊚The paper used in this publication meets the minimum requirements of the American National Standard for Information Sciences—Permanence of Paper for Printed Library Materials, ANSI Z39.48-1992.

Manufactured in the United States of America
Library of Congress Cataloging-in-Publication Data

Novas, Fernando E.
 The age of dinosaurs in South America / Fernando E. Novas.
   p. cm. — (Life of the past)
 Includes bibliographical references and index.
 ISBN 978-0-253-35289-7 (cloth : alk. paper)
 1. Dinosaurs—South America. I. Title.
 QE861.4.N69 2009
 567.9098—dc22
                                         2008042734

1 2 3 4 5 14 13 12 11 10 09

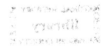

This book is dedicated to Argentine paleontologists

Osvaldo A. Reig (1929–1992)

José Bonaparte (1928–)

Rodolfo Casamiquela (1932–2008)

who envisioned the importance of exploring and studying the Mesozoic in South America

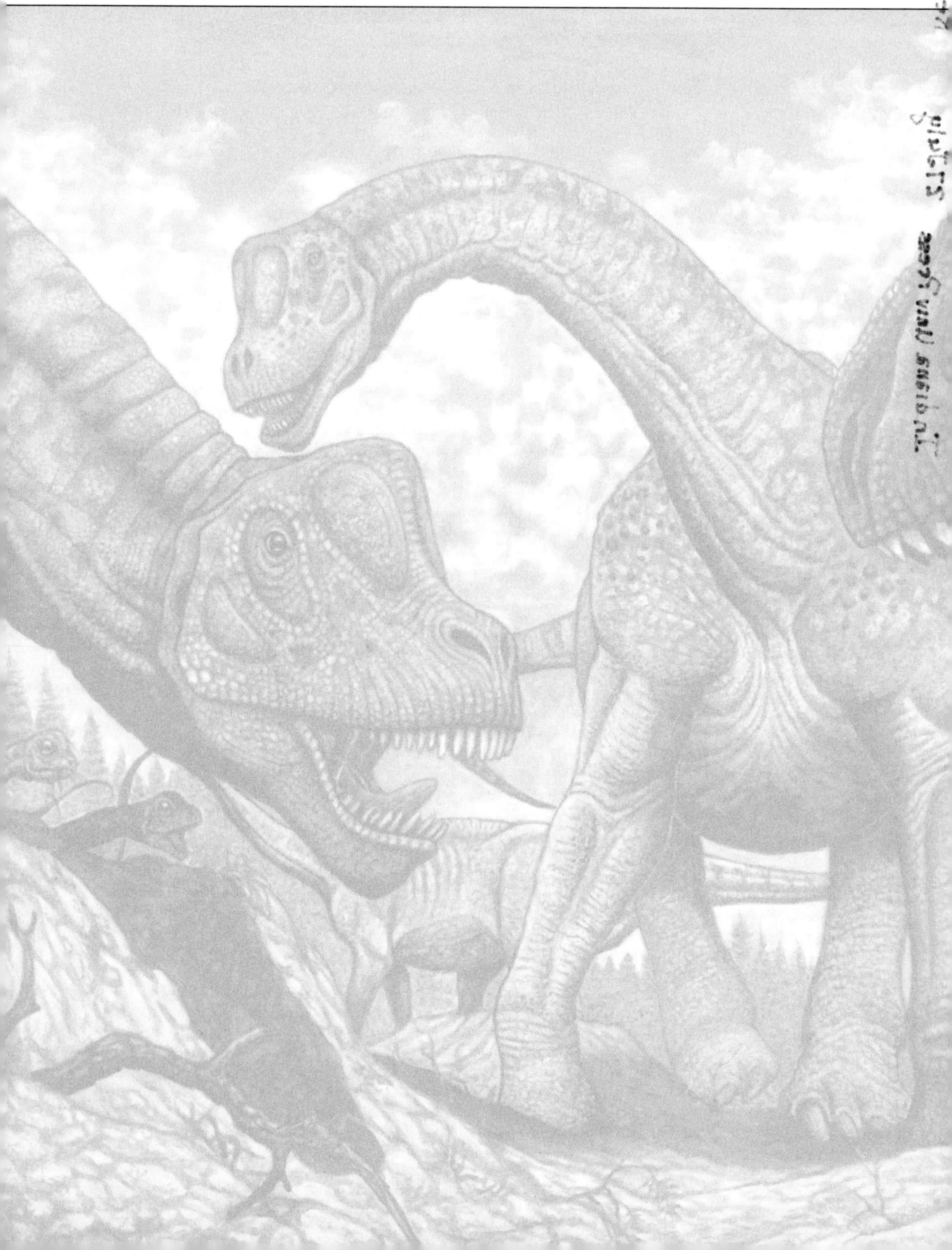

CONTENTS

ACKNOWLEDGMENTS *XI*

LIST OF ANATOMICAL ABBREVIATIONS USED IN THE FIGURES *XIII*

INTRODUCTION *XIX*

1 **An Overview of the Anatomy and Phylogenetic Relationships of Dinosaurs** *3*

What Are Dinosaurs? *3*

Dinosaur Skeletal Anatomy *4*

A Brief Historical Account of the Systematics of Dinosauria *11*

Dinosaur Forerunners *13*

Characteristics That Distinguish Dinosauria from Other Archosaurs *17*

2 **Triassic Dinosaurs** *25*

Triassic Dinosaur-Bearing Beds in South America *26*

The Triassic Dinosaur Record in South America *35*

Dinosauria *41*

Saurischia *41*

Theropoda *54*

Sauropodomorpha *57*

Ornithischia *71*

Ichnological Evidence *76*

Triassic Dinosaurs: An Overview *80*

3 **Jurassic Dinosaurs** *87*

Jurassic Dinosaur-Bearing Beds in South America *88*

The Jurassic Dinosaur Record in South America *100*

Sauropoda *100*

Neosauropods *108*

Theropoda *114*

Ornithischia *123*

Jurassic Dinosaurs: An Overview *128*

4 **The Fossil Record of Cretaceous Dinosaurs in South America** *135*

The Cretaceous Dinosaur Record in South America *138*

5 **Cretaceous Sauropods** *167*

Diplodocimorpha *169*

Titanosauria *181*

Cretaceous Sauropod Eggs and Nests *236*

Cretaceous Sauropod Footprints *238*

6 **Cretaceous Theropods** *243*

Ceratosauria *246*

Abelisauroidea *249*

Abelisauridae *272*

Tetanurae *286*

Cretaceous Theropod Footprints and Eggs *338*

7 Cretaceous Ornithischians *343*

Stegosauria *345*

Ankylosauria *346*

Ornithopoda *352*

Ceraptopsia *379*

8 A Summary of the Cretaceous Dinosaurs *385*

Dinosaur Diversification in the Cretaceous of South America *386*

WORKS CITED *405*

INDEX *439*

ACKNOWLEDGMENTS

I would like to express my gratitude to Pat Rich, who suggested that I submit a book proposal on South American dinosaurs to Indiana University Press, and to Jim Farlow and Bob Sloan for their kind encouragement. Many thanks to all of them for their trust in the results of this project. Jim made clever comments and suggestions about how to improve my original manuscript, but I am particularly grateful to him for his encouraging words and support.

Several colleagues and friends from Argentina, Brazil, Bolivia, Canada, Chile, England, Italy, Uruguay, the United States, and Venezuela contributed with information and corrections to the present volume. Thanks are due to José Bonaparte, Jaime Powell, Jorge Calvo, Leonardo Salgado, Diego Pol, Teresa Manera de Bianco, Rubén Martínez, Ricardo Martínez, Olga Giménez, Bernardo González Riga, Luis Chiappe, Andrea Cambiaso, Silvina De Valais, Andrea Arcucci, Sergio Martín, Alejandro Kramarz, Eduardo Bellosi, Rubén Juárez Valieri, Juan Porfiri, Sebastián Apesteguía, Ismar de Souza Carvalho, César Schultz, Manuel Medeiros, Alexander Kellner, Max Langer Cardoso, Luis Borges Ribeiro, Roberto Candeiro, Marcelo Sánchez-Villagra, Matías Soto, Mario Suárez Riglos, Saswati Bandyopadhyay, Fabbio Dalla Vechia, Paul Barrett, Phil Currie, Cathy Forster, Mike Brett-Surman, Martin Lockley, Dave Krause, and my brilliant students Federico Agnolín, Juan Canale, Martín Ezcurra, Augusto Haro, Ariel Méndez, Diego Pais, and Agustín Scanferla for sharing their knowledge with me.

All my gratitude to Fernando Spinelli, Stella Alvarez, Gastón Lo Coco, Federico Agnolín, Martín Ezcurra, Diego Pais, Agustín Martinelli, Rodrigo Vega, and Marina Caporale for their valuable help in preparing the illustrations and text. Kate Babbitt's detailed edit improved the manuscript. The book has benefited from the artwork done by Jorge González, Carlos Papolio, Maurilio Oliveira, Orlando Grillo, and Gabriel Lio. Special thanks to the skillful technicians Marcelo Isasi, Pablo Puerta, and Santiago Reuil for their discoveries in the field and for the preparation of fossil specimens. I would like to thank the Argentine institutions Consejo Nacional de Investigaciones Científicas y Técnicas (CONICET) and Agencia Nacional de Promoción Científica y Técnica (ANPCyT) for the support that enabled me to conduct research in my country, and the National Geographic Society (Washington) and the Jurassic Foundation (Calgary) for support for field work.

Finally, my gratitude to my beloved family—my wife, Liliana, and my sons, Mariano and Francisco—for their help and patience.

ANATOMICAL ABBREVIATIONS USED IN FIGURES

ac	acetabulum	bt	biceps tubercle of the coracoid
af	antorbital fenestra	c	cannaliculii
afa	articular facet of the fibula	ca	capitulum
alr	alveolar ridge	cal	calcaneum
amb	scar for M. ambiens	cap	caudal ascending process of the astragalus
an	angular		
ap	acromial process	car	carpal bones
apa	ascending process of the astragalus	cas	cranial articular surface
apl	articular process of the lacrimal	cb	cancellous bone
apm	ascending process of the maxilla	cd1	caudal 1
apsq	articular process of the squamosal	cds1	caudosacral 1
ar	articular	cdpl	centrodiapophyseal lamina
arc	articular cavity of the quadrate	cf	coracoid foramen
arsc	articular surface of the scapula	cmtc	carpometacarpus
as	astragalus	cn	cnemial crest
asa	articular surface of the astragalus	cor	coracoid
asi	articular surface of the ilium	cp	coronoid process
at	anterior trochanter	cpi	caudal process of the ilium
ati	antitrochanter	clp	collateral ligament pit
atl	atlas	cr	coronoid
ax	axis	cs	conical scute
b	bulk	ct	calcaneal tuberosity
be	blind excavation	ctf	crista tibiofibularis
bf	brevis fossa	cuf	cuppedicus fossa
bk	beak	cv1–9	cervical vertebrae 1–9
blp	bilobate process	cvr	cervical rib
bpt	basipterygoid tuberosity	d	dentary
bptr	basipterygoid recess	d1–d15	dorsal vertebra 1–15
brt	brachial tubercle	db	distal bevel on the pubis
bs	brevis shelf	dca	distal carpals
bsp	basisphenoid	dcr	diagonal crest on the dorsal surface of the astragalus
bspw	basisphenoidal wing		
bst	basisphenoid tubera	dgv	distal groove

dif	distal ischiac foot	hh	humeral head
dp	diapophysis	hs	core for the horny spike
dpc	deltopectoral crest	hyf	hyposphene
dpf	distal pubic foot	hyp	hypantrum
dpr	depression	hyt	hypotarsus
dr	dorsal rib	idp	interdental plate
ds	dentary simphysis	il	ilium
dsa	dorsosacral vertebra	ilp	iliac pedicle
dt 3–4	distal tarsals 3–4	im	internal malleolus
ect	ectopterygoid	indl	infradiapophyseal lamina
ed	elliptical depressions	is	ischium
elp	extensor ligament pit	isl	interspinal ligament scar
ep	epipophysis	isp	ischiadic pedicle
epr	epipophyseal-prezygapophyseal ridge	it	internal tuberosity
fa	"foot" for anchoring the neural spine	j	jugal
4t	fourth trochanter	k	frontal knob
f	fossa	knp	knob-like nasal projection
fap	facet for the ascending process of the astragalus	lac	lacrimal
		lb	lateral bulge of proximal femur
fc	fibular condyle	lct	lateral condyle of proximal tibia
fcr	fibular crest	lgp	ligament pit
fe	femur	libl	left iliac blade
fh	femoral head	ll	lateral lamina of the neural spine
fib	fibula	ln	left nasal
flp	flexor ligament pit	lp	lateral pit
flt	flexor tubercle	lpr	lateral primary ridge of the tooth
fm	foramen magnum	lrp	lateral ridge of the pubis
fo	foot	ltf	lateral temporal fenestra
fr	frontal	maw	medial acetabular wall
fs	surface for fibular articulation	mbf	tuberosity for insertion of M. biceps femoris
fse	fan-shaped expansion		
fu	ventral furrow	mcfb	scar of insertion of M. caudifemoralis brevis
fur	furcula (fused clavicles)		
g	collateral groove	mdc	mediodistal crest of the femur
ga	gastralia	mf	mandibular fenestra
gl	glenoid cavity	mo	mosaics
gm	guilliotine-like (cutting) margin	mps	site of origin of M. pseudotemporalis
gt	greater trochanter	ms	muscle scar on the distal femur
h	humerus	mtcI–V	metacarpals I–V
ha	haemal arch	mttI–V	metatarsals I–V

mvr	median vertical ridge on the distal tibia	pnc	pneumatic cavity
mx	maxilla	pnf	pneumatic foramen
mxf	maxillary fenestra	po	postorbital
n	notch on the distal tibia	poaw	postacetabular wing
na	nasal	poc	paroccipital process
nc	neural canal	posf	postspinal fossa
no	nasal openning	posl	postspinal lamina of the neural spine
ns	neural spine	poz	postzygapophysis
o	orbit	pp	parapophysis
o1–o4	dermal ossification type 1–4	pphp	proximopalmar hook-like process
obp	obturator process	ppp	prepubic process
oc	occipital condyle	pra	prearticular
od	odontoid process of the axis	prf	prefrontal
of	obturator foramen	prl	prespinal lamina of the neural spine
ol	olecranon process	pro	prootic
om	outer malleolus of the distal tibia	prom	prominence for muscular attachment on the scapula
op	opisthotic		
osl	ossified ligament	prsf	prespinal fossa
p	parietal	prz	prezygapophysis
pad	process for articulation of the dentary	prztu	prezygapophyseal protuberance
pal	palpebral bone	ps	pubic shaft
papr	pubic apron	pt	pterygoid
parpr	parasphenoidal recess	ptq	pterygoid ramus of the quadrate
pas	proximal articular surface of the phalanx	pu	pubis
		pup	pubic pedicle
paw	preacetabular wing	pv	posteroventral margin of the ilium
pd	predentary	pvh	posteroventral heel of the pedal phalanx
pdl	posterodorsal lip of the ungual		
pdp	posterodorsal process of the ischium	pvqp	posteroventral quadrangular process of the pedal ungual
peg	peg in a socket		
ph I–IV	phalanx I–IV	q	quadrate
phs	phalanges	qf	quadrate foramen
pia	pubic ischiac articulation	qj	quadratojugal
pl	palatine	r	rib
plr	posterolateral ridge (or *crista lateralis plantaris*)	ra	radius
		re	raised eminence
pma	protuberance for muscle attachment on the distal pubis	rg	rugosity
		ri	ridge
pmx	premaxila	ribl	right iliac blade
pmxf	promaxillary fenestra	rn	right nasal

rpmx	rostromedial process of the maxilla	t	tooth
rtp	retroarticular process	tar	tarsal bones
s	sacrum	tc	tibial condyle
s1–s8	sacrals 1–8	tib	tibia
sa	surangular	tp	transverse process
sac	supracetabular crest	trf	trochanteric fossa of the proximal femur
sap	supraacetabular process		
sc	scapula	trs	trochanteric shelf
scb	scapular blade	ts	tibial articulation surface
scl	sclerotic ring	tu	tuberculum
scr	subcondylar recess	tub	tuberosity
se	tooth serrations	tyr	tympanic recess
sl	ligament groove on the proximal femur	u	ungual phalanx
		ul	ulna
sms	sulcus for M. supracoracoidei	v	vomer
soc	supraoccipital	vk	ventral keel
sp	splenial	vnt	ventral notch
sq	squamosal	vs	ventral surface
sr	sacral rib	w	wrinkles
st	spine table	wf	wear facet
ste	sternal plates	I–V	first to fifth digit of manus or pes
stf	supratemporal fenestra	1–4	first to fourth phalanges
stp	stapes		
strp	supratrochanteric process		

INTRODUCTION

South America has been revealed as one of the most important continents for dinosaur discoveries. Although numerically less impressive than that of North America, the fossil record of South America is highly important for understanding the evolution of dinosaurs: their origin in Triassic times; the acquisition of gigantic size by some species; the effects of the geographic fragmentation of Gondwana on the evolution of dinosaur communities. The fossil record of South America includes many "unique pieces" that offer a glimpse into the evolutionary history of dinosaur clades not recorded in other parts of the world.

After dinosaurs were first documented in South America in 1883, knowledge of their evolutionary history remained fragmentary and restricted to a handful of poorly understood species. Although some discoveries were published prior to 1930, little progress in understanding dinosaur evolution in South America was made during the first half of the twentieth century. A major change in fossil documentation occurred at the end of the 1950s, when exploration of the Triassic beds of the Ischigualasto Formation, which crop out in northwest Argentina, resulted in remarkable discoveries. This exploration represented a pivotal moment in the history of dinosaur paleontology in South America because it involved the participation of the first Argentine specialists—Osvaldo Reig, José Bonaparte, and Rodolfo Casamiquela—who were committed to studying Triassic fossils in different parts of the country. The work done by these Argentine pioneers, in particular José Bonaparte, became the starting point for a remarkable increase in our knowledge of Jurassic and Cretaceous dinosaur faunas and inspired a new generation of researchers working mainly under Bonaparte's direction. Among this group of active younger paleontologists, Jaime Powell, Luis Chiappe, Guillermo Rougier, Leonardo Salgado, Jorge Calvo, Rodolfo Coria, Rubén Martínez, Oscar Alcober, Andrea Arcucci, and Bernardo González Riga figure prominently.

In contrast to Argentina, the development of vertebrate paleontology in general and dinosaur studies in particular was slower in other South American countries. However, Brazil has seen a sustained increase in discoveries and research, including descriptions of several outstanding discoveries of Cretaceous dinosaurs published by Llewellyn Ivor Price between 1948 and 1969. Innovative studies of South American ichnology were done by Giuseppe Leonardi and important advances in dinosaur studies are currently being developed by Alexander Kellner, Ismar de Souza Carvalho, and Max Langer.

Activities carried out by foreign paleontologists also contributed to better knowledge of dinosaur faunas, mainly in Argentina, including pioneering studies by Richard Lydekker and Friedrich von Huene and the sustained work by Alfred S. Romer in Triassic rocks of northwest Argentina. In more recent years, Argentine paleontology has benefited considerably from the activities conducted by Philip Currie in Cretaceous beds of northwest Patagonia, by Paul Sereno in outcrops of the Ischigualasto Formation, by Patricia and Thomas Rich in Jurassic rocks of Chubut, by Oliver Rauhut in Jurassic and Cretaceous formations in the same Argentine province, and by Kenneth Lacovara and Matthew Lamanna in Cretaceous fossil beds of central Patagonia.

Work carried out in Mesozoic outcrops of South America has resulted in a remarkable string of discoveries, including more than 50 dinosaur species (diagnosable on the basis of derived osteological features), which have revealed that a rich and complex evolutionary history took place in this southern continent. The fossil record of South American dinosaurs covers the Triassic, Jurassic, and Cretaceous periods, revealing remarkable aspects of their origin and later diversification. We now know that South America was once populated by dinosaurs of different pedigree, some of them sharply different from their Laurasian counterparts. In comparison with dinosaurs from other southern continents, those from South America are the best represented in terms of number of specimens, quality of preservation, and diversity of species. The fossil evidence amassed in Argentina alone (consisting of skeletons, footprints, eggs, embryos, nests, and skin impressions belonging to the main evolutionary streams of Saurischia and Ornithischia) constitutes a comprehensive dataset that is taxonomically more diverse, chronologically better documented, and phylogenetically more informative than the fossil record obtained in other continents that formed part of Gondwana (Africa, Madagascar, India, Antarctica, and Australia).

Deficits of the South American fossil record mainly relate to basic taxonomical problems (for example, questions about the validity of species founded on partial evidence, in particular Cretaceous titanosaurs, or clades such as stegosaurs and ankylosaurs that are known only from incomplete specimens). There are geographic sampling problems as well. Dinosaur discoveries are almost unknown from South American regions other than Patagonia, northwest Argentina, and southern Brazil. The stratigraphic representation is also patchy, and the most productive sedimentary units of South America are still underrepresented in the number of discovered taxa compared with formations of similar ages of other parts of the world. For example, the intensively worked Campanian–Maastrichtian Dinosaur Park Formation of western Canada has yielded remains of nearly 30 different genera of dinosaurs (Weishampel et al. 2004), double the number of taxa documented in the Bajo de la Carpa and Anacleto formations, two of the most productive beds of South America.

The purpose of this book is to offer an up-to-date and comprehensive review of the anatomy, systematics, and evolution of South American dinosaurs within paleogeographic and paleoecological contexts. To carry out this task, I review information from a variety of sources, most of them published in Spanish and Portuguese. Firsthand observations of dinosaur specimens from South America form an important part of the book, as well as comparisons with forms discovered in other Gondwanan landmasses. Current understandings of phylogeny, paleoecology, and paleobiogeography are considered in light of current controversies. Main geological events (e.g., diastrophism, volcanism, continental breakup) as well as faunal and floristic changes that occurred in South America are also briefly examined.

THE AGE OF DINOSAURS IN SOUTH AMERICA

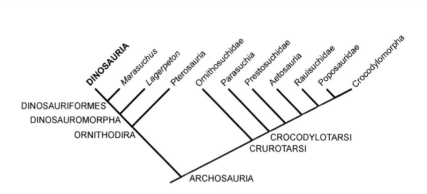

Fig. 1.1. Cladogram depicting the phylogenetic relationships of main groups of Archosauria. *Redrawn from Parrish (1997).*

AN OVERVIEW OF THE ANATOMY AND PHYLOGENETIC RELATIONSHIPS OF DINOSAURS

1

This chapter provides the reader with some basic knowledge about how dinosaurs are phylogenetically related to other reptiles, emphasizing features of skeletal construction. It will also offer some generalities about dinosaur anatomy to familiarize the reader with the main morphological details this book will frequently use in its descriptions of South American dinosaurs. Finally, the chapter presents a brief historical account of the systematic interpretations of Dinosauria that highlights the impact of discoveries in South America on the generation of currently accepted phylogenetic hypotheses.

Dinosaurs are a diverse lineage of vertebrates that dominated terrestrial landscapes during most of the Mesozoic Era, a time span of approximately 135 million years. Because birds are a subgroup of living dinosaurs, the group as a whole has lasted some 200 million years. Dinosaurs are tetrapods, the group of vertebrates that are four-footed land animals. Dinosaurs are amniotes, a subgroup of four-footed vertebrates characterized by internal fecundation and reproduction with a shelled egg. Amniota includes two main evolutionary streams: Synapsida (the group leading to the mammals) and Reptilia (the lineage that includes living turtles, lizards, crocodiles, and birds plus a wide variety of extinct members). The first members of Reptilia lived during the Carboniferous Period—320 million years ago—near the end of the Paleozoic Era.

Within Reptilia, dinosaurs belong to a subgroup known as Archosauria, which is currently defined by modern systematics to include the last common ancestor of the two extant groups of archosaurs, the crocodilians and the birds, and all of the descendants of that common ancestor (e.g., Sereno 1991a; Parrish 1997; Fig. 1.1). Archosauria originated during the Triassic Period, and the group is subdivided into two main evolutionary streams: Crurotarsi (a name that refers to a complex construction of ankle bones involving a movable articulation between the astragalus and the calcaneum) and Ornithodira (a word meaning "bird neck," referring to the S-shaped curvature of the neck). Crurotarsan archosaurs include living crocodiles plus a wide spectrum of extinct crocodile-like creatures, most of them quadrupedal, that prospered during the Triassic Period (e.g., ornithosuchians, parasuchians, aetosaurs, rauisuchids). On the other side, Ornithodira is the archosaur lineage that joins pterosaurs (extinct flying archosaurs that proliferated during Triassic, Jurassic, and Cretaceous times) and dinosauromorphs (the clade that gathers Mid-Triassic

What Are Dinosaurs?

3

dinosaur forerunners such as *Lagerpeton* and *Marasuchus* and dinosaurs themselves, including birds).

Dinosaurs and their kin underwent important transformations of their locomotor apparatus when they acquired an upright posture of their hindlimbs, resembling more the shape and posture of the hindlimbs of living birds than the sprawling posture of living crocodiles or lizards. In sharp distinction to these other reptile groups, the hindlimbs of a dinosaur were vertically oriented below the body during both slow walking and fast running. In other words, dinosaurs carried their bellies far from the ground. In addition, while in the above-cited living reptiles the foot contacts the ground with its entire sole, in dinosaurs only the toes rested on the ground, as in living birds (footprints left by early dinosaurs that lived around 200 million years ago are similar to the footprints that a modern chicken leaves in the mud). Dinosaurs and their more immediate relatives (*Lagerpeton, Marasuchus*) were among the first reptiles that abandoned crawling. Dinosaurs preceded by several tens of millions of years other vertebrates (for example, jerboa rats, kangaroos, hominids) that would evolve a bipedal carriage.

Bipedalism, which implies freedom of the forelimbs from their ancestral locomotory function, evolved in the immediate forerunners of dinosaurs. The forelimbs of early dinosaurs presumably participated in tasks other than supporting body weight (for example, handling food items). Although later in their history many groups of plant-eating dinosaurs regained a quadrupedal style of locomotion, other forms (in particular, the meat-eating dinosaurs) retained freed forelimbs, which had enormous adaptive consequences for the evolution of wings and flying capabilities in birds.

The word dinosaur means "fearfully great lizard," and this image applies well to ferocious creatures like a *Tyrannosaurus* and *Gigantosaurus* or even bigger beasts like *Brachiosaurus* and *Argentinosaurus*, which weighed more than 50 tonnes. Nevertheless, many dinosaurs were tiny animals, not fearsome giants. *Scipionyx* from Italy measured roughly 40 centimeters long, while *Microraptor*, from Cretaceous beds in China, was just 25 centimeters from its nose to the tip of its tail.

Dinosaur Skeletal Anatomy

For better comprehension of the anatomical descriptions of South American dinosaurs, I offer some general comments about the skeletal morphology of these animals. The main sources of information for this part of the book are Romer (1956) and Holtz and Brett-Surman (1997).

The skeleton of a tetrapod, dinosaurs included, can be divided into the following sections from a topological point of view: the skull, the vertebral column (including the ribs and haemal arches), the pectoral girdle and forelimbs, and the pelvic girdle and hindlimbs. All the elements of the skeleton, with the exception of the skull ("cranium" in Latin), are collectively called the postcranium (the part of the skeleton that is posterior to the cranium).

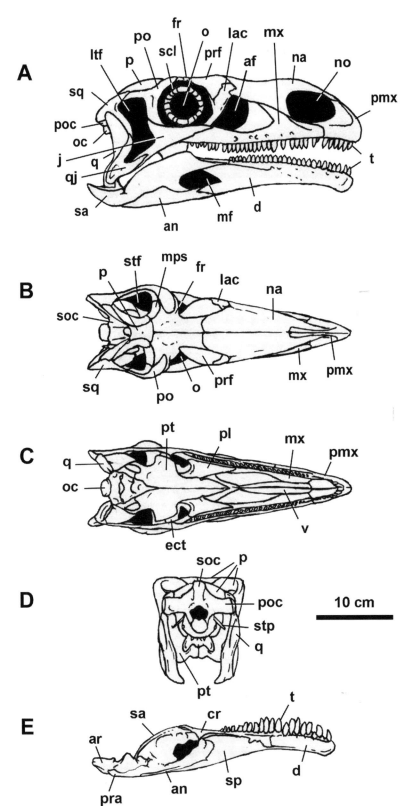

Fig. 1.2. Sauropodomorph dinosaur *Plateosaurus*. Skull and jaw in (A) lateral, (B) dorsal, (C) ventral, and (D) occipital views. (E) Left jaw in medial view. *Redrawn from Galton and Upchurch (2004a).*

The skull is composed of paired bones repeated as mirror images of each other on both sides of the axial plane (Fig. 1.2). There are also single bones that lie along the midline of the skull. The skull houses the main sense organs (nose, eyes, ears) and the brain. Two bones of the ventral margin of the skull may have teeth: the premaxilla and the maxilla (together forming the upper jaw). In some dinosaurs (e.g., ornithischians, titanosaurid sauropods, ornithomimid theropods, and living birds), teeth are absent on premaxillae, on maxillae, or on both and a horny beak covered the margins of the mouth.

The premaxilla is ventral to the nasal opening, and the maxilla is ventral to the antorbital fenestra, an opening that is present in most dinosaurs (and other archosaurian reptiles). Some dinosaurs (the ornithischians) had reduced or lost the antorbital opening, while in others (particularly in theropod dinosaurs) the antorbital fenestra is wide and was associated with other pneumatic openings (the maxillary and the promaxillary openings). The paired premaxillae and the maxillae articulate dorsally with the paired nasals (sometimes fused in a single piece of bone) to surround the nasal passage.

The eye socket, or orbit, is surrounded ventrally by the jugal, rostrally (toward the snout) by the lacrimal, caudally (toward the tail) by the postorbital, and dorsally by the frontal. The frontal is a paired bone that meets on the midline of the skull, roofing part of the brain cavity. The frontals articulate rostrally with the nasals, caudally with the parietals (which also contribute to the walls of the braincase), and laterally with the bones that encase the orbit (e.g., the lacrimal and postorbital). In the rear end of the skull, a pair of openings, the temporal fenestrae, that housed the jaw muscles: the lateral temporal fenestrae (located on both sides of the skull) and the supratemporal fenestrae (lying on the dorsal surface of the skull). These pairs of fenestrae evolved in the common ancestor of diapsids, the group of reptiles that includes lepidosaurs (e.g., lizards and snakes) and archosaurs (e.g., crocodiles, pterosaurs, and dinosaurs, including birds).

In the posterior end of the skull there is a left and right quadrate, a large bone that served for articulation with the lower jaw, which may have teeth or a horny sheath. The tooth-bearing bone of the mandible is the dentary, and the left and right dentaries form the symphysis that unites the two sides of the jaws. Dinosaur teeth, when present, were always growing and could always be replaced. They have different morphologies and sizes depending on the dinosaur group. In theropods, teeth are conical, transversely compressed, serrated, and caudally curved. In sauropodomorphs, teeth are leaf-shaped with margins that might or might not be serrated, although some derived sauropods had pencil-like teeth. Ornithischians had a variety of types of teeth. Basal ornithischians had a spatulate and serrated tooth but more-derived lineages evolved a prismatic-shaped tooth that had no serrations.

The dentary articulates medially (toward the midline of the animal) with the splenial and caudally with three bones: dorsally with the

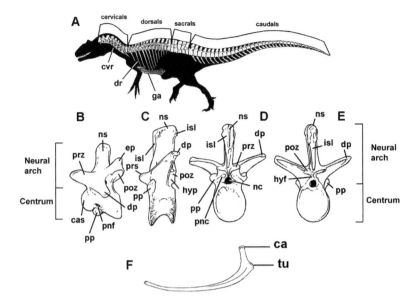

Fig. 1.3. Theropod dinosaur *Allosaurus*: (A) Silhouette of the body indicating regions of the vertebral column and thorax. (B) Cervical 7 in lateral view. (C–E) Dorsal 10 in (C) left lateral, (D) cranial, and (E) caudal views. (F) Dorsal rib in cranial view. (B–F) Redrawn from Madsen (1976).

surangular and ventrally with both the angular (on the outer side of the mandible) and the prearticular (on the inner side of the jaw). These three bones meet caudally with the articular, the bone that has a hinge-joint articulation with the quadrate bone of the skull. The angular, surangular, prearticular, and articular bound a wide fossa (open space) that housed the adductor muscles that closed the jaw.

The braincase is made up of several tightly sutured bones. The braincase proper lies beneath the roof bones of the outer skull. The spinal cord exits from the braincase through a great opening called the foramen magnum located on the posterior end of the skull. The foramen magnum is dorsally and ventrally bounded by unpaired bones, the supraoccipital and the basioccipital, respectively, and laterally by the paired exoccipitals. The basioccipital forms a bulbous prominence, the occipital condyle, for articulation with the vertebral column (i.e., the atlas and axis).

The vertebral column of dinosaurs can be divided into cervical (neck), dorsal (trunk), sacral (hip), and caudal (tail) vertebrae (Fig. 1.3A). Each of these sections is composed of vertebrae of different morphologies and the number of vertebrae in each module is variable, depending on the group of dinosaur considered. In general, cervicals are low (with a relatively small top-to-bottom dimension) and elongate (from front to back). Dorsal vertebrae are more robustly constructed. The dinosaur sacrum consists of two or more vertebrae, sometimes fused into a single element. The sacrum attaches on either side to the left and right ilia, which constitute the upper bones of the pelvic girdle (see below). The caudal series is composed of several vertebrae that tend to diminish in size toward the tailtip.

A single vertebra is formed by a large spool or cylindrical structure, the centrum, which supports the neural arch dorsally (Fig. 1.3B–E). Both the centrum and the neural arch bound the neural canal, which houses the spinal cord. The centrum has cranial and caudal surfaces for articulation with contiguous centra. Articular surfaces may be flattened (amphyplatyan) or slightly concave (amphicoelous), although in some special cases the cranial surface is concave but the caudal end is convex (procoelous, as in the tail vertebrae of titanosaurid sauropods) or the reverse condition occurs (the cranial surface is convex and the caudal end is concave, a condition termed opisthocoelous, as in the neck vertebrae of some derived theropod and sauropod dinosaurs). The lateral surface of the centra may be perforated by pneumatic pores known as pleurocoels. Such foramina connect with the interior of the vertebral centrum and may have served as passage for air sacs. Such perforations occur in the cervical vertebrae of most saurischian dinosaurs and are also present in the dorsal, sacral, and caudal vertebrae of some derived theropods and sauropods. Centra that are perforated by pneumatic pores are internally hollow.

From the neural arch projects a dorsal neural spine, which offers surfaces of attachment for muscles and tendons of muscles running over the vertebral column and ligamentous connections between successive neural spines. Both cranially and caudally there are pairs of finger-like projections called zygapophyses. The prezygapophyses (i.e., the anterior zygapophyses) have articular surfaces oriented upward and slightly inward and serve as support of the postzygapophyses (i.e., the posterior zygapophyses), which have articular surfaces oriented downward and slightly outward. The trunk vertebrae of saurischian dinosaurs evolved additional surfaces for vertebral support: the hypantrum and the hyposphene. The hyposphene is a posteriorly directed median wedge projecting below the postzygapophyses that inserts into the hypantrum, a narrow space between the prezygapophyses.

The cervical vertebrae of most dinosaurs exhibit prominences on the dorsal surface of the postzygapophyses, called epipophyses, to which neck muscles attach.

The cervical and dorsal vertebrae have conspicuous apophyses projecting from both sides of the neural arch and centrum. Such projections are termed diapophyses (extending laterally from the neural arch) and parapophyses (present on the cranial margin of cervical and dorsal vertebrae). These paired apophyses served for articulation with the double-headed cervical and dorsal ribs, which themselves proximally split into a dorsal process, the tubercle, to attach to the diapophysis, and a ventral process, the capitulum, to attach to the parapophysis (Fig. 1.3F). In archosaurian reptiles, including dinosaurs, the parapophysis migrates upward and backward from the centrum to the neural arch from one vertebra to the next along the presacral series, finally lying close to the diapophysis so that diapophysis and parapophysis may fuse together into a single-headed structure. Ribs articulated to the dorsal vertebrae form a ribcage (thorax) around the vital organs. Along the belly of some dinosaurs are gastralia, or "belly ribs," which strengthened the ventral side of the animal and acted as a girdle to hold in the viscera but probably also contributed dynamically in the breathing process.

Fig. 1.4. Girdles and limb bones of several dinosaurian taxa. (A) *Allosaurus fragilis*, scapular girdle and forelimbs in cranial view. (B) *Brachiosaurus brancai*, trunk and sacral vertebrae articulated with pectoral and pelvic girdles and fore- and hindlimbs in left lateral view. (C, D) Pelvic girdles in left lateral view of (C) the basal saurischian *Plateosaurus* and (D) the basal ornithischian *Lesothosaurus*. (E) *Allosaurus fragilis*, pelvic girdle and hindlimbs in cranial view. *(A) Redrawn from Carpenter (2002). (B) Redrawn from Christiansen (2000). (E) Composite reconstruction based on Gilmore (1920) and Madsen (1976).*

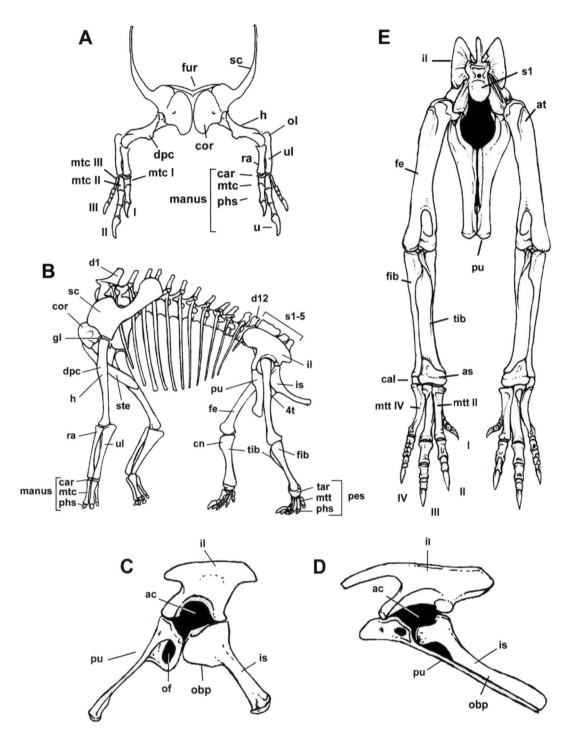

The caudal vertebrae exhibit laterally projected transverse processes from the neural arch that do not articulate distally with ribs (which are absent in the caudal series). Ventral to the caudal vertebrae are the chevrons, or haemal arches. These are Y-shaped in cranial aspect and articulate below the connection between two contiguous caudal centra.

Some dinosaurs have ossified tendons running along different portions of the vertebral column. This is the case for most ornithischians, in which a net of ossified tendons runs across the neural spines of the dorsal, sacral, and caudal vertebrae as well as across the haemal arches of the tail. Ossified tendons also evolved in the neck of certain titanosaurid sauropods and in the tail of some clades of theropod dinosaurs (e.g., abelisauroids, dromaeosaurids).

The pectoral (or shoulder) girdle (Fig. 1.4A and B) attached through ligament and muscle connections to the anterior region of the ribcage. It is composed of a somewhat strap-like elongate scapula that articulates with an elliptical or rectangular-shaped ventral element, the coracoid. The junction of the scapula and coracoid (along their caudal margin) forms the glenoid cavity, a deep concavity for articulation with the proximal element of the forelimb (the humerus). Running over the cranial edges of both scapulae and coracoids were the paired clavicles, which in most predatory dinosaurs fused into a single bone, the furcula (the avian "wishbone"). The coracoids of both sides connect along the midline of the body with a paired sternum, or "breastbone," the shape and size of which varies among groups of dinosaurs.

Dinosaur forelimbs (Fig. 1.4A and B) are made up of three main bones: the humerus, the ulna, and the radius. The proximal humerus joins the two forearm bones at the elbow. In most dinosaurs, the humerus has a pronounced and cranially directed deltopectoral crest, which served for the attachment of pectoral muscles that originated from the animal's chest. The ulna is larger and more posterior than the radius. It has a proximal prominence, the olecranon process, which is variably developed for attachment of extensor muscles that straightened the limb. Both the ulna and the radius articulate with the hand, or manus. The proximal components of the manus are a series of small carpal (wrist) elements articulated distally to elongate metacarpals, which are designated by Roman numerals I to V. I is the most medial (inner) and V is the most lateral (outer). Each digit is numbered to match its corresponding metacarpal. The individual bones of the digits are the phalanges. The distalmost (toward the fingertip) of the phalanges are the unguals, which supported horny claws or hooves. Primitively, dinosaurs were bipedal reptiles with elongate and slender forelimbs about half as long as the hindlimbs and hands suited for manipulation rather than for progression. As a quadrupedal posture was acquired, mainly by massively constructed dinosaurs (such as sauropods, ankylosaurs, stegosaurs), the forelimb bones became robust but the hands retained an almost unreduced digit count (Fig. 1.4B). The ungual phalanges of quadrupedal dinosaurs were lost or were transformed into hooves. In dinosaurs that retained a bipedal posture, mainly the meat-eating theropods, the number of outer digits of the manus was reduced and the unguals became proportionally large and trenchant. Theropod forelimbs manifested dissimilar trends; in tyrannosaurids and abelisaurids they were strongly reduced, while in some derived avian-like theropods (deinonychosaurs), particularly flying birds, the forelimb was elongated.

The pelvic girdle on each side of the body is composed of three bones: a dorsal bone (the ilium); an anteroventral bone (the pubis); and a posteroventral bone (the ischium) (Fig. 1.4C and D). At the junction of these three pelvic bones is a perforated depression, the acetabulum, which receives the femoral head. The left and right ilia are connected to the sacral vertebrae. The ilium, the pubis, and the ischium served as expanded surfaces for muscle attachment, mainly the origin of several hindlimb protractors (to pull the limb forward) and retractors (to pull the limb backward) and the muscles of the trunk and tail. The distal extremities of both pubes and ischia may have served as points of support of the body while the animal was in a resting position.

Most of the saurischian dinosaurs retained from their archosaur ancestors a "triradiate," or propubic, pelvis in which the pubis was projected forward and downward (Fig. 1.4C). In contrast, from their early origins, ornithischian dinosaurs manifested a backward-oriented (opisthopubic) pelvis (Fig. 1.4D). However, in derived saurischians such as the coelurosaurian therizinosaurids, mononykids, dromaeosaurids, and birds, the pubis also became retroverted, independently acquiring a condition that superficially resembles that of the ornithischian dinosaurs.

The hindlimb bones are the femur, the tibia, and the fibula (Fig. 1.4B, E). The femur articulates proximally with the pelvic acetabulum through a medially offset femoral head and joins distally at the knee to the tibia and the fibula. The dinosaur femur has different prominences for muscle attachment: the anterior trochanter proximally (to attach fibers of protractor muscles originated on the ilium) and the fourth trochanter at caudal midshaft (to attach retractor muscles that originated from the ilium and the tail). The tibia constitutes the larger and more medial bone of the lower leg and the fibula is the thinner and more lateral bone. Proximally, the tibia has a rectangular-shaped cranially projecting cnemial crest. Distally, both the tibia and the fibula articulate with two tarsals (the bones of the ankle), respectively the astragalus and calcaneum, which contact distally with a small number of disc-shaped distal tarsals. The tarsal bones articulate with the elongate metatarsals (numbered I to V in medial-to-lateral fashion), and the metatarsals articulate distally with their respective digits. The digits, metatarsals, and tarsals are collectively called the foot, or pes.

A Brief Historical Account of the Systematics of Dinosauria

English naturalist Richard Owen coined the name Dinosauria in 1841 for a group of land-living reptiles of large size that exhibited features that were more advanced than extant reptiles in the construction of the vertebral column and hindlimbs. After that time, dinosaur discoveries multiplied considerably and it became clear that Dinosauria did not constitute a homogeneous group but was a deeply diversified taxon with a wide variety of adaptations. Paleontologists became interested in bringing order to this diversity and proposed numerous classifications. In 1887, British paleontologist Harry Seeley made the case that dinosaurs could be sorted into two different taxa: Saurischia for dinosaurs with a cranioventrally projected pubis, and Ornithischia for dinosaurs with a caudoventrally oriented pubis. These important distinctions in pelvic construction led Seeley to argue that Saurischia and Ornithischia were not closely related to each other. Seeley reinterpreted the features that Owen recognized as characterizing Dinosauria as superficial resemblances due to functional similarities related to hindlimb posture. Over the course of the twentieth century, Seeley's interpretations gained wide acceptance and the name Dinosauria began to be used as an informal term for a polyphyletic assemblage of archosaurian reptiles. Thus, the respective origins of saurischians and ornithischians were sought in totally different ancestral groups.

The sharp morphological gap between dinosaurs and basal archosaurs as well as the lack of a truly phylogenetic methodological approach created confusion about dinosaur origins and interrelationships. Moreover, during the 1970s the idea of a marked polyphyletism became predominant among authors to the point of almost destroying not only Dinosauria but also Saurischia and Theropoda as monophytetic groups (e.g., Thulborn 1975). However, this changed during that decade, mainly due to the availability of new fossil material from the Triassic of Argentina. Discoveries of dinosaur-like archosaurs such as *Lagerpeton chanarensis* and *Marasuchus lilloensis* (Romer 1971, 1972; Bonaparte 1975, 1984a) made it possible to study, for the first time, advanced reptiles that were potential ancestors of saurischians and ornithischians. Furthermore, more information on early dinosaurs, such as *Pisanosaurus* from Argentina, *Anchisaurus* from the United States, and *Syntarsus* from Zimbabwe, became available (Raath 1969; Galton 1976; Bonaparte 1976). This new evidence provided sufficient grounds for reevaluating the long-held belief that Dinosauria was not a natural group (e.g., Seeley 1887, 1888; Baur 1891; von Huene 1914; Colbert 1964; Romer 1971).

Robert Bakker and Peter Galton (1974) provided the concept that saurischians and ornithischians were closely related phylogenetically and supported a hypothesis that both lineages descended from a common archosaurian ancestor. In sharp contrast with Seeley, whose methodological procedure emphasized the anatomical differences between saurischians and ornithischians, Bakker and Galton paid special attention to the shared characteristics of these two groups. These authors (like Owen before them) observed that early saurischians and ornithischians were bipedal and that their hindlimbs adopted a vertical posture, constituting features that were absent in other archosaurian reptiles. Additional characteristics Bakker and Galton listed in support of their hypothesis included a completely open acetabulum, a femoral head that was distinct from the rest of the bone, a reduced fibula, a mesotarsal type of tarsus, and a foot with a digitigrade stance.

In 1975, Argentine paleontologist José Bonaparte published an extensive study of *Marasuchus* (following Sereno and Arcucci [1994], the name *Lagosuchus talampayensis* is regarded as a nomen dubium and the name *Marasuchus lilloensis* is used here to refer the holotype of "*Lagosuchus*"

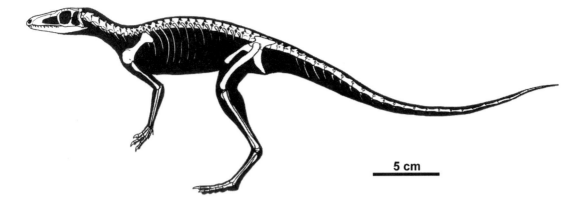

Fig. 1.5. Reconstructed skeleton of *Marasuchus lilloensis*. Illustration by J. González.

lilloensis; Fig. 1.5) that demonstrated its resemblances to saurischians and ornithischians, thus filling an important morphological gap between early dinosaurs, which were mostly bipedal and digitigrade, and basal archosaurs, which were mostly quadrupedal and plantigrade. Bonaparte's research offered strong support to Bakker and Galton's hypothesis of the common origin of Saurischia and Ornithischia.

The hypothesis that dinosaurs were monophyletic gained adherents during the 1980s, and at present no alternative hypothesis (e.g., that dinosaurs are not a clade) has been offered on the basis of derived characteristics. In recent years cladistic analysis has been extensively applied to the archosaur systematic. Gauthier (1986) offered an exhaustive analysis of characteristics of saurischian dinosaurs, supporting the idea that Dinosauria, Saurischia, and Theropoda were monophyletic. He argued that the clade Ornithodira encompassed the common ancestor of Pterosauria and Aves and all of its descendants. Novas (1992a) found evidence in support of a monophyletic group composed of *Marasuchus* plus Dinosauria, for whom the name Dinosauriformes was coined.

Dinosaur Forerunners

Gauthier (1986) and Sereno (1991a) studied in depth the monophyly of Ornithodira (a group including Pterosauria, *Lagerpeton*, and Dinosauriformes and their most recent common ancestor). Sereno and Novas (1990), Sereno (1991a), and Novas (1992a) found evidence to support the notion that Pterosauria were outgroups of a clade composed of *Marasuchus* plus Dinosauria (see Fig. 1.1). Also, the phylogenetic placement of two taxa was clarified: the bizarre archosaur *Lagerpeton* (Arcucci 1986, 1997; Sereno and Arcucci 1993) is now considered the sister taxon of the Dinosauriformes (Sereno and Novas 1990; Sereno 1991a), and *Pseudolagosuchus* (formerly placed within "Lagosuchidae"; Arcucci 1987) was more recently interpreted as the immediate sister group of Dinosauria (Novas 1992a). Pterosauria and *Lagerpeton* are now interpreted as successive ornithodiran outgroups of Dinosauriformes (e.g., Sereno 1991a).

The clade Dinosauriformes (Novas 1992a) is defined to include the common ancestor of *Marasuchus* and Dinosauria and all the descendants of that common ancestor (Fig. 1.1). Several synapomorphies concerned

Fig. 1.6. Cervical and dorsal vertebrae of *Marasuchus lilloensis*. From Bonaparte (1975). Reprinted with permission of the author and Fundación Miguel Lillo.

Fig. 1.7. The hindlimb of dinosauriforms. (A–G) Proximal end of femora of several archosaur taxa: (A, D) Right femur of *Lagerpeton*. (B, E) Left femur (reversed) of *Marasuchus*. (C, F, G) Right femur of *Herrerasaurus*, in (A–C) proximal, (D–F) caudal, and (G) cranial views. (H–J) Proximal end of left tibia in lateral view of several dinosauriform taxa: (H) *Marasuchus*, (I) *Herrerasaurus*, and (J) *Dryosaurus*. (K–M) Distal end of left tibia in lateral view of several dinosauriform taxa: (K) *Pseudolagosuchus*, (L) *Herrerasaurus*, and (M) *Dryosaurus*. (N–P) Distal view of left tibia of several dinosauriform taxa: (N) *Pseudolagosuchus*, (O) *Herrerasaurus*, and (P) *Dryosaurus*. (Q–T) Left astragalus in proximal view of several dinosauriform taxa: (Q) *Marasuchus*, (R) *Pseudolagosuchus*, (S) *Herrerasaurus*, and (T) *Scutellosaurus*. Modified from Novas (1996a). © Copyright 1996 the Society of Vertebrate Paleontology. Reprinted and distributed with permission of the Society of Vertebrate Paleontology.

with neck anatomy and hindlimb morphology have been recognized to support the hypothesis that Dinosauriformes are monophyletic. *Marasuchus* shows an incipient sigmoid curvature of the neck but it does not have the strongly S-shaped neck of dinosaurs, including birds (Bonaparte 1975; Gauthier 1986; Fig. 1.6).

The hindlimbs of *Marasuchus* are notably similar to those of primitive dinosaurs, including a trochanteric fossa on the proximal articular surface of the femoral head (Fig. 1.7A–F), which served for sliding over a prominent pelvic antitrochanter (located on the posterior portion of acetabulum; Fig. 1.8A–C). *Marasuchus* and Dinosauria also share an anterior trochanter on the femur, which conforms to a subvertical prominence that is continuous distally with a subhorizontal ridge (the trochanteric shelf; Fig. 1.8F and G). The trochanteric shelf constitutes a prominent posterodistally oriented ridge on the lateroproximal surface of the femur that is uniquely documented among dinosauriform archosaurs (e.g., *Marasuchus, Pseudolagosuchus, Silesaurus, Herrerasaurus, Syntarsus*).

Marasuchus and early dinosaurs also share important similarities in the shape of the tibia: on the proximal end of this bone, a cnemial crest developed that is square-shaped in side view (Fig. 1.7H). In addition, the distal end of the tibia is quadrangular and is laterally notched by a groove that runs longitudinally along the distal portion of the tibial shaft (Fig. 1.7K–O). This groove is expressed in distal aspect as a U-shaped notch that is bounded by the facet for the ascending process of the astragalus and the posterior descending process of the tibia. In dinosaurs this notch on the tibia correlates with the differentiation of a more complex dorsal surface of the astragalus (Fig. 1.7L, O, and S).

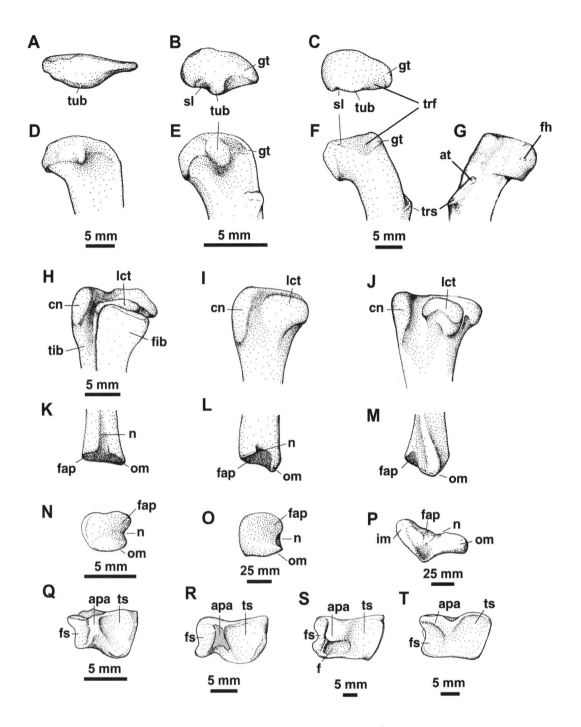

Some dinosauriforms (e.g., the Mid-Triassic *Pseudolagosuchus* from Argentina and the Late Triassic *Silesaurus* from Poland; Dzik 2003) exhibit features that are closer to the dinosaurian condition than those present in *Marasuchus*. For example, dinosaurs share with *Pseudolagosuchus* and *Silesaurus* an elongate pubis that is equivalent to 70 percent or more of femoral length, in contrast with more basal ornithodirans (e.g., Pterosauria, *Lagerpeton*, *Marasuchus*), in which the pubis is shorter (nearly

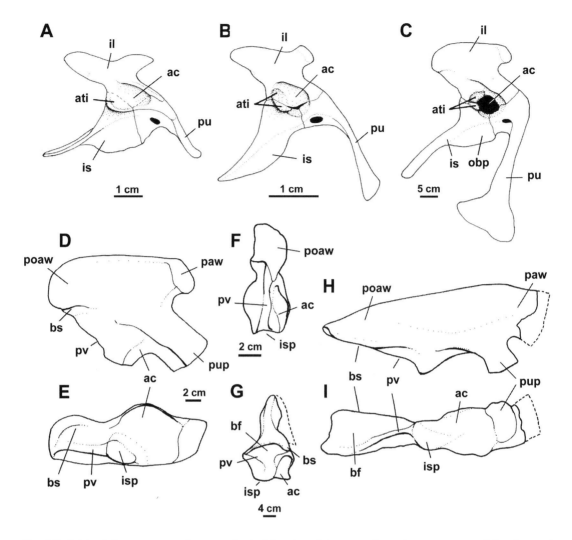

Fig. 1.8. Pelvic girdles of dinosauromorph taxa: (A) *Lagerpeton*. (B) *Marasuchus*. (C) *Herrerasaurus*, in right lateral view. (D–F) Right ilium of *Herrerasaurus* in lateral (D), ventral (E), and caudal (F) views. (G–I) Right ilium of *Torvosaurus* in (G) caudal, (H) lateral, and (I) ventral views. Modified from Novas (1996b). © Copyright 1996 the Society of Vertebrate Paleontology. Reprinted and distributed with permission of the Society of Vertebrate Paleontology.

as long as the ischium) and is approximately 50 percent of the femoral length (Fig. 1.8A–C).

Pseudolagosuchus and Dinosauria also developed changes in astragalar anatomy, consisting of the acquisition of a pyramid-shaped ascending process and a caudal subvertical facet and an elliptical depression behind the process (Fig. 1.7S). In Archosauria ancestrally the ascending process of the astragalus constitutes a craniocaudally oriented ridge that separates the tibial and fibular facets (Novas 1989a). This condition was retained in *Lagerpeton* and *Marasuchus* (Fig. 1.7Q). *Pseudolagosuchus*, in contrast, has a pyramid-shaped ascending process resulting from development of a posterior, subvertical, non-articular surface. Behind this surface is an elliptical depression on the proximal face of the astragalus. Dinosaurs are distinguished from other dinosauriforms in that the posterior surface of the ascending process articulates with the posterior process of the tibia, the latter being transversely broader than in basal Dinosauriformes. In addition, the posterior process of the tibia conceals the elliptical depression present on the dorsal astragalar surface. In

Dinosauria, the ascending process of the astragalus shows important morphological variation: in ornithischians (except *Pisanosaurus*; Bonaparte 1976; Novas 1989a) the ascending process is dorsoventrally reduced and anteroposteriorly flattened (e.g., *Scutellosaurus*, *Dryosaurus*; Colbert 1981; Galton 1981), while in tetanurine theropods the ascending process forms an anteroposteriorly flattened, tall, and broad wedge (Gauthier 1986). Nevertheless, all dinosaurs share a caudal facet on the ascending process that articulates with the cranial surface (i.e., the astragalar surface) of the distal tibia (Novas 1989a).

Characteristics That Distinguish Dinosauria from Other Archosaurs

Dinosaurian diagnostic features (i.e., derived features that evolved in the common ancestor of Saurischia and Ornithischia but are absent in their immediate sister groups) include several anatomical transformations in the head, the neck, and the pelvic girdles.

In the back of the skull, pseudotemporal muscles spread beyond the boundaries of the supratemporal openings, partially invading the dorsal surface of the frontal bones (Fig. 1.2B, mps), a modification that may have increased the force of jaw closure. The neck vertebrae of dinosaurs manifested a more pronounced inclination of the articular surfaces of the cervical centra, conferring a distinctive sigmoid curve to the neck, thus resembling the avian condition more closely. At least some of the cervical vertebrae developed epipophyses above the postzygapophyses.

With the exception of *Staurikosaurus*, *Guaibasaurus*, and *Saturnalia*, in which the sacrum is apparently formed by two sacrals, most of the early dinosaurs increased the sacral count to three by incorporating a new vertebra from the presacral series (e.g., Welles 1984; Novas 1996a; Fig. 1.9). The incorporation of a new sacral (from the dorsal vertebrae) involved craniocaudal shortening of the last dorsal vertebrae and original sacrals 1 and 2. This resulted in placement of the last dorsal vertebra clearly behind the tip of the pubic pedicle of the ilium. The first sacral ribs accompanied the craniocaudal shortening of its corresponding centrum, articulating in a more caudal position than ancestrally. As a result of this transformation, the last trunk vertebra was incorporated into the sacrum (it became the first dorsosacral) and the perpendicular orientation of the primordial sacral ribs allowed the articulation of the dorsosacral ribs to the ilium. Axial shortening of the last dorsal and primordial sacral vertebrae probably constituted the most important process that made incorporation of more vertebrae to the sacrum possible, at least during early dinosaur evolution. The sacral count increased independently in Theropoda, Ornithischia, and Sauropodomorpha, in all of which the sacrum has at least five vertebrae.

Dinosaurs have a deltopectoral crest (Fig. 1.4A and B) with the distal corner positioned down the shaft away from the head of the humerus at a distance greater than 30 percent of the maximum length of the humerus (e.g., Bakker and Galton 1974; Sereno and Novas 1992; Sereno, Forster, Rogers, and Monetta 1993; Sereno 1994).

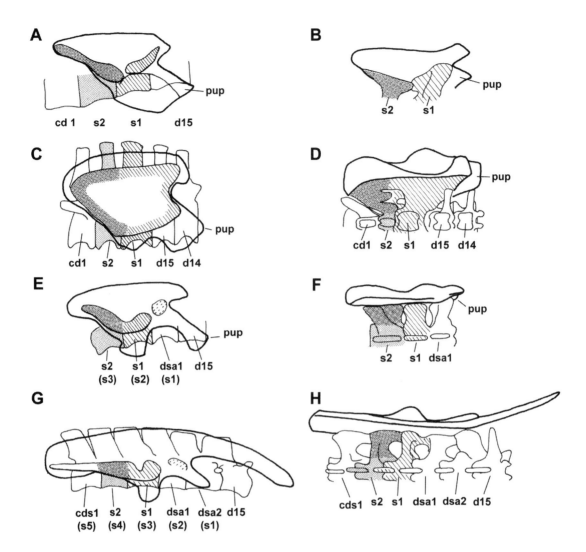

Fig. 1.9. Ornithodiran sacra showing relative position of the dorsal, sacral, and caudal vertebrae with respect to the ilium in several dinosauriform taxa. (A, B) *Marasuchus*. (C, D) *Herrerasaurus*. (E, F) *Riojasaurus*. (G, H) *Hypsilophodon*. (A, C, E, G) lateral view of sacral, dorsal, and caudal vertebrae (silhouette of ilium indicated by solid line). (B, D, F, H) dorsal view of sacral, dorsal, and caudal vertebrae (only the left ilium and left half of vertebral column is indicated).

Dinosaurs are distinguished from other dinosauriforms by a perforated acetabulum (Fig. 1.8C, shown in black), the result of the opening of the medial acetabular walls of the ilium, pubis, and ischium. The opening of the acetabulum is relatively small in basal dinosaurs (e.g., *Guaibasaurus, Saturnalia, Staurikosaurus, Herrerasaurus, Lesothosaurus*) but is larger in more-advanced dinosaurs (e.g., sauropodomorphs, theropods). The dinosaur pelvis is also characterized by a brevis shelf that consists of a distinct and prominent shelf on the posterolateral margin of the iliac blade that flares laterally above the posteroventral iliac margin (Fig. 1.8D–I). Both the brevis shelf and the posteroventral iliac margin bound a ventral fossa that cradled the caudifemoralis brevis muscle, a femoral retractor. Most theropods, ornithischians, and early sauropodomorphs exhibit a well-developed brevis shelf and a transversely wide and dorsoventrally deep brevis fossa. Only Dinosauria among ornithodirans have a brevis shelf and a fossa, since in basal forms (e.g., Pterosauria, *Lagerpeton, Marasuchus*) the postacetabular portion of the iliac blade is transversely compressed, with the posteroventral margin running from the ischiadic peduncle toward the posterior end of the blade (Novas 1992a, 1996a).

The dinosaur ischium has a slender shaft and a ventral keel (obturator process) that is limited to the proximal third of the bone (Figs. 1.4C–D and 1.8A–C). In basal archosaurs the ischium forms a plate-like structure that is distal to the acetabular region. This condition was retained in Ornithodira ancestrally because it is present in Pterosauria, *Lagerpeton*, and *Marasuchus*, in which the ischium has a proximodistally expanded and ventrally projected obturator process. In contrast, ornithischians, sauropodomorphs, and theropods share a distally elongate ischiadic shaft, with the obturator process that is limited almost entirely to the proximal third of the bone.

The hindlimb bones of early dinosaurs have several modifications compared to more-basal dinosauromorphs. The proximal femur of crurotarsan archosaurs and basal ornithodirans (e.g., *Lagerpeton*, *Marasuchus*, and *Pseudolagosuchus*) shows a well-developed tuberosity on the caudal surface of the head (Fig. 1.7A, B, D, and E). This tuberosity externally bounds the groove for a ligament attached to the femoral head and in basal archosaurs has a rounded and transversely wide prominence. The tuberosity is placed nearly halfway between the medial and lateral margins of the proximal end of the femoral head. In Dinosauria, this tuberosity is significantly reduced and has a slight prominence that is medially placed (Fig. 1.7C and F). Another modification in the proximal end of the dinosaurian femur is related to the development of a prominent proximally projected anterior trochanter to attach femoral protractors.

Interesting modifications are detected in the tibial-astragalar articulation of basal dinosaurs. In basal archosaurs the tibia articulates with the proximal surface of the astragalus but does not extend laterally beyond the ascending process of the astragalus. Consequently, there is no clear distinction between the posterior process of the tibia and the remainder of the distal articular surface of the bone. This condition is retained in Dinosauriformes ancestrally, since it is seen in *Marasuchus* and *Pseudolagosuchus*, in which the distal end of the tibia is transversely convex and posterodistally inclined (Fig. 1.10). In contrast, the tibia in dinosaurs, although quadrangular in basal dinosaurs (e.g., *Herrerasaurus*, *Pisanosaurus*; Novas 1989a, 1994), covers the ascending process posteriorly, resulting in the articulation of the latter beneath the craniolateral corner of distal articular surface of the tibia (Fig. 1.7L, O, and S). The distal tibial articular surface becomes topographically complex and it is possible to distinguish an anterolateral facet for the ascending process of the astragalus and a ventrally projected posterior process. The latter covers posteriorly the ascending process of the astragalus and its dorsal surface behind the ascending process. The dorsal astragalar surface that is posterior to the ascending process becomes enlarged toward the rear, and the posterior margin of the bone becomes straight in dorsal view (Fig. 1.7S and T). By contrast, in *Pseudolagosuchus* this portion of the astragalus

Not to scale. The counting of dorsosacral vertebrae (dsa1, dsa2) starts with the vertebrae attached to the ilium that articulates with the original sacral 1 (s1). Counting of caudosacral vertebra (cds1) starts with the vertebra attached to the ilium that articulates with the original sacral 2 (s2). The count of sacrals based on vertebrae that articulate with the ilium is indicated parenthetically below the nomenclature (i.e., original sacrals 1 and 2, dorsosacrals 1 and 2, and caudosacral 1). Light stippling marks the centrum of s2; dark stippling marks the transverse process and rib of s2 and its place of articulation with the ilium. Cross lines mark s1 and its corresponding transverse process, rib, and place of attachment to the ilium. Dashed lines indicate the articulation area of the first dorsosacral rib. *Modified from Novas (1996a). © Copyright 1996 the Society of Vertebrate Paleontology. Reprinted and distributed with permission of the Society of Vertebrate Paleontology.*

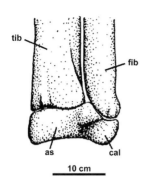

Fig. 1.10. Right tibia, fibula, and tarsus of *Marasuchus* in caudal aspect. *From Novas (1989a). Reprinted with permission of The Paleontological Society and the Society for Economic Paleontology and Mineralogy.*

Fig. 1.11. Fibulae, tarsals, and metatarsals of several ornithodiran taxa: (A) Right fibula and calcaneum of *Marasuchus* in lateral view. (B) Right fibula of *Marasuchus* in lateral view. (C) Right calcaneum of *Pseudolagosuchus* in lateral view. (D) Right calcaneum of *Herrerasaurus* in lateral view. (E) Left distal tarsals 3 and 4 of the basal pterosaur *Dimorphodon* in proximal view. (F) Left distal tarsals and metatarsals of *Lagerpeton* in proximal view. (G) Left distal tarsal 4 of *Lagerpeton* in caudal view. (H–J) Left distal tarsals and metatarsals of *Lagosuchus* in (H) proximal, (I) lateral, and (J) caudal views. (K–M) Right (reversed) distal tarsal 4 and metatarsal IV of *Herrerasaurus,* in (K) proximal, (L) lateral, and (M) caudal

is deeply notched in dorsal view and is a non-articular surface. Also, in the dinosaur astragalus, a deep elliptical basin is present immediately posterior to the ascending process (Fig. 1.7S).

Two more modifications occurred in the tarsal bones of basal dinosaurs. The proximal surface of the calcaneum became concave, creating a tighter articulation with the distal end of fibula (Fig. 1.11D). In contrast, in *Pseudolagosuchus* and *Marasuchus* the calcaneum has a hemicylindrical proximal condyle for articulation with the fibula (Fig. 1.11A and C). Also, distal tarsal 4 of dinosaurs acquired a proximodistally flattened condition and a triangular-shaped contour in proximal view (Fig. 1.11K–M), differing from the more block-like distal tarsal 4 of more basal ornithodirans (Fig. 1.11E–J).

From the available evidence, the early evolution of the Ornithodira manifested a sustained improvement in locomotor capabilities. Bipedal and digitigrade postures were acquired by the ancestral ornithodiran species (Gauthier 1986) and were inherited by dinosauromorphs, including Dinosauria ancestrally. Many of the synapomorphies of Ornithodira (Sereno 1991a) pertain to the hindlimb. Dinosauromorphs inherited such adaptations, evolving transformations in the pelvic and hindlimb skeleton that are also related to locomotion. On the basis of studies made by Sereno (1991a) and Sereno and Arcucci (1994), features of the tarsus and pes constitute 70 percent of the apomorphies diagnostic of Dinosauromorpha.

Moreover, although the synapomorphies that diagnose Dinosauriformes include a strong sigmoid neck curvature, most of the evolutionary novelties pertain to the locomotor apparatus. In early dinosauriforms, some outstanding skeletal modifications include the remodeling of the femoral head as a trochanteric fossa developed, which probably restricted femoral movement in the acetabular socket more than in Archosauria ancestrally. The presence of the anterior trochanter and trochanteric shelf suggests greater development of or differentiation of both the femoral protractors and the retractors. Also, the proximal and distal ends of the tibia exhibit interesting modifications: a cnemial crest on the proximal tibia, which suggests differentiation and enlargement of the protractive (extensor) muscles of the lower leg, and flexor muscles of the ankle.

In contrast to the transformations noted above, modifications of pelvic elements were delayed for the hindlimb bones during the early evolution of dinosauriforms. Since pelvic bone morphology retained the ancestral archosaurian condition (e.g., a craniocaudally short ilium and a pubis and ischium that were not ventrally elongated), it is assumed that no major modifications occurred in the placement of origin of the pelvic musculature. This implies that bone morphology, mainly the acetabular-femoral and tibiofibular-tarsal articulations (but not muscular rearrangement at the area of origin), were major constraints of hindlimb movements in a parasagittal plane in the early evolution of ornithodirans. Modifications occurred at the insertion areas of the protractive and retractive muscles of the femur, probably allowing an increase in the power

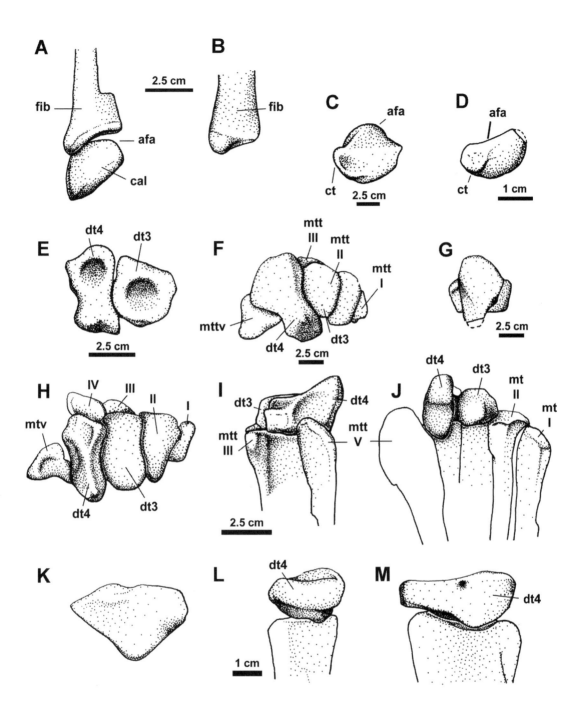

stroke of the hindlimb. The lengthening of the pubis and ischium and consequent enlargement of the origin area for the femoral protractors occurred later in dinosauriform evolution (e.g., in the common ancestor of *Pseudolagosuchus* and Dinosauria).

Diagnostic features of Dinosauria pertain to the skull, the neck vertebrae, the forelimbs, the pelvic girdle, and the hindlimbs. Those characteristics pertaining to the skull are equivocal synapomorphies of

views. *Modified from Novas (1996a). © Copyright 1996 the Society of Vertebrate Paleontology. Reprinted and distributed with permission of the Society of Vertebrate Paleontology.*

Dinosauria, since the skull is unknown in the immediate sister taxa of Dinosauria (*Pseudolagosuchus, Marasuchus, Lagerpeton*). Seventy-five percent of the postcranial adaptive innovations hypothesized in the common dinosaurian ancestor pertain to the pelvic girdle and hindlimb, including a strengthening of the articulation of the pelvis to the vertebral column (with the addition of a new sacral vertebra), the development of prominences (e.g., trochanters and shelves), and a widening of pelvic bone surfaces for muscular attachment. Of the hindlimb adaptive innovations, 40 percent pertain to the proximal femur and the other 60 percent to the articulation of the distal tibia and the tarsals. Apparently, the lateral surface of the iliac blade is wider in dinosaurs than in early dinosauromorphs (e.g., *Lagerpeton, Marasuchus*). Since the pelvis is not completely known in *Pseudolagosuchus*, it is not certain that the increase in the size of the ilium occurred concomitantly with the lengthening of the pubis and ischium. Except for the development of the brevis shelf and fossa, the ilium of the ancestral dinosaurs retained the brachyiliac condition. The main modification expected in the dinosaur iliac musculature was an increase in volume. However, the ancestral dinosaur modified the pelvis much more than other ornithodirans ancestrally. It is interesting that after dinosaurs differentiated, no major modifications occurred in hindlimb anatomy: saurischians are diagnosed by nine unequivocal synapomorphies (Novas 1994), none of which pertains to the hindlimb and only one of which pertains to the pelvis (e.g., an ischiadic shaft that is rod-like and triangular in cross-section). Ornithischians, in contrast, were more "progressive" than saurischians with reference to the pelvic girdle and hindlimb morphology: they show a few transformations in hindlimb anatomy (three synapomorphies have been recognized in the femur and pes by Sereno [1986]), but the pelvis was highly modified, accumulating 11 derived features (Sereno 1986).

Most of the synapomorphies diagnostic of Ornithodira, Dinosauromorpha, Dinosauriformes, and Dinosauria are involved with hindlimb anatomy. Thus, it is reasonable to conclude that the early evolution of Ornithodira was characterized by improved locomotor capabilities (e.g., erect hindlimbs, bipedality, digitigrady). Although new discoveries may demonstrate that other portions of the skeleton changed extensively during the early evolution of ornithodirans, it is clear that they improved locomotion. Modifications in their pelvic girdle and hindlimb anatomy were more profound than in their scapular girdle and forelimbs. Only pterosaurs and theropods manifested extensive transformations in the latter skeletal elements (Gauthier 1986; Sereno 1991a; Novas 1994).

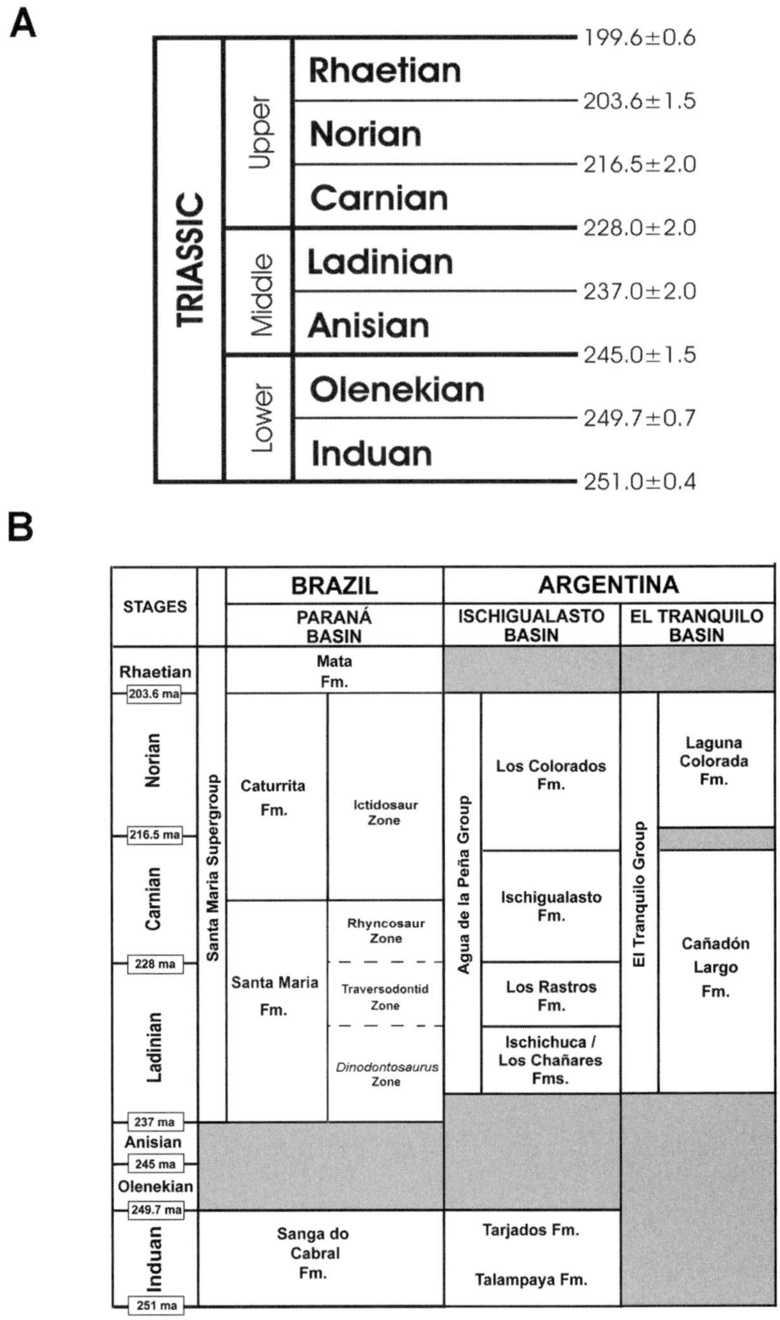

Fig. 2.1. (A) Geologic scale of the Triassic. (B) Stratigraphy of dinosaur-bearing Triassic formations of Brazil and Argentina.

A

TRIASSIC			
	Upper	Rhaetian	199.6±0.6
		Norian	203.6±1.5
		Carnian	216.5±2.0
			228.0±2.0
	Middle	Ladinian	237.0±2.0
		Anisian	
			245.0±1.5
	Lower	Olenekian	249.7±0.7
		Induan	
			251.0±0.4

B

STAGES		BRAZIL		ARGENTINA	
		PARANÁ BASIN		ISCHIGUALASTO BASIN	EL TRANQUILO BASIN
Rhaetian 203.6 ma		Mata Fm.			
Norian 216.5 ma	Santa Maria Supergroup	Caturrita Fm.	Ictidosaur Zone	Los Colorados Fm.	Laguna Colorada Fm.
Carnian 228 ma		Santa Maria Fm.	Rhyncosaur Zone	Ischigualasto Fm.	Cañadón Largo Fm.
Ladinian			Traversodontid Zone	Los Rastros Fm.	
			Dinodontosaurus Zone	Ischichuca / Los Chañares Fms.	
237 ma Anisian 245 ma Olenekian 249.7 ma					
Induan 251 ma		Sanga do Cabral Fm.		Tarjados Fm. Talampaya Fm.	

Agua de la Peña Group; El Tranquilo Group

TRIASSIC DINOSAURS   2

Dinosaurs originated, radiated, and became the dominant group of large animals in continental tetrapod communities of the world during the Triassic, a period that lasted from 251 through 199 million years ago that was a time of considerable faunal and floral change (Fig. 2.1A).

At the onset of the Triassic, landmasses were assembled into a single large supercontinent called Pangea that had considerable latitudinal spread and was positioned nearly symmetrically about the Equator. A single vast ocean (Panthalassa—the forerunner of the modern Pacific) covered the rest of the globe and had a westward-extending arm, the Tethys Sea. The Triassic was a time of great continental emergence due to a combination of widespread epeirogenic uplift and a relatively low sea level. Marine environments were mostly confined to the Tethys and the circum-Pacific (Lucas and Orchard 2004).

This peculiar geographic configuration of the Earth facilitated widespread faunal and floral exchange, so continental Triassic animals and plants are quite similar across Pangea. However, with the onset of rifting in the Gulf of Mexico basin in Late Triassic times, this supercontinent started to separate into two landmasses, Gondwana to the south and Laurasia to the north.

The climate during Triassic times was characterized by increased warmth with relatively wide subtropical desert belts roughly at 10 degrees to 30 degrees latitude. Climate models suggest that seasonality was monsoonal with only two seasons, wet and dry, but with little annual temperature fluctuation. Abundant rainfall occurred during the summer months (Wing and Sues 1992). The vegetation of Gondwana during the Triassic was dominated by the pteridosperms (e.g., *Dicroidium*), an extinct group of basal gymnosperms informally known as seed ferns.

The portion of Gondwana corresponding to South America was mainly continental, with the exception of a shallow sea that covered most of Perú, central Bolivia, and northern Chile and a small portion of northern Argentina (Fig. 2.2A; Bossi 1990; Uliana and Biddle 1988). Other marine incursions were located in Ecuador and Colombia. A Late Permian marine transgression that covered the eastern side of South America (e.g., large parts of Brazil, Uruguay, and Argentina) retreated at the beginning of the Triassic, followed by deposits of red beds and evaporites (Bossi 1990). The Triassic sedimentary fills of the Paraná Basin in southernmost Brazil are the continental red beds of the Santa Maria Supergroup (e.g., Barberena, Araujo, and Lavina 1985).

The Triassic sedimentary accumulations in central Chile and west-central and southern Argentina were deposited in a complex system of

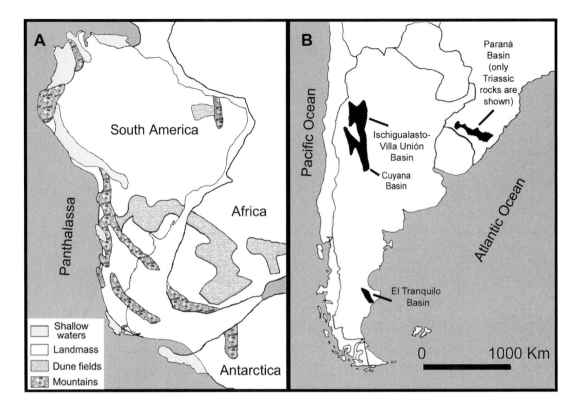

Fig. 2.2. (A) Paleogeographic map of South America during the Triassic. (B) Map of southern South America indicating Triassic sedimentary basins of Argentina and Brazil that have yielded dinosaur remains. *(A) Redrawn from Bossi (1990).*

northwest-trending narrow and elongated troughs that developed as result of the breakup of the Pangean supercontinent (Fig. 2.2B; Uliana and Biddle 1988). Most of the southwestern Gondwana Triassic basins, such as the extensional basins of Chile and western Argentina and the Karoo Basin of southern Africa, were directly related to the Sierra de la Ventana–Cape Fold Belt, which was part of a Gondwanan "cordillera" that diagonally crossed Argentina and southern Africa (Bossi 1990; Zerfass, Lavina, Schultz, Vasconcellos Garcia, Faccini, and Chemale 2003).

In central-western Argentina (e.g., Mendoza, San Juan, and La Rioja provinces), two large depocenters, the Triassic Cuyana and Ischigualasto–Villa Unión basins, were filled by thick, predominately fluvial, lacustrine, and volcaniclastic deposits (e.g., pyroclastic deposits and lava flows).

Although Triassic rock successions in South America are predominantly non-marine, only those from the El Tranquilo (northeastern Patagonia), Ischigualasto–Villa Unión (northwestern Argentina), and Paraná (southern Brazil) basins have yielded both dinosaur remains and abundant fossils of other vertebrates and plants (Fig. 2.1B). These sedimentary basins are examined below.

Triassic Dinosaur-Bearing Beds in South America

Sedimentary rock sequences rich in terrestrial fossil remains are exposed in northwestern Argentina and Southern Brazil that accumulated almost without interruption from Mid-Triassic times (e.g., Ladinian, 237 million years ago) through the end of the Triassic (Late Norian, 203 million years ago). Thus, the Triassic record of South America reveals approximately 34 million years of archosaur evolution, illustrating the sequence of morphological transformations that began with primitive, quadrupedal, sprawling, and plantigrade reptiles and led to small, bipedal, erected-limbed, digitigrade early dinosaurs.

The oldest known dinosaurs and their most closely related outgroups have been documented in Triassic beds in Argentina and Brazil. They include a still-growing list of forms, most of which are represented by several specimens, some of them fairly complete, belonging to different dinosaurian subclades. These basal dinosaurs (and their kin) provide the best available information about early dinosaur anatomy, and an extensive literature exists on this topic (Romer 1971; Reig 1963; Casamiquela 1967; Colbert 1970; Bonaparte 1975, 1976; Bonaparte, Ferigolo, and Ribeiro 1999; Bonaparte, Brea, Schultz, and Martinelli 2007; Galton 1977; Novas 1986, 1992b, 1994; Sereno 1994; Sereno and Novas 1992, 1994; Sereno and Arcucci 1993, 1994; Sereno, Forster, Rogers, and Monetta 1993). Interpretations of the phylogenetic relationships and the timing and mode of the early evolutionary radiation of Dinosauria mainly rely on such South American discoveries.

1. Argentina

1.A. ISCHIGUALASTO–VILLA UNIÓN BASIN

This basin is located in northwestern Argentina, at the border between San Juan and La Rioja provinces (Stipanicic and Bonaparte 1979; Figs. 2.2B and 2.3). It is filled by an almost continuous 4,000-meter thick sequence of sediments of Early, Middle, and Late Triassic ages, constituting the Early Triassic Talampaya and Tarjados formations, which are overlain by the Agua de la Peña Group (Fig. 2.1B). The highly fossiliferous Agua de la Peña Group is composed (from bottom to top) of the Los Chañares–Ischichuca (approximately Early Ladinian), Los Rastros (approximately Late Ladinian), Ischigualasto (Carnian), and Los Colorados (possibly Late Norian) formations.

Of these, the units that have yielded fossil remains of several archosaurs of principal interest for the early evolution of dinosaurs are the Los Chañares, Ischigualasto, and Los Colorados formations (Bonaparte 1982).

1.A.1. THE LOS CHAÑARES FORMATION   This unit has good exposures in La Rioja Province at the Los Chañares site and it is stratigraphically equivalent to the Ischichuca Formation, which crops out in San Juan Province. These beds were explored by Alfred Romer and later by José Bonaparte and his assistants in 1964 (Fig. 2.4A). The Los Chañares Formation, whose age is estimated to be Early Ladinian, is made up of grey tuffaceous fluvial and lacustrine sandstones and siltstones. A wide diversity of tetrapods has been recovered from these beds (Romer 1972; Bonaparte 1975; Arcucci 1986, 1987; Sereno and Arcucci 1993, 1994).

Fig. 2.3. Map depicting Triassic fossil localities of Argentina (shown in black in upper map). The lower map shows the provinces of Argentina. The two northern provinces (La Rioja and San Juan) and the southern province (Santa Cruz) that have yielded dinosaurs are shown in black. Sites that have produced particular dinosaur finds are indicated in the individual provincial maps.

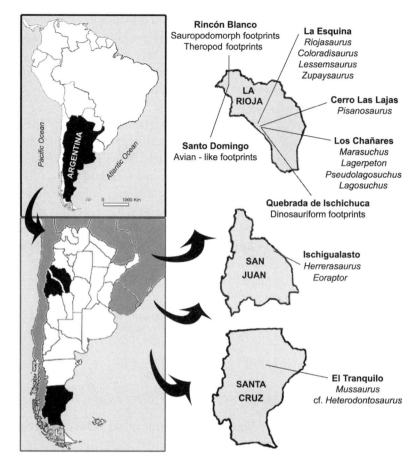

Among synapsids, three family-level groups of cynodonts were reported: Traversodontidae (including *Massetognathus major*, *Massetognathus pascuali*, *Massetognathus terugii*, and *Megagomphodon oligodens*), Chiniquodontidae (represented by *Probelesodon lewisi* and *Probelesodon minor*), and Probainognathidae (represented by *Probainognathus jenseni*). The dicynodonts from the Los Chañares Formation (Fig. 2.4B) include members of the family Stahleckeriidae (i.e., *Dinodontosaurus brevirostris*, *Dinodontosaurus platygnathus*, and *Chanaria platyceps*) and an indeterminate kannemeyerid. Among Archosauria, several proterochampsid genera were recovered from these beds, such as *Chanaresuchus bonapartei*, *Gualosuchus reigi*, and *Tropidosuchus romerii*, as well as the crocodylotarsians *Gracilisuchus stipanicicorum* and *Luperosuchus fractus*. The oldest dinosauromorphs were found in the Los Chañares Formation, including *Lagosuchus talampayensis*, *Pseudolagosuchus major*, *Marasuchus lilloensis*, *Lagerpeton chanarensis*, and probably *Lewisuchus admixtus* (interpreted by Arcucci [1997, 2005] as a possible senior synonym of *Pseudolagosuchus major*).

1.A.II. THE LOS RASTROS FORMATION This unit (whose name means "the tracks," in reference to the archosaur footprints found in its exposures) is formed by lacustrine black shales and deltaic sandstones

Fig. 2.4. (A) Outcrops of the Los Chañares Formation, La Rioja Province, northwest Argentina; Early Middle Triassic Los Chañares fauna. (B) A group of *Marasuchus* investigates a carcass of the cow-sized dicynodont *Dinodontosaurus*. *(A) Photograph courtesy of José Bonaparte. (B) Illustration by J. González.*

with intercalations of coal beds and abundant plant remains. The Los Rastros Formation has yielded the remains of a temnospondyl tetrapod interpreted as Chigutisauridae indet. as well as tridactyl footprints belonging to bipedal animals, probably dinosauriforms or dinosaurs (Marsicano, Arcucci, Mancuso, and Caselli 2004).

1.A.III. THE ISCHIGUALASTO FORMATION   Extensive outcrops of this paleontologically rich and important formation are in Ischigualasto, a place informally known as the Valley of the Moon in San Juan Province (Plate 1). Early explorations mainly recovered plant remains of the *Dicroidium* flora, but successive field trips, starting with the pioneering work of Alfred Romer and Osvaldo Reig, discovered a wide diversity of fossil vertebrates that were thoroughly reviewed by José Bonaparte in

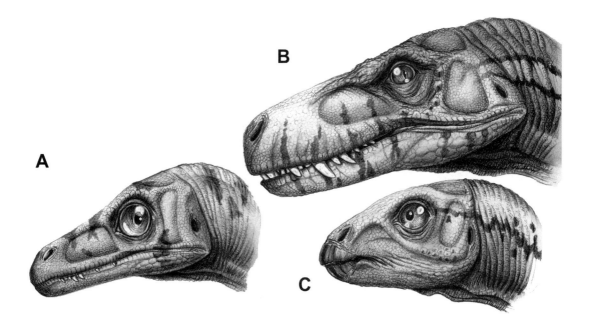

Fig. 2.5. Reconstructed heads of dinosaurs from the early Upper Triassic Ischigualasto fauna. Not to scale. (A) *Eoraptor* (skull length 13 cm). (B) *Herrerasaurus* (skull length 32 cm). (C) *Pisanosaurus* (skull length approximately 12 cm). Illustrations by J. González.

several authoritative publications (1973, 1982, 1997a). The age of the Ischigualasto beds has been widely accepted as Carnian, an interpretation supported by isotopic dating of a bentonite bed in the lower part of the formation dated as earliest Carnian (ca. 228 million years; Rogers, Swisher, Sereno, Monetta, Forster, and Martínez 1993). The Ischigualasto Formation is made up of fluvial sandstones, mudstones, and tuffs that are predominantly gray in color.

The tetrapods from the Ischigualasto beds constitute one of the most diversified Triassic faunas worldwide (Bonaparte 1973, 1982; Langer 2005b), at present represented by amphibian mastodonsaurids (*Promastodonsaurus bellmani*) and chigutisaurids (*Pelorocephalus ischigualastensis*), large dicynodonts (*Ischigualastia jenseni*), a wide variety of cynodonts (*Ecteninion lunensis*, cf. *Probainognathus* sp., *Chiniquodon sanjuanensis*, *Exaeretodon frenguellii*, and *Ischignathus sudamericanus*), and a notable abundance but low diversity of rhynchosaurs (*Hyperodapedon sanjuanensis* and *H. mariensis*), represented by adult, juvenile, and hatchling specimens. The Ischigualasto Formation is also famous for its set of archosaurs, comprised of several disparate clades that include proterochampsids (*Proterochampsa barrionuevoi*, cf. *Chanaresuchus*), probable sphenosuchids (*Trialestes romeri*), aetosaurs (*Aetosauroides scagliai*), ornithosuchids (*Venaticosuchus rusconii*), rauisuchians (*Saurosuchus galilei* and *Sillosuchus longicervix*), basal saurischians (*Eoraptor lunensis* and *Herrerasaurus ischigualastensis*), and possible ornithischians (*Pisanosaurus mertii*) (Fig. 2.5).

1.A.IV. THE LOS COLORADOS FORMATION This unit is widely exposed in La Rioja Province, close to the border with San Juan Province (Fig. 2.6). The Los Colorados Formation is composed of fluvial red

Fig. 2.6. Outcrops of the Los Colorados Formation, San Juan Province, northwest Argentina. *Photograph by Fernando Novas.*

beds that are possibly Late Norian in age. The upper levels of this unit have yielded highly prolific tetrapod remains (Fig. 2.7), particularly prosauropods, which were collected thanks to the intensive work done by José Bonaparte in several field trips during the 1960s and 1970s. Fossils from this unit constitute three species of sauropodomorphs (*Riojasaurus incertus*, *Coloradisaurus brevis*, and *Lessemsaurus sauropoides*), the coelophysoid theropod *Zupaysaurus rougieri*, the remains of a small indeterminable theropod, the crocodylomorphs *Hemiprotosuchus leali* and *Pseudohesperosuchus jachaleri*, the aetosaurian *Neoaetosauroides engaeus*, the ornithosuchian *Riojasuchus tenuiceps*, the rauisuchian *Fasolasuchus tenax*, the derived cynodont *Chaliminia musteloides*, and the chelonian *Paleochersis talampayensis* (Bonaparte 1972, 1979a, 1982, 1999c; Rougier, de la Fuente, and Arcucci 1995; Arcucci and Coria 2003; Ezcurra and Novas 2005, 2007). However, at the base of the Los Colorados Formation, close to the boundary with the underlying Ischigualasto beds, Bonaparte recovered the skull of the dicynodont *Jachaleria colorata*. To this body of fossils are added some chirotheroid footprints (Arcucci, Marsicano, and Caselli 2004).

This rich tetrapod fauna has been interpreted as representing a unique transitional assemblage that has elements typical of both the Late Triassic and the Early Jurassic. Some have challenged this interpretation, arguing instead that the Los Colorados fauna are a mixture of horizons of different ages (e.g., Olsen and Sues 1986; Shubin and Sues 1991; Benton 1994). However, recent stratigraphic analysis of the levels of the Los

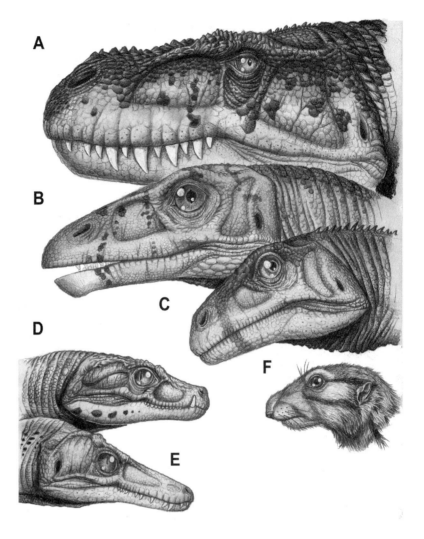

Fig. 2.7. Reconstructed heads of some typical tetrapods from the late Upper Triassic Los Colorados fauna. Not to scale. (A) Rauisuchid archosaur *Fasolasuchus tenax* (skull length 75 cm). (B) Prosauropod *Riojasaurus incertus* (skull length 25 cm). (C) Prosauropod *Coloradisaurus brevis* (skull length 20 cm). (D) Early crocodyliform *Hemiprotosuchus leali* (skull length 8 cm). (E) Sphenosuchian archosaur *Pseudohesperosuchus jachaleri* (skull length 15 cm). (F) Derived cynodont *Chaliminia musteloides* (skull length 3 cm). *Illustrations by J. González.*

Colorados that include tetrapods (Caselli, Marsicano, and Arcucci 2001; Arcucci, Marsicano, and Caselli 2004) confirms the transitional nature of the association. Typical Triassic taxa (e.g., aetosaurs) have been found in the same beds with advanced theropod dinosaurs and cynodonts, which are known from Early Jurassic levels in other Gondwanan and Laurasian areas. This faunal association is not based on isolated remains (e.g. teeth) that were perhaps moved after death but on almost complete and articulated skeletons (Bonaparte 1972).

1.A.V. THE SANTO DOMINGO FORMATION This unit also belongs to the Ischigualasto–Villa Unión Basin but crops out in La Rioja Province in the northwestern corner of the basin. This Norian–Early Jurassic unit is important because it yielded bird-like footprints (Melchor, de Valais, and Genise 2002).

1.B. THE EL TRANQUILO BASIN

The southernmost Triassic deposits of South America are those that fill the El Tranquilo Basin in northeastern Santa Cruz Province (Fig. 2.3). These beds constitute the El Tranquilo Group, which is divided into the lower Cañadón Largo Formation (considered to be Ladinian–Carnian in age) and the overlying Laguna Colorada Formation. Both units preserve abundant remains of *Dicroidium* flora, but the Laguna Colorada Formation has also yielded abundant tetrapod remains, including the prosauropod dinosaur *Mussaurus patagonicus*, which is known by adult individuals, post-hatchling juveniles, eggs, and nests (Bonaparte and Vince 1979), and an fragmentary heterodontosaurid ornithischian collected from the same locality as the articulated juveniles of *Mussaurus* (Báez and Marsicano 2001).

The age of the Laguna Colorada Formation is usually considered to be Late Triassic (Norian) based on the *Dicroidium* flora found below and above the horizon where the *Mussaurus* remains were found (Fig. 2.1B). Furthermore, the El Tranquilo Group is intruded upon by the granitoids of the La Leona Formation. Radiometric dating of these granitoids yielded an age of 203 million years (Rapela and Pankhurst 1996), providing an upper bound for the age of El Tranquilo Group (Stipanicic and Marsicano 2002).

2. Brazil

The Triassic rocks of the Paraná Basin in the state of Rio Grande do Sul in the south of Brazil (Fig. 2.8) are represented by the Early Triassic Sanga do Cabral Formation and the overlying Ladinian to Rhaetian Santa Maria Supergroup comprised of the Santa Maria, Caturrita, and Mata formations (see Rubert and Schultz 2004, Langer 2005a, and Langer, Ribeiro, Schultz, and Ferigolo 2007 for an updated and detailed review of the Triassic beds and faunas from Brazil). Along most of its outcrop area, these Triassic beds are directly overlain by the Late Jurassic to Early Cretaceous Botucatu (aeolian sandstones) and Serra Geral (lava floods) formations, the latter forming an extensive plateau (the Planalto Sul-Rio–Grandense) that limits the outcrop area of the Triassic units (Zerfass, Lavina, Schultz, Vasconcellos Garcia, Faccini, and Chemale 2003).

The beds of the Santa Maria Supergroup are restricted to a sinuous belt running roughly 500 kilometers west to east in the State of Río Grande do Sul west of the city of Porto Alegre. The most relevant Triassic localities of the Paraná Basin are Santa Maria, Candelaria, and Chiniquá, which have been intensively worked beginning with the explorations carried on by the great German paleontologist Friedrich von Huene at the end of the 1920s.

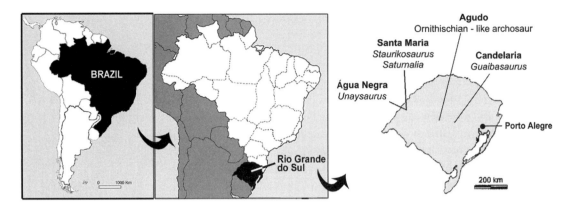

Fig. 2.8. Map depicting Triassic fossil localities of Brazil (shown in black in left map). The map in the center shows the Brazilian state of Rio Grande do Sul (shown in black) that has yielded dinosaurs. The map on the right indicates sites that have produced particular dinosaur finds in the state of Rio Grande do Sul.

2.A. THE SANTA MARIA FORMATION

This formation is subdivided into the following biostratigraphic subunits (from older to younger): the Eoladinian *Dinodontosaurus* Zone (or "Therapsid Cenozone"; Barberena, Araujo, and Lavina 1985; Schultz and Langer 2007), the Neoladinian Traversondontid Zone (Schultz and Langer 2007), and the Carnian Rhynchosaur Zone (Fig. 2.1B).

In the city of Candelária, the lower levels of the Santa Maria Formation (i.e., the *Dinodontosaurus* Zone), yielded remains of dicynodonts, cynodonts, chanaresuchids, and large archosaurs that are taxonomically similar to those recovered from the Los Chañares Formation in Argentina (Barberena, Araujo, and Lavina 1985) as well as procolophonids and rhynchosaurs (Schultz and Langer 2007) that have not yet been recorded in the Los Chañares Formation. No remains of dinosaurians or Dinosauriformes have been recovered so far from this biostratigraphic unit. From these beds, which crop out close to the Chiniquá region, came the partial skeleton of *Spondylosoma absconditum*, which for a long time was considered a saurischian dinosaur (von Huene 1942; Romer 1956; Colbert 1970) but was recently recognized as a rauisuchian archosaur (Galton 2000).

The upper levels of the Santa Maria Formation (i.e., the Rhynchosaur Zone) crop out mainly near the cities of Santa Maria and the district of Inhamandá. The Rhynchosaur Zone is almost equivalent in age with the Ischigualasto Formation of northwestern Argentina because of the presence of abundant rhynchosaur remains (*Hyperodapedon*). From the layers of the Rhynchosaur Zone the following animals have been discovered (Schultz and Langer 2007): the cynodonts *Exaeretodon riograndensis, Gomphodontosuchus brasiliensis, Charruodon tetracuspidatus, Therioherpeton cargnini,* and *Prozostrodon brasiliensis*; the rhynchosaurs *Hyperodapedon mariensis, Hyperodapedon sanjuanensis,* and *Hyperodapedon huenei*; the crurotarsan archosaurs *Cerritosaurus binsfeldi, Proterochampsa nodosa, Hoplitosuchus raui, Rauisuchus tiradentes, Rhadinosuchus gracilis,* and *Aetosauroides* sp.; and the basal saurischian dinosaurs *Staurikosaurus pricei* and *Saturnalia tupiniquim*.

2.B. THE CATURRITA FORMATION

This unit overlies the Santa Maria Formation; it is seen, for example, near Mount Botucaraí in Candelária (Rubert and Schultz 2004; Langer 2005a). The Caturrita beds yielded fossils of the dicynodont *Jachaleria candelariensis* (Araújo and Gonzaga 1980), the occurrence of which seems to mark a faunal stage almost equivalent with that of the base of the Los Colorados Formation of northwestern Argentina (i.e., Early Norian). Additional material from the same outcrop includes vertebrae and partial pelvic elements of a possible herrerasaurid dinosaur (Kischlat and Barberena 1999) as well as a fragmentary mandible symphysis with phytosaur affinities (Kischlat and Lucas 2003). Aside from Mount Botucaraí, two other productive localities of this formation were recently discovered, namely Faxinal do Soturno (Rubert and Schultz 2004; Fig. 2.9) and Água Negra (Langer 2005a; Leal, Azevedo, Kellner, and da Rosa 2004).

To date, the fossil content of the Caturrita beds includes the bizarre dinosauriform *Sacisaurus agudoensis* (Ferigolo and Langer 2007); the basal saurischian *Guaibasaurus candelariensis* (Bonaparte, Ferigolo, and Ribeiro 1999; Bonaparte, Brea, Schultz, and Martinelli 2007); the prosauropod dinosaur *Unaysaurus tolentinoi*, a possible herrerasaurid (Kischlat and Barberena 1999); small cynodonts such as *Riograndia guaibensis, Irajatherium hernandezi, Brasilodon quadrangularis,* and *Brasilitherium riograndensis* (Bonaparte, Martinelli, Schultz, and Rubert 2003; Rubert and Schultz 2004; Langer 2005a); the dicynodont *Jachaleria candelariensis* (Araújo and Gonzaga 1980); the sphenodontian *Clevosaurus brasiliensis* (Bonaparte and Sues 2006), a possible phytosaur (Kischlat and Lucas 2003); and the procolophonid *Soturnia caliodon* (Cisneros and Schultz 2003).

The faunal assemblage of the Caturrita Formation does not conform well with the Ischigualasto or the Los Colorados faunas but seems to represent a transitional stage in Triassic tetrapod evolution. It includes basal saurischians that are more derived than herrerasaurids, and prosauropods that are more advanced than *Saturnalia*.

The Triassic Dinosaur Record in South America

Paleontological exploration of Triassic beds in South America started as early as 1928–1929, when German paleontologist Friedrich von Huene (Fig. 2.10), from the University of Tübingen, visited the outcrops of the Santa Maria Formation in southeastern Brazil and made a quick inspection of the Ischigualasto fossil locality in northwestern Argentina. He described remains of a wide variety of fossil tetrapods, including rauisuchians, rhynchosaurs, and dicynodonts. Von Huene (1942) also interpreted one of his discoveries, the archosaur *Spondylosoma absconditum*, as an early representative of the saurischian dinosaurs. Some years later, in 1936, an expedition from the Museum of Comparative Zoology of Harvard, which included Brazilian paleontologist Llewellyn Ivor Price, spent many months making systematic collections of Triassic reptiles around

Fig. 2.9. Outcrops of the Caturrita Formation exposed at Faxinal do Soturno, State of Río Grande do Sul, southeast Brazil. (A) General aspect of the outcrops. (B) The richness of this new fossil site is expressed by the abundance of nicely preserved synapsid jaws. *Photographs courtesy of José Bonaparte.*

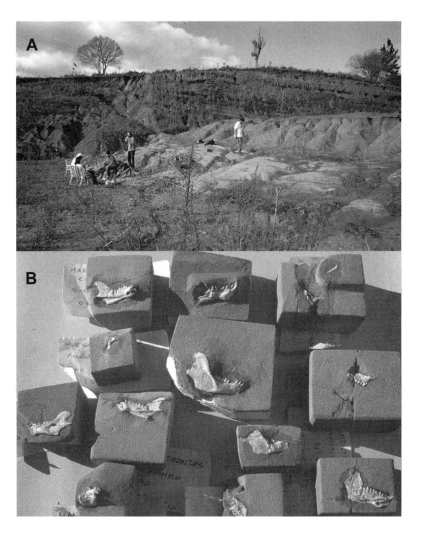

Santa Maria city, resulting in the discovery of *Staurikosaurus pricei*, a basal saurischian dinosaur described and named by Edwin Colbert in 1970. Important fossil tetrapod collections were made through the efforts of Professor Mario Costa Barberena and his assistants from the Federal University of Río Grande do Sul in Porto Alegre. Although Barberena did not describe dinosaur remains, he and his crew provided valuable biostratigraphical information for understanding the Permian–Triassic faunal succession in Brazil (e.g., Barberena, Araujo, and Lavina 1985).

In 1958, Alfred S. Romer, a renowned professor from Harvard University, conducted the first exploration to Ischigualasto in San Juan Province in northwestern Argentina in search of Triassic fossil reptiles (Fig. 2.11). Romer collected a formidable amount of fossil vertebrates (most of them are housed at Harvard University), but he did not describe any of the material excavated in that trip. Although only a few of the specimens Romer collected were left in Argentine museums, his visit triggered the deep interest of Argentine paleontologist Osvaldo A. Reig (and later José F. Bonaparte and Rodolfo Casamiquela; Fig. 2.12) in digging in these

Fig. 2.10. Portrait of the German paleontologist Friedrich von Huene, taken around the end of the 1930s. *Archives of the Institut für Palaeontologie, Tübingen University. Courtesy of Michael Maisch.*

wonderful and prolific localities of Argentina and in carrying out their own pioneering studies of dinosaur origins. Soon after Alfred Romer left the Ischigualasto Valley in 1958, the location was explored with great success by Osvaldo A. Reig (at that time a professor at the University of Tucumán) and José F. Bonaparte. Later, Reig reported the first Triassic dinosaurs from South America, which for long time constituted the oldest known examples in the world. In 1963, Reig named *Herrerasaurus ischigualastensis*, *Ischisaurus cattoi* (later interpreted as a junior synonym of *Herrerasaurus*; Novas 1994), and *Triassolestes romerii* (later renamed *Trialestes romerii* and reinterpreted as a sphenosuchian crocodylimorph; Bonaparte 1972).

Rodolfo Casamiquela, a multitalented naturalist at that time working at the Museum of Natural Sciences in La Plata, Argentina, exhumed prosauropod skeletons at El Tranquilo farm in several explorations from 1962 to 1968, which he originally described as an unknown species of

Fig. 2.11. American paleontologists Alfred Romer and Bryan Patterson. *Photograph of the Museum of Comparative Zoology, Harvard University.* © Copyright President and Fellows of Harvard College, Museum of Comparative Zoology, Harvard University.

the European genus *Plateosaurus* (Casamiquela 1977). Casamiquela also published the first description of *Pisanosaurus mertii*, a specimen from the Ischigualasto beds, widely considered to be the oldest known ornithischian dinosaur (Casamiquela 1967).

Romer returned to Argentina in 1964 to conduct field exploration in La Rioja Province, where he and his crew discovered the new and spectacular Triassic fossil site named Los Chañares. This site yielded remains of dicynodonts, a variety of cynodonts, and different kinds of archosaurs, including the long-limbed, bipedal, and small-sized *Lagerpeton chanarensis*, *Lagosuchus talampayensis*, and *Lagosuchus lilloensis*, which have been very relevant in the elucidation of dinosaur origins.

Following Reig, José F. Bonaparte spent most of his energies in the 1960s and 1970s on excavations in different Triassic outcrops of Argentina, including Ischigualasto, Los Chañares, Los Colorados, and El Tranquilo. He and his group of collaborators from the Instituto Miguel Lillo in Tucumán made interesting discoveries of more-complete specimens of *Lagosuchus talampayensis* (described by Bonaparte in 1975) as well as several prosauropod dinosaurs, including *Riojasaurus incertus* (Bonaparte 1972), *Mussaurus patagonicus* (Bonaparte and Vince 1979), *Coloradisaurus brevis* (Bonaparte 1979a), and *Lessemsaurus sauropoides* (Bonaparte 1999c).

However, after the highly productive "Triassic rush" of Romer, Reig, Casamiquela, and Bonaparte in the 1960s (which resulted in a long series of seminal publications on Triassic tetrapods), explorations in the Triassic beds of Argentina were considerably curtailed. In 1988, a joint exploration

Fig. 2.12. A meeting of vertebrate paleontologists in 1968. From left to right: Alfred Romer, José Bonaparte, William Sill, Rodolfo Casamiquela, Rosendo Pascual, and Osvaldo Reig. *Photograph courtesy of Rosendo Pascual.*

carried on by the Universidad Nacional de San Juan, the Museo Argentino de Ciencias Naturales, and the University of Chicago became interested once again in the search for the oldest dinosaurs (Fig. 2.13). Discoveries during these joint explorations included complete skeletons of *Herrerasaurus ischigualastensis*, which was described in several publications by Paul Sereno and the present writer, and *Eoraptor lunensis*, which improved our knowledge of early dinosaurs and closely related creatures (e.g., Novas 1992b, 1994, 1996a, 1997b; Sereno and Novas 1992, 1994; Sereno 1994; Sereno, Forster, Rogers, and Monetta 1993).

Another hiatus in the discovery and study of Triassic dinosaurs and dinosaur forerunners in Argentina followed the publication of *Eoraptor* in 1993, but a wealth of very interesting findings began to be made in South Brazil in 1999 through the work of José Bonaparte and young scientists from different Brazilian institutions, in particular Max Cardoso Langer, resulting in the discovery of *Guaibasaurus candelariensis* (Bonaparte, Ferigolo, and Ribeiro 1999), *Saturnalia tupiniquim* (Langer, Abdala, Richter, and Benton 1999), and *Unaysaurus tolentinoi* (Leal, Azevedo, Kellner, and da Rosa 2004), as well as intriguing new archosaurs with ornithischian-like features in the skull (Langer and Ferigolo 2005; Ferigolo and Langer 2007).

At present, seven species of basal dinosaurs (*Herrerasaurus ischigualastensis*, *Staurikosaurus pricei*, *Guaibasaurus candelariensis*, *Eoraptor lunensis*, *Saturnalia tupiniquim*, *Unaysaurus tolentinoi*, and *Pisanosaurus mertii*) have been documented in Carnian and Early Norian beds in South America, represented by almost-complete skeletons and in some

Fig. 2.13. Field exploration carried out in 1988 by the Universidad Nacional de San Juan, the Museo Argentino de Ciencias Naturales, and the University of Chicago. (A) Carrying a jacket containing a skeleton of *Herrerasaurus* (from left to right: Cathy Forster, Jorge Garay, Fernando Novas, Oscar Alcober, Paul Sereno, Yu Chao, Bill Stevens, and José Gómez). (B) The same jacket during preparation. *Photographs by Liliana R. Lo Coco.*

cases more than one specimen. Thus, the South American fossil record of basal dinosaurs currently constitutes the best source for anatomical, phylogenetic, and paleoecological study. Moreover, this core of early dinosaurs is chronostratigraphically preceded by a set of less-derived archosaurs (i.e., *Lagerpeton*, *Marasuchus*, and *Pseudolagosuchus*) that illustrate the evolutionary transformations leading to Dinosauria. The sequence of Mid-Triassic dinosaur forerunners and early dinosaurs of Carnian age is followed by Late Norian units (e.g., the Los Colorados and El Tranquilo formations) that have yielded remains of sauropodomorphs (e.g., *Mussaurus*, *Coloradisaurus*, *Riojasaurus*, and *Lessemsaurus*) and basal theropods (i.e., the coelophysoid *Zupaysaurus*).

However, the phylogenetic relationships of these animals with other saurischians and ornithischians continue to be debated (see Langer 2004 for a comprehensive review).

Dinosauria

TEYUWASU BARBERENAI

Among the several archosaurs Friedrich von Huene described from Santa Maria is the rauisuchian archosaur *Hoplitosuchus raui*, recovered from levels of the Alemoa Member cropping out at Sanga Grande. More recently, Kischlat (1998) pointed out that some hindlimb elements (i.e., right femur and tibia) of *Hoplitosuchus*-type material correspond in fact to a robust dinosaur, for which he coined the name *Teyuwasu barberenai* (Kischlat 1998). Although referring hindlimb bones to Dinosauria (or, more cautiously, to Dinosauriformes) might be correct (for example, the tibia has a proximal cnemial crest and the distal articular surface is rounded and has a developed distal process for articulation with the ascending process of astragalus), Kischlat listed no autapomorphic attributes in support of his new taxon. He mentioned that the femur of *Teyuwasu* lacks a trochanteric shelf and has only a pair of parallel and proximodistally extended ridges, but this condition is variably present among basal dinosaurs.

In sum, the present information available only permits us to consider *Teyuwasu* as a possible member of Dinosauria, but in the absence of autapomorphic features the validity of this taxon remains doubtful.

Saurischia

Saurischia has been defined as a stem-based taxon that includes all Dinosauria that are closer to *Allosaurus* than to *Stegosaurus* (Gauthier 1986; Padian and May 1993; Sereno 1999; Langer 2004). Saurischia is composed of two major lineages, the plant-eating Sauropodomorpha and the (primitively) meat-eating Theropoda (Fig. 2.14). However, some taxa that are consistently interpreted as saurischians (e.g., Herrerasauridae, *Eoraptor*, and *Guaibasaurus*) do not have a clear phylogenetic allocation within these two main dinosaur branches. Recent cladistic analysis carried out by Max Langer (2004) depicted Herrerasauridae and *Eoraptor* as the sister taxa of a clade composed of Sauropodomorpha plus Theropoda (=Eusaurischia; sensu Langer 2004). For the sake of clarity, I adopt this later phylogenetic hypothesis here, although it is necessary to note that alternative trees depicting herrerasaurs and *Eoraptor* as basal theropods differ in just a few evolutionary steps.

Herrerasauridae

Herrerasauridae is a clade of basal saurischians known from Ischigualastian (Carnian) beds of South America (Reig 1963; Colbert 1970; Benedetto 1973; Novas 1986, 1992b, 1994, 1996a; Sereno 1994; Sereno and Novas 1994) and Carnian and Norian beds from North America (Murry and Long 1989). South American herrerasaurids include *Staurikosaurus pricei* (Colbert 1970; Galton 1977), from the Santa Maria Formation of southwestern Brazil, and *Herrerasaurus ischigualastensis* (Reig 1963), from the

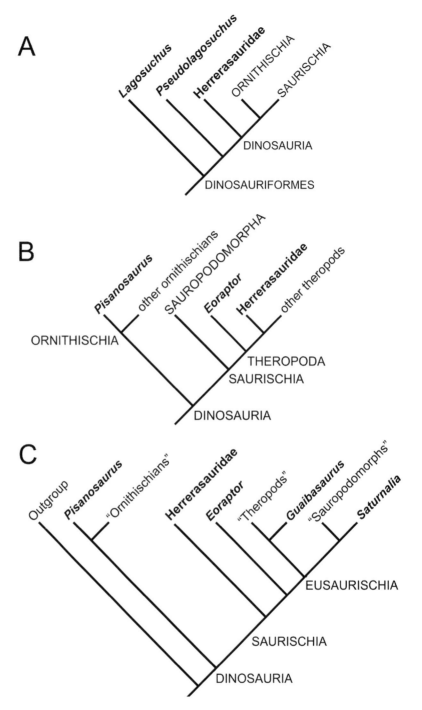

Fig. 2.14. Phylogenetic hypotheses depicting relationships among Saurischia: (A) Hypothesis defended by Novas (1992b). (B) Hypothesis defended by Sereno, Forster, Rogers, and Monetta (1993). (C) Hypothesis defended by Langer (2004). Genera discovered in South America are indicated in bold.

Fig. 2.15. *Herrerasaurus ischigualastensis:* (A) Reconstructed skeleton. (B–E) Skull and jaws in (B) left lateral, (C) right lateral, (D) ventral, and (E) dorsal views. (F) Dorsal vertebra 9 in cranial view. (G) Sequence of articulated dorsal vertebrae 9 through 15 in left lateral view. (H) Caudal vertebrae 1 through 7 in left lateral view. (I) Distal caudal vertebrae 27 through 29 in left lateral view. *(A) From Novas (1997b). Reprinted from* Encyclopedia of Dinosaurs, *Academic Press, San Diego. Eds. Philip Currie and Kevin Padian. Article:* Herrerasauridae. *Author: Fernando Novas. Pages 303–311. © Copyright 1997. With permission from Elsevier. (B–E) From Sereno and Novas (1994). © Copyright 1994 the Society of Vertebrate Paleontology. Reprinted and distributed with permission of the Society of Vertebrate Paleontology. (F–I) From Novas (1994). © Copyright 1994 the Society of Vertebrate Paleontology. Reprinted and distributed with permission of the Society of Vertebrate Paleontology.*

Ischigualasto Formation of San Juan, Argentina. A probable North American representative of this clade is *Chindesaurus briansmalli*, from the Late Carnian–Early Norian Chinle Formation (Murry and Long 1989).

Herrerasaurids ranged from nearly 2 meters (e.g., *Staurikosaurus*) to 5 meters (e.g., *Chindesaurus* and *Herrerasaurus*) in length (Fig. 2.15A). They represent the largest, most abundant, and earliest meat-eating

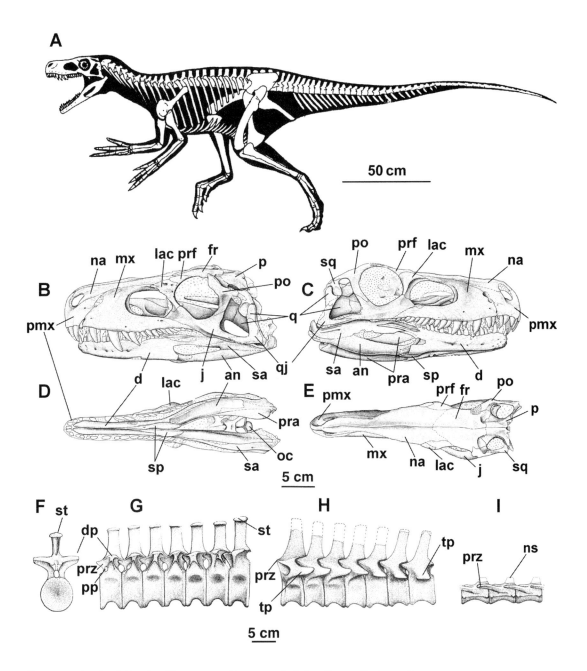

dinosaurs in the Triassic. Herrerasaurids preceded the first radiation of true theropods, Coelophysidae, which dominated during Norian and Early Jurassic times. During the Carnian, herrerasaurids played the role of large predators and were surpassed in size only by the rauisuchians *Saurosuchus*, *Prestosuchus*, and *Postosuchus*, which ranged from 4 to 6 meters in length. South American herrerasaurids evolved in faunas that were numerically dominated by rhynchosaurs and traversodonts, which were very likely their principal prey. Evidence for this is the find of a juvenile rhynchosaur *Hyperodapedon* within the rib cage of one specimen of *Herrerasaurus*, direct evidence of the diet of this animal.

Fig. 2.16. *Herrerasaurus ischigualastensis:* (A) Left scapula in side view. (B–C) Left manus in (B) dorsal and (C) medial views. (D–F) Pelvic girdle and sacrum in (D) dorsal, (E) right lateral, and (F) cranial views. (G–I) Right femur in (G) cranial, (H) lateral, and (I) distal views. (J–K) Right tibia in (J) lateral and (K) distal views. (L) Right foot in dorsal view. *(A–C) From Sereno (1994). © Copyright 1994 the Society of Vertebrate Paleontology. Reprinted and distributed with permission of the Society of Vertebrate Paleontology. (D–L) From Novas (1994). © Copyright 1994 the Society of Vertebrate Paleontology. Reprinted and distributed with permission of the Society of Vertebrate Paleontology.*

Herrerasaurids have a long and low skull that is almost the same length as the femur. The teeth are conical, trenchant, and serrated. *Herrerasaurus* has large "caniniform" teeth in a robust and dorsoventrally deep maxilla (Novas 1986). This set of characteristics, in conjunction with a deep jugal, suggests a powerful buccal apparatus. Also interesting is the presence in herrerasaurids of an intramandibular joint, which means that both the dentary and splenial could slide against the postdentary bones (e.g., surangular and angular, respectively; Fig. 2.15B–D). This kind of mandibular articulation, which is also present in theropod dinosaurs, allowed the toothed anterior segment of the jaw to flex around struggling prey, preventing them from escaping (Sereno and Novas 1994).

Herrerasaurids are readily distinguished from other dinosauriforms (e.g., *Marasuchus, Guaibasaurus, Saturnalia,* and *Coelophysis*) in that the dorsal vertebrae are axially shortened, not only the centra but also the transverse processes and neural spines (Fig. 2.15F and G). The stout neural spines are tall and axially very short. In the anterior dorsal vertebrae, the neural spines are subrectangular in cross-section, but they are squared in the posterior dorsals, the top of which expands into a spine table. Both *Herrerasaurus* and *Staurikosaurus* have accessory intervertebral articulations along the dorsal series, a feature that among dinosauriforms only occurs in saurischian dinosaurs (Gauthier 1986; Novas 1994).

In despite of the derived characteristics of the presacral column, the sacrum is comprised of only two sacrals, the lowest number recorded among dinosaurs (Fig. 2.16D). Originally, I interpreted this characteristic as support for placing Herrerasauridae outside the remaining dinosaurs (Novas 1992a, 1992b; see Fig. 2.15A). However, it is very likely that other basal saurischians (e.g., *Guaibasaurus, Saturnalia*) also had two sacrals. Despite the reduced number of sacrals, the sacrum is widely attached to the ilia through very robust sacral ribs.

The total number of caudals was probably close to 50. The prezygapophyses become longer, so much so that behind caudal 35 the prezygapophyses overlap the posterior half of the preceding vertebra (Fig. 2.15I). This characteristic, which is widely distributed among theropod dinosaurs, probably acted as a stabilization device for running and leaping.

The pectoral girdle and forelimbs of herrerasaurids are also peculiar in the highly derived characteristics they exhibit (Fig. 2.16A–C). For example, the scapular shaft is narrow and distally unexpanded, a condition resembling that of maniraptoran theropods such as dromaeosaurids and birds. The acromial process of the scapula is fairly prominent, which is indicative of a strongly developed M. deltoides clavicularis, a humeral protractor and elevator muscle.

Herrerasaurids were obligatory bipeds, a condition that is widely distributed among primitive dinosaurs. The forelimbs are less than half the length of the hindlimbs, but they have an elongate manus (Sereno and Novas 1992), a condition again resembling that of derived coelurosaurian theropods (Gauthier 1986). Consistent with the predaceous habits inferred from the mandibular and teeth morphology, the forelimbs have

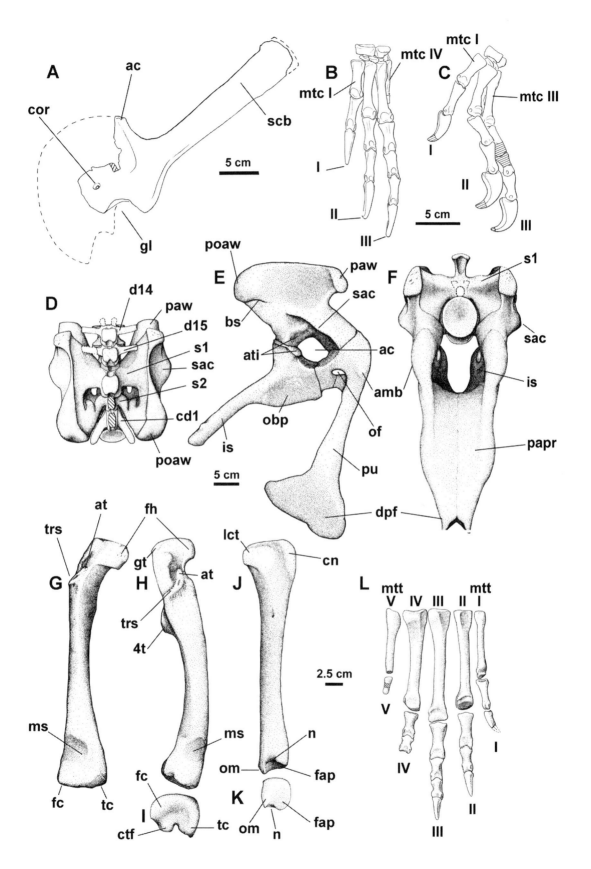

specialized features for capturing prey (Fig. 2.16C). Metacarpals IV and V are splint-like and are considerably more reduced than in any other Triassic dinosaur, including theropods (e.g., *Coelophysis*; Colbert 1989), but digits I, II, and III have elongated phalanges and large and trenchant claws that were adapted for grasping (Sereno and Novas 1992).

The mosaic of derived and primitive features is even more impressive in the pelvic girdle and hindlimb (Fig. 2.16D–F). The ilium is brachyiliac, the acetabulum is bony and is perforated only by a relatively small and elliptical fenestra, and the brevis fossa is represented by a horizontal furrow in the posterolateral surface of that bone, in contrast with the transversely wide fossa present in saurischians and ornithischians (Novas 1992b). The pubis, instead, looks highly derived, with its distal half anteroposteriorly expanded, ending distally in a robust pubic foot that is more than 25 percent of the length of the pubis.

The hindlimb can be described as an enlarged and more robust version of the hindlimb of the basal dinosauriform *Marasuchus* (Bonaparte 1975; Sereno and Arcucci 1994), especially because of trochanteric shelf on the lateral surface of the proximal femur, a modest cnemial crest on the proximal tibia, and the square-shaped distal articular surface of the tibia. Interestingly, the morphology of hindlimb bones of herrerasaurids is very similar to that of the basal sauropodomorph *Saturnalia* (Langer 2003), indicating that it represents the basal pattern for Dinosauria (Fig. 2.16G–L).

The tarsus of herrerasaurids resembles that of basal dinosauriforms such as *Pseudolagosuchus* and may represent the kind of tarsus from which the remaining ankle types of dinosaurs evolved (Novas 1989a). The herrerasaurid foot, known only in *Herrerasaurus*, is constituted by five metatarsals and a phalangeal formula of 2-3-4-5-1. Digits I to IV have slightly curved, non-raptorial ungual phalanges, the longest and most recurved of which corresponds to the second digit.

The anatomy of herrerasaurids is peculiar for the mixture of plesiomorphic and highly derived characteristics. For example, herrerasaurids have two sacral vertebrae associated with an enlarged pubic foot resembling that of the Jurassic *Allosaurus*. This melange of traits created (and still generates) disagreements among paleontologists about the phylogenetic relationships of herrerasaurids. These dinosaurs have been variously considered as theropods of uncertain relationships (e.g., Reig 1963; Benedetto 1973), carnivorous sauropodomorphs (Galton 1977), the sister group of the clade Saurischia + Ornithischia (Gauthier 1986; Brinkman and Sues 1987; Sereno and Novas 1992; Novas 1992b), and basal theropods (Novas 1994; Sereno and Novas 1992, 1994). This latter hypothesis is based on numerous derived features of the skull, vertebral column, forelimb, and pelvic girdle. More recently, Max Langer (2004) suggested that Herrerasauridae are saurischians that occupy a basal position with respect to the more-derived eusaurischians (i.e., Sauropodomorpha plus Theropoda).

Two South American herrerasaurids are currently known, *Herrerasaurus ischigualastensis* and *Staurikosaurus pricei*.

HERRERASAURUS ISCHIGUALASTENSIS

This taxon, briefly described by Osvaldo Reig in 1963, is one of the best-known Triassic dinosaurs, and detailed anatomical information has become available in the last several years (e.g., Novas 1992b, 1994; Sereno and Novas 1992, 1994; Sereno 1994). The nearly ten specimens known of *Herrerasaurus ischigualastensis* come from the lower third of the Ischigualasto Formation, which is exposed at the Hoyada de Ischigualasto in San Juan Province, Argentina. Two other species have been described from the same formation: *Ischisaurus cattoi* (Reig 1963) and *Frenguellisaurus ischigualastensis* (Novas 1986). However, their respective type specimens reveal the same synapomorphies recognized for *Herrerasaurus ischigualastensis*, and because of this they were considered junior synonyms of the latter taxon (Novas 1994).

The pubis has the most remarkable autapomorphies of *Herrerasaurus*: it is proximally curved, resulting in an almost ventral and slightly posterior orientation of the distal end of the bone (Reig 1963). This condition emerged convergently in Late Mesozoic segnosaurian and paravian theropods (Gauthier 1986; Paul 1988). Also, the distal end forms an extensive enlarged foot that is even more enlarged than in *Staurikosaurus*.

Also curious is the existence of a definite and unusual subcircular muscle scar on the distal anterior surface of the femur, a characteristic reported only in *Herrerasaurus* among ornithodirans, although it calls to mind a smaller muscle scar present in the sauropodomorph *Saturnalia* (Langer 2003). This scar may represent peripheral areas of the origin of muscles related to tibial protraction.

STAURIKOSAURUS PRICEI

Staurikosaurus pricei comes from the Alemoa Member (Rhynchosaur Zone) of the Santa Maria Formation. This herrerasaurid (Fig. 2.17) is more gracile, smaller (2 meters long), and anatomically more primitive than *Herrerasaurus*. For example, in *Staurikosaurus* the pubis is cranioventrally oriented (instead of caudoventrally as in *Herrerasaurus*) and has a straight external margin, as in most basal dinosauriforms (instead of a sigmoid margin, as in *Herrerasaurus*; compare Figures 2.16F and 2.17F). As in Dinosauriformes ancestrally, the tibia of *Staurikosaurus* is longer than the femur and the trunk vertebrae lack the dorsal table that characterize *Herrerasaurus*.

One potential autapomorphy of *Staurikosaurus* is a distal bevel on the anterior margin of pubis (Fig. 2.17E and F), a feature that is apparently absent in other dinosaurs and immediate outgroups, in which the anterodistal surface is flat (e.g., *Marasuchus* and early sauropodomorphs) or slightly convex (e.g., *Herrerasaurus* and *Coelophysis*).

Some particular aspects of the anatomy of *Staurikosaurus* require brief comment: it was said (e.g., Colbert 1970) that the scapular shaft of this dinosaur retained a primitive shape because it is craniocaudally wide,

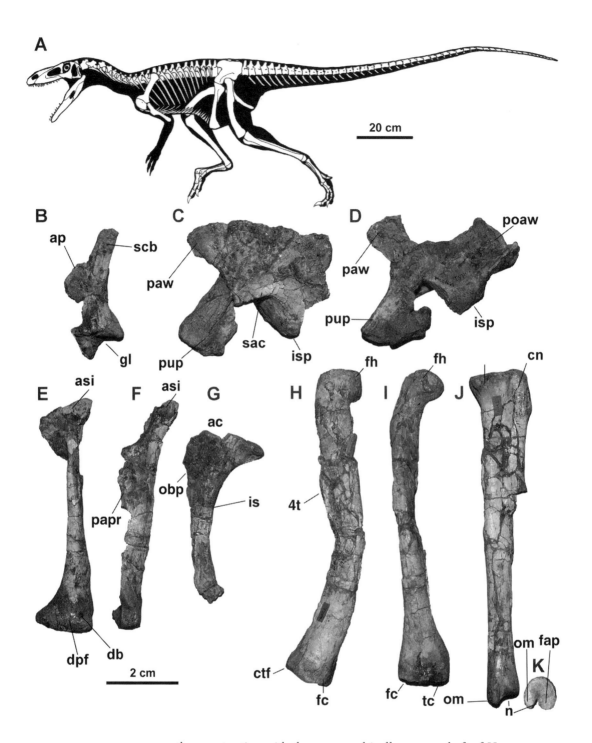

thus contrasting with the apomorphically narrow shaft of *Herrerasaurus* (Novas 1992b; Sereno 1994). However, the available scapula (which has been frequently misidentified as an ischium; see Galton 2000 for a review of this problem) matches the scapula of *Herrerasaurus* in that it has a narrow scapular shaft and a prominent acromial process (Fig. 2.17B).

The ilium of *Staurikosaurus* has been described (and illustrated) as having a caudally truncated postacetabular wing (e.g., Colbert 1970), leading some authors (Rauhut 2003a) to assume that this is an autapomorphic trait of this animal. However, examination of the opposite (and more complete) ilium reveals that it was caudally rounded (Fig. 2.17C and D). Aside from that, it is evident that this pelvic bone was craniocaudally short, resembling that of *Herrerasaurus*, and different from the more elongate condition present in more-derived basal saurischians (e.g., *Saturnalia* and *Guaibasaurus*).

Non-Herrerasaurid Basal Saurischians

EORAPTOR LUNENSIS

Eoraptor lunensis is known from an almost complete holotypic skeleton (Fig. 2.18A) and several unpublished referred individuals (Sereno 2007), corresponding to a probable carnivorous animal nearly 1 meter long. Found in the Ischigualasto Formation (Sereno, Forster, Rogers, and Monetta 1993), *Eoraptor lunensis* increases our knowledge of dinosaur groups present during the Carnian, presenting a new species of carnivorous dinosaur that did not pertain to the herrerasaurid radiation. *Eoraptor* lacks the derived characteristics recognized for the Herrerasauridae (Novas 1992b, 1994).

On the basis of published information (Sereno 2007; Sereno, Forster, Rogers, and Monetta 1993), *Eoraptor* is distinguished from other dinosaurs by a posterolateral process on the premaxilla and a heterodont dentition that consists of leaf-shaped premaxillary and anterior maxillary teeth. These contrast with the remaining teeth, which are blade-like and caudally curved (Fig. 2.18B). The lower jaw resembles that of neotheropods in that it has an intramandibular joint between the splenial and angular bones (Sereno 2007). In addition to this derived feature, *Eoraptor* has two rows of rudimentary palatal teeth on the pterigoid, a remarkably primitive feature recorded for the first time within Dinosauria (Sereno 2007).

The vertebral column is made up of 24 presacrals, 3 sacrals, and just over 40 caudals. The neural spines of the mid- to posterior dorsals are axially expanded, a condition frequently found among basal dinosauriforms (e.g., *Marasuchus*, sauropodomorphs, ornithischians, and coelophysids), with the exception of herrerasaurids, in which the neural spines are anteroposteriorly narrow (Novas 1992b). The scapular blade is broad, resembling that of *Marasuchus*, ornithischians, coelophysids, and sauropodomorphs. The forelimb, nearly half the length of the hindlimb, has well-developed digits I, II, and III, but digits IV and V are strongly reduced, a condition that *Eoraptor* shares with *Herrerasaurus* and more-derived theropods. The manual unguals are not trenchant (Sereno, Forster, Rogers, and Monetta 1993).

Fig. 2.17. *Staurikosaurus pricei:* (A) Reconstructed skeleton. (B) Left scapula in side view. (C) Left ilium in lateral view. (D) Right ilium in medial view. (E, F) Right pubis in (E) lateral and (F) cranial aspects. (G) Left ischium in lateral view. (H, I) Right femur in (H) lateral and (I) cranial views. (J, K) Right tibia in (J)lateral and (K) distal views. *(A) From Novas (1997b). Reprinted from* Encyclopedia of Dinosaurs, *Academic Press, San Diego. Eds.: Philip Currie and Kevin Padian. Article:* Herrerasauridae. *Author: Fernando Novas. Pages 303–311. © Copyright 1997. With permission from Elsevier. (B–K) Photographs by Fernando Novas.*

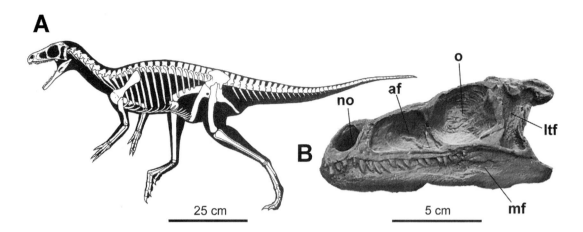

Fig. 2.18. *Eoraptor lunensis:* (A) Reconstructed skeleton. (B) Skull in lateral view. *(A) From Novas (1997b). Reprinted from* Encyclopedia of Dinosaurs, Academic Press, San Diego. Eds.: Philip Currie and Kevin Padian. *Article:* Herrerasauridae. *Author: Fernando Novas. Pages 303–311. © Copyright 1997. With permission from Elsevier. (B) Photograph courtesy of Hernán Canutti.*

The pelvic girdle and hindlimb bones are rather primitive, but they have derived features that are diagnostic of Dinosauria (e.g., a perforate acetabulum, brevis fossa on the posteroventral ilium, an ischium with a slender shaft; Novas 1996a).

Sereno, Forster, Rogers, and Monetta (1993) concluded that the available specimen of *Eoraptor* represents an adult individual on the basis of the closure of sutures in the vertebral column and partial fusion of the scapulocoracoid. However, fusion of the neural arches and the centra and of the pelvic and scapular girdle bones in herrerasaurids and basal sauropodomorphs is not always correlated with large size (Novas 1994). Moreover, Bonaparte (1996a) has emphasized that the proportionally large orbital opening of *Eoraptor* may be a juvenile trait. Nevertheless, at present several specimens of this dinosaur have been recovered from the same spot and stratigraphic level, and they are closely similar in size. Although seems obvious that these were not hatchlings, their ontogenetic status remains controversial.

Sereno, Forster, Rogers, and Monetta (1993) described *Eoraptor* as anatomically close to the predicted structure and size of the common dinosaurian ancestor, although they listed 15 synapomorphic traits present in *Eoraptor* that are absent in the common dinosaur ancestor and that support the placement of this taxon within Theropoda. Some researchers have disagreed with this interpretation (e.g., Padian and May 1993; Langer and Benton 2006), but an extensive cladistic analysis on this dinosaur is still lacking.

GUAIBASAURUS CANDELARIENSIS

Two specimens of *Guaibasaurus* were collected from levels of the Early Norian Caturrita Formation, which is exposed in the vicinity of the cities of Candelária and Faxinal do Soturno in the state of Río Grande do Sul, Brazil. The holotype specimen of *Guaibasaurus* (Fig. 2.19A–N), which measures around 2 meters long, is composed of several vertebral elements, a scapular and a pelvic girdle, and several hindlimb elements (Bonaparte, Ferigolo, and Ribeiro 1999). A second specimen of

Guaibasaurus consists of a fairly complete articulated skeleton lacking the skull, jaws, and neck (Brea, Bonaparte, Schultz, and Martinelli 2005; Bonaparte, Brea, Schultz, and Martinelli 2007).

The scapular shaft (Fig. 2.19D) is typically dinosaurian in that it is proportionally elongate and craniocaudally narrower than in basal dinosauriforms (e.g., *Marasuchus*). The humerus is short and more robust than in *Marasuchus*, herrerasaurids, and *Coelophysis*, thus resembling more the condition present in sauropodomorphs (Bonaparte, Brea, Schultz, and Martinelli 2007). The ulna and radius are also robust and proportionally short with respect to the humerus, contrasting with the elongate forearm bones of *Marasuchus*, *Silesaurus*, and herrerasaurids. In the hand of *Guaibasaurus*, digit III is shorter than II, a derived condition that is closer to that of theropods and sauropodomorphs than to that of herrerasaurids, which plesiomorphically retained a manual digit III that is longer than digit II.

The sacrum is represented by 2 vertebrae (Fig. 2.19B), of which sacral 1 is larger than sacral 2. Because this pattern resembles the two-sacral sacrum of *Staurikosaurus* (Galton 1977), I assume that *Guaibasaurus* also had a sacrum comprised of only 2 vertebral elements. This interpretation counters that of Bonaparte, Brea, Schultz, and Martinelli (2007), which concluded that *Guaibasaurus* had a three-sacral sacrum based on the morphology of the scars on the medial side of the iliac blade.

The ilium of *Guaibasaurus* lacks a well-developed brevis shelf (Fig. 2.19E and F). In this regard, the caudal extremity of the postacetabular wing is transversely narrow and vertically oriented, unlike that of sauropodomorphs and theropods, in which the brevis shelf is well developed and its fossa is more deeply excavated ventrally. In *Guaibasaurus*, the supraacetabular crest is notably large and projects laterally, apparently more so than in other dinosaurs. The medial acetabular aperture is closed, as in basal dinosauriforms, although this feature in *Saturnalia* supports the idea that the acetabulum was closed in the ancestral dinosaur and that it opened independently in different dinosaurian subclades. The ischium of *Guaibasaurus* (Fig. 2.19I) has a more robust shaft than in *Saturnalia* and Herrerasauridae, and its distal end has a large and triangular distal expansion. The ischial shaft is triangular-shaped in cross-section (it is dorsally flattened and the apex is ventrally oriented), as in prosauropod dinosaurs.

Because the pubis of *Guaibasaurus* is shorter than the ischium (Fig. 2.19G and H), I suspect that it is distally incomplete. If this is the case, we do not know if the pubis was distally expanded. Enough of the bone is preserved to show that it was not axially expanded, as characteristically occurs in herrerasaurids. Bonaparte, Ferigolo, and Ribeiro (1999) mention that the cranial surface of the femur is transversely narrow, making a broad crest that runs distally from the anterior trochanter. Such a crest is similar to that of *Herrerasaurus* (Novas 1994) and the basal sauropodomorph *Saturnalia* (Langer 2003). The anterior trochanter of the femur is quite small, and a trochanteric shelf is absent (Fig. 2.19J and K). The

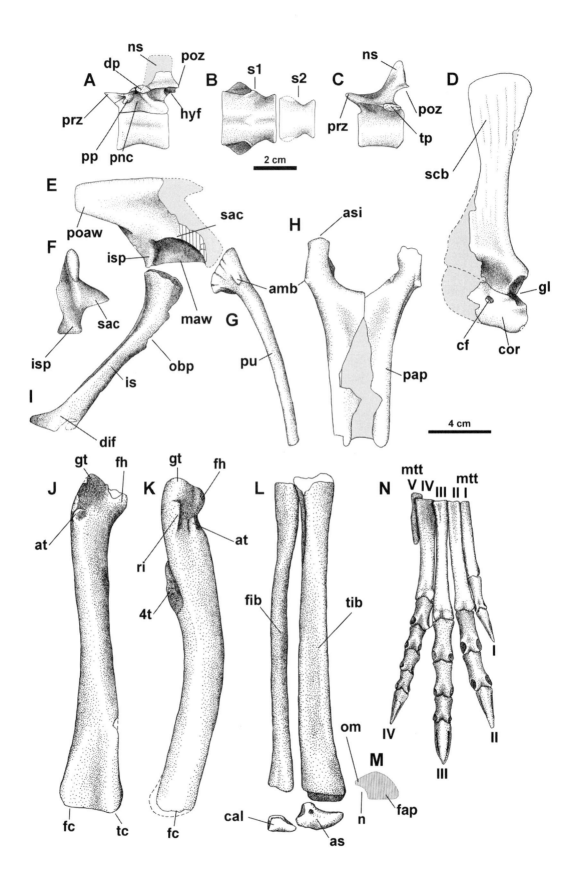

tibia, which is as long as the femur, has a small cnemial crest and the distal articular end is quadrangular. Metatarsal V is greatly reduced, probably lacking phalanges, as indicated by the three articulated feet available (Fig. 2.19N).

Bonaparte, Ferigolo, and Ribeiro (1999) claimed that a combination of several postcranial features distinguishes *Guaibasaurus* from all other basal dinosaurs. Some of these features are neural arches of the midcranial dorsal vertebrae with large pneumatic cavities, a parapophyseal-prezygapophysial lamina with a fossa below it, an undivided infradiapophyseal lamina, an acetabular area of ilium that is closed or nearly so with only an incipient reduction of its medial wall, a lateroposterior ridge lying above the level of the anterior trochanter, and the absence of a trochanteric shelf. However, all these features are also present in a variety of basal saurischians, and they can not be interpreted as synapomorphies of *Guaibasaurus*. Hindlimb morphology, especially tarsal anatomy, is rather uniform among basal dinosaurs, including *Guaibasaurus*, and distinctions in foot structure appear to be related to differences in metatarsal and phalanx lengths rather than structural modifications. Recognition of autapomorphic features for this early dinosaur is not easy.

Bonaparte and colleagues (Bonaparte, Ferigolo, and Ribeiro 1999; Bonaparte, Brea, Schultz, and Martinelli 2007) considered *Guaibasaurus* to be closely related with *Saturnalia* on the basis of general resemblances in the postcranial skeleton that unite them within the family Guaibasauridae. Although this hypothesis might be right, it still needs to be cladistically tested on the basis of apomorphic features. Phylogenetic relationships of *Guaibasaurus* with the remaining dinosaurs are far from clear: Bonaparte and colleagues (Bonaparte, Ferigolo, and Ribeiro 1999; Bonaparte, Brea, Schultz, and Martinelli 2007) made the rather confusing statement that *Guaibasaurus* is a possible member of Theropoda, at the same time indicating that it is more primitive than herrerasaurids (the theropod affinities of which are debatable) and hypothesizing that *Guaibasaurus* is morphologically suitable to be an ancestor of Sauropodomorpha and Theropoda. Langer (2004) presented a cladistic analysis depicting *Guaibasaurus* as the sister group of Theropoda but listed no features to support this hypothesis. In my view, *Guaibasaurus* shows the hyposphene-hypantrum complex in its dorsal vertebrae, thus supporting saurischian affinities, a conclusion that is in accordance with that of other authors (Bonaparte, Ferigolo, and Ribeiro 1999; Langer 2004). Moreover, this Brazilian taxon seems more derived than Herrerasauridae and *Eoraptor* because of a distally expanded ischium, a feature it shares with eusaurischians (=Theropoda plus Sauropodomorpha). On the basis of the morphology of the ischia, the robust condition of the forelimbs, and pedal unguals that are proportionally large, I think that this saurischian dinosaur may be more closely related to sauropodomorphs than with Theropoda, countering recent opinions expressed by Bonaparte and colleagues (Bonaparte, Ferigolo, and Ribeiro 1999; Bonaparte, Brea, Schultz, and Martinelli 2007) and Langer (2004).

Fig. 2.19. *Guaibasaurus candelariensis:* (A) Dorsal vertebra in lateral view. (B) Articulated centra of sacral vertebrae in ventral aspect. (C) Mid-caudal vertebrae in lateral view. (D) Left scapulocoracoid in side view. (E, F) Right ilium in (E) lateral and (F) caudal views. (G) Right pubis in lateral view. (H) Articulated paired pubes in cranial aspect. (I) Right ischium in lateral view. (J, K) Right femur in (J) cranial and (K) lateral views. (L) Right tibia, fibula, astragalus, and calcaneum in cranial aspect. (M) Right tibia in distal view. (N) Right foot in cranial aspect. From Bonaparte, Ferigolo, and Ribeiro (1999). Reprinted with permission of the authors and National Science Museum Monographs, Tokyo.

Theropoda

At present, Carnian rocks of South America have yielded herrerasaurs, early sauropodomorphs (i.e., *Saturnalia*), and basal saurischians (e.g., *Eoraptor*, *Guaibasaurus*) but no derived theropods (e.g., coelophysoids). This group of carnivorous dinosaurs has been reported from upper levels of the Norian Los Colorados Formation.

Coelophysoidea

The coelophysoid theropods were the dominant group of predatory dinosaurs during the Norian, Rhaetian, and Early Jurassic times. Coelophysoids were medium-sized animals about 2 to 6 meters long. The group is characterized by a bizarre snout with a subnarial gap (interrupting the upper tooth row) that received a dentary fang-like tooth when the mouth was closed. The skull of these theropods is characteristically elongated and low and has a high number of maxillary and dentary teeth (more than 18) and heterodonty between the premaxillary and maxillary teeth. The cervical vertebrae are slender and long and have low neural spines and blind fossae that excavate both sides of the centra. The pelvic girdle and hindlimbs have derived features diagnostic of Theropoda, sharply distinguishing them from other basal saurischians. For example, the preacetabular wing of the ilium is well developed, the cnemial crest of the proximal tibia is prominent, the outer malleolus of the distal tibia is laterally expanded and craniocaudally thick, the ascending process of the astragalus is more laminar and higher than in other dinosaurs, and metatarsal I is reduced to its distal half.

The Coelophysoidea are currently interpreted as the sister group of Ceratosauria plus Tetanurae (Carrano, Sampson, and Forster 2002; Rauhut 2003a; Ezcurra and Novas 2007; Fig. 2.20). The earliest record of coelophysoid theropods comes from the Late Carnian of North America with the fragmentary material of *Camposaurus arizonensis* (Hunt, Lucas, Heckert, Sullivan, and Lockley 1998). In the Norian, the coelophysoids had a flourish of species diversity. Basal members of the Coelophysoidea include the Norian *Liliensternus liliensterni* from Germany (Rowe and Gauthier 1990), *Gojirasaurus quayi* from the United States (Rauhut 2003a), and *Zupaysaurus rougieri* from South America (Arcucci and Coria 2003; Ezcurra and Novas 2005). Coelophysidae includes the North American taxa *Coelophysis bauri* (Norian), *Syntarsus kayentakatae* (Sinemurian–Pliensbachian), and *Segisaurus halli* (Pliensbachian–Toarcian) and the South African *Coelophysis rhodesiensis* (Hettangian–Sinemurian?) (Rowe, Tykoski, and Hutchinson 1997; Sereno 1999).

The South American record of coelophysoid theropods is represented by the recently reassigned *Zupaysaurus rougieri* (sensu Ezcurra and Novas 2005, 2007) from the Los Colorados Formation (Norian). Putative South American members of the Coelophysoidea also include fragmentary remains reported by Bonaparte (1972).

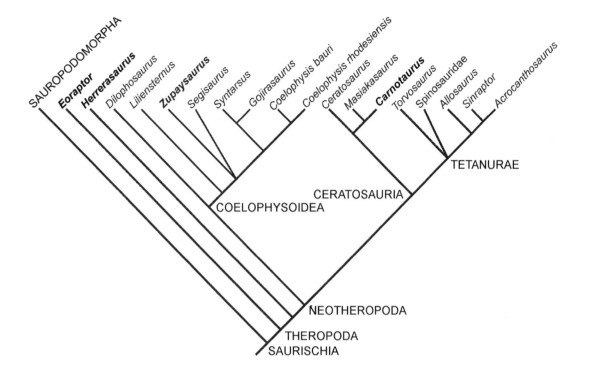

Fig. 2.20. Phylogenetic hypothesis depicting relationships among basal Theropoda. Genera discovered in South America are indicated in bold.

ZUPAYSAURUS ROUGIERI

This taxon comes from the Los Colorados Formation (Late Triassic, La Rioja Province, northwestern Argentina). It was originally described by Andrea Arcucci and Rodolfo Coria (1997, 1998, 2003), who interpreted *Zupaysaurus* as the oldest known member of Tetanurae. However, a review of the available materials of *Zupaysaurus* carried on by Martín Ezcurra and the present author (Ezcurra and Novas 2005, 2007) demonstrates its membership in the basal neotheropod clade Coelophysoidea (Fig. 2.20). The Argentine taxon has distinctive features of this clade, such as a skull length more than three times the height of the caudal skull, a ventral margin of the maxilla that is curved dorsally on its rostral portion, a minimum length of the internal antorbital fenestra that is 25 percent longer or more than the maximum skull premaxilla-quadrate length, a lacrimal rostral ramus that is longer than the ventral ramus, fang-like teeth at the rostral end of the dentary, and deep fossa on the cranial surface of the ascending process of the astragalus. Moreover, *Zupaysaurus* shares the following apomorphic features of more than 18 maxillary teeth, and maxilla with a lateral alveolar ridge above the tooth row with a coelophysoid subgroup made up of *Liliensternus* + (*Coelophysis* + *Syntarsus*). *Zupaysaurus* shares with *Coelophysis* and *Syntarsus* a square-shaped rostral margin of the maxillary antorbital fossa and a rostrocaudally compressed infratemporal fenestra. In particular, the Argentine theropod *Zupaysaurus* apomorphically resembles *Syntarsus* in the three-pronged condition of the rostral process of the jugal (Ezcurra and Novas 2005, 2007; Ezcurra 2007). Within Coelophysoidea, the Argentine taxon seems to be more

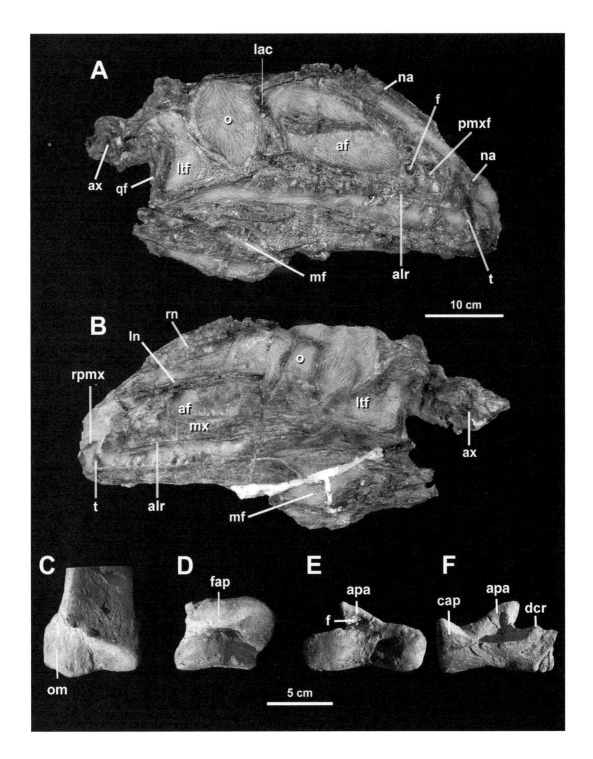

closely related to Coelophysidae (= *Coelophysis* + *Syntarsus*) than to the basal coelophysoid *Dilophosaurus*. The phylogenetic analysis recently performed by Ezcurra and Novas (2007) depicts *Zupaysaurus* within a trichotomy formed by *Zupaysaurus*, *Segisaurus*, and the Coelophysidae.

The distal end of the tibia and tarsus of *Zupaysaurus* are almost identical to those of the coelophysoid *Liliensternus* (Ezcurra and Novas 2005, 2007). Both taxa share an outer malleolus of the distal tibia that is polygonal-shaped, contrasting with the lobular-shaped structure present in the remaining saurischians (Fig. 2.21C–F).

Another point of discussion concerns with the purported presence of parasagittal crests in the skull roof of *Zupaysaurus*. Arcucci and Coria interpreted this as a diagnostic feature of this taxon. Close examination of the holotype specimen carried on by Ezcurra and Novas (2005, 2007) reveals that the right nasal is rotated outward from its natural position, thus looking like a longitudinal crest (Fig. 2.21A and B). The absence of a parasagittal crest in *Zupaysaurus* invites review of the skull of *Syntarsus kayentakatae*, for which these crests were also described (Rowe 1989).

Fig. 2.21. *Zupaysaurus rougieri:* (A, B) Skull and jaws in (A) right lateral and (B) left lateral views. (C, D) Distal end of right tibia in (C) cranial and (D) distal views. (E, F) Right astragalus-calcaneum in (E) cranial and (F) caudal views. *Photographs by Fernando Novas.*

THEROPODA INDET. FROM THE LOS COLORADOS FORMATION

Bonaparte (1972) described the postcranial remains of a small-sized theropod from the Los Colorados Formation (Norian). The materials include a cervical vertebra, a femur, the proximal half of both a tibia and a fibula, and a distal tibia articulated with the astragalus. Bonaparte (1972) pointed out some resemblances with coelophysoids, but the fragmentary material does not have clear tetanuran or ceratosaurian traits. So I prefer to consider it as Theropoda indet.

Sauropodomorpha

Sauropodomorpha is the saurischian clade that includes the well-known Jurassic and Cretaceous sauropods as well as their Upper Triassic (Carnian)–Early Jurassic (Toarcian) forerunners known as prosauropods. Sauropodomorphs constituted the first plant-eating dinosaurs and successfully diversified and populated all continents during the Triassic and Early Jurassic.

No consensus exists among authors about whether Prosauropoda represents a monophyletic taxon (as defended by Sereno [1999] and more recently by Galton and Upchurch [2004a]) or whether prosauropods constitute a paraphyletic assemblage of taxa, some of which are more closely related to Sauropoda (Fig. 2.22). I follow this last view, so I use the name "prosauropod" in this book in an informal way to refer to all these dinosaurs at a similar evolutionary grade (i.e., non-sauropod sauropodomorphs).

Prosauropod dinosaurs are known from Brazil and Argentina. The earliest and morphologically most primitive is the Carnian *Saturnalia tupiniquim*. More-derived forms are known from Early Norian beds of Brazil (*Unaysaurus tolentinoi*) and Late Norian rocks of Argentina (*Coloradisaurus brevis, Riojasaurus incertus, Mussaurus patagonicus,* and *Lessemsaurus sauropoides*).

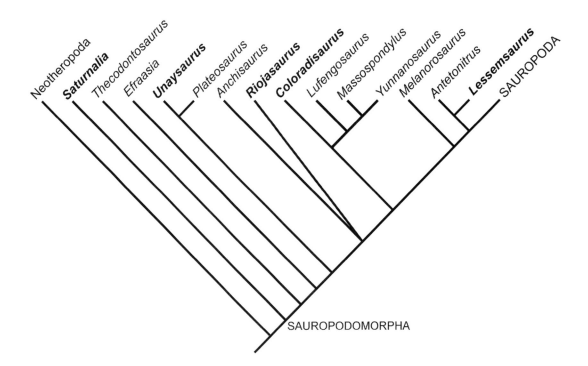

Fig. 2.22. Phylogenetic relationships among basal Sauropodomorpha. Genera discovered in South America are indicated in bold.

SATURNALIA TUPINIQUIM

Saturnalia is known by three semi-articulated skeletons in excellent preservation, which come from the Carnian Alemoa Member (Rhynchosaur Zone) of the Santa Maria Formation in Rio Grande do Sul (Langer, Abdala, Richter, and Benton 1999; Langer 2003). Although most of the skeleton of this creature is described in only a preliminary manner (Langer, Abdala, Richter, and Benton 1999), the pectoral and pelvic girdles and forelimb and hindlimb bones have been described and compared in considerable detail (Langer 2003; Langer, Franca, and Gabriel 2007).

Saturnalia was a gracile animal (Fig. 2.23A) measuring approximately 1.5 meters long that retained a primitive construction of the hindquarters (for example, the femur, tibia, fibula, tarsus, and foot are closely similar to those of *Herrerasaurus*; see Fig. 2.23C–N) but also shows derived traits present only among sauropodomorph dinosaurs (Langer, Abdala, Richter, and Benton 1999). For example, the teeth of *Saturnalia* are prosauropod-like, being small, lanceolate, and coarsely serrated. In *Saturnalia* there is some degree of heterodonty (the anterior teeth are taller), a suggested apomorphy of Sauropodomorpha (Langer, Abdala, Richter, and Benton 1999). Furthermore, the vertebral column of *Saturnalia* also shows similarities with those of basal sauropodomorphs, including a penultimate cervical vertebra that is much longer than the trunk vertebrae. The humerus of *Saturnalia* is transversely more expanded than in other basal dinosauriforms (e.g. *Marasuchus*, *Herrerasaurus*, *Lesothosaurus*), resembling the condition of sauropodomorphs.

Langer, Abdala, Richter, and Benton (1999) and Langer (2003) postulated that the sacrum of *Saturnalia* is composed of three vertebrae and that the vertebra added to the sacrum is of caudal origin. However, two main vertebrae form the sacrum of this dinosaur, and they are firmly attached to the ilia by means of massive ribs and transverse processes (Langer 2003). Two more vertebrae (of caudal origin) are placed within the limits of the iliac blades, but based on the illustrations offered by Langer (2003), the caudal expansion of the transverse processes of the second sacral does not leave space for articulation (on the medial surface of the iliac blade) with either of the first two caudals. I think that *Saturnalia* retained the primitive count of two sacral vertebrae, as in herrerasaurids and probably also in *Guaibasaurus*.

Bonaparte and colleagues (2007) interpreted *Saturnalia* as closely related with *Guaibasaurus*. Similarities are seen in the low and elongate construction of the dorsal vertebrae, the robust proportions of both the humerus and the ulna, the plate-like pubes, and the medially closed acetabulum. However, most of these resemblances are widely distributed among basal dinosauriforms, and unfortunately, no information is currently available on the skull, teeth, and neck of *Guaibasaurus*. These are valuable elements that may shed light on the phylogenetic relationships of this taxon with *Saturnalia*.

Saturnalia is known by its jaw, teeth, and cervical vertebrae, the anatomy of which supports its sauropodomorph affinities. Recent cladistic analyses carried out by Yates (2003), Yates and Kitching (2003), and Langer (2004) depicted *Saturnalia* as the sister taxon of all other known sauropodomorphs, a condition that accords with the Carnian age of the animal. However, another phylogenetic hypothesis was presented by Galton and Upchurch (2004a), who considered *Thecodontosaurus* to be the most basal member of a monophyletic Prosauropoda.

UNAYSAURUS TOLENTINOI

This prosauropod dinosaur was recently described by Leal, Azevedo, Kellner, and da Rosa (2004) from the Caturrita Formation (Early Norian), which crops out in Água Negra, Brazil, 13 kilometers north of Santa Maria. The type material of *Unaysaurus tolentinoi* is represented by a semi-articulated skeleton of an animal around 2.5 meters long comprised of dorsal and caudal vertebrae, both scapula girdles and forelimbs, portions of a hindlimb, and a skull and jaws that are almost complete (Fig. 2.24A). Gnaw marks made by small animals (presumably carnivorous cynodonts) are present on several bones, suggesting that the carcass of *Unaysaurus* was exposed for some time before burial (Leal, Azevedo, Kellner, and da Rosa 2004).

The skull of *Unaysaurus* is around 15 centimeters long, about the same length as the humerus (Fig. 2.24B). The rostral portion of the skull is deep, thus differing from other South American prosauropods (e.g., *Coloradisaurus*, *Riojasaurus*), for which this region of the skull is more

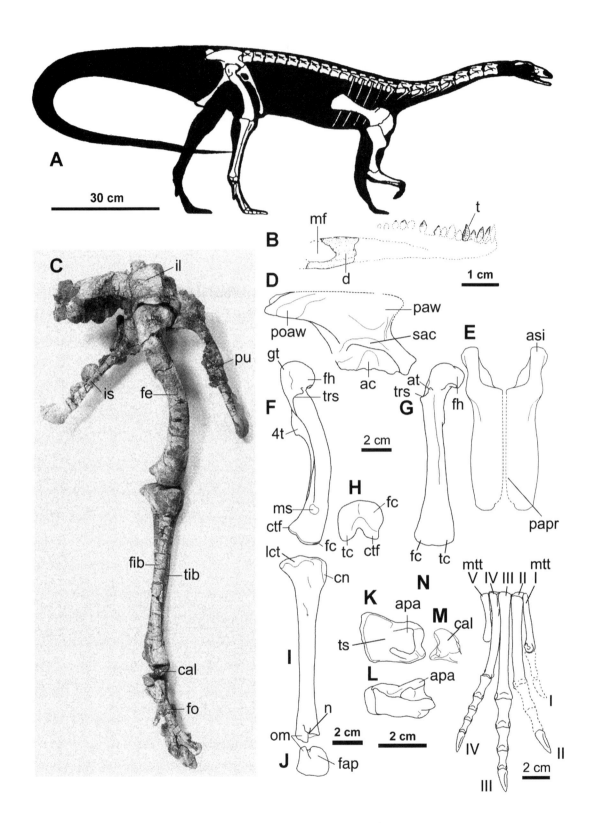

depressed (Leal, Azevedo, Kellner, and da Rosa 2004). *Unaysaurus* differs from other prosauropod dinosaurs in the following features: a well-developed laterodorsally oriented process formed by the frontal and parietal, two wide excavations on the lamina that medially bounds the antorbital fossa of the maxilla, and a conspicuous blunt ridge running on the lateral surface of the deltopectoral crest of the humerus.

Leal, Azevedo, Kellner, and da Rosa (2004) concluded that *Unaysaurus* shows some apomorphic traits shared by *Thecodontosaurus* and more-derived prosauropods (with the exception of *Saturnalia*), such as a first dentary tooth inset from the dentary tip, maxillary and dentary teeth with coarse serrations angled at 45 degrees from the tooth margins, tooth crowns that are not curved, and a proximal metatarsal II with a concave medial margin for the reception of metatarsal I. However, *Unaysaurus* is phylogenetically more derived than *Thecodontosaurus* because of enlarged caudal vascular foramina on the maxilla, a dentary that is downturned below the alveolar margin, a caudolateral shelf on the dentary, centra of the middle caudal vertebrae that are less than twice as long as high, and a proximal width of metacarpal I that is more than 65 percent of its length (Fig. 2.24C). The presence of a deep cleft on the basal tubera of the basicranium constitutes a derived condition shared by *Unaysaurus*, *Plateosaurus*, and more-derived sauropodomorphs (exclusive of *Efraasia*, *Thecodontosaurus*, and *Saturnalia*). Moreover, the tibia and astragalus of *Unaysaurus* are more robust than in *Saturnalia*, *Anchisaurus*, *Efraasia*, and *Thecodontosaurus* (Fig. 2.24F–J), a bulkier condition shared with derived prosauropods (Leal, Azevedo, Kellner, and da Rosa 2004). Although a cladistic analysis carried out by Leal, Azevedo, Kellner, and da Rosa (2004) depicts *Unaysaurus* as a sister taxon of the European prosauropod *Plateosaurus*, these authors emphasized that no unequivocal features unite both taxa.

MUSSAURUS PATAGONICUS

Mussaurus patagonicus is represented by small skeletons (25–30 centimeters long; Fig. 2.25A and B), found by José Bonaparte and assistants in levels of the Late Norian El Tranquilo Formation, which crops out in La Colorada Lake on the Cañadón Largo farm in north-central Santa Cruz Province (Bonaparte and Vince 1979). A series of intermediate-sized skeletons link these juveniles to larger individuals (3 meters long) from the same site that Rodolfo Casamiquela (1977, 1980b) briefly described as "*Plateosaurus* sp." It is currently thought that all the prosauropod specimens collected in the El Tranquilo area (juveniles and adults) belong to the same taxon, *Musssaurus patagonicus*. Recent studies carried on by Diego Pol and Jaime Powell (2005, 2007a) allowed identification of several autapomorphic features shared by juveniles and adults, including an anterior margin of the premaxilla that is posterodorsally directed (forming an angle of 45 degrees with the alveolar margin), a thin ridge along the entire ventral ramus of the lacrimal, a dorsally projected peg-like process on the anterior tip of the dentary, and a dorsoventrally deep anterior end

Fig. 2.23. *Saturnalia tupiniquim:* (A) Reconstructed skeleton. (B) Lower jaw in right lateral view. (C) Articulated pelvic girdle and right hind limb in lateral view. (D) Right ilium in lateral aspect. (E) Paired pubes (reconstructed) in cranial aspect. (F–H) Right femur in (F) lateral, (G) cranial, and (H) distal views. (I, J) Right tibia in (I) lateral and (J) distal views. (K, L) Right astragalus in (K) dorsal and (L) caudal views. (M) Right calcaneum in dorsal view. (N) Left foot in dorsal view. (A, D, N) From Langer (2003). Reprinted with permission of the author. (B, C) From Langer, Abdala, Richter, and Benton (1999). Reprinted from Sciences de la terre et des planètes, Comptes Rendus de l'Académie des Sciences, Paris, 329: 511–517, p. 512, fig. 2 and p. 513, fig. 3. © Copyright 1999. With permission from Elsevier.

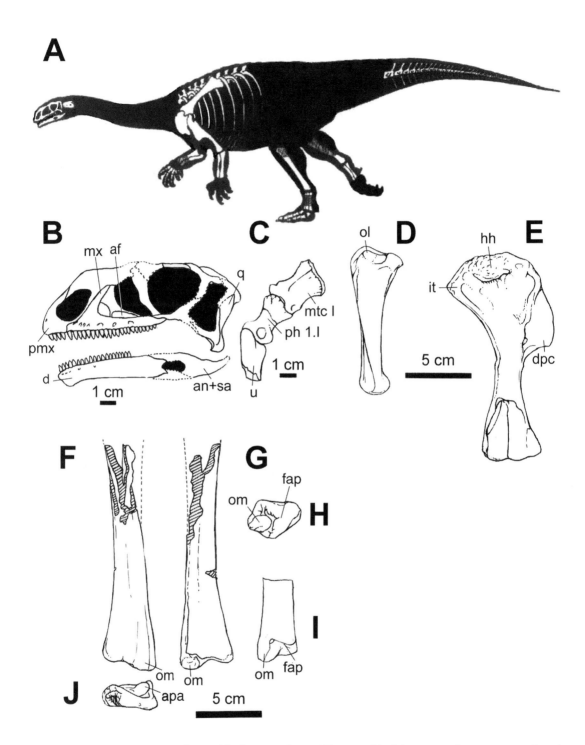

of mandibular symphysis (Fig. 2.25C). Moreover, the metacarpal and manual phalanges of *Mussaurus* are notably more robust and transversely wider than in the European *Plateosaurus*. *Mussaurus* differs from other South American prosauropods (e.g., *Riojasaurus, Unaysaurus, Saturnalia*) in the remarkable transverse expansion of the distal articular end of the tibia (Fig. 2.25G; Novas 1989a).

As Pol and Powell (2005, 2007a) have noted, the skull anatomy of *Mussaurus* shows derived characteristics shared with some basal sauropodomorphs and eusauropods that are absent in other basal sauropodomorphs (e.g. *Thecodontosaurus*, *Plateosaurus*). These include an extension of the infratemporal fenestra that is ventral to the orbit, dorsal and anterior rami of the quadratojugal that are subperpendicular to each other, dorsoventral expansion of the dentary at mandibular symphysis, and slightly procumbent teeth with broad serrations restricted to the apical region (these are absent in some teeth).

The discovery of *Mussaurus* skeletons and eggs is still the only available evidence of nesting behavior in Triassic dinosaurs. Seven juvenile individuals of *Mussaurus* at a similar stage of ontogenetic development were found near two fossil eggs (Bonaparte and Vince 1979). However, since erosion had acted on the original surface of the fossil site, Bonaparte and Vince considered it probable that components of the original nest were more numerous than those that were recovered. The discovery of juveniles in association with eggs and eggshell fragments suggests that the young stayed in the nesting area for a certain period after hatching. Moreover, Bonaparte and Vince (1979) estimated that the juvenile specimens of *Mussaurus* were some weeks old when buried because the largest of the collected eggs could not hold an embryo of the size of the specimens collected.

The preservation of individuals in different ontogenetic stages invites examination of the morphological and proportional changes that occurred in the development of these dinosaurs. Bonaparte and Vince (1979) masterfully interpreted some cranial and vertebral features of *Mussaurus* in both ontogenetic and phylogenetic contexts. For example, the skull of *Mussaurus* is short and high with large orbits and a short snout, features that indicate its juvenile condition. The maxilla and premaxilla are very high and axially short, different from the low and elongate skull of certain prosauropod dinosaurs (e.g., *Anchisaurus*; Fig. 2.25D). In contrast, the skull of the juvenile *Mussaurus* recalls the skull type of the Late Jurassic sauropod *Camarasaurus*. The antorbital fenestra of the juvenile *Mussaurus*, the major axis of which is directed dorsoventrally, likewise recalls this sauropod genus.

More recently, Pol and Powell (2007a) published a detailed description of the skull of *Mussaurus* and offered new interpretations about the acquisition of evolutionary novelties in basal sauropodomorphs through heterochronic processes. The skull of *Mussaurus* shows remarkable changes in the rostral and temporal region, which commonly vary during the ontogeny of dinosaurs and other archosaurs. The rostrum in the subadult specimens is relatively longer than in the post-hatchling specimens, and the premaxilla, maxilla, and nasal bones are relatively large compared to the total skull length. Similarly, the relative size of the external nares and the antorbital fenestra increased during the ontogeny of *Mussaurus*, as was recently noted for the Early Jurassic basal sauropodomorph *Massospondylus carinatus* (Reisz, Scott, Sues, Evans, and Raath 2005; Pol and Powell 2007a).

Fig. 2.24. *Unaysaurus tolentinoi:* (A) Reconstructed skeleton. (B) Reconstructed skull in left lateral view. (C) Right manual digit 1 and its metacarpal in medial view. (D) Right ulna in lateral view. (E) Right humerus in caudal aspect. (F–I) Right tibia in (F) caudal, (G) cranial, (H) distal, and (I) lateral views. (J) Right astragalus in cranial view. *Redrawn from Leal, Azevedo, Kellner, and da Rosa (2004).*

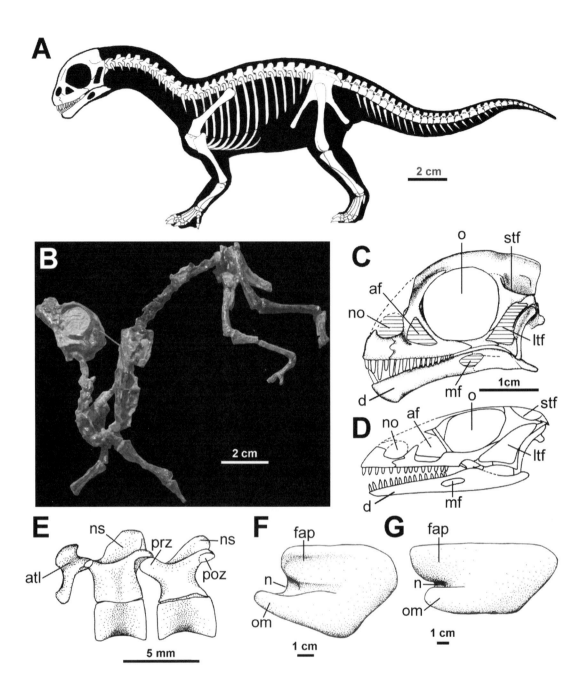

Other changes, however, may be more specific to the ontogenic development of *Mussaurus* (Pol and Powell 2007a), such as changes on the anterior margin of the premaxilla, which is remarkably more vertical in the post-hatchling specimens than in the subadults (it changes from a range of 63 to 71 degrees to 45 degrees). Similarly, the ascending ramus of the maxilla changes its orientation (changing the angle with the alveolar margin from 90 to 75 degrees).

Pol and Powell (2007a) emphasized that numerous skull characteristics of *Mussaurus* were subject to strong ontogenetic change. Some of these characteristics show a sauropod-like condition in the post-hatchling specimens but a plesiomorphic condition in the subadult specimens (e.g., the anterior margin of premaxillae is subvertically oriented, the antorbital opening is proportionately short and high, and the infratemporal fenestra projects extensively underneath the orbit). These characteristics fit the hypothesis that major features of the skull morphology of the sauropod could have been originated through paedomorphic processes, as first proposed by Bonaparte and Vince (1979) because of some similarities in the skull morphology of the *Mussaurus* holotype and *Camarasaurus* (e.g. short premaxilla and maxilla, a dorsoventrally high antorbital fenestra, an expanded mandibular symphysis, and enlarged teeth).

However, Pol and Powell (2007a) indicated that other skull characteristics seem to have an opposite developmental trend, showing a plesiomorphic-like condition in the post-hatchlings and a more-derived condition in the subadult skulls, such as a relative enlargement of the descending process of the postorbital, an increase in the posterior extension of the posterodorsal process of the prefrontal along the orbital rim, a decreased anteroposterior extension and increased dorsoventral depth relative to the anteroposterior extension of the infraorbital region of the jugal, and an increased dorsoventral height of dentary at the mandibular symphysis relative to the anteroposterior extension of this element.

The cervical vertebrae are proportionally short and tall in the juvenile *Mussaurus* (Fig. 2.25E), a morphology that resembles that of the basal dinosauriform *Marasuchus*, but is in turn distinct from the low and elongated neck vertebrae of adult prosauropods, in which the cervicals became more gracile as they are closer to the skull (Bonaparte and Vince 1979). The forelimb bones of juvenile *Mussaurus* resemble those of other prosauropods except for their proportions with respect to the hindlimb bones: the humerus is proportionally longer than in other prosauropods, suggesting that a quadrupedal pose was preferred in the initial stages of development (Reisz, Scott, Sues, Evans, and Raath 2005).

As seen, the skull changed proportionally during ontogeny; the size of the braincase and orbit became relatively smaller during the first life stages. In contrast, the cervical vertebrae became proportionally longer, thus achieving the characteristically low and elongate shape of adult sauropodomorphs.

Bonaparte and Vince (1979) concluded that the sauropod-like characteristics that *Mussaurus* shows in the skull, mandibular symphysis, teeth, and forelimb proportions as well as its juvenile condition suggest the possibility that Sauropoda differentiated from prosauropods because of the persistence of some juvenile characteristics in the adult state (neoteny). This interpretation has been more recently supported by Reisz, Scott, Sues, Evans, and Raath (2005) in a study of juvenile specimens of *Massospondylus*.

Pol and Powell (2005) studied the phylogenetic relationships of *Mussaurus* in a comprehensive cladistic analysis of basal Sauropodomorpha. The most parsimonious trees they recovered in their phylogenetic

Fig. 2.25. *Mussaurus patagonicus:* (A) Reconstructed skeleton. (B) Skeleton of holotype specimen as preserved. (C, D) Skulls of (C) juvenile *Mussaurus* compared with skull of (D) an adult specimen of *Anchisaurus*. (E) Atlantal neural arch, axis, and cervical 3 in left lateral view. (F, G) Right tibiae of (F) *Riojasaurus* and (G) *Mussaurus* (adult specimen) in distal view. *(A) Illustration by J. González. (C–E) From Bonaparte and Vince (1979). Reprinted with permission of* Ameghiniana. *(F, G) From Novas (1989b). Reprinted with permission of The Paleontological Society.*

analysis depicted *Mussaurus* as more closely related to "melanorosaurids" and more-derived sauropodomorphs than to the more plesiomorphic forms from the Late Triassic of Europe (e.g., *Plateosaurus engelhardti* and closely related forms). Nevertheless, Pol and Powell (2007a) cautiously pointed out that cranial morphology of the adult stage in *Mussaurus* is currently unknown and that new materials are needed in order to understand the late ontogenetic changes of this taxon. Such remains will also be crucial for appropriately testing the phylogenetic relationships of *Mussaurus patagonicus*, since many of the characteristics subject to ontogenetic change in this taxon have been previously interpreted as varying phylogenetically among basal sauropodomorphs and Sauropoda.

COLORADISAURUS BREVIS

The type specimen of *Coloradisaurus brevis* was collected by José Bonaparte and his assistants from the upper levels of the Late Norian Los Colorados Formation near La Esquina in La Rioja Province. *Coloradisaurus* preserves the cervical vertebrae, the pectoral girdle, incomplete forelimb bones, the astragalus, the pedal phalanx, and a skull and jaws that are almost complete (Fig. 2.26). Bonaparte (1979a) pointed out several cranial features that distinguish *Coloradisaurus* from other prosauropods, including frontals that are proportionally elongate, an antorbital opening partially below the orbit, and an incipient rostral projection of maxilla (Fig. 2.26A and B). Furthermore, *Coloradisaurus* differs from the European *Plateosaurus* and the Chinese *Lufengosaurus* in that it has a shorter antorbital region. The Argentine taxon is further distinguished from *Plateosaurus* in the dorsoventrally depressed condition of both the mandibular symphysis and the caudal half of the jaw (Fig. 2.26C).

The short skull of *Coloradisaurus* has sharp distinctions with the elongate and low skull of *Riojasaurus*. The available cervicals of *Coloradisaurus* are more delicate than the larger and more robustly constructed ones of *Riojasaurus* (Fig. 2.26D and E).

Recent cladistic analysis (e.g., Galton and Upchurch 2004a; Yates and Kitching 2003) supports the original interpretation of Bonaparte (1979a), placing *Coloradisaurus* within plateosaurid prosauropods.

RIOJASAURUS INCERTUS

This is by far the best-represented Triassic dinosaur from South America, due to the large number of specimens that are mainly housed in the Instituto Miguel Lillo, Tucumán. The name *Riojasaurus incertus* was coined by Bonaparte in 1967 on the basis of several postcranial remains that lacked the skull, the jaws, axial elements, the scapulae, the ischia, and some manual elements (Fig. 2.27A). This holotype specimen was collected in La Esquina from upper levels of the Los Colorados Formation (Late Norian). Originally, Bonaparte (1967) described the prosauropod taxon *Strenusaurus procerus* from the same locality and stratigraphic levels, which he later synonymized with *Riojasaurus*

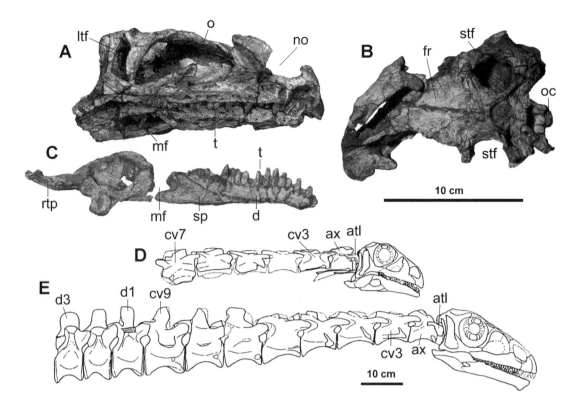

Fig. 2.26. *Coradisaurus brevis:* (A, B) Skull in (A) lateral and (B) dorsal views. (C) Lower jaw in medial aspect. (D, E) Skull and cervical vertebrae of (D) *Coloradisaurus brevis* and (E) *Riojasaurus incertus* compared. (A–C) Photographs by Fernando Novas. (D, E) Redrawn from Bonaparte and Pumares (1995).

incertus, arguing that morphological differences between both taxa were due to ontogenic variation.

In recent years, Bonaparte and Pumares (1995) reported the discovery of a new specimen of *Riojasaurus* that includes an almost-complete skull (Fig. 2.27B–D). They pointed out distinctive features of *Riojasaurus*, including a wide antorbital depression that is larger than that in any other known prosauropod as well as cervical vertebrae that are more robust than in the plateosaurids *Coloradisaurus* and *Plateosaurus*.

The available skull of *Riojasaurus incertus*, which measures 25 centimeters long, accords well with the cranial morphology of other prosauropods, although it is dorsoventrally lower and more elongate than in *Thecodontosaurus*, *Coloradisaurus*, *Massospondylus*, and *Yunnanosaurus*, particularly in the preorbital region (Bonaparte and Pumares 1995). In *Riojasaurus*, the jaw articulation is level with the alveolar row of the maxilla, a difference from *Plateosaurus* and the majority of prosauropods, in which that region of the jaw is much lower.

A notable feature that Bonaparte and Pumares (1995) described for the first time for prosauropod dinosaurs is the possible presence of a beak in the rostral region of the skull. In *Riojasaurus*, the premaxilla shows a slight lateral expansion that is well defined by a dorsoventrally oriented border, in front of which is a rugose surface that was probably covered by horny material (Fig. 2.27B). This particular morphology of the premaxilla also corresponds with similar features in the symphysial region of the lower jaw, which apparently was also covered by a horny sheath.

Fig. 2.27. *Riojasaurus incertus:* (A) Reconstructed skeleton. (B–D) Skull in (B) lateral, (C) dorsal and (D) occipital views. (E) Series of selected cervical, dorsal, and sacral vertebrae of the holotype specimen. (F) Left scapulocoracoid in lateral aspect. (G) Left humerus in cranial view. (H) Right manus in dorsal view. (I) Left ilium in lateral aspect. (J) Paired pubes (reconstructed) in cranial view. (K, L) Left femur in (K) caudal and (L) medial views. (M) Right tibia in lateral view. (N) Left distal tibia, fibula, and tarsal bones and foot in dorsal aspect. *(A) Illustration by J. González. (B–D) From Bonaparte and Pumares (1995). Reprinted with permission of* Ameghiniana. *(E–N) From Bonaparte (1972). Reprinted with permission of the author and Fundación Miguel Lillo.*

Also remarkable is that the dentary tooth row is medially set off with respect to the external surface of the jaw, thus resulting in a well-defined cheek. Galton (1976) indicated that this trait is not common among prosauropods, so the state of this characteristic in *Riojasaurus* is undoubtedly derived (Bonaparte and Pumares 1995).

The dentition of *Riojasaurus* is not of the type present in *Plateosaurus* or other prosauropods, which have transversely flattened leaf-shaped teeth with a constriction between the crown and the root. In *Riojasaurus*, the teeth are subcylindrical and slightly curved and have very small serrations. Bonaparte and Pumares (1995) believe that the kind of teeth described for *Riojasaurus* could be the primitive condition for Sauropodomorpha.

In *Riojasaurus*, the length of the humerus and the radius combined is approximately 67 percent of the femur and the tibia combined. Thus, the size disparity between forelimbs and hindlimbs is less marked than in plateosaurid prosauropods (Bonaparte and Pumares 1995). Based on forelimb-to-hindlimb proportions as well as the robustness of the vertebral column and limb bones, Bonaparte (1972) concluded that *Riojasaurus* is a member of Melanosauridae. Recent cladistic analysis (Galton and Upchurch 2004a) supports this view but interprets *Riojasaurus* as the most basal member of this subgroup of prosauropod dinosaurs (Fig. 2.22).

LESSEMSAURUS SAUROPOIDES

Lessemsaurus is a larger sauropodomorph than *Riojasaurus*. It was collected by José Bonaparte and his crew at the La Esquina fossil site in southwestern La Rioja Province, Argentina, from levels of the Norian Los Colorados Formation. The type specimen of this sauropodomorph is composed of several axial elements, but the postcranial remains of three other specimens of different size were collected in association with the holotype (Fig. 2.28). Bonaparte (1999c) pointed out the resemblance of *Lessemsaurus* to sauropod dinosaurs in the construction of the presacral vertebral column: the mid- to posterior cervicals and dorsal vertebrae have neural arches that are proportionally higher than in other prosauropods and the pre- and postzygapophyseal articulations are located high and are separated from the base of the neural arch, a condition resembling that of sauropod dinosaurs. Furthermore, the high position of the postzygapophyses creates a wide space below them that is accompanied by a strong transverse compression of the neural arch, both features that also occur in Jurassic sauropods such as *Patagosaurus*, *Diplodocus*, and *Camarasaurus*. Also, the cervical vertebrae of *Lessemsaurus* have deeply excavated postspinal fossae that are limited by sharply defined crests, as occurs in sauropods, thus contrasting with less-derived prosauropods (e.g., *Riojasaurus*, *Plateosaurus*) in which the postspinal fossa is shallower and the bounding crests are considerably less well developed. The morphology of the dorsal vertebrae of *Lessemsaurus* also resembles that of sauropods more than that of prosauropods, especially in the shape of the neural spines, which are dorsoventrally deeper than they are craniocaudally wide.

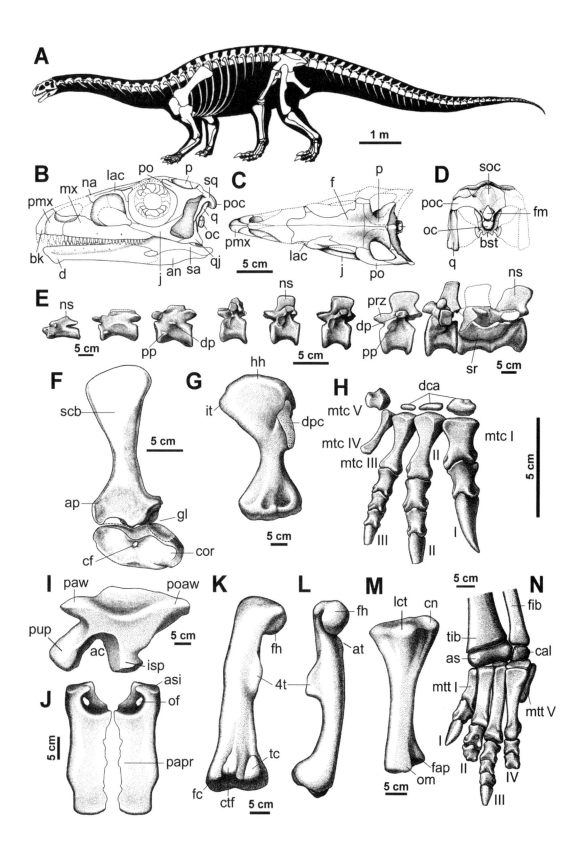

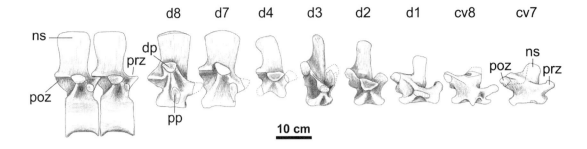

Fig. 2.28. *Lessemsaurus sauropoides*: Cervicals 7–8 and dorsals 1–4, 7, and 8 of the holotype specimen and two posterior dorsals of a referred specimen. From Bonaparte (1999c). Reprinted with permission of Ameghiniana.

In a recent paper, Pol and Powell (2007b) offered an extensive description of the pectoral girdle, forelimb, pelvis, and hindlimb of *Lessemsaurus*, also emphasizing, as did Bonaparte (1999c), the intermediate condition of this taxon between prosauropods and early sauropods. For example, the scapula has strongly expanded proximal and distal ends, which is different from the more-slender scapulae of earlier prosauropods (e.g. *Riojasaurus*, *Lufengosaurus*, *Plateosaurus*, *Massospondylus*) but is similar to the expanded scapula of *Antetonitrus ingenipes*, from the Norian Elliot Formation of South Africa (Yates and Kitching 2003). The humerus of *Lessemsaurus* has a low deltopectoral crest, which contrasts with the condition of most prosauropods, in which this crest is much more prominent. The ungual of manual digit I is poorly curved and has a reduced flexor tubercle (Pol and Powell 2007b), suggesting that the forelimb served as support rather than as the weapon that is indicated by the strongly curved manual ungual of most basal prosauropods.

Bonaparte (1999c) pointed out that *Lessemsaurus* represents a derived evolutionary state of prosauropod dinosaurs toward the acquisition of typical sauropod morphologies. Beyond the apomorphic resemblance *Lessemsaurus* and sauropods share in the morphology of the vertebrate column, Bonaparte (1999c) considered this Triassic dinosaur to be a member of the family Melanosauridae, a phylogenetic hypothesis that was supported by a recent study by Galton and Upchurch (2004a), who depicted *Lessemsaurus* as the most-derived member of the Melanosauridae (Fig. 2.22). However, Galton and Upchurch (2004a) recognized Prosauropoda as a monophyletic group within Sauropodomorpha and consequently interpreted the sauropod features present in *Lessemsaurus* as autapomorphies of this taxon that were acquired independently from Sauropoda. But in phylogenetic contexts in which Prosauropoda is not identified as a natural assemblage but some of its members are instead regarded as more proximately related to Sauropoda (e.g., Yates 2003; Fig. 2.22B), *Lessemsaurus* appears to be more proximately linked to basal sauropods. Moreover, *Lessemsaurus* shares exclusively with *Antetonitrus* a scapula with a broad and proportionally short blade, a metacarpal I with a distal lateral condyle that is dorsoventrally taller than the medial condyle, and a manual phalanx 1.I that is slightly wider than it is proximodistally long. This suite of derived features, which are uniquely shared by *Lessemsaurus* and *Antetonitrus*, suggests that these two taxa may have

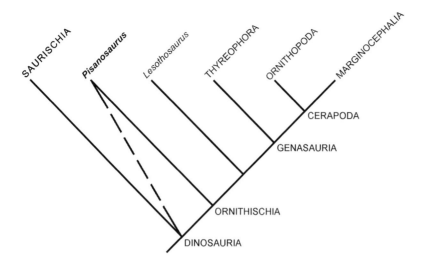

Fig. 2.29. Phylogenetic hypothesis depicting relationships among main clades of Ornithischia. *Pisanosaurus* is depicted as a basal member of this clade, but the dashed line indicates the possibility that it constitutes an earlier evolutionary branch of Dinosauria.

formed part of a lineage of highly derived prosauropods that populated Gondwana at the end of the Triassic (Pol and Powell 2007b).

Ornithischia

Ornithischia is defined as a stem-based taxon that includes all dinosaurs that are closer to *Iguanodon* than to the sauropod *Cetiosaurus* (Norman, Witmer, and Weishampel 2004). Researchers agree that the Early Jurassic *Lesothosaurus* represents the sister taxon of all the remaining ornithischians, which are grouped in the clade Genasauria (the name applied to ornithischians with maxillary and dentary cheeks by virtue of the medial inset of the tooth rows with respect to the external surface of the skull). Genasauria includes two main branches, Cerapoda (which gathers famous ornithischians such as ornithopods, ceratopsians, and pachycephalosaurs) and Thyreophora (the clade of armored ornithischians, including stegosaurs and ankylosaurs, among the most conspicuous members of the group) (Fig. 2.29).

Triassic archosaurs found in South America that have been referred as to Ornithischia are *Pisanosaurus mertii* and cf. *Heterodontosaurus* from Argentina, and *Sacisaurus agudoensis* from Brazil.

PISANOSAURUS MERTII

The remains of this animal were collected from the middle section of the Ischigualasto Formation in the locality of Hoyada del Cerro Las Lajas in La Rioja, northwestern Argentina (Casamiquela 1967). *Pisanosaurus* is currently represented by an incomplete maxilla and dentary with teeth, impressions of the partial pelvic girdle, and some bones of the hindlimb (Fig. 2.30A). This taxon has been frequently considered to be a heterodontosaurid ornithopod (e.g., Casamiquela 1967; Bonaparte 1976) and more recently as the basalmost ornithischian (Weishampel and Witmer 1990; Sereno 1991b).

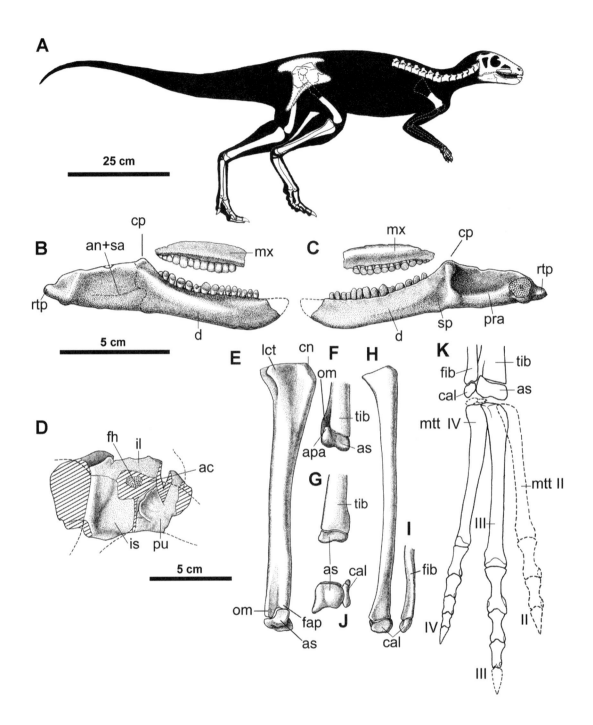

Most derived features of *Pisanosaurus* correspond to those of the lower jaw, which has cheeks resulting from the medial inset of the tooth row (Fig. 2.30B and C). Maxillary and dentary teeth are subcylindrical and tightly grouped, thus forming a contiguous occlusal surface, which probably was adapted for chewing tough plant material. Recently, Norman, Witmer, and Weishampel (2004) reviewed the several synapomorphies of Ornithischia that are present in *Pisanosaurus* (e.g., separation of the crown and root of the teeth by a neck, a maximal tooth size near the middle of the tooth row, and a dentary that forms the rostral portion of the coronoid process). Furthermore, these authors listed for *Pisanosaurus* some features diagnostic of Genasauria (e.g., a pronounced buccal emargination of the maxillary, dentary tooth rows, and the development of extensive and apparently confluent wear facets between blades formed by the maxillary and dentary teeth). In sum, on the basis of cranial and dental features, *Pisanosaurus* could be considered a member of Genasauria and even of Cerapoda (Norman, Witmer, and Weishampel 2004).

In his review of this taxon, Bonaparte (1976) offered a sketch depicting the original position of the partial skeleton, indicating that the specimen was found articulated in a much more complete condition than it is currently available. When found, the holotype specimen of *Pisanosaurus mertii* consisted of parts of the skull and jaws, articulated cervical and dorsal vertebrae, some manual elements, the hindlimb bones, and some impressions of the pelvis and sacral vertebrae (Bonaparte 1976). Although parts of the articulated skeleton are now lost or badly damaged, the original association negates the suspicion of Norman, Witmer, and Weishampel (2004) that *Pisanosaurus* is represented by two different individuals, perhaps belonging to two different taxa. As shown by Bonaparte (1976), *Pisanosaurus* is represented by pieces belonging to only one individual, and this association of cranial and postcranial elements is crucial for addressing its phylogenetic affinities.

First of all, the proportionally small size of the fragmentary scapula and metacarpals (Fig. 2.30A) are indicative of reduced forelimbs and thus a bipedal stance of *Pisanosaurus*, a condition that is in accordance with that of other basal dinosauromorphs. Second, the morphology of the preserved pelvic girdle and hindlimb is notably primitive for an ornithischian dinosaur: the pelvis of *Pisanosaurus*, represented by molds of the internal surface of a partial ilium, ischium, and pubis around the acetabulum (Fig. 2.30D), has dinosaurian traits (e.g., an open acetabulum) but lacks ornithischian features (Sereno 1991b). For example, the proximal end of the pubis suggests that this bone was cranioventrally oriented instead of being caudally projected, as characteristically occurs in Ornithischia, and the proximal end of the pubis is dorsoventrally deep and devoid of an obturator foramen, in contrast to basal ornithischians (e.g., *Lesothosaurus*, *Heterodontosaurus*), in which the proximal extremity of the bone is dorsoventrally depressed and deeply notched.

Contrary to the assertion of Norman, Witmer, and Weishampel (2004) that the hindlimb elements of *Pisanosaurus* suggest an animal of non-dinosaurian affinities, the tibia and proximal tarsals (Fig. 2.30E–K) are clearly dinosaurian in a cnemial crest, a distal end of the tibia that is quadrangular with a notch for reception of the ascending process of the astragalus, and a well-developed posterodistal flange (Bonaparte 1976; Novas 1989a). Norman, Witmer, and Weishampel (2004) cited as an autapomorphic feature of *Pisanosaurus* a recess of the distal tibia to receive the ascending process of the astragalus, but this feature is synapomorphic

Fig. 2.30. *Pisanosaurus mertii:* (A) Reconstructed skeleton. (B, C) Right maxilla and dentary in (B) lateral and (C) medial views. (D) Left pelvic girdle in medial aspect. (E) Right tibia and astragalus in lateral view. (F, G) Distal end of right tibia and astragalus in (F) cranial and (G) caudal views. (H, I) Right fibula and calcaneum in (H) lateral and (I) cranial views. (J) Right astragalus and calcaneum in distal aspect. (K) Reconstructed right foot in dorsal view. (A) Illustration by J. González. (B–K) From Bonaparte (1976). Reprinted with permission of The Paleontological Society and the Society for Economic Paleontology and Mineralogy.

of Dinosauria (or even of a more inclusive group that includes *Silesaurus* and probably *Pseudolagosuchus*), and thus it can hardly be considered diagnostic of *Pisanosaurus*. In a previous article (Novas 1989a), I suggested that a laterally expanded postfibular flange and a transversely compressed calcaneum represent ornithischian synapomorphies (Fig. 2.30F and J), but a similarly outward extension of the postfibular flange has been reported in the dinosaurian sister taxon *Silesaurus* (Dzik 2003), and the transverse compression of the calcaneum is so intense that it is distinctive of *Pisanosaurus* (Sereno 1991b).

Discordant characteristics between highly derived cranial and tooth morphologies, on the one hand, and plesiomorphic postcranial anatomy, on the other, create uncertainties about the phylogenetic location of *Pisanosaurus*. Although it is unanimously accepted that *Pisanosaurus* is a member of Dinosauria, its position within Ornithischia automatically creates homoplasies—evolutionary reversions when *Pisanosaurus* is nested within Genasauria and Cerapoda or convergences with Cerapoda if *Pisanosaurus* is located as a sister group of the remaining ornithischians.

As Parker, Irmis, Nesbitt, Martz, and Browne (2005) have correctly pointed out, the crurotarsan *Revueltosaurus* and the dinosauriform *Silesaurus* hint that the diversity of Triassic archosaurs with similar tooth morphology is poorly understood. By virtue of the recent discoveries of these Late Triassic archosaurs with highly derived masticatory features reminiscent of ornithischian dinosaurs, it is neither improbable nor illogical that *Pisanosaurus* constitutes, in fact, another case of basal dinosaurs that developed special adaptive devices for processing plant material convergently with ornithischian dinosaurs. In other words, *Pisanosaurus* may be an "herbivorous experiment" of early dinosaurs or ornithischians, just as herrerasaurids may be considered "predatory experiments" of basal saurischians. In both cases, plesiomorphic features are associated with surprisingly derived traits.

In sum, I interpret *Pisanosaurus* to be a member of Dinosauria (as demonstrated by the pelvic and hindlimb elements) and (cautiously) as a basal Carnian representative of Ornithischia (as indicated by some cranial and dental traits).

CF. *HETERODONTOSAURUS*

Fragmentarily preserved cranial remains of an ornithischian dinosaur were described from the Norian Laguna Colorada Formation in Santa Cruz Province (Báez and Marsicano 2001). The cranial elements consist of a partial maxillary with natural external molds of teeth and an isolated caniniform tooth with a serrated cutting edge (Fig. 2.31). Báez and Marsicano (2001) also reported an isolated vertebra and unidentified bone fragments but offered no descriptions of these elements. They recognized high-crowned maxillary teeth with wear facets lying in a single plane and caniniform tooth with serrated cutting edges to be diagnostic features of Heterodontosauridae. Báez and Marsicano pointed out that the Laguna

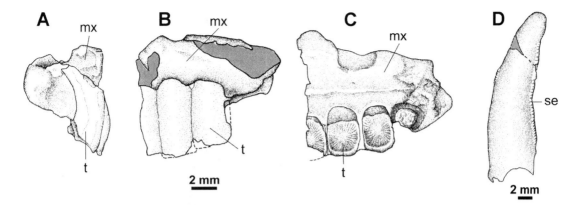

Fig. 2.31. cf. *Heterodontosaurus:* (A–C) Caudal fragment of left maxilla in (A) rostral, (B) lateral, and (C) ventral views. (D) Caniniform tooth in lateral view. *From Báez and Marsicano (2001). Reprinted with permission of* Ameghiniana.

Colorada heterodontosaurid shares the following features with the Early Jurassic *Heterodontosaurus* from South Africa: columnar maxillary teeth (owing to the absence of a distinction between crown and root), adjacent crown margins that are squared-shaped (thus allowing close apposition between successive teeth), and contiguous crowns that are practically touching each other throughout their lengths.

As Báez and Marsicano (2001) pointed out, this heterodontosaurid constitutes the first South American record of this group of small ornithischians, hitherto known from southern Africa, England, North America, and (possibly) China. In addition, this Late Triassic record from Patagonia extends the Early Jurassic stratigraphic range of heterodontosaurids farther back in time (Báez and Marsicano 2001).

SACISAURUS AGUDOENSIS

Max Langer and Jorge Ferigolo recently described remains of this Late Triassic archosaur (Langer and Ferigolo 2005; Ferigolo and Langer 2007) as a probable ornithischian dinosaur. The specimens were found in a bone bed belonging to the Caturrita Formation (Early Norian; Langer 2005a), which crops out in the locality of Agudo, Río Grande do Sul, southeastern Brazil. The specimens include several skull bones, some teeth, some vertebrae, a scapula, a well-preserved pelvic girdle, and hindlimb bones (Fig.2.32). The rostral portion of the jaw is edentulous, thus suggesting a horned beak resembling that of the Carnian basal dinosauriform *Silesaurus opolensis* from Poland (Dzik 2003; Fig.2.32C). Langer and Ferigolo (2005) pointed out some diagnostic features of Ornithischia, such as presence of a predentary bone (albeit paired rather than single as in Ornithischia), maxilla and a dentary with larger teeth located on the caudocentral portion of the dental series, tooth crowns that are low and triangular in lateral profile, a basal cingulum that is lingually more expanded so that the tooth is asymmetrical in mesial and distal views, tooth carina composed of large denticles, and a distal tibia with a laterally projected postfibular flange. However, the Brazilian taxon shows outstanding plesiomorphic characteristics unknown among ornithischians, such as a large antorbital fenestra, a caudal ramus of the maxilla that is

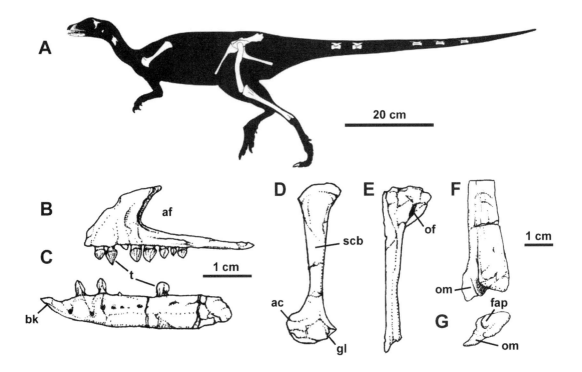

Fig. 2.32. *Sacisaurus agudoensis*: (A) Reconstructed skeleton. (B) Left maxilla in lateral view. (C) Left dentary in lateral view. (D) Left scapula in lateral aspect. (E) Left pubis in lateral view. (F, G) Distal end of right tibia in (F) cranial and (G) distal views. *Redrawn from Ferigolo and Langer (2007).*

narrow and has an oblique dorsal margin, a reduced number of teeth, two sacral vertebrae, a propubic pelvis, an almost fully closed acetabulum, a short preacetabular wing of the ilium, a medially laminar pubic shaft, a knob-like anterior trochanter of femur, and a non-pendant femoral fourth trochanter. Langer and Ferigolo (2005) and Ferigolo and Langer (2007) considered this curious archosaur from Brazil to be a basal ornithischian, but the reduced set of ornithischian-like features enumerated above for the dentary, dentition, and distal tibia are also documented in the non-dinosaurian dinosauriform *Silesaurus*. Thus, it is not improbable that *Sacisaurus* is instead a case of convergent evolution with more derived plant-eating dinosaurs (i.e., ornithischians).

Ichnological Evidence

Tetrapod footprints have been known in southern South America since 1931, when von Huene (1931a) described a large trackway of a quadrupedal reptile (*Rigalites ischigualastianus*) from the Triassic Ischigualasto–Villa Unión Basin. Since then, archosaurian footprints of Middle to Late Triassic age of South America have been described from basins in Argentina and Brazil (Leonardi 1989; Marsicano and Barredo 2004; Marsicano, Arcucci, Mancuso, and Caselli 2004; Marsicano, Domnanovich, and Mancuso 2005; Melchor, de Valais, and Genise 2002; Da Silva, Souza Carvalho, and Schwanke 2007).

Some trackways and isolated footprints from the Middle Triassic Los Rastros Formation have been referred to theropod dinosaurs (Stipanicic and Bonaparte 1979; Arcucci, Forster, Abdala, May, and Marsicano 1995).

More recently, large tridactyl tracks of a medium- to large-sized biped animal (footprint length 42 centimeters) were reported from the same beds cropping out in the Quebrada de Ischichuca, La Rioja Province (Marsicano, Arcucci, Mancuso, and Caselli 2004). The pes prints show a functionally tridactyl digitigrade pes with a long anteriorly directed digit III. The trackway has a very high pace angulation, with the footfall pattern on a nearly straight line. As noted by Marsicano and collaborators, obligate bipedalism and digitigrade pedal posture characterize the Ornithodira, and within this group, a symmetric tridactyl pes distinguishes the members of the dinosauriform clade. However, poor preservation of the material and the fact that the trackway consists of only one clearly preserved print prevented Marsicano and colleagues from referring this large biped footprint to a particular clade within Dinosauriformes (Marsicano, Arcucci, Mancuso, and Caselli 2004).

More recent discoveries of Los Rastros ichnites in the Ischichuca area (Marsicano, Domnanovich, and Mancuso 2005) have revealed a more diverse ichnofauna than was previously suspected. The ichnological association includes several tracks and trackways of bipeds with very high pace angulations, as the footfall pattern is on a nearly straight line. The prints, which are of variable size (15 to 42 centimeters long), show a functionally tridactyl digitigrade foot with an anteriorly directed digit III that is much longer than digits II and IV and the well-developed claws. Previously, the large tridactyl footprints from the Ischichuca area were attributed to theropod dinosaurs (Arcucci, Forster, Abdala, May, and Marsicano 1995), although no synapomorphies are preserved in the three-toed footprints that might discriminate among theropods, basal sauropodomorphs, and basal ornithischians as possible trackmakers (Marsicano, Domnanovich, and Mancuso 2005). This ichnological evidence from South America coupled with information of dinosauriform or dinosaurian tridactyl footprints known from the Middle Triassic of France and Germany supports an earlier appearance and broader geographic distribution of dinosaurs and their close relatives than the Carnian record of dinosaurian body fossils (Novas 1996a; Marsicano, Domnanovich, and Mancuso 2005).

A new tetrapod track assemblage has been recently described by Marsicano and Barredo (2004) from the Carnian Portezuelo Formation in Rincón Blanco, San Juan Province. The tracks from the Portezuelo Formation include non-mammalian synapsids and a rather diverse archosaur fauna made up by crurotarsal archosaurs and dinosaurs (Marsicano and Barredo 2004; Fig. 2.33A–C). The trackways show a relatively high pace angulation and narrow pace width, suggesting a narrow gait and parasagittal posture for their makers. All this evidence suggests that the trackmakers are related to Dinosauria. Moreover, the relatively large size of one of the Portezuelo trackmakers (estimated height at the hip approximately 1.65 meters) and its footprint shape suggest that its producer was a medium-sized facultatively bipedal prosauropod dinosaur (Fig. 2.33A and B). Nevertheless, facultative bipedal posture may have also occurred in basal sauropods, which are secondarily obligate quadrupeds (Wilson

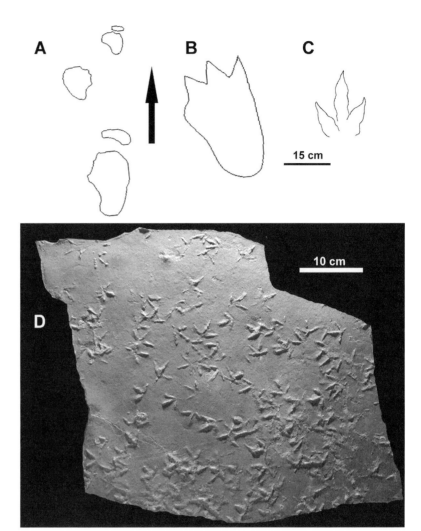

Fig. 2.33. Footprints of Triassic dinosaurs. (A, B) Sauropodomorph trackways and (C) theropod footprint from the Carnian Portezuelo Formation. (D) Bird-like theropod footprints from the Upper Triassic Santo Domingo Formation. *(A–C) Redrawn from Marsicano and Barredo (2004). (D) Photograph courtesy of Silvina de Valais.*

and Sereno 1998). Because the earliest record of Sauropoda is from the Upper Triassic of Laurasia (Thailand; Buffetaut et al. 2000; Buffetaut, Suteethorn, le Loeuff, Cuny, Tong, and Khansubha 2002; Gillette 2003), basal sauropod affinities for this type of facultatively bipedal trackmaker cannot be dismissed (Marsicano and Barredo 2004). As the latter authors remark, this hypothesis is supported by the semi-plantigrade condition of the hindfoot prints, the size disparity between the manus and pes, and the relatively equal length of the digits.

Footprints of fully bipedal animals have also been described from the Portezuelo Formation. One of them has been identified as a large prosauropod and another as a small- to medium–sized (height at hip approximately 75 centimeters) animal that could be allied to Theropoda (Fig. 2.33C).

The presence in the Rincón Blanco tetrapod footprint assemblage of putative basal members of Sauropodomorpha (perhaps both prosauropods and basal sauropods) of probably Carnian age implies that these dinosaurian herbivores were distributed in western Gondwana earlier than was previously documented by body fossils. As Marsicano and Barredo (2004) have argued, evidence afforded by the Portezuelo Formation strongly supports previous assumptions that the initial radiation of sauropodomorph dinosaurs and close relatives might have occurred during the Middle Triassic (Sereno 1999; Carrano and Wilson 2001).

Recently, Da Silva, Souza Carvalho, and Schwanke (2007) reported from the Carturrita Formation (Early Norian) in Novo Treviso, southern Brazil, the discovery of circular and concave structures of various sizes, ranging from a few to 50 centimeters. They show features that are congruent with those of prosauropod dinosaurs (e.g., large size of the footprints, a tridactyl pattern, footprints that are oval in outline, and impressions of the hands that are comparatively smaller than the impressions of the hindfeet). Da Silva, Souza Carvalho, and Schwanke concluded that ichnofossils from the Caturrita beds were left by large-sized animals, probably prosauropods, a conclusion that is congruent with the record of the prosauropod *Unaysaurus tolentinoi* in the same beds (Leal, Azevedo, Kellner, and da Rosa 2004).

The kind of dinosaurian footprints recovered from the Portezuelo, Los Rastros, and Caturrita formations are morphologically congruent with other footprints and body fossils from the Late Triassic. However, the recent discovery of well-preserved and abundant footprints with clearly avian characteristics in Late Triassic red beds from northwestern Argentina constitutes exceptional evidence that supports the unexpected presence of highly derived, bird-like theropods. The footprints have been described by Melchor, de Valais, and Genise (2002) from the Late Triassic Santo Domingo Formation, Santo Domingo River, La Rioja Province (Fig. 2.33D). These footprints, which were left in an environment researchers have interpreted as small ponds, document the activities of an unknown group of Late Triassic small theropods that had some avian characteristics. The tracks are bipedal, digit III is around 3 to more than 4 centimeters long, and the tracks display a high pace angulation and a straight or slightly curved path. However, the Santo Domingo tracks have several features that characterize bird footprints: they are wider than they are long, the digit impressions are slender, the angle between digits II and IV is wide, and the hallux impression is posterior or posteromedial (Melchor, de Valais, and Genise 2002). Moreover, the high footprint density and absence of preferred orientation of the Santo Domingo footprints resemble occurrences of tracks of modern water birds and waders. Bird-like footprints are rare in the Mesozoic record, occurring predominantly in Cretaceous strata, and the "aviform" pre-Cretaceous ichnological evidence, especially that of Triassic age, is sparser and shows few avian characteristics and is not comparable with the footprints of Santo Domingo. As Melchor, de Valais, and Genise (2002) have pointed out, whatever the ichnotaxonomic affinities of the Santo Domingo footprints, their producers are unknown from Late Triassic skeletal remains. No Late Triassic theropods presently recorded show evidence of an avian-like reversed hallux. In sum, bird-like footprints of the Late Triassic Santo Domingo Formation are attributed to an unknown group of theropods showing some avian characteristics (Melchor, de Valais, and Genise 2002).

Triassic Dinosaurs: An Overview

The Pattern of Early Dinosaur Diversification

Until recently, the sister taxa of Dinosauria were restricted to Ladinian forms (e.g., *Lagerpeton, Marasuchus, Pseudolagosuchus*), chronologically preceding the record of the earliest known dinosaurs, which were of Carnian age. Primitive dinosaurs were viewed as representative of the main two lineages into which Dinosauria is divided, Saurischia and Ornirthischia. The resulting pattern of evolution of Dinosauriformes was depicted as a stepwise process in which dinosaurs diversified after the extinction of basal dinosauriforms without any temporal overlap between true dinosaurs and their forerunners. However, discoveries made in the last years have deeply altered this rather simple pattern of diversification, showing a much richer and more complicated picture of early dinosaur evolution (Novas and Ezcurra 2005). Different lines of evidence, in support of this new view, are summarized below as general statements:

1. Dinosaurs probably radiated in Mid-Triassic times. Body fossils document early members of Theropoda (e.g., "*Camposaurus arizonensis*" from North America, which may represent the oldest known coelophysoid theropod; Hunt, Lucas, Heckert, Sullivan, and Lockley 1998; Irmis, Nesbitt, and Parker 2005; Nesbitt, Irmis, and Parker 2007) and Sauropodomorpha (e.g., *Saturnalia tupiniquim*) in Carnian beds. However, large theropod-like footprints belonging to bipedal dinosaurs (or dinosauriforms) are recorded from Ladinian rocks of the Los Rastros Formation (Marsicano, Arcucci, Mancuso, and Caselli 2004; Marsicano, Domnanovich, and Mancuso 2005), and large tracks (presumably belonging to Sauropoda) have been documented in the Carnian Portezuelo Formation (Marsicano and Barredo 2004). As already mentioned, body fossils and footprints strongly suggest that the initial radiation of dinosaurs might have occurred during the Middle Triassic. Moreover, footprints discovered in the Santo Domingo Formation indicate that by Upper Triassic times, theropod dinosaurs had acquired anatomical and functional features present in living birds.

2. Some Carnian–Norian dinosaurs do not fit comfortably within Ornithischia, Sauropodomorpha, and Theropoda. The available fossil evidence indicates that along with the Carnian sauropodomorphs and theropods just mentioned, dinosaurs existed (e.g., *Herrerasaurus, Staurikosaurus, Alwalkeria, Eoraptor, Agnosphitys, Pisanosaurus*) that do not fit comfortably within Saurischia and Ornithischia. Moreover, members of this basal dinosaurian radiation survived into the Early Norian, including the herrerasaurid *Chindesaurus briansmalli* (Murry and Long 1989; Hunt, Lucas, Heckert, Sullivan, and Lockley 1998) and the basal saurischian *Guaibasaurus*, indicating that basal representatives of Dinosauria exist with members of Sauropodomorpha (e.g., *Unaysaurus*) and Theropoda (e.g., *Coelophysis*).

3   Non-dinosaurian dinosauriforms survived into Carnian and Norian times. Recent discoveries and studies demonstrate an unexpected survival of non-dinosaurian dinosauriforms well into the Late Triassic: this is the case for *Silesaurus opolensis*, from Carnian beds of Poland (Dzik 2003), as well as for *Eucoelophysis baldwini* (Early Norian, United States), originally thought to be a coelophysoid theropod (Sullivan and Lucas 1999) but currently reinterpreted as a basal dinosauriform (Irmis, Nesbitt, and Parker 2005; Ezcurra 2006; Nesbitt, Irmis, and Parker 2007). The presence of *Silesaurus*, *Sacisaurus*, and *Eucoelophysis* in Upper Triassic rocks demonstrates that basal Dinosauriformes survived into the Late Triassic and were coeval with the oldest known dinosaurs (Novas and Ezcurra 2005; Ezcurra 2006).

4   Herbivory evolved among basal dinosauriforms before the origin of sauropodomorphs and ornithischians. *Silesaurus* has notable cranial and dental apomorphies (e.g., leaf-shaped teeth, a beaked jaw) as well as quadrupedal posture, demonstrating that herbivory was acquired among basal dinosauriforms independently of sauropodomorphs and ornithischians. Homoplasies of *Silesaurus* with ornithischians include a beak-like structure, a feature also reported in *Sacisaurus*, a possible dinosauriform from the Santa Maria Formation of Brazil (Langer and Ferigolo 2005; Ferigolo and Langer 2007). Similarly, *Pisanosaurus* may constitute another example of an early dinosaur that developed highly derived adaptations for oral plant processing prior to the origin of ornithischians.

5   "Dinosaur-mimic" archosaurs were contemporaneous with true dinosaurs. A topic of high interest is the fact that some purported dinosaurs from the Carnian–Norian of North America, Africa, and Madagascar have been recently reinterpreted as basal archosaurs that convergently acquired dinosaurian traits. The list of Triassic archosaurs that were once considered basal dinosaurs includes *Revueltosaurus callenderi* (Norian Chinle Formation, United States), *Azendohsaurus laarousii* (Carnian Argana Formation, Morocco), "prosauropods" from the Makay Formation (Isalo beds) of Madagascar, and *Shuvosaurus inexpectatus* (Norian Cooper Canyon Formation, United States). *Revueltosaurus* was originally described as a basal ornithischian (Hunt 1989), but newly discovered skeletal remains clearly indicate that it is a crurotarsan archosaur (Parker, Irmis, Nesbitt, Martz, and Browne 2005). *Azendohsaurus* has some derived features in the dentition that resembles those of basal ornithischians (Gauffre 1993; Galton and Upchurch 2004a), but recent discovery of a more-complete postcranium of this animal reveals that it lacks synapomorphies of Dinosauria (Jalil and Knoll 2002). Consequently, *Azendohsaurus* is considered to be outside this clade. The cranial and dental remains found in the Makay Formation (Isalo beds) of Madagascar are in a similar situation. Initially they were thought at first to be Carnian prosauropods (Flynn, Parrish, Rakotosamimanana, Simpson Whatley, and Wyss 1999), but this interpretation has been abandoned and some authors think that the Isalo "prosauropods" correspond in fact to herbivorous reptiles that are phylogenetically close to the unusual Moroccan archosaur *Azendohsaurus* rather than to dinosaurs (Langer 2005b). Another putative dinosaur is the bizarre archosaur *Shuvosaurus*, once interpreted either as an ornithomimosaur (Chatterjee 1993) or as a coelophysoid (Rauhut 2003a), which is currently reinterpreted as a member of the crurotarsan clade of archosaurs (Nesbit and Norell 2005).

In sum, the documented morphological diversity of Carnian dinosaurs and their allies suggests that an extensive evolutionary radiation occurred before the Late Triassic. Dinosaurian diversification was preceded and partially accompanied by a proliferation of non-dinosaurian dinosauriforms (e.g., *Silesaurus, Eucoelophysis*) as well as by bizarre dinosaur-mimic crurotarsan archosaurs (Isalo "prosauropods," *Azhendohsaurus, Revueltosaurus, Shuvosaurus*). Also, the first theropods and sauropodomorphs were contemporary with basal dinosaurs such as *Pisanosaurus* and herrerasaurids that evolved features convergently with derived ornithischians and theropods, respectively. It seems obvious that the Late Norian dominance of coelophysoids, basal sauropodomorphs, and basal ornithischians was preceded by a rich (but still poorly known) evolutionary radiation of basal dinosauriforms and dinosaur-mimic crurotarsan archosaurs.

Dinosaur Domination: Good Competitors or Victors by Default?

From a paleoecological point of view, no consensus exists about how dinosaurs attained numerical dominance during the Late Triassic. Some researchers (i.e., Bonaparte 1982) argued that the numerical preponderance of dinosaurs was reached thanks to their "competitive superiority" relative to other contemporary tetrapods. Other authors (e.g., Benton 1991, 1994), felt that the evidence demonstrates that the evolutionary radiation of dinosaurs occurred fortuitously in an "empty ecospace" after the extinction of synapsids, basal archosaurs, and rhynchosaurs. However, the evolutionary radiation of dinosauriforms was much more complex than was initially thought. Their explosive diversification needs to be evaluated in a way that takes into account the general statements listed above.

The early radiation of dinosauriforms was characterized by a sustained increase in body length, from 0.50 meters in mid-Triassic dinosauriforms (e.g., *Marasuchus lilloensis*) to up to 5 meters in Carnian dinosaurs (e.g., *Herrerasaurus ischigualastensis*; Novas 1994, 1997b, 1997c). This increase in body size also involved a shift from insectivorous toward megapredatory feeding habits and was accompanied by an increase in numerical abundance in Carnian times. In the Early Late Triassic, herrerasaurids entered the terrestrial biotas as large and highly predatory forms, sharing with rauisuchid archosaurs the role of superpredators. In the Ischigualasto Formation, herrerasaurids are numerically more abundant than other contemporary terrestrial carnivores (e.g., *Saurosuchus galilei, Venaticosuchus rusconii, Chiniquodon theotonicus*). Clearly, the herrerasaurid radiation did not wait for the extinction of non-dinosaurian tetrapods (Novas 1997c). Moreover, the increase in body size and numerical importance of basal dinosaurs occurred almost in tandem with a comparable phenomenon in other plant-eating tetrapods, such as rhynchosaurs and traversodontid cynodonts. The later ones were well diversified and numerically abundant at the time of deposition of the Early Ladinian Los Chañares Formation and manifested a significant increase in size and body mass in the Carnian, from cat-size forms (*Massetognathus*) to animals the size of wild boars (e.g., *Exaeretodon*). This allows the conclusion that herrerasaurids and other meat-eating dinosaurs (e.g., *Eoraptor*) co-evolved with their plant-eating prey (Novas 1997c). On the other hand, no clear pattern of competitive exclusion exists between basal meat-eating dinosaurs and other predatory archosaurs, such as ornithosuchids and rauisuchids, which survived until the Norian (Bonaparte 1972, 1982; Long and Murry 1995).

In consequence, the available Triassic record of South America does not entirely support the interpretation that dinosaur dominance was reached in an empty ecospace after several non-dinosaurian groups (e.g., rhynchosaurs and dicynodont, traversodont, and chiniquodontid synapsids) became extinct (contrary to Benton 1991, 1994). On the contrary, dinosauriforms co-evolved during the Middle to Early Late Triassic with the above-mentioned non-dinosauriform tetrapods, amplifying the range of body size, species diversity, and feeding habits. In sum, the ecological diversification of the dinosauriforms documented in the Carnian can be seen as part of an uninterrupted co-evolutionary process that began at least in the Mid Triassic (Novas 1997c).

The radiation of the herbivorous dinosaurs that became dominant in Late Triassic–Early Jurassic times (e.g., sauropodomorphs and ornithischians) was preceded by the radiation of small-sized (~1 meter long) plant-eating dinosauriforms (e.g., *Silesaurus*, *Sacisaurus*) and basal dinosaurs (e.g., *Pisanosaurus*). In other words, the acquisition of plant-eating habits occurred more than twice among basal dinosauriforms.

During Ischigualastian times, basal and small-sized sauropodomorphs are present (*Saturnalia*), but they constitute a secondary element within faunas numerically dominated by rhynchosaurs. However, ichnological evidence from the Carnian Portezuelo Formation (Marsicano and Barredo 2004) reveals that prosauropod dinosaurs coexisted with sauropodomorphs that were more evolved toward the sauropod condition in pedal morphology and quadrupedal pose and walk, indicating that these plant-eating saurischians were deeply diversified (albeit not numerically dominant) during the Carnian.

In Late Norian beds of South America (e.g., Los Colorados, El Tranquilo), sauropodomorphs constitute the terrestrial tetrapods that are numerically most important (Bonaparte 1972, 1982), represented by a variety of genera such as the plateosaurids *Coloradisaurus brevis* and *Mussaurus patagonicus*, the melanorosaurid *Riojasaurus incertus*, and the possible sister group of Sauropoda, *Lessemsaurus*.

The "explosive evolution" (e.g., rapid increase in numerical abundance and body size) manifested by prosauropod dinosaurs during the Late Norian was interpreted as the result of opportunistic evolution after the extinction of several non-dinosauriform tetrapods (e.g., rhynchosaurs and traversodonts; Benton 1991). However, sauropodomorphs were present in the Carnian (*Saturnalia*, Portezuelo footprints), and their rarity in Ischigualastian times can be explained by successful competitors (e.g., rhynchosaurs and traversodonts).

Competition and opportunism constitute two ecological processes that are mutually exclusive in current interpretations about the early evolution of dinosaurs. Probably a combination of both processes played out in the structure of tetrapod communities during the Mid- to Late Triassic, but it is not possible to say among which species and with what intensity such ecological relationships took place. Perhaps it will be more adequate, in accordance with available evidence, to consider that the explosive dominance of rhynchosaurs and the deep diversification of archosaurs and synapsids resulted from the same eco-evolutionary processes that began in Mid-Triassic times and that affected the tetrapod fauna as a whole.

Fig. 3.1. Geological time scale of Jurassic Period.

JURASSIC	Upper	Tithonian	145.5±4.0
			150.8±4.0
		Kimmeridgian	
			155.5±4.0
		Oxfordian	
			161.2±4.0
	Middle	Callovian	
			164.7±4.0
		Bathonian	
			167.7±3.5
		Bajocian	
			171.6±3.0
		Aalenian	
			175.6±2.0
	Lower	Toarcian	
			183.0±1.5
		Pliensbachian	
			189.6±1.5
		Sinemurian	
			196.5±1.0
		Hettangian	
			199.6±0.6

JURASSIC DINOSAURS 3

The Jurassic is the second period of the Mesozoic, extending from 199 through 145 million years ago (Fig. 3.1). The paleogeographic configuration was similar to the Triassic, with landmasses assembled into the single supercontinent of Pangea. However, in contrast with earlier times, during the Jurassic, new seaways developed and the continent began to fragment (Fig. 3.2). The rifting of Pangea commenced in the Late Triassic and continued slowly through the Early Jurassic, but the separation of the major continental areas by growing ocean basins took place during the Middle or Early Late Jurassic (Wing and Sues 1992). In the absence of barriers to dispersal, Jurassic faunas and floras included numerous cosmopolitan elements (Wing and Sues 1992).

During the Jurassic, dinosaurs diversified considerably; this is the period when the first really big representatives of Sauropoda and Theropoda and smaller-sized ornithischians radiated. The Triassic pteridosperms (the main elements of the *Dicroidium* flora) were replaced by a flora dominated by a mixture of woody gymnospermous groups (mainly large conifers such as araucariaceans and smaller bennettitaleans) and herbaceous pteridophytes (Wing and Sues 1992).

The climate during the Jurassic was similar to that of the Triassic, characterized by warm temperatures and the absence of glacial accumulations at high latitudes. Seasonally dry conditions were widespread during the Early and Middle Jurassic, and increasing aridity in the Late Jurassic is thought to have resulted from a breakdown of the monsoonal circulation that had characterized the interior of Pangea during the Triassic and the beginning of the Jurassic. Strong latitudinal variation in rainfall continued throughout most of the Jurassic (Wing and Sues 1992).

South America remained joined with Africa and Antarctica, but some isolation from North America had begun by Middle Jurassic times (Fig. 3.2), when continental separation of North America from Africa created the North Atlantic, thus allowing the interchange of invertebrate and vertebrate marine faunas of the Tethys Sea and the American margin of the Pacific Ocean (Bonaparte 1981; Gasparini 1992).

The west coast of South America was cyclically flooded by the waters of the Pacific, as evidenced by thick marine deposits of Jurassic age along the Andes from Colombia to southern Patagonia (Zambrano 1981). The marine conditions in most of Perú and Chile that had been initiated in the Triassic continued into the Jurassic. Some regions of Patagonia were also covered by Pacific waters because of regional faulting that began in Triassic times. Important magmatic activity occurred along the Pacific margin of South America, and a submarine island arc complex developed that finally coalesced as an essentially continuous volcanic feature (Uliana and Biddle 1988).

Vast lava flows and thick ash falls occurred during the Jurassic over most of the South American territory. Inland regions such as the Amazonas and Parnaíba basins in northern Brazil; the Paraná and Chaco basins in southern Brazil and northeastern Argentina, respectively; and most of Patagonia experienced intense and recurrent volcanic events related to the fragmentation of western Gondwana. The most spectacular lava flows were those of the Upper Jurassic–Lower Cretaceous Serra Geral Basalts. These were widely represented in the Paraná Basin, which together with its counterpart in western Africa formed the Paraná-Etendeka volcanic province (Fig. 3.2).

Intermittent igneous activity in Patagonia occurred through most of the Jurassic, a time interval of around 60 million years, and abundant volcanic tuffs (resulting from ash falls) accumulated, interstratified with less common lava flows, fluvial sandstones, and lacustrine mudstones. These later sedimentary deposits, which were exposed in the Deseado Massif of southern Patagonia, yield the fossil remains of plants, frogs, insects, conchostracan arthropods, and fresh-water mollusks (Stipanicic and Reig 1956; Casamiquela 1964a). This set of volcanic and sedimentary rocks is widespread in this southern region of South America, from the Atlantic to the Andes, and constitutes the most relevant geological feature of this age.

The Paraná and Chaco basins continued to receive sediments under desert conditions dominated by aeolian regimes that had begun in Triassic times. Toward the end of the Jurassic and into the beginning of the Cretaceous, thick sand deposits of the Botucatu Formation accumulated, a record of the earth's largest paleodesert during Mesozoic times.

Jurassic Dinosaur-Bearing Beds in South America

The record of Jurassic dinosaurs in South America (Figs. 3.3–3.5) is limited to a few taxa and is restricted geographically mostly to the southern extremity of the continent (e.g., Argentina), thus contrasting with the more abundant and diverse record of Triassic and especially Cretaceous faunas of this continent. With the exception of sporadic finds from Venezuela, Colombia, and Brazil, the most complete and informative remains come from Patagonia. The known taxa of Jurassic Theropoda currently include the large *Piatnitzkysaurus floresi* (Bonaparte 1986b) and *Condorraptor currumilli* (Rauhut 2005a), both of the Callovian Age, and footprints of small theropods, the ichnotaxa *Sarmientichnus scagliai* and *Wildeichnus navesi* (Casamiquela 1964a), both from the Bathonian-Oxfordian Age. Several sauropodomorph taxa have been found in South America, among the most important of which are the Hettangian *Massospondylus* sp. from northwest Argentina (R. N. Martínez 1999), the Bajocian *Amygdalodon patagonicus* (Cabrera 1947), the Callovian *Volkheimeria chubutensis* and *Patagosaurus fariasi* (Bonaparte 1979b, 1986a, 1986c), and the Kimmeridgian–Tithonian *Tehuelchesaurus benitezii* (Rich, Vickers-Rich,

Fig. 3.2. Paleogeographic map of South America and surrounding continents during the Early Cretaceous (143 million years

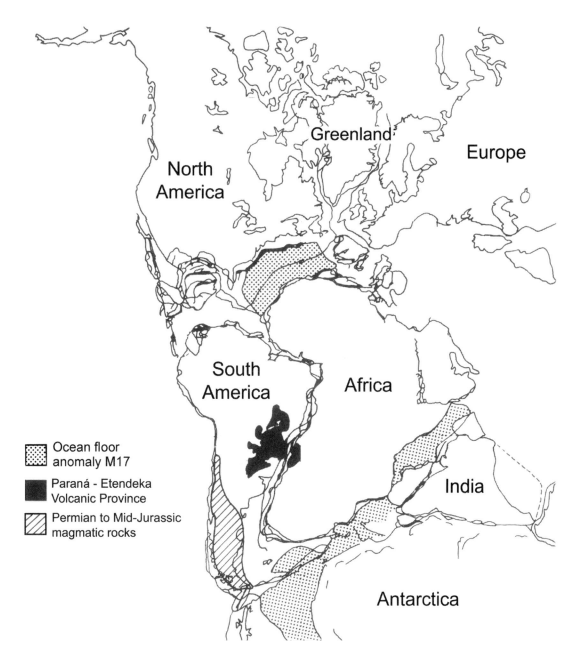

Giménez, Cúneo, Puerta, and Vacca 1999) and *Brachytrachelopan mesai* (Rauhut, Remes, Fechner, Cladera, and Puerta 2005). In contrast, only a few ornithischian taxa are currently known from Jurassic beds of this continent, represented by remains of a tiny animal collected in the La Quinta Formation of Venezuela (Russell, Odreman Rivas, Battail, and Russell 1992; Barrett et al. 2008) and footprints of small- to medium-sized quadrupedal and bipedal ornithopods from Argentina (*Delatorrichnus goyenechei*), Chile, and Brazil (e.g., Casamiquela 1964a; Leonardi 1989; Moreno and Rubilar 1997; Moreno and Benton 2005; Moreno, Rubilar, and Blanco 2000; Rubilar 2003).

ago), indicating magmatic activity that occurred during the Jurassic (included within the Gondwanan magmatic rocks) and Cretaceous (Paraná–Etendeka Volcanic Province and oceanic bottom anomaly M17). *Redrawn from Rapela and Llambías (1999).*

Fig. 3.3. (A) Map of Colombia, Venezuela, and Chile indicating fossil localities that have yielded Jurassic dinosaurs. (B) Stratigraphy of dinosaur-bearing Jurassic formations of Colombia, Venezuela, Chile, and Brazil.

Because of strong geographical and taxonomical biases, Jurassic dinosaurs from South America are poorly known, in contrast with the contemporary faunas from North America, Africa, and Asia, about which we know much (Weishampel 1990). Even so, extensive and rich outcrops of Jurassic age in South America remain to be prospected and exploited, and it is not improbable that very relevant findings could emerge in the near future.

1. Colombia

Jurassic sediments of continental origin are widespread in the Eastern Cordillera of Colombia (Mojica and Dorado 1987). Several sedimentary units have yielded plant remains in Colombia, but dinosaur remains are currently restricted to an isolated sauropod vertebra reported from the Upper Jurassic Girón Formation, which crops out in the Department of Magdalena (Fig. 3.3; Langston and Durham 1955; Mojica and Dorado 1987).

2. Venezuela

The La Quinta Formation is widely exposed in the Venezuelan Andes (Andes de Mérida, western Venezuela, near the border with Colombia; Fig. 3.3). The fossil flora and invertebrates at the La Quinta Formation indicate continental and freshwater/brackish conditions, but different facies identified in the La Quinta Formation indicate distinct paleoenvironments that include alluvial plains, swamps, freshwater lagoons, and semi-arid facies (Barrett et al. 2008).

The La Quinta Formation of the Venezuelan Andes yields a dinosaur fauna that is of Early or Middle Jurassic age (Schubert 1986; Russell, Odreman Rivas, Battail, and Russell 1992; Sánchez-Villagra and Clark 1994; Moody 1997). Because of its age and geographical location, it has the potential to provide an important window on dinosaur evolution and paleobiogeography (Barrett et al. 2008). Previous authors have reported fragmentary ornithischian material (Russell, Odreman Rivas, Battail, and Russell 1992; Sánchez-Villagra and Clark 1994) and isolated indeterminate theropod teeth (Moody 1997) from this unit. However, a large sample of additional dinosaur specimens from the La Quinta Formation is now available (Sánchez-Villagra and Clark 1994). Although some of the remains were originally attributed to a small ornithischian dinosaur, recent work suggests that more than one dinosaur taxon is represented in the sample (Barrett et al. 2008).

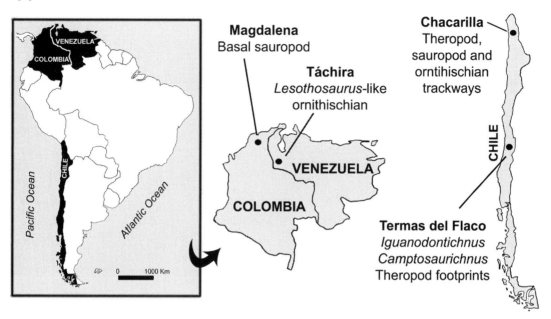

3. Chile

Jurassic dinosaurs from Chile are currently limited to some badly preserved trackways from the Tithonian Baños del Flaco Formation. This unit, which is exposed on the northern side of the Tinguirica River near Termas del Flaco in central Chile (Fig. 3.3), yielded trackways corresponding to sauropods (*Iguanodonichnus frenki*; Moreno and Benton 2005), ornithopods (*Camptosaurichnus fasolae*; Moreno and Rubilar 1997; Rubilar, Moreno, and Vargas 1998), and medium- to small-sized theropods (Moreno and Rubilar 1997; Rubilar, Moreno, and Vargas 1998; Rubilar 2003). The Termas del Flaco tracksite reveals a dinosaur ichnocoenosis in a carbonate platform environment (Moreno and Benton 2005) that is taxonomically similar (to the extent that such determinations can be made from footprints) to Early–Middle Jurassic faunas from Argentina known from body fossils of sauropods, medium-sized theropods, and small ornithopods (the last known only by footprints).

The Chacarilla Formation (Upper Jurassic–Lower Cretaceous?) crops out in I and II Regions of northern Chile (Fig. 3.3). It contains a large number of trackways corresponding to theropods, sauropods, ornithopods, and possibly stegosaurians, indicating that a diverse dinosaur fauna inhabited the area by the end of the Jurassic (Moreno, Rubilar, and Blanco 2000; Rubilar, Moreno, and Blanco 2000).

4. Brazil

Two large basins accumulated sediments and volcanic rocks during most of the Jurassic: the Paraná Basin in southeastern Brazil and the Parnaíba Basin in northeastern Brazil. The rock successions are similar in both basins; they are composed of Upper Jurassic through Lower Cretaceous sandstones of the Botucatu Formation and intercalated Serra Geral basalts.

The Botucatu Formation has yielded an ichnocoenosis composed of tracks of theropods, ornithischians, and mammals (e.g., Leonardi and Oliveira 1990). This unit is primarily comprised of fine- to medium-grained reddish well-sorted sandstones that have sedimentological characteristics of migrating dunes. The Botucatu sandstones are intercalated with basalts of the Serra Geral Formation and their age is estimated by stratigraphic relationships rather than by fossils. Bigarella and Salamuni (1961) suggested that the aeolian sandstones accumulated between the Late Jurassic and Early Cretaceous (Fig. 3.3B). According to Scherer, Faccini, and Lavina (2000), the youngest limit of the Botucatu Formation's age corresponds to the Hauterivian, contemporaneous with the most recent lava flows of the Serra Geral Formation (Fernandes, Bueno dos Reis, and Souto 2004). The Botucatu Formation covers most of the Paraná Basin in Brazil, eastern Paraguay, and northeastern Argentina. Dinosaur footprints were found at several sites in the state of São Paulo (Fig. 3.4),

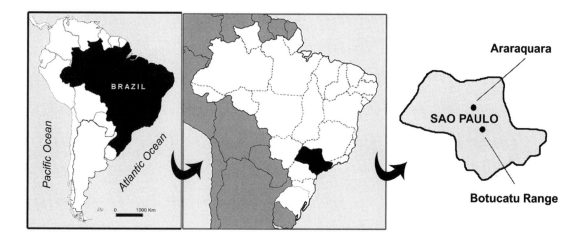

Fig. 3.4. Map depicting Jurassic fossil localities of Brazil (shown in black on the left map). The map of the center shows in black the Brazilian state of São Paulo that has yielded Jurassic dinosaurs. The map on the right indicates sites producing particular dinosaur finds in the State of São Paulo.

such as Araraquara, Jurucé, Rifaina, and the Botucatu Range and two more localities in Rio Grande do Sul (Tramandaí and Santa Cruz do Sul). Giuseppe Leonardi (1980a) presented an account of the sedimentology, paleoenvironment, and paleoichnology of the Araraquara fossil site. Leonardi (1981, 1989), Leonardi and Godoy (1980), and Leonardi and Oliveira (1990) reported about eight or nine forms of dinosaur trackways, including small theropods and ornithopods, all of them bipedal. The known dinosaurian trackways represent about 40 individual theropods and around 12 ornithopods. These tracks are associated with those of tiny mammals (*Brasilichnium elusivum*), which are numerically predominant. More recently, Fernandes and Carvalho (2007) reported the discovery of bipedal trackways of a large ornithopod dinosaur.

5. Argentina

The most productive Jurassic fossil sites in this country are in Patagonia (Fig. 3.5). However, some new but still barely prospected localities are in Neuquén Province in southern Argentina and San Juan Province in northwestern Argentina.

5.A. THE ISCHIGUALASTO–VILLA UNIÓN BASIN

Ricardo N. Martínez (1999) reported the discovery of well-preserved prosauropods in the Mogna Range, 60 kilometers north of the city of San Juan in northwestern Argentina. The find was made in beds of the Cañón del Colorado Formation and consists of several individuals of different sizes, including a skull and several postcranial skeletons. Martínez (1999) referred the specimens to the African sauropodomorph *Massospondylus*, from the Upper Elliot and Clarens formations of Zimbabwe. Based on *Massospondylus*, Martínez inferred that the Cañón del Colorado Formation dates to the Hettangian to Pliensbachian age. This find constitutes the first record of Lower Jurassic dinosaurs for southern South America. I will not include anatomical and phylogenetic considerations of the

Jurassic Dinosaurs   93

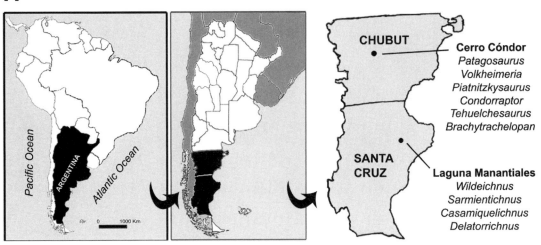

Massospondylus sp. from the Mogna Range in this book, but I look forward to more-detailed descriptions and illustrations of this interesting discovery.

5.B. THE SOMUNCURA MASSIF

Fossil-bearing beds that were deposited under continental conditions and correspond to different stages of the Jurassic period are in this extensive region of central Patagonia (Fig. 3.6). From bottom to top, these units are the Cerro Carnerero (Toarcian–Aalenian), Cañadón Asfalto (Callovian), and Cañadón Calcáreo (Kimmeridgian–Tithonian) formations (Fig. 3.5B).

The Cerro Carnerero Formation is comprised of tuffs and conglomerates and constitutes the basal unit of the Lonco Trapial Group, which unconformably overlies marine beds of Early Toarcian age (Rauhut 2005a). The Cerro Carnerero Formation is overlain by the volcanic Cañadón Puelman beds, possibly of Aalenian age (Page et al. 1999). The Cerro Carnerero Formation has yielded remains of fossil plants as well as remains of the basal sauropod *Amygdalodon patagonicus*.

The Cañadón Asfalto Formation, which is dated as Callovian, lies over the Lonco Trapial Group. This formation, which has extensive exposures in the central-western part of Chubut Province along the Chubut River (Fig. 3.6A), consists of a series of mainly lacustrine sediments with frequent basaltic intercalations. Cerro Cóndor is a Chubutian village famous for the abundance of dinosaur remains excavated mainly by José Bonaparte and his assistants during the period 1976–1985. The Cañadón Asfalto Formation has yielded a wealth of invertebrate, plant, and vertebrate fossils. Lacustrine algal limestone is very abundant, mainly formed by stromatolites (Cabaleri and Armella 1999). The most frequent invertebrates are conchostracans, but bivalves are also common in some layers. Apart from dinosaurs, the rich vertebrate fauna includes the remains of fishes, anurans, turtles, lepidosaurs, crocodiles, pterosaurs, and mammals (Rauhut, López Arbarello, and Puerta 2001; Rauhut, Martin, Ortiz-Jaureguizar, and Puerta 2002). Dinosaur taxa recorded in this unit are the theropods *Piatnitzkysaurus floresi* (Bonaparte 1979b, 1986b) and *Condorraptor currumili* (Rauhut 2005a) and the sauropods *Patagosaurus fariasi* and *Volkheimeria chubutensis* (Bonaparte 1979b, 1986a, 1986c) (Fig. 3.7). The record of Mid-Jurassic sauropods and theropods from the Cañadón Asfalto beds is highly relevant because it documents the taxonomic composition of dinosaur faunas from this part of the globe, which are approximately 15 million years older than the well-known dinosaur faunas from the Morrison Formation of western North America and the Tendaguru beds of Eastern Africa, both of which are assigned to the latest Jurassic (Bonaparte 1979b).

The Cañadón Calcáreo Formation (Kimmeridgian–Tithonian) overlies the Callovian Cañadón Asfalto beds; its sediments are dominated by fluvial sandstones (Proserpio 1987). Not far from the village of Cerro

Fig. 3.5. (A) Map depicting Jurassic fossil localities of Argentina (shown in black in left map). The map of the center shows the southern provinces of Argentina (Chubut and Santa Cruz; shown in black), that have yielded dinosaurs. Sites producing particular dinosaur finds are indicated in the individual provincial maps. (B) Stratigraphy of dinosaur-bearing Jurassic formations in Argentina.

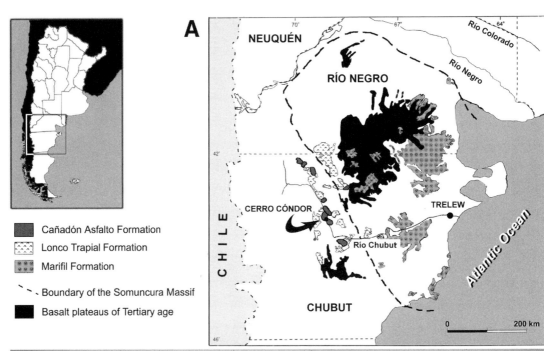

Fig. 3.6. (A) Somuncura Massif, in central Patagonia, showing distribution of sedimentary and volcanic rocks. (B) Volcanic rocks of the Marifil Formation, Cóndor, the Cañadón Calcáreo Formation has yielded abundant fishes as well as skeletons of the bizarre dicraeosaurid *Brachytrachelopan mesai* (Rauhut, Remes, Fechner, Cladera, and Puerta 2005) and the basal titanosauriform *Tehuelchesaurus benitezii* (Rich, Vickers-Rich, Giménez, Cúneo, Puerta, and Vacca 1999).

5.C. THE DESEADO MASSIF

This massif constitutes a topographically high region of southern Patagonia (Fig. 3.8) that is comprised of resistant bedrock, which contrasts with the subsiding regime of surrounding areas, such as the San Jorge and Austral basins (Barrio, Panza, and Nullo 1999). During the Jurassic, most of Patagonia experienced extensive deformation of the earth's crust, and volcanic events were of great importance in the geologic evolution of the Deseado Massif. The La Matilde Formation is a fossil-bearing unit exposed on the northeastern corner of Santa Cruz Province (almost coincident with the northern limit of the Deseado Massif). This is one of the most remarkable Jurassic formations of South America because of sedimentological conditions that preserved a diversity of fossils, making this unit especially amenable to paleoecological research. The formation is composed mainly of tuffs and sandstones that laterally and vertically interdigitate with the volcanic Chon Aike Formation, which together constitute the Bahía Laura Group. The La Matilde Formation has been interpreted as representing accumulation in a low-energy fluvial system, including associated shallow ponds. No sand dunes like those reported for the Botucatu Formation have been identified in La Matilde beds (Bonaparte 1996a). Sedimentation of the latter unit was accompanied by intense volcanic activity whose products were commonly reworked and diluted by fluvial currents (Barrio at al. 1999; De Valais, Melchor, and Genise 2003). The age of the La Matilde Formation is estimated as Bathonian–Oxfordian on the basis of radiometric dating of volcanics of the interdigitated Chon Aike Formation (Barrio, Panza, and Nullo 1999; Fig. 3.5B).

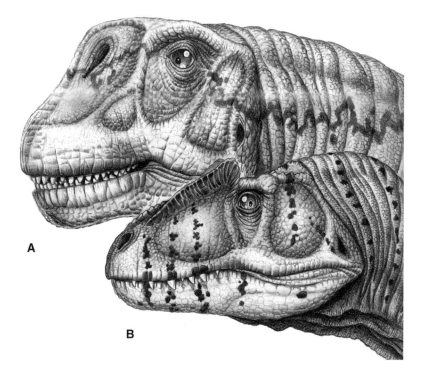

Fig. 3.7. Reconstructed heads of two dinosaurs recorded in the Callovian Cañadón Asfalto Formation, Central Patagonia. Not to scale. (A) Basal eusauropod *Patagosaurus fariasi* (skull length approximately 60 cm). (B) Basal tetanuran *Piatnitzkysaurus floresi* (skull length approximately 50 cm). *Illustrations by J. González.*

which crops out in central Chubut. *(A) Redrawn from Page et al. (1999). (B) Photograph by Fernando Novas.*

Jurassic Dinosaurs

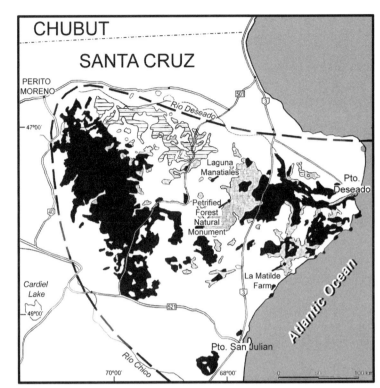

At Estancia Laguna Manantiales, the La Matilde Formation has yielded a diverse group of tetrapod tracks and some invertebrate trace fossils in addition to abundant permineralized plant remains. The Petrified Forest National Monument at Madre e Hija Hill also rests on beds of the La Matilde Formation and contains huge petrified araucarian logs (Melchor, de Valais, and Genise 2004).

Several fossil localities corresponding to the La Matilde Formation exist in the northeastern corner of Santa Cruz Province between the Deseado and Chico rivers (Fig. 3.8A). In the outcrops located closer to the Atlantic coast (including the type locality), the unit contains interbedded lacustrine shales (e.g., Mazzoni, Spalletti, Iñiguez Rodriguez, and Teruggi 1981) that have yielded abundant fossil remains. The body fossils recorded include silicified conifer logs, cones, seeds, seedlings and in situ stumps; leaves; fungi; ostracods; conchostracans; insects; mollusks; and the basal frog *Notobatrachus degiustoi*, represented by numerous articulated specimens (e.g., Stipanicic and Reig 1957; Báez and Nicoli 2004).

The Laguna Manantiales farm is an outstanding fossil site of the La Matilde Formation; it is located close to the Madre e Hija Hill, where petrified plant remains (araucarians) are abundant. Trace fossils from Laguna Manantiales Farm are diverse and include well-preserved tracks and trackways of insects (De Valais and Melchor 2003), mammals, and small dinosaurs (Casamiquela 1964a). Notably, some sites of the La Matilde Formation contain leaf litter horizons covered by tuff laminae that have tetrapod footprints as well as raindrop imprints (Melchor, de Valais, and Genise 2004). Rodolfo Casamiquela (1964a) named and described the trace fossils as *Ameghinichnus patagonicus* (mammal; Fig. 3.8B), *Wildeichnus navesi* and *Sarmientichnus scagliai* (small theropods), and *Delatorrichnus goyenechei* (presumably a small quadrupedal ornithischian). De Valais and Melchor (2003) mentioned the discovery of more footprints of these animals, including some specimens they identified as cf. *Grallator*, represented by tridactyl tracks with digital pads 69 millimeters long and 42 millimeters wide (De Valais and Melchor 2003). More recently, Coria and Paulina Carabajal (2004) described a purported new theropod footprint taxon, *Casamiquelichnus navesorum*.

5.D. THE NEUQUINA BASIN

This sedimentary basin, which constitutes a large part of northwestern Patagonia, is especially famous for its rich dinosaurian fossil assemblage of Cretaceous age (see chapter 4). In contrast, highly fossiliferous beds of the Jurassic period of the Neuquina Basin were deposited under marine conditions, so dinosaur remains are poorly known from levels of this age. Currently, some isolated sauropod records that have yielded little information have come from the Aalenian Calquenque and Kimmeridgian Tordillo formations (García, Salgado, and Coria 2003; Salgado and Gasparini 2004).

Fig. 3.8. (A) Deseado Massif, Santa Cruz province, southern Patagonia, Argentina, showing distribution of sedimentary and volcanic rocks. (B) Intensive quarrying in the Bathonian–Oxfordian La Matilde Formation, which crops out in southern Patagonia, has yielded great numbers of well preserved trackways of a tiny mammal (ichnogenus *Ameghinichnus*), as seen in this photo, and small dinosaurs. (A) Redrawn from Barrio, Panza, and Nullo (1999). (B) Photograph courtesy of José Bonaparte.

The Jurassic Dinosaur Record in South America

Perhaps the first references to Jurassic dinosaurs from South America are those from Friedrich von Huene, who in 1931 published a study of invertebrate and vertebrate footprints from the Botucatu Formation in Brazil (von Huene 1931b). Later, discoveries of body fossils and trackways were made in Argentina, including the early sauropod *Amygdalodon patagonicus*, which Ángel Cabrera studied in 1947; dinosaur footprints discovered in southern Patagonia, which Rodolfo Casamiquela studied in 1964; and numerous well-preserved sauropod and theropod skeletons that José Bonaparte and his crew excavated in Mid-Jurassic beds of central Patagonia. More recently, important finds resulted from explorations carried out in Lower Jurassic beds exposed in the Andes of Venezuela, where hundreds of bones of small-sized ornithischians were found (Russell, Odreman Rivas, Battail, and Russell 1992), and in the Andean Precordillera of northwestern Argentina, where several specimens of a prosauropod closely similar to *Massosponylus* were collected (R. N. Martínez 1999). Finally, Oliver Rauhut's fruitful explorations in the Tithonian beds cropping out in central Patagonia have resulted in the discovery of short-necked sauropods (*Brachytrachelopan mesai*).

Sauropoda

Sauropoda is phylogenetically defined as sauropodomorphs that are more closely related to *Saltasaurus* than to *Plateosaurus* (Wilson and Sereno 1998; Upchurch, Barrett, and Dodson 2004; Fig. 3.9). Among the most primitive sauropod is *Blikanasaurus cromptoni*, from the Lower Elliot Formation (Norian) of South Africa (Upchurch, Barrett, and Dodson 2004). Sauropods, as exemplified by *Patagosaurus*, are united by a suite of derived postcranial features, including modifications in the vertebral column (e.g., opisthocoelous cervical and cranial trunk vertebrae; a sacrum composed of four or more vertebrae) and hindquarters (e.g., an ilium with a pubic peduncle much longer than the ischiadic articulation, which is virtually absent; a columnar orientation of principal limb bones; a femur that is nearly straight in lateral view; metatarsals that are shorter than metacarpals; an enlarged and sickle-shaped pedal digit I ungual).

The clade Eusauropoda is composed of sauropods more related to *Saltasaurus* than to *Vulcanodon* (Wilson and Sereno 1998; Fig. 3.9). Among the best-represented primitive eusauropods are *Barapasaurus tagorei* from the Early Jurassic of India (Jain, Kutty, Roy-Chowdury, and Chatterjee 1979) and *Patagosaurus fariasi* from the Middle Jurassic of Patagonia. Eusauropods are diagnosed on the basis of a number of synapomorphic features: posterodorsally retracted external nares, the absence of antorbital fossa, broadly laterally exposed supratemporal fossa, a craniocaudally reduced orbital ventral margin, a craniocaudally shortened temporal bar, spatulate tooth crowns (Fig. 3.10A–D), opisthocoelous cervical centra, dorsal neural spines dorsoventrally elongated and broader transversely than craniocaudally, a femur lacking a lesser trochanter and asymmetrical distal condyles, a reduced manual phalangeal formula, and hoof-like pedal ungual phalanges, among many more features (Wilson and Sereno 1998).

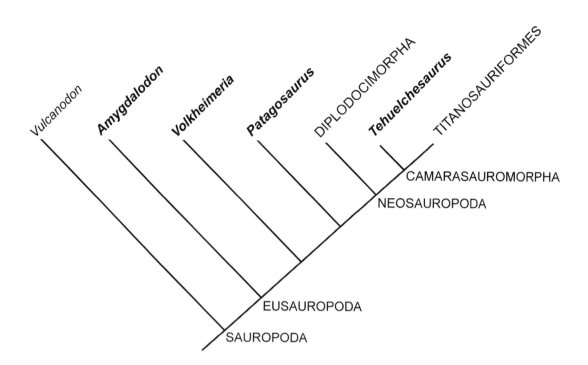

Fig. 3.9. Cladogram depicting possible phylogenetic relationships among basal sauropods. Taxa recorded in South America are indicated in bold.

The fossil record of Jurassic sauropods from South America, albeit far from complete, includes early representatives of Eusauropoda, such as *Amygdalodon* and *Volkheimeria*, as well as more-derived sauropods such as *Patagosaurus* (resembling the Mid-Jurassic *Cetiosaurus* from Europe), some sister taxa of Titanosauria (i.e., *Tehuelchesaurus*), and peculiar members of Diplodocoidea (i.e., *Brachytrachelopan*).

As Rauhut and colleagues (2005) pointed out, although the Late Jurassic is usually regarded as the heyday of sauropod evolution, our knowledge of sauropods from this interval is based on only a few individual formations, almost all of which are in the northern hemisphere. The highly productive Late Jurassic Tendaguru beds of Tanzania, which have yielded abundant sauropod remains, and sauropod discoveries made in the Cañadón Asfalto and Cañadón Calcáreo formations (Bonaparte 1979b, 1986a, 1986c; Rauhut, Remes, Fechner, Cladera, and Puerta 2005) constitute some of the few sources for understanding the evolutionary diversification and paleobiogeography of Jurassic sauropod dinosaurs in Gondwana.

Basal Eusauropoda

AMYGDALODON PATAGONICUS

In 1947, Ángel Cabrera coined the name *Amygdalodon patagonicus* for a sauropod taxon collected in Pampa de Agnia (west of Chubut Province) from levels of the Cerro Carnerero Formation, a unit currently thought to be Toarcian to Aalenian in age (Page et al. 1999). Rodolfo Casamiquela

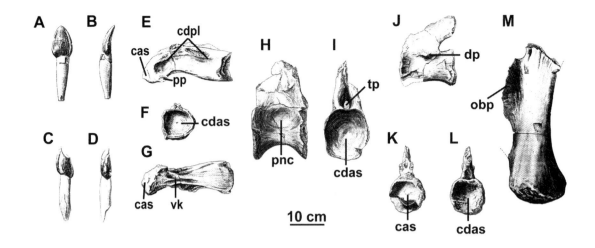

Fig. 3.10. *Amygdalodon patagonicus:* (A, B) Isolated tooth in (A) front and (B) side views. (C, D) Another isolated tooth in (C) front and (D) side views. (E–G) Anterior cervical vertebra in (E) lateral, (F) caudal, and (G) ventral views. (H, I) Mid-dorsal vertebra in (H) lateral and (I) caudal views. (J–L) Mid-caudal vertebra in (J) lateral, (K) proximal, and (L) distal views. (M) Right pubis in lateral view. *From Cabrera (1947). Reprinted with permission of Museo de La Plata, Argentina.*

(1963) redescribed the original material and referred to the same individual several other elements that had been collected previously, apparently from the same fossil site. Thus, the available remains of *Amygdalodon* consist of several teeth, incomplete cervical and posterior dorsal vertebrae, sacral and caudal vertebrae, ribs, a partial pubis, and the distal end of a tibia. More recently, Oliver Rauhut (2003a) published a review of this dinosaur in which he argued that the supposed single specimen of *Amygdalodon patagonicus* is a composite of more than one skeleton, a conclusion based mainly on size incongruence among some postcranial bones. However, he also noted that the consistency in phylogenetic information of all the material originally referred to *Amygdalodon* suggests that it may belong to the same species. This situation, in which the assemblage of bones belong to individuals of the same species of different ontogenetic ages, is not uncommon in the sauropodomorph record.

The crowns of the teeth of *Amygdalodon* are spatulate with a well-developed apical wear facet and several longitudinal ridges and grooves; the roots are considerably longer and narrower than the crown (Fig. 3.10A–D). The preserved portions of a cervical vertebra indicate that it was opisthocoelous and that a well-developed bony strut connected the centrum with the diapophysis (i.e., the centrodiapophyseal lamina). On the other hand, the centrum is poorly pneumatized and devoid of pleurocoels (Fig. 3.10E–G). Although incomplete, the dorsal vertebrae have neural arches that exceed the height of the centrum, although the neural arches are devoid of lateral excavations and extensive pneumatization (Fig. 3.10H and I).

These features of the morphology of *Amygdalodon* teeth and its cervical and dorsal vertebrae are also found among Eusauropoda (e.g., Bonaparte 1986a; Rauhut 2003b), but the absence of extensive pneumatization in the cervical and dorsal vertebrae (which is characteristic of the derived neosauropods) supports the interpretation that *Amygdalodon* represents one of the most basal known members of Eusauropoda (Rauhut 2003b; Fig. 3.9).

VOLKHEIMERIA CHUBUTENSIS

Bonaparte (1979b) coined the genus and species *Volkheimeria chubutensis* based on an incomplete skeleton composed of some cervicals, some dorsals and sacral vertebrae, pelvic bones, and a femur and tibia (Fig. 3.11). The specimen comes from Cerro Cóndor, in the center of Chubut Province, from outcrops of the Cañadón Asfalto Formation.

Bonaparte (1979b, 1986a, 1986c, 1999c) considered *Volkheimeria* to be one of the basal-most sauropods, resembling other primitive sauropods (such as the Bathonian *Lapparentosaurus* from Madagascar) in having neural arches of dorsal vertebrae taller than in prosauropods and well-developed dorsoventral crests on the sides of the neural spines (Fig. 3.11 A–D). McIntosh (1990) concluded that *Volkheimeria* may be related to Brachiosauridae, but, as Bonaparte pointed out (1999c), no single feature supports this suspicion. Recently, Upchurch and colleagues (2004) considered the position of *Volkheimeria* as uncertain within Sauropoda, although this Patagonian form could at least be included within Eusauropoda on the basis of its opisthocoelous cervical centra, the absence of the femoral anterior trochanter, and a laterally projecting cnemial crest of the tibia. However, *Volkheimeria* retains several plesiomorphic traits within Eusauropoda, such as dorsal neural spines that are broader craniocaudally than transversely. Additionally, the slender ischial shaft and an elongate pubis with a dorsoventrally reduced pubic plate (Fig. 3.11E) are reminiscent of prosauropods and *Vulcanodon*, suggesting a basal position for *Volkheimeria* within Eusauropoda (Fig. 3.9).

Volkheimeria is smaller than the larger specimens of *Patagosaurus*. For example, the length of the ilium of *Volkheimeria* is 45 centimeters, contrasting with that of *Patagosaurus*, which is 95 centimeters long.

PATAGOSAURUS FARIASI

Patagosaurus constitutes the best-known Jurassic sauropod thus far discovered in South America. Eleven specimens of this animal were excavated by José Bonaparte and collaborators in the Cañadón Asfalto Formation, which crops out in the village of Cerro Cóndor, close to the Chubut River (Fig. 3.12). *Patagosaurus* was a large animal of up to 14 meters long (Fig. 3.13A) that was found in the same bed rocks (but from a different fossil spot) as the smaller-sized *Volkheimeria*. However, *Patagosaurus* distinguishes from this dinosaur in that the neural arches of the dorsal vertebrae are taller than in *Volkheimeria* and the neural spines are morphologically more complex (Bonaparte 1979b, 1986a, 1986c, 1999c).

Some skull bones and teeth of *Patagosaurus* have been recovered (Bonaparte 1986c; Rauhut 2003c; Fig. 3.13B–D): a premaxilla, maxillae, and a juvenile dentary, the morphology of which match fairly well with the skull of the Late Jurassic *Camarasaurus*. *Patagosaurus* has a robust premaxilla that is rostrocaudally shortened and dorsoventrally deep. The maxilla has a thin and hooked nasal process and a well-delimited antorbital fenestra. The dentary is relatively short and deep, and the rostral

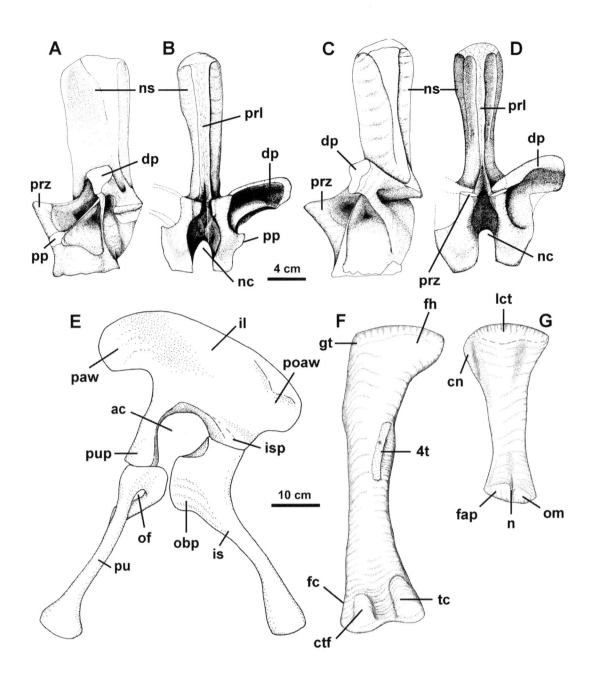

Fig. 3.11. *Volkheimeria chubutensis:* (A, B) Posterior dorsal vertebra in (A) lateral and (B) cranial views. (C, D) Another posterior dorsal vertebra in (C) lateral and (D) cranial views. (E) Pelvic girdle in left lateral view. (F) Left femur in caudal aspect. (G) Left tibia in lateral view. *From Bonaparte (1986c).*

portion is dorsoventrally expanded. The maxillary teeth are spatulate with a constricted neck between crown and root, and the crown is devoid of serrations. The dentary teeth, in contrast, have small marginal denticles on the proximal portion of the crown on both the rostral and caudal margins of the teeth (Rauhut 2003c).

As Bonaparte noted (1986c), the cervical vertebrae of *Patagosaurus* strongly resemble those of *Cetiosaurus* from the Middle Jurassic of Europe. They are elongate and have low neural spines (Fig. 3.13 E–G). The lateral surface of each cervical centra is deeply excavated, but a pleurocoel (e.g., a pneumatic foramen perforating the sides of the centra)

Fig. 3.12. Field work at Cerro Cóndor, Chubut Province, resulted in the discovery of abundant bone remains. (A) José Bonaparte handling a femur of a juvenile *Patagosaurus*. (B) Transporting a large plaster jacket containing sauropod bones down a hill, January 1979 (from left to right: Orlando Coria, Martín Vince, Tomás Fasola, Juan Leal, and Fernando Novas). *Photographs courtesy of José Bonaparte.*

is absent. The dorsal vertebrae of *Patagosaurus* are craniocaudally short and amphiplatyan and have poorly developed lateral excavations (Fig. 3.13H–J). The base of the neural arch is strongly constricted transversely, and the zygapophyses are placed well above the dorsal margin of the centrum. The neural spines of the dorsal vertebrae have well-developed dorsoventral laminae that give the spine an X-shaped contour in cross-section (Bonaparte 1986a, 1986c). The sacrum is composed of five strongly fused vertebrae (Fig. 3.13K).

The morphology of the scapula and coracoid closely resembles that of other basal eusauropods such as *Barapasaurus* and *Cetiosaurus* (Bonaparte 1986c); it has a proximodistally shortened scapular blade and a poorly developed acromial depression. The outline of the coracoid is elliptical to circular. The humerus and radius-ulna are elongate and have poorly expanded extremities. The ilium of *Patagosaurus* has a large pubic peduncle and an iliac blade that is anteriorly well projected (Fig. 3.13L). The femur is nearly straight and has a well-developed fourth

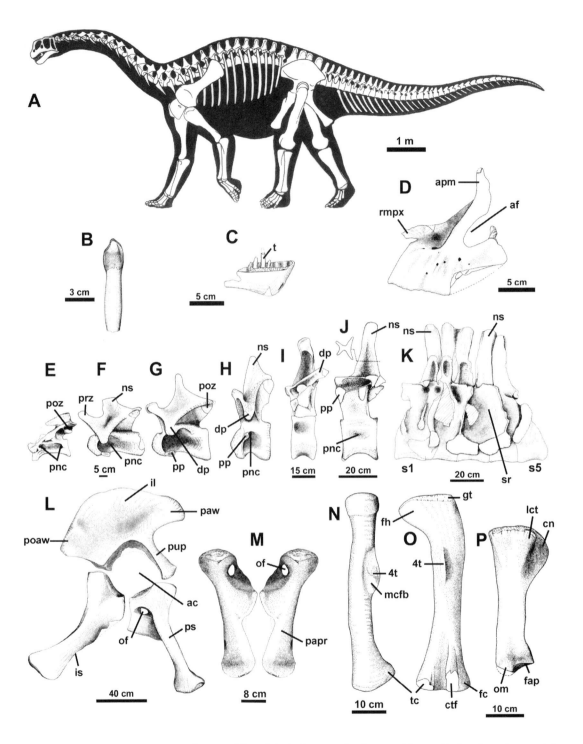

trochanter. The tibia is short and robust, with a well-developed cnemial crest (Fig. 3.13N–P).

Patagosaurus is considered a derived eusauropod, and it is here depicted as the sister taxon of Neosauropoda (Fig. 3.9). Upchurch and colleagues (2004) depicted the Indian *Barapasaurus*, the European

Cetiosaurus, and the Patagonian *Patagosaurus* as forming a monophyletic clade, Cetiosauridae, that splits at the base of the eusauropods.

A monospecific assemblage formed by five specimens of *Patagosaurus fariasi* was collected in the Cerro Cóndor quarry. Two specimens represent adult individuals, but the other three are juveniles (Bonaparte 1986c; Coria 1994b). The bone assemblage, which occurred at the same stratigraphic level and were in a similar state of preservation, has a relatively high proportion of juveniles compared to adults. All these factors suggest that burial of specimens occurred during or after a catastrophic event (Coria 1994b). On the basis of this assumption, Coria (1994b) speculated that *Patagosaurus* had a relatively complex social behavior, with parental care extended to the young of several age classes.

The discovery of individuals of *Patagosaurus* belonging to different stages of maturity allow us to characterize some ontogenetic changes of the postcranial skeleton: in juveniles the dorsal vertebrae have proportionally low dorsal spines with excavations on their sides that are restricted to the proximal half of the spines. Adult individuals, instead, have neural spines that are proportionally taller than in juveniles, and the lateral depressions extend toward the tip of the spines (Bonaparte 1996c).

Fig. 3.13. *Patagosaurus fariasi:* (A) Reconstruction of skeleton. (B) Isolated teeth in lingual view. (C) Juvenile left dentary in medial view. (D) Left maxilla in lateral aspect. (E–K) Vertebrae in left lateral view. (E) Axis. (F) Anterior cervical vertebra. (G) Posterior cervical vertebra. (H) Anterior dorsal vertebra. (I) Mid-dorsal vertebra. (J) Posterior dorsal vertebra; inset shows cross-sectional shape of neural spine. (K) Sacrum. (L) Pelvic girdle in right lateral view. (M) Reconstructed paired pubes in cranial aspect. (N, O) Right femur of juvenile specimen in (N) medial and (O) caudal views. (P) Right tibia of juvenile specimen in lateral view. *(A) Illustration by J. González. (B–P) From Bonaparte (1986c).*

Poorly Known Jurassic Sauropods from South America

Salgado and Gasparini (2004) described a fragmentary sacrum from marine beds of the Aalenian (Middle Jurassic) Calquenque Formation that is exposed at Tricolor Hill, southwest of Mendoza Province. The sacrum has a marked lateral constriction as well as internal spongy tissue, features that allow us to identify the specimen as an indeterminate sauropod. This fossil, albeit highly fragmentary, is one of the oldest records of Jurassic dinosaurs in Patagonia (Salgado and Gasparini 2004).

García and colleagues (2003) reported the discovery of fragmentary hindlimb bones from the Kimmeridgian Tordillo Formation near Chos Malal in Neuquén Province. The specimen consists of the distal extremity of a femur and the proximal ends of both a tibia and a fíbula. It was referred to Sauropoda on the basis of the shape of the cnemial crest as seen from the side, which is symmetrical in contour and has a distally placed point of maximum development. The specimen may be allocated to Eusauropoda because the cnemial crest is laterally projected, as is characteristic of this sauropod group (Wilson and Sereno 1998). However, the ovoid shape of the tibia in cross-section excludes the possibility that the specimen is a member of Neosauropoda, in which the tibia is subcircular in transverse section. This represents an interesting discovery that demonstrates that non-neosauropod dinosaurs survived in Gondwana together with more-derived sauropod groups (Salgado and Coria 2005).

The only Jurassic sauropod documented in a South American country other than Argentina was found in northern Colombia in the Department of Magdalena (Fig. 3.3). The material, which Langston and

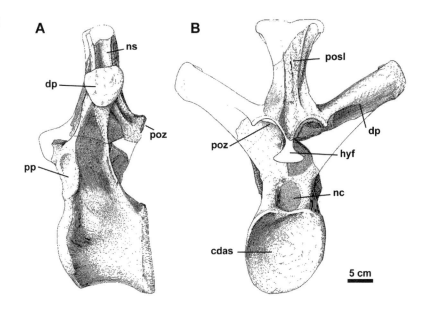

Fig. 3.14. Sauropod dorsal vertebra from Colombia. (A) Left lateral and (B) caudal views. *From Langston and Durham (1955). Reprinted with permission of The Paleontological Society and the Society for Economic Paleontology and Mineralogy.*

Durham described in 1955, is a single dorsal vertebra (Fig. 3.14) presumably excavated from beds of the Upper Jurassic Girón Formation (Mojica and Dorado 1987). The vertebra has diagnostic features of Eusauropoda (Bonaparte 1986a; Rauhut 2003b), such as a neural arch that exceeds the height of the centrum and a neural spine that is transversely broader than craniocaudally long. Moreover, this isolated vertebra resembles cranial dorsals of *Patagosaurus* in having elongate centropostzygapophyseal and postzygodiapophyseal laminae, features that Wilson (2002) has interpreted as autapomorphic of the Patagonian taxon.

Neosauropods

The name Neosauropoda was coined by José Bonaparte (1986a), and is currently applied to the clade formed by the common ancestor of Diplodocidae and Camarasauromorpha and all of its descendants (Wilson and Sereno 1998, Fig. 3.15). The most noteworthy features that diagnose Neosauropoda are the presence of pleurocoels on the presacral vertebrae, an enclosed external mandibular fenestra, the absence of marginal tooth denticles, chevrons that are dorsally opened, and a count of two or less carpal bones.

Camarasauromorpha

This clade of neosauropod dinosaurs was first defined by Salgado, Coria, and Calvo (1997; Fig. 3.15) to include conspicuous sauropods such as *Camarasaurus*, from the Late Jurassic of North America, and the Titanosauriformes, a prolific group of Late Jurassic and Cretaceous sauropods that will be examined in more detail in chapter 5. *Tehuelchesaurus benitezii* is a basal representative of Camarasauromorpha.

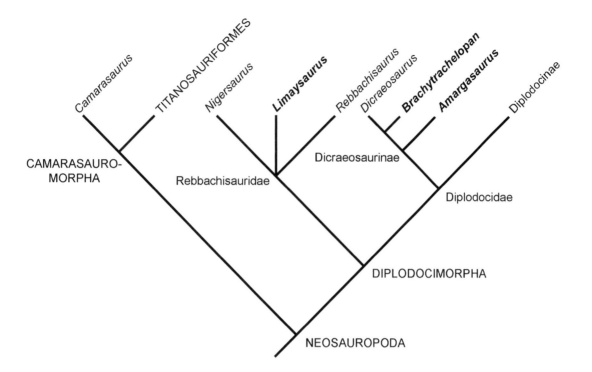

Fig. 3.15. Cladogram depicting possible phylogenetic relationships among diplodocimorph sauropods. Genera discovered in South America are indicated in bold.

TEHUELCHESAURUS BENITEZII

In 1999, Thomas Rich and coauthors described the sauropod *Tehuelchesaurus benitezii* from levels of the Kimmeridgian–Tithonian Cañadón Calcáreo Formation from a fossil site located 27 kilometers north of where *Patagosaurus fariasi* was excavated. *Tehuelchesaurus* was a large animal (albeit slightly smaller than *Patagosaurus*); its humerus was 114 centimeters long and its femur reached 158 centimeters in length.

Rich, Vickers-Rich, Giménez, Cúneo, Puerta, and Vacca (1999) considered *Tehuelchesaurus* to be closely similar to the genus *Omeisaurus*, from the Middle Jurassic of China, which is interpreted as the sister group of Neosauropoda (Wilson and Sereno 1998). In fact, *Tehuelchesaurus* resembles *Omeisaurus* in deep elliptical pneumatic excavations (i.e., pleurocoels) on the sides of the vertebral centrum as well as in a strong deltopectoral crest on the humerus. Rich and his colleagues pointed out that *Omeisaurus* and *Tehuelchesaurus* have a condition of their dorsal vertebrae that seems to be evolutionarily intermediate between the solid dorsal centra of more-primitive sauropods and the highly excavated condition found in neosauropod dinosaurs. Nevertheless, *Tehuelchesaurus* has features that support a higher position within Sauropoda rather than a position outside Neosauropoda. In this regard, the Patagonian genus has pleurocoels on the presacral vertebrae and a coracoid that is subquadrangular in contour, features that support affinities to the neosauropods (Upchurch 1998). Moreover, Salgado and Coria (2005) indicated that the opisthocoelous condition of the dorsal vertebrae of *Tehuelchesaurus* suggests its affiliation with Camarasauromorpha (Salgado, Coria, and

Calvo 1997). Rauhut (2002) went further, presenting evidence strengthening the hypothesis that *Tehuelchesaurus* is a camarasauromorph but also enumerating features that unite this genus with Titanosauriformes (e.g., a dorsoventrally enlarged pubic plate, a dorsoventrally deep pubo-ischiadic articulation, dorsal ribs with pneumatic cavities, plate-like cranial dorsal ribs, a femoral shaft with a lateral bulge on its proximal portion).

Interestingly, the holotype specimen of *Tehuelchesaurus benitezii* includes skin impressions (Fig. 3.16); it is one of the few sauropods in which this kind of anatomical evidence is recorded (Czercas 1997). The skin impressions of *Tehuelchesaurus* (Rich, Vickers-Rich, Giménez, Cúneo, Puerta, and Vacca 1999; Giménez 1996, 2007) come from different parts of the body, including patches that correspond to the thoracic region and a larger portion found close to the articulated dorsal vertebrae. Skin impressions found close to the back consist of non-imbricated flattened scales that are hexagonal in outline and have a rosette-like pattern. The scales are 3 centimeters in maximum diameter and are surrounded by scales of smaller size (2 centimeters in maximum diameter). Patches from the thoracic region are composed of diminutive rhomboidal scales 1 to 3 millimeters wide and 2 to 4 millimeters long (Giménez 1996, 2007). The scales of a basal sauropod from the Jurassic of Europe were of similar size and formed a similar rosette-like pattern (Czercas 1997).

Diplodocimorpha

Diplodocimorphs are one of the most famous dinosaur groups, including very well-known taxa such as *Diplodocus* and *Apatosaurus* from the Late Jurassic of North America (Figs. 3.9 and 3.15). In South America, diplodocimorphs have been recently recorded from Jurassic beds from Central Patagonia (Rauhut, Remes, Fechner, Cladera, and Puerta 2005). Diplodocimorphs were extraordinarily successful at the end of the Jurassic, especially in North America, where they flourished during a short period (Kimmerdigian).

Diplodocimorphs are distinguished by their pencil-like teeth as well as by a group of features related to a major rearrangement of the skull, involving a ventral displacement of the occiput and retraction and confluence of the external nares, which are characteristically located on the top of the skull (Fig. 3.17A and B). Salgado and Calvo (1997) interpreted the retraction of the external nares in the skull roof in adult diplodocimorphs as a consequence of the strong posterior inclination of the braincase as a whole. This cranial rearrangement is also expressed by the slit-like infratemporal fenestra, resulting from the strong anteroventral inclination of the quadrate bone and the consequent rostral shifting of the quadratojugal bone beyond the rostral border of orbit. Also, the basipterygoid processes are directed forward and the basioccipital condyle is directed downward. Diplodocimorphs are also characterized by very deep neural arches on the posterior dorsal and proximal caudal vertebrae, which maximally are

Fig. 3.16. Detail of the skin of *Tehuelchesaurus benitezii*. Courtesy of Olga Giménez.

nearly three times deeper than the corresponding vertebral centrum. The tail is distally narrow, conferring an ability to use it like a whip. In contrast to the long-armed brachiosaurs, diplodocimorphs have proportionally short forearms; the humerus is less than 70 percent of femur length.

Diplodocimorpha is subdivided into the diverse and highly derived Diplodocidae and the anatomically more primitive Cretaceous Rebbachisauridae (Fig. 3.15). Diplodocidae is a distinctive clade of diplodocimorphs containing two branches: the Laurasian Diplodocinae and the mainly Gondwanan Dicraeosaurinae (Calvo and Salgado 1995). Diplodocinae includes the five well-known Jurassic genera: *Diplodocus*, *Apatosaurus*, *Barosaurus*, *Amphicoelias*, and *Supersaurus*. Dicraeosaurinae includes the African *Dicraeosaurus* and the South American *Brachytrachelopan* (Jurassic) and *Amargasaurus* (Cretaceous). Diplodocidae are characterized by the shared presence of bifid neural spines in the cervical and anterior dorsal vertebrae, slightly procoelous anterior caudal vertebrae, mid-caudal chevrons with fore- and aft-directed processes, a prominence for the ambiens muscle on the anterior part of pubis, ischia that are distally expanded, and a metatarsal I with its distal articulation laterally projected (Salgado 1999). Dicraeosaurines were comparatively small diplodocids. An adult *Amargasaurus* reached around 9 meters long, in contrast with the *Diplodocus*, which was 25 meters long (Salgado 1999).

Dicraeosaurinae

Up to now, this sauropod clade has been recorded only in Gondwana. Its taxa include *Dicraeosaurus hansemanni* and *Dicraeosaurus sattleri*, both

Fig. 3.17. Diplodocimorph sauropods. (A, B) Skull of *Diplodocus carnegii* in (A) lateral and (B) dorsal views. (C–I) *Brachytrachelopan mesai*: (C) Reconstruction of skeleton. (D–G) Vertebrae in lateral view. (D) Cervical 5. (E) Cervical 8. (F) Centrum of cervical 10. (G) Dorsal 2. (H) Dorsal 7 in anterior view (broken parts of centrum and neural arch indicated in shadow). (I) Dorsal 11 in lateral view. *Redrawn from Rauhut, Remes, Fechner, Cladera, and Puerta (2005).*

from the Upper Jurassic Tendaguru Formation in Tanzania (Janensch 1914, 1929, 1935–36); *Brachytrachelopan mesai* from the Tithonian Cañadón Calcáreo Formation (Rauhut, Remes, Fechner, Cladera, and Puerta 2005); and *Amargasaurus cazaui*, from the Lower Cretaceous La Amarga Formation of Patagonia (Salgado and Bonaparte 1991). Recent discoveries and studies carried on by Leonardo Salgado, Jorge Calvo, José Bonaparte and Oliver Rauhut have contributed to a better knowledge of dicraeosaurine anatomy, shedding light on their phylogenetic relationships.

Dicraeosaurines are characterized by the bifurcation of neural spines, at least through presacral 6, transverse processes of dorsal vertebrae that project dorsoventrally at a 45° angle above the horizontal, anterior caudals with both neural arches and neural spines dorsoventrally deeper than in other diplodocids, and the absence of pneumatic cavities in the centra of both cervical and dorsal vertebrae.

Notably, dicraeosaurines were among the smaller diplodocimorphs, and some authors believe that they tended to reduce their size over geologic time. For example, Russell and colleagues (1980) noted that *Dicraeosaurus sattleri* is somewhat smaller than *Dicraeosaurus hansemanni* and that *Amargasaurus cazaui* is even smaller (Salgado and Bonaparte 1991). Moreover, it seems that a shortening in the backbone occurred in the Cretaceous dicraeosaurids in comparison with their Jurassic relatives, because *Amargasaurus cazaui* has 23 presacral vertebrae, one fewer than in *Dicraeosaurus hansemanni* (Salgado 1999).

BRACHYTRACHELOPAN MESAI

This taxon was described by the German paleontologist Oliver Rauhut and colleagues in 2005 and constitutes one of the most notable dinosaur finds in recent years. *Brachytrachelopan* is known from a beautifully preserved and articulated partial vertebral column, ribs, and some other isolated postcranial bones (Fig. 3.17C). The specimen comes from the Upper Jurassic (Tithonian) Cañadón Calcáreo Formation, which crops out 25 kilometers northeast of the village of Cerro Cóndor in Chubut Province. *Brachytrachelopan* was a small sauropod, as was usual among dicraeosaurids; its estimated length was 10 meters.

Analysis of *Brachytrachelopan*'s characteristics places it as the sister taxon of the almost contemporary African genus *Dicraeosaurus* rather than with the highly derived *Amargasaurus* from the Early Cretaceous of Patagonia. The main similarities between *Brachytrachelopan* and *Dicraeosaurus* occur in the morphology of the cervical vertebrae, which have well-developed ventral keels and strongly constricted centra. The cervical vertebrae are strongly opisthocoelous, whereas the dorsals and sacrals are amphiplatyan (Fig. 3.17D). The dorsal neural spines are elongate and deeply bifurcated up through dorsal 6. The neural spine of dorsal 7 is broad and petal-shaped, as seen from the front. The neural arches of the sacral vertebrae form a continuous sheet of bone over the sacrum (Fig. 3.17C).

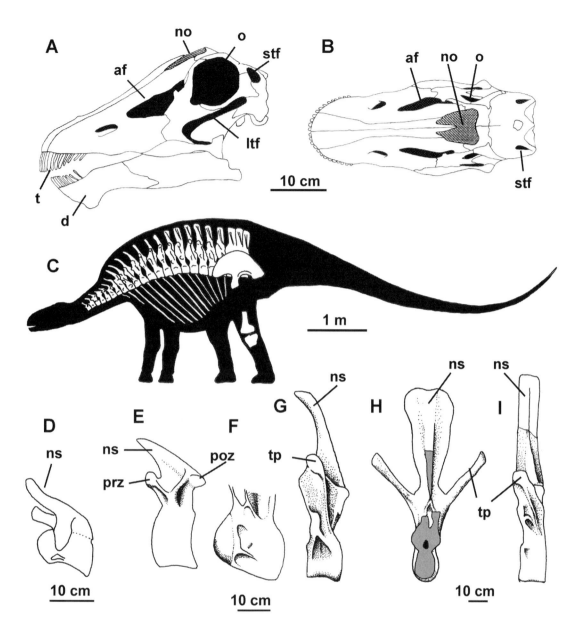

The exclusive traits of *Brachytrachelopan* include a pronounced anterior inclination of the mid-cervical neural spines, with the tip of the spines extending beyond the anterior end of the centrum. Moreover, dorsal vertebrae 1 to 6 have neural spines with cranially flexed tips (Fig. 3.17C–G). However, the most striking characteristic of this dicraeosaurid is a very short neck with cervical centra that are as long as (or shorter than) they are high, thus contrasting with the general evolutionary tendency manifested among sauropods toward an increase in neck length (Rauhut, Remes, Fechner, Cladera, and Puerta 2005). Neck elongation involves an increase in the number of cervical vertebrae and the length of individual elements, or both. Other diplodocoids (including the dicraeosaurids *Dicraeosaurus* and *Amargasaurus*) have elongate necks. The neck length of *Brachytrachelopan*, in contrast, is only 75 percent or less of that of the dorsal vertebral column, thus representing an autapomorphic feature of this sauropod, probably constituting an adaptation for low browsing (Fig. 3.17C). The structure of the cervical neural arches in *Brachytrachelopan* would seriously have restricted dorsal flexion of the neck (Rauhut, Remes, Fechner, Cladera, and Puerta 2005). On the basis of the available evidence, it seems that this animal was specialized for feeding on plants 1–2 meters high. Thus, *Brachytrachelopan* may have had the same ecological role as large low-browsing iguanodontians in Late Jurassic ecosystems of the northern hemisphere (Rauhut, Remes, Fechner, Cladera, and Puerta 2005).

Sauropod Footprints

IGUANODONICHNUS FRENKI

This ichnotaxon, recorded in the Tithonian Baños del Flaco Formation of central Chile, was originally described by Casamiquela and Fasola (1968) as a bipedal ornithopod dinosaur, probably a member of Iguanodontidae. However, subsequent reviews reinterpreted these footprints as belonging to a medium-sized sauropod (see Moreno and Benton [2005] for a detailed analysis of *Iguanodonichnus frenki*). The footprints are longer than they are wide and are 50–70 centimeters long. The pes claw impression corresponding to digit I is long and narrow. Moreno and Benton (2005) calculated that the hip of the trackmaker would be approximately 3 meters high, and because of the narrowness of the hips (as suggested by the small space between the trackway midline and the inside margins of the pes tracks), these authors attributed *Iguanodonichnus* to basal Sauropoda, diplodocoids, or basal macronarians, which are narrow-hipped sauropods.

From the Upper Jurassic–Lower Cretaceous? Chacarilla Formation in northern Chile, Moreno and colleagues (2000) and Rubilar and colleagues (2000) reported the discovery of superbly preserved sauropod footprints that they referred to *Brontopodus*.

Theropoda

The Jurassic record of theropod dinosaurs in South America is currently restricted to two closely related basal tetanurans. Ichnological evidence from the La Matilde beds, however, demonstrates that by Mid- to Late Jurassic times, Patagonia was populated by small-sized theropods, some of which have a highly specialized construction of the foot (Casamiquela 1964a; Coria and Paulina Carbajal 2004).

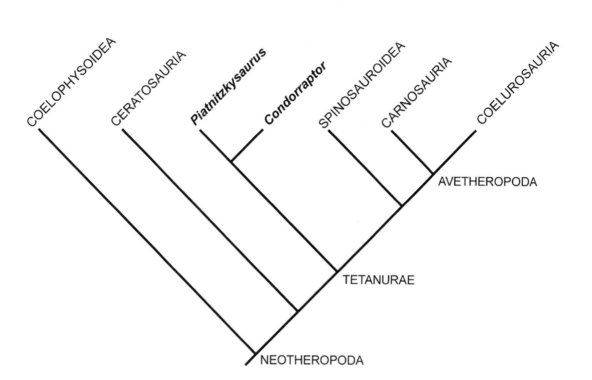

Fig. 3.18. Cladogram depicting possible phylogenetic relationships among basal theropods. Genera recorded in South America are indicated in bold.

Tetanurae

This theropod group was recognized by Jacques Gauthier in 1986 and is currently defined to include all taxa that share a more recent common ancestor with birds than with Ceratosauria (Fig. 3.18; Holtz, Molnar, and Currie 2004). Tetanurines are theropods recorded in Jurassic and Cretaceous rocks around the world and include small forms such as *Microraptor* as well as large animals around 14 meters long such as *Giganotosaurus*. Birds are considered to be a tetanurine lineage that survived the K-T mass extinction. Tetanurines are distinguished by a maxillary fenestra, a single pair of pleurocoels in the cervical vertebrae, prezygapophyses of the cranial cervicals that are entirely lateral to the neural canal, a ventral keel on the cranial dorsals, a shelf-like supraacetabular crest of the ilium, and a lateral ridge for contact with the fibula on the tibia that is offset from the proximal end of the tibia (Rauhut 2003a, 2005a).

The basal tetanurans recorded in Jurassic rocks of Argentina are reviewed below.

PIATNITZKYSAURUS FLORESI

This theropod is known from two partial skeletons, including parts of the skull (Bonaparte 1979b, 1986b), and is therefore the most completely known theropod from the Middle Jurassic of the southern hemisphere (Fig. 3.19A). *Piatnitzkysaurus* is one of the most primitive members of Tetanurae (Novas 1989b, 1992a; Holtz 2000; Rauhut 2003a) and is thus of greatest importance for our knowledge of evolution of characteristics

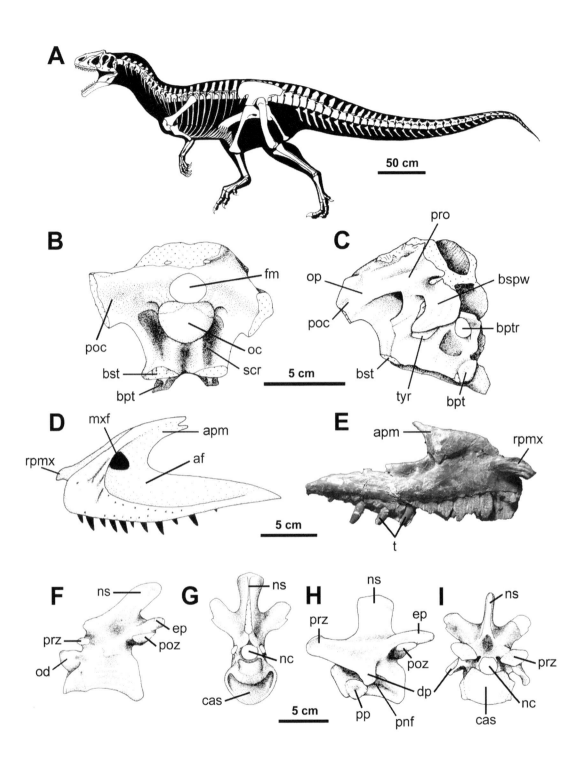

at the base of this group of theropods (Novas 1989b; Rauhut 2004). José Bonaparte provided an extensive description and illustration of this theropod taxon in several articles (Bonaparte 1979b, 1986b, 1996a), which together represent one of the best sources of information about basal tetanuran anatomy. Bonaparte interpreted *Piatnitzkysaurus* as a member of Allosauridae on the basis of its general similarity to *Allosaurus*, from the Upper Jurassic of North America. Later, Novas (1989b, 1992a) and Rauhut (2003a, 2005a) considered *Piatnitzkysaurus* to be a sister taxon of more-derived tetanurans (e.g., *Allosaurus* plus coelurosaurians), but more recent authors (e.g., Sereno 1997; Holtz, Molnar, and Currie 2004) have interpreted the Patagonian taxon as a member of Spinosauroidea (i.e., the tetanuran taxon including *Spinosaurus aegyptiacus* and all taxa sharing a more recent common ancestor with it than with *Passer domesticus*; Holtz, Molnar, and Currie 2004).

The braincase of *Piatnitzkysaurus* has been reviewed in detail by Rauhut (2004), whose article constitutes one of the few detailed accounts of braincase morphology in basal theropods. *Piatnitzkysaurus* has several probable autapomorphic characteristics, including extremely short and narrow basipterygoid processes, parasphenoid recesses that seem to communicate with each other, and basipterygoid recesses that are approximately as long anteroposteriorly as they are high (Fig. 3.19B and C). As Rauhut observed, a noteworthy aspect of the braincase of *Piatnitzkysaurus* is the large number of accessory pneumatic cavities, which indicates that the tympanic recess is plesiomorphic for tetanurans; it is possible that the subcondylar and basipterygoid recesses are as well. In contrast, the bones enclosing the brain (e.g., laterosphenoid, prootic, supraoccipital) are very massive and lack the internal pneumatic chambers that are present in later tetanurans. Below the occipital condyle are three well-excavated pneumatic cavities (subcondylar recesses; Fig. 3.19B). The basipterygoid processes in *Piatnitzkysaurus* are short and broad, a condition shared with other basal tetanurans (Rauhut 2004). Another noteworthy aspect of the braincase of *Piatnitzkysaurus* is a prominent hook-like basisphenoidal wing (Fig. 3.19C), a feature also distributed among other theropods such as *Ceratosaurus*, abelisaurids, *Allosaurus*, and tyrannosaurs (Rauhut 2004).

The frontal bone of *Piatnitzkysaurus* is considerably longer than wide and rectangular in shape, unlike the relatively much shorter and broader bone in *Allosaurus*, for example. The maxilla (Fig. 3.19D and E) has a large and laterally exposed maxillary fenestra, a feature that *Piatnitzkysaurus* shares with more-derived tetanurans (e.g., *Allosaurus*, *Compsognathus*, Tyrannosauridae, maniraptorans). However, *Piatnitzkysaurus* retains an unfenestrated medial surface of the maxilla, a primitive condition that distinguishes its maxilla from the medially fenestrated bone of *Allosaurus* and higher coelurosaurians (Fig. 3.19E; Novas 1989b).

The vertebral column of *Piatnitzkysaurus* is represented by several elements (Bonaparte 1986b). The axis (Fig. 3.19F and G) has no pneumatic foramina (e.g., pleurocoels), a plesiomorphic feature that distinguishes this genus from *Allosaurus* and coelurosaurians, in which this

Fig. 3.19. *Piatnitzkysaurus floresi:* (A) Reconstructed skeleton. (B, C) Braincase in (B) caudal and (C) rostro-lateral views. (D) Left maxilla (reconstructed) in lateral view. (E) Left maxilla in medial aspect. (F, G) Axis in (F) lateral and (G) cranial views. (H, I) Cervical 3 in (H) left lateral and (I) cranial views. *(A) Illustration by J. González. (B, C, F–I) From Bonaparte (1986b).*

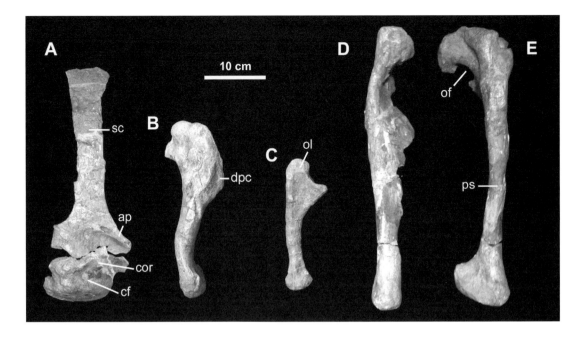

Fig. 3.20. *Piatnitzkysaurus floresi:* (A) Right scapula and coracoid in side view. (B) Right humerus in lateral view. (C) Right ulna in lateral view. (D, E) Right pubis in (D) cranial and (E) lateral views. *Photographs by Fernando Novas.*

neck vertebra is fenestrated. However, the remaining neck vertebrae (Fig. 3.19H and I) have pneumatic foramina, a condition that is shared with the remaining tetanurans. The sacrum of *Piatnitzkysaurus* is composed of five vertebrae that are craniocaudally shortened and describe a ventral arch in side view (Novas 1989b).

The scapular blade (Fig. 3.20A) is proximodistally shorter and craniocaudally wider than in more-derived tetanurans (e.g., *Allosaurus*, coelurosaurians). The humerus of *Piatnitzkysaurus* represents 50 percent of the length of the femur, a primitive condition present among basal theropods (e.g., *Dilophosaurus, Syntarsus, Eustreptospondylus*). The humerus is robust and strongly sigmoid in lateral view, and its distal end is cranially flexed (Fig. 3.20B). The relative lengths of the ulna (Fig. 3.20C) of *Piatnitzkysaurus* with respect to that of the humerus (77 percent) and the femur (40 percent) are similar to those of basal theropods (e.g., *Dilophosaurus, Syntarsus*). Thus, the forelimbs of *Piatnitzkysaurus* are proportionally longer than in *Allosaurus*.

The pelvic girdle of *Piatnitzkysaurus* is morphologically intermediate between that of basal theropods (e.g., *Dilophosaurus*) and derived tetanurans (e.g., *Allosaurus*, coelurosaurians; Bonaparte 1986b). For example, the pubic pedicle of the ilium is craniocaudally wide, as in *Allosaurus*, but the proximal portion of the pubis looks more primitive than in *Allosaurus* in that it has an obturator fenestra that is enclosed by bone (Fig. 3.20E) instead of being ventrally fully open, as in *Allosaurus* and coelurosaurians. Also primitive for a dinosaur is the dorsoventrally wide contact between pubis and ischium (Fig. 3.20E), a condition that is present in *Piatnitzkysaurus* and is sharply different from that of *Allosaurus* and more-derived tetanurans, in which this contact is reduced. The pubis in *Piatnitzkysaurus* has a distal foot that is more modestly developed than in *Allosaurus*, at least. Finally, the robust construction of the ischiadic shaft of *Piatnitzkysaurus* contrasts with the slender shaft that characterizes *Allosaurus* and coelurosaurians.

The hindlimb of *Piatnitzkysaurus* is well known (Bonaparte 1986b) and has tetanurine features in all its components. For example, the femoral head is well developed and transversely expanded and the anterior trochanter is wing-like and proximally projected, albeit not as much as in *Allosaurus* and coelurosaurians. The tibia of *Piatnitzkysaurus* has a pronounced cnemial crest as well as a well-developed fibular crest that is separated from the outer condyle of the proximal tibia by a notch, a feature that characterizes tetanurans (Rauhut 2005a). In addition, the distal end of the tibia is craniocaudally compressed and the surface for articulation with the ascending process of the astragalus is deeper than in coelophysoids and ceratosaurians. The metatarsals of *Piatnitzkysaurus* are elongate, representing 54 percent of tibial length, proportions that this taxon inherited from basal dinosaurs and theropods.

CONDORRAPTOR CURRUMILLI

Oliver Rauhut (2005a) recently described this taxon based on several pieces of the postcranial skeleton (Fig. 3.21). *Condorraptor* was found in levels of the Callovian Cañadón Asfalto Formation that are exposed in the nearby village of Cerro Cóndor.

Condorraptor strongly resembles *Piatnitzkysaurus*, which is found in almost the same beds and fossil locality. Rauhut thinks, however, that the two taxa are different and that their close resemblance is due to their similar position within theropod phylogeny. Rauhut found some distinctions between the two tetanuran taxa, such as a less-well-developed cnemial crest in *Condorraptor* (different from the rectangular-shaped and cranially projected cnemial crest of *Piatnitzkysaurus*; Fig. 3.21I) and a first sacral vertebra with a shallower infraprezygapophyseal fossa (different from the very deep and tunnel-like fossa of *Piatnitzkysaurus*). Even so, in my view, such differences are better interpreted as the result of individual variation within the population of a single theropodan species (e.g., *Piatnitzkysaurus floresi*).

Theropod Footprints

Ichnological evidence of Jurassic theropods from South America comes from Brazil (Botucatu Formation), Chile (Baños del Flaco, San Salvador, and Chacarilla formations), and Argentina (La Matilde Formation), revealing a variety of forms that considerably amplify the size range and taxonomic diversity, evidenced solely on the basis of body fossils.

Moreno, Blanco, and Tomlinson (2004) described theropod tracks from the Upper Jurassic San Salvador Formation, including tridactyl footprints with slender digits corresponding to a small animal (the length of the ichnite is 12 centimeters). Rubilar (2003) cited medium- to small-sized

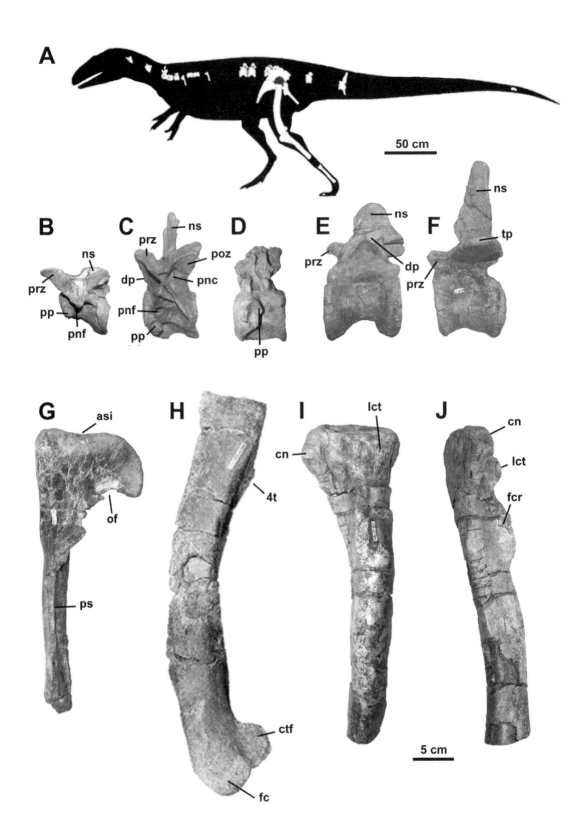

theropod footprints from the Tithonian Baños del Flaco Formation in central Chile (Fig. 3.3). Rubilar estimated that the hip of the trackmaker was approximately 0.80–1.1 meters high. The discovery of several theropod trackways was reported from the Chacarilla Formation (Upper Jurassic–Lower Cretaceous?), which crops out in northern Chile (Fig. 3.3), including small forms (ichnites 10 centimeters long) alongside large footprints (65 centimeters long) that correspond to a big animal that probably reached a height of 2.9 meters at its hips. These constitute the largest known theropod pedal footprints recorded in South America (Rubilar, Moreno, and Blanco 2000; Rubilar 2003).

The Botucatu Formation yielded unnamed theropod footprints of small size (Leonardi 1989; Leonardi, Carvalho, and Fernandes 2007; Fig. 3.22). Although most of the impressions are not well preserved, thus precluding a more precise taxonomic assignment, they demonstrate that the Botucatu dinosaur fauna apparently included small animals similar to those reported from the La Matilde beds of southern Patagonia rather than large theropods (Fig. 3.22D–F).

In Argentina, the only known Jurassic locality that has yielded dinosaur footprints is the La Matilde Formation, which crops out at Laguna Manantiales in Santa Cruz Province (Figs. 3.5 and 3.8), where abundant and nicely preserved trackways of small theropods have been found (Casamiquela 1964a; Fig. 3.23A and B). They correspond to the ichnotaxa *Wildeichnus navesi*, *Sarmientichnus scagliai*, and *Casamiquelichnus navesorum*.

Fig. 3.21. *Condorraptor currumili:* (A) Reconstruction of skeleton. (B–F) Vertebrae in left lateral view. (B) Anterior cervical. (C) Last cervical. (D) Dorsal 2. (E) Posterior dorsal. (F) Anterior caudal. (G) Proximal half of right pubis in medial view. (H) Left femur (lacking head and neck) in lateral aspect. (I–J) Left tibia in (I) lateral and (J) cranial views. *Photographs courtesy of Martín Ezcurra.*

WILDEICHNUS NAVESI

Casamiquela (1964a) coined this ichnotaxon on the basis of tridactyl footprints of a gracile theropod the size of a hen (the central axis is around 4 centimeters in maximum length). The morphology of these ichnites (Fig. 3.23B) corresponds to a conservative pattern in theropods, both in the relative lengths of the digits and the acute angles of divergence between digits (Coria and Paulina Carabajal 2004).

SARMIENTICHNUS SCAGLIAI

This is the most interesting ichnotaxon of the La Matilde footprint assemblage: it consists of monodactylous impressions that were probably left by the central digit (Fig. 3.23A). Because these tracks are found in the same levels as (and sometimes in association with) other dinosaur tracks that left imprints of three digits, I assume that the monodactylous condition of *Sarmientichnus* is not an artifact attributable to substrate conditions but reflects a true morphological characteristic. Casamiquela (1964a) drew attention to the peculiar claw print of this digit: it is ventrally slightly convex (in both directions, transversely and proximodistally), a shape that is different from a raptorial claw mark, which is ventrally concave (at least proximodistally). The shape of the largest claw mark differs from that of the acuminate claw marks of meat-eating theropods. Whether the

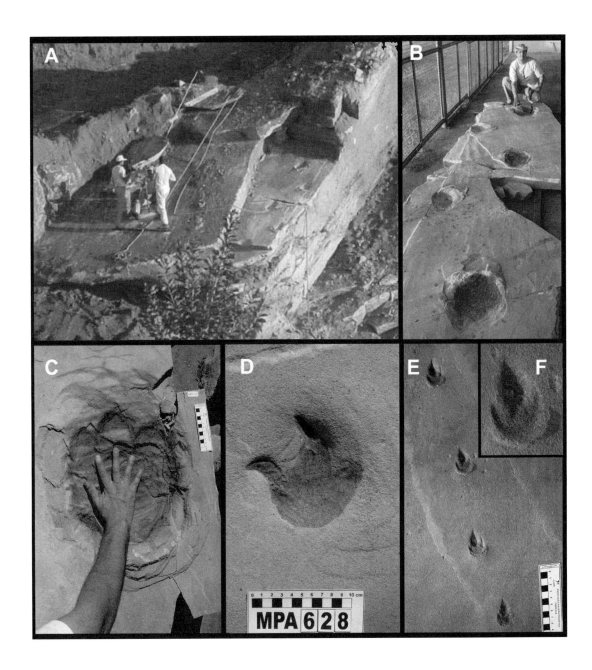

Sarmientichnus-maker was omnivorous is a matter of discussion, but an ostrich-like diet cannot be dismissed. The size of each *Sarmientichnus* print ranges from 10 to 14 centimeters, roughly the size of the greater rhea (*Rhea americana*). The theropod that produced these peculiar footprints constitutes the largest animal of the assemblage. *Sarmientichnus* is represented by numerous tracks, which indicates that it was a common member of the La Matilde fauna.

CASAMIQUELICHNUS NAVESORUM

Coria and Paulina Carabajal (2004) named this ichnotaxon (documented from the same fossil site from which *Sarmientichnus* was recovered) on the basis of functionally monodactyl or didactyl pedal prints produced by a small bipedal dinosaur. The impression of the central digit is narrow and elongate with a maximum length of 5.9 centimeters and a width of 1.8 centimeters. Digits II and IV were strongly reduced, as is evidenced by the small marks they left. Notably, each of these footprints are associated with an elongate discontinuous groove oriented in the same direction as the track. Coria and Paulina Carabajal assumed that the grooves were left by one of the pedal claws of both feet.

Coria and Paulina Carabajal felt that no ichnotaxon (even from the same fossil site at Laguna Manantiales) shares similar features with *Casamiquelichnus*. However, *Casamiquelichnus* is closely similar to *Sarmientichnus* in the functionally monodactyl or didactyl foot and the slight medial curvature of the footprint, so it is obvious that these two ichnotaxa were produced by phylogenetically related theropods, if not the same kind of dinosaur. Coria and Paulina Carabajal indicated that *Casamiquelichnus* differs from *Sarmientichnus* because it is smaller, lacks a rugose surface around the caudal end of the footprint (a feature that is present in all the available ichnites of *Sarmientichnus*), and produced footprints accompanied by a straight groove (a feature that Coria and Paulina Carabajal did not observe in the specimens of *Sarmientichnus* they studied). However, Silvina de Valais (pers. comm.) studied footprints of *Sarmientichnus* (Fig. 3.23A) that combine large size (a footprint length of 14 centimeters), a rugose surface around the caudal end of the footprint, and a straight groove that accompanies each of the footprints. In this context, distinctions in absolute size may reflect different ontogenetic stages of footprint producers. All of this indicates that *Casamiquelichnus navesorum* represents a junior synonym of *Sarmientichnus scagliai*.

Small-Sized Ornithischian from Venezuela

A rich collection of bones and teeth belonging to small-sized dinosaurs was made in the Lower Jurassic La Quinta Formation, cropping out in the state of Táchira in western Venezuela (Fig. 3.3). This discovery, which was briefly described in preliminary reports (e.g., Russell, Odreman Rivas, Battail, and Russell 1992; Sánchez-Villagra 1994; Sánchez-Villagra and Clark 1994), may include the only body fossil record of ornithischians in the Jurassic of South America (Fig. 3.24). Work carried out by John Moody (at the time working at the Museum of Biology of the University of Zulia, Maracaibo, Venezuela) resulted in a large collection of dental, cranial, and postcranial elements of small dinosaurs from the same locality.

The dinosaur specimens are disarticulated and associations of the elements are rare, but at least two distinct taxa appear to be present (Barrett

Fig. 3.22. Dinosaur footprints from the Botucatu Formation. (A) Rock quarry at Araraquara, São Paulo State, southern Brazil. Note the ornithopod trackway at the right of the image. (B) Trackway left by a large ornithopod. (C) Isolated ornithopod footprint. (D) Isolated theropod footprint. (E) Theropod trackway. (F) Detail of one of the footprints of figure. *Courtesy of Marcelo Adorna Fernandes.*

Ornithischia

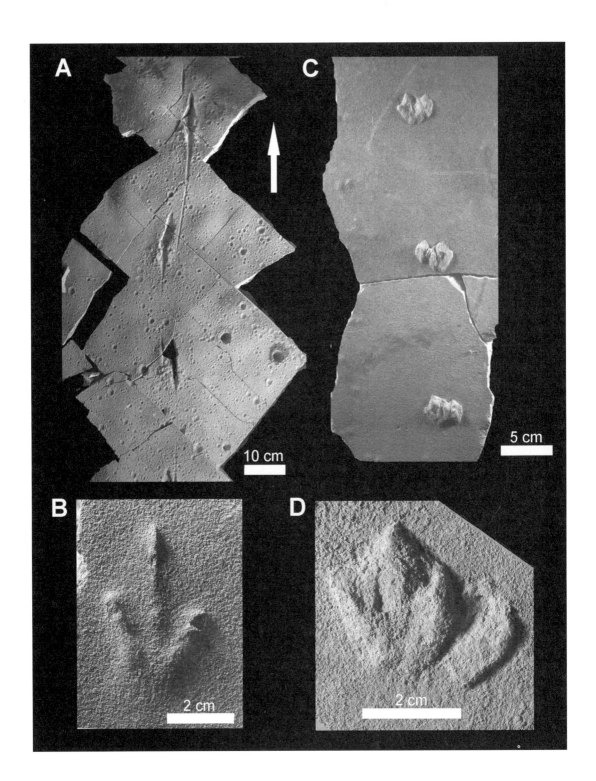

et al. 2008). Ornithischian dinosaurs are identified on the basis of isolated teeth and a distal tibia. The teeth collected from La Quinta beds (Fig. 3.24A and B) strongly resemble those of basal ornithischians (e.g., *Lesothosaurus*; Sereno 1991b). They have a large number of features that are often considered synapomorphic for Ornithischia (Barrett et al. 2008), including low crowns that have sub-triangular outlines in labial view, crowns that are expanded labiolingually above the roots to form a cingulum supported mesially and distally by ridges, a neck between the crown and the root, crowns with enlarged denticles on the mesial and distal margins, and crowns with an associated interdental pressure facet that suggests an en echelon arrangement in the jaw. Although some of these features occur in non-dinosaurian Triassic archosaurs (e.g., *Silesaurus*; Dzik 2003), the majority of these features have not yet been documented in non-ornithischian Jurassic archosaurs. In sum, the teeth from the La Quinta Formation are identified as pertaining to ornithischian dinosaurs.

However, the La Quinta teeth can be distinguished from those of other ornithischians (e.g., *Lesothosaurus*, *Scutellosaurus*; Sereno 1991b; Colbert 1981) because they have apicobasally extending ridges on the labial and lingual surfaces. Additionally, at least some of the crowns have apicobasal heights that significantly exceed mesiodistal widths. The unique combination of characteristics present in the La Quinta teeth suggests that they represent a taxon that is distinct from known ornithischian dinosaurs. It must be emphasized that previous reports of the basal ornithischian *Lesothosaurus* sp. from the La Quinta Formation cannot be substantiated on the basis of available data (Barrett et al. 2008). Additional and detailed comparisons are needed to clarify the phylogenetic allocation of the new South American form.

The distal end of the tibia (Fig. 3.24C and D) is strongly expanded transversely relative to the shaft and is triangular in distal view with a relatively flat anterior surface and a convex posterior surface, the apex of which extends proximally along the posterior margin of the shaft as a sharp and well-defined ridge. The outer malleolus extends further distally and is transversely broader and craniocaudally narrower than the inner malleolus. A broad beveled surface extends across the cranioventral surface of distal tibia and represents the articular surface for the astragalus. Proximally, this surface is bounded by a low horizontal ridge that extends transversely across the medial malleolus. The nearly horizontal inclination of this low ridge indicates that the ascending process of the astragalus was low and weakly developed. The flat anterior surface of the distal tibia is interrupted by a distinct raised eminence that is positioned at the lateral edge of the medial malleolus. Comparison with basal ornithischians (e.g., *Lesothosaurus*; Barrett et al. 2008) suggests that this low eminence articulated with the weakly developed ascending process of the astragalus and the distal end of the fibula.

The strong transverse expansion of the distal end of distal tibia is seen in ornithischians and a variety of theropods (Novas 1989a; Rauhut 2003a; Langer and Benton 2006). As in ornithischians, the ridge that

Fig. 3.23. Dinosaur footprints from the La Matilde Formation. (A) *Sarmientichnus scagliai*. (B) *Wildeichnus navesi*. (C, D) *Delatorrichnus goyenechei*, showing (C) a trackway and prints of a pes and manus. White arrow indicates the direction of steps. *Courtesy of Silvina de Valais.*

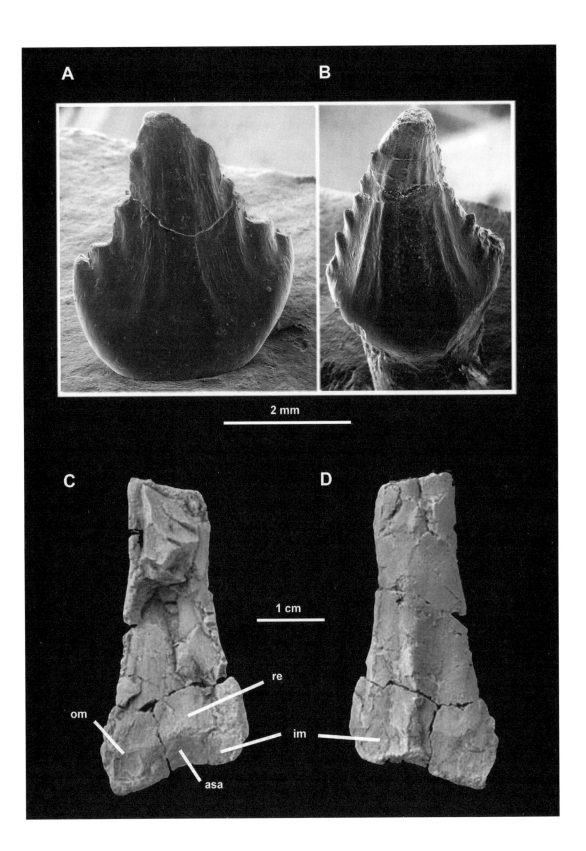

proximally delimits the articular facet for the astragalus is horizontal, and the corresponding ascending process of the astragalus would have been low and poorly developed. This contrasts with the general theropod condition in which the corresponding ridge extends proximolaterally from the mediodistal corner of the tibia across the anterior surface of the shaft and there is a corresponding well-developed ascending process of the astragalus. In addition, the raised eminence at the anterolateral edge of the medial malleolus is present in basal ornithischians but not in theropods. In light of these similarities, Barrett and his coauthors identify this tibia as an indeterminate ornithischian dinosaur.

Fig. 3.24. Selected elements of basal small ornithischian from the Lower Jurassic of Venezuela. (A, B) Two isolated teeth in side view (height of tooth crown approximately 5 mm). (C, D) Distal end of right tibia in (C) cranial and (D) caudal views. *Courtesy of Marcelo Sánchez-Villagra, Paul Barrett, and Jim Clark.*

Ornithischian Footprints

Rubilar (2003) described ornithopod footprints of large size (50 centimeters long) from the Chacarilla Formation (Upper Jurassic–Lower Cretaceous?), which is exposed at Quebrada de Chacarilla (I Región of northern Chile; Fig. 3.3). Casamiquela and Fasola (1968) described from these beds a set of footprints that they named *Iguanodonichnus frenki* and identified as belonging to Iguanodontidae. However, recent reviews reassigned *Iguanodonichnus frenki* to Sauropoda (see above).

Camptosaurichnus fasolae was reported from the Baños del Flaco Formation in central Chile (Fig. 3.3; Casamiquela and Fasola 1968; Moreno and Rubilar 1997; Rubilar, Moreno, and Vargas 1998; Rubilar 2003), a trace fossil that corresponds to a medium-sized ornithopod that walked both bipedally and quadrupedally and to other footprints of smaller bipedal ornithopods with an estimated hip height not exceeding 1 meter.

Casamiquela (1964a) described the ichnotaxon *Delatorrichnus goyenechei* based on the tracks of a quadrupedal small-sized dinosaur (Fig. 3.23C and D). The footprint, not exceeding 3 centimeters long, is tridactyl with broad toe marks, while the handprint is elliptical and lacks clear prints of digits. *Delatorrichnus* was originally considered to be a possible trail of a quadrupedal theropod dinosaur by Casamiquela, an interpretation Thulborn (1990) followed, who also noted the resemblances of *Delatorrichnus* to *Atreipus*, a quadrupedal trackway recorded in the Upper Triassic of North America and Europe. Instead, Bonaparte (1978) supported ornithischian affinities for *Delatorrichnus*, an interpretation that I follow. In fact, the shape of the ichnites that correspond to the foot matches well with those of ornithopod dinosaurs.

From the Late Jurassic–Early Cretaceous Botucatu Formation, Leonardi and Oliveira (1990) reported the discovery of a dinosaurian ichnite 14 centimeters long that they assigned to Ornithischia. However, its poor preservation prevents a detailed account of its shape and certain assignment to this group of dinosaurs. There are reports of the abundant presence of mammalian footprints (*Brasilichnium elusivum*) in company with bipedal trackways of small-sized theropods and ornithischians (Leonardi,

Carvalho, and Fernandes 2007). More recently, Marcelo Adorna Fernandes and Ismar de Souza Carvalho (2007) announced the discovery of the ichnites of a large ornithopod. The trackways were found in a rock quarry at the city of Araraquara in the state of São Paulo in Brazil and consist of a sequence of five tridactyl footprints with pes imprints of up to 35 centimeters long (Fig. 3.22B and C) that correspond with a trackmaker of approximately 5 meters long. The digits are short and rounded, and the double pace (i.e., the distance between successive footprints of the same foot) is 2 meters. This discovery presents the largest footprints recorded so far in this sedimentary unit.

Also from the Botucatu Formation, Fernandes, Bueno dos Reis, and Souto (2004) reported fossil evidence of liquid waste (urolite) consisting of several trace fossils 50–75 centimeters long in an excavation crater associated with the lines of gravitational streams. The size of the traces indicates that they would have been produced by the biggest animals documented in the Botucatu beds, ornithopod dinosaurs. Fernandes and colleagues interpreted this evidence to assume that at least some dinosaurs urinated.

In sum, footprint evidence indicates that during the Jurassic, South America was populated by small to large ornithopods. Ichnological records of ornithischian groups other than Ornithopoda (e.g., ankylosaurs, stegosaurs) remain to be discovered.

Jurassic Dinosaurs: An Overview

Available evidence indicates that sauropods constituted a common element in the Jurassic dinosaur faunas of South America. They were widely distributed latitudinally—from Colombia to Patagonia with some records in Chile, although they were apparently absent or numerically scarce from the eastern half of South America (i.e., the Paraná Basin), as suggested by the ichnofaunal assemblage in the Botucatu Formation. During the Early through Mid-Jurassic, sauropods were represented by basal forms (e.g., *Amygdalodon*, *Volkheimeria*, *Patagosaurus*), but toward the end of the period they included higher neosauropods such as camarasauromorphs (*Tehuelchesaurus*) and bizarre diplodocimorphs (*Brachytrachelopan*), indicating that a profound evolutionary diversification was in progress during Mid- to Late Jurassic times. The other plant-eating dinosaurs, the ornithischians, are represented in the Jurassic by small- to mid-sized forms (e.g., the tiny basal ornithischian from the Early Jurassic of Venezuela, the quadrupedal *Delatorrichnus* from the Mid-Jurassic of southern Patagonia, and *Camptosaurichnus* from the Upper Jurassic of Chile). The presence of large ornithopods, however, is evidenced by footprints as long as 50 centimeters from the Chacarilla Formation of northern Chile. Clearly ornithischians were present in South America during the Jurassic, although not as abundantly as the sauropods. Ornithischians are numerically underrepresented even compared to theropod dinosaurs. In this respect, South American dinosaur faunas contrast with other Jurassic dinosaur assemblages, such as in the Tendaguru beds of Africa, where the stegosaur *Kentrosaurus* and the iguanodontian *Dryosaurus* are represented by large numbers of individuals (Russell, Béland, and McIntosh 1980), and the Morrison Formation of the western United States, where Jurassic dinosaurs are represented variously by stegosaurs, ankylosaurs, and ornithopods. Of course, the intensity of exploration and digging in Jurassic beds of South America is less than that in the Tendaguru beds and the Morrison Formation, so these differences may be due, in part, to sample bias. Jurassic meat-eating dinosaurs are represented by an array of forms from reptiles the size of a pigeon (*Wildeichnus*) to big animals that probably were as tall as 2.9 meters at the hips (based on the footprints 65 centimeters long reported from the Chacarilla Formation of northern Chile; Rubilar, Moreno, and Blanco 2000; Rubilar 2003). The producer of such footprints was even larger than the tetanurans *Piatnitzkysaurus* and *Condorraptor*, which are known by bone remains.

Some Jurassic dinosaurs from South America are similar to those of other parts of the world—for example, the Early Jurassic *Massospondylus* sp. from northwestern Argentina. A comparable situation also applies for Mid-Jurassic times, as evidenced by the discovery of sauropods (e.g., *Volkheimeria, Patagosaurus*) and theropods (*Piatnitzkysaurus, Condorraptor*), which have notable morphological similarities with contemporary relatives from England and India (Bonaparte 1986b, 1986c; Rauhut 2005a). Although the available evidence demonstrates that a relatively uniform dinosaur fauna was widely distributed throughout the world, some peculiar cases indicate that some regionalization in dinosaur distribution was controlled by geographic and/or environmental conditions. In this regard, the fact that dicraeosaurids in the Late Jurassic are found in both Africa (in the Tendaguru beds) and South America (in the Cañadón Calcáreo Formation) and their absence in rocks of the same age in the northern hemisphere supports the belief that this clade of sauropods radiated and dispersed rapidly in the southern landmasses after the separation of Laurasia and Gondwana in the latest Middle Jurassic (Rauhut, Remes, Fechner, Cladera, and Puerta 2005). Following this line of interpretation, Bonaparte (1981) proposed that terrestrial faunas from Gondwana started to differentiate from those of Laurasia after Mid-Jurassic times, when the Tethys Sea connected with the Pacific Ocean and severed terrestrial connections between Laurasia and Gondwana.

The presence of some bizarre forms in Patagonia also speaks in favor of the theory that dinosaur faunas differentiated regionally: the dicraeosaurid *Brachytrachelopan* is distinguished by its shortened neck, an unknown condition among sauropod dinosaurs, which sharply distinguishes it from its African relatives (e.g., *Dicraeosaurus*). Furthermore, the producer of the *Sarmientichnus* trackways was a small theropod with a monodactylous feet, a highly derived construction this is difficult to match among Jurassic and Cretaceous coelurosaurs known from bone remains. In sum, *Brachytrachelopan* and *Sarmientichnus* may be viewed as evidence that some degree of endemism affected Jurassic dinosaur faunas, at least in Patagonia.

However, the most impressive evidence that, in my opinion, suggests strong regional differentiation is provided by the La Matilde and Botucatu ichnofaunas. These ichnofaunas from Brazil and Patagonia lack large dinosaurs, irrespective of their phylogenetic relationships. Leonardi (1989) has pointed out that the numerically rare "giants" that left their footprints in the Botucatu sandstones are ornithopods that could have been no more than 5 meters long. Although dinosaurs recorded in the La Matilde beds and the Botucatu sandstones are not the same, they correspond to a "dwarf fauna" composed of small-sized dinosaurs in association with tiny mammals (*Ameghinichnus* in Patagonia and *Brasilichnium* in Brazil; Leonardi and Oliveira 1990; Leonardi, Carvalho, and Fernandes 2007). Footprints left in the Botucatu sandstones have a simple rounded or elliptical cavity that furnished no morphological details, contrasting with examples from La Matilde, which include footprints left in a soft substrate as well as detailed ichnites imprinted in wetter sands (Leonardi and Oliveira 1990). The observed differences can be explained by the paleolatitudinal gradation of humidity, from desert conditions in the Paraná Basin to greater humidity in the Deseado Massif. Bonaparte (1996a) interpreted the La Matilde environment as marginal to the paleodesert of Botucatu and suggested that the sands accumulated in an arid environment with prolonged dry seasons. These dry intervals alternated with wetter periods during which river floods were more common.

It is noteworthy that such ichnofaunas, mostly made by small creatures, occur in parts of South America that are about 2,800 kilometers apart (i.e., the distance from the Laguna Manantiales farm to the southernmost record of the Botucatu ichnofaunas in Tramandaí in the state of Rio Grande do Sul). Moreover, footprints corresponding to this peculiar faunal assemblage are also recorded within the Paraná Basin, some 1,500 kilometers away (the distance from Tramandaí in the state of Rio Grande do Sul to Rifaina in the state of São Paulo). I believe that sampling of these beds is reasonably good and that they cover a considerable region and that this indicates that the absence of sauropods, for example, is not an accident. Also interesting is the fact that footprints of turtles, lepidosaurs, and crocodyliforms remain undocumented in both the Botucatu and La Matilde formations.

The causes of the remarkable absence of large animals (so common in the Jurassic and Lower Cretaceous strata of Patagonia and the rest of the world) are not well understood. However, climatic stress due to arid environments may be suspected (cf. Leonardi 1989). Volkheimer (1969) pointed out that a latitudinally extensive arid zone developed in South America during the Late Jurassic to Early Cretaceous (Fig. 3.25), as indicated by thick anhydrite deposits that accumulated along the Pacific margin from Patagonia to southern Perú that are almost equivalent in time (and located approximately at the same paleolatitudes) as the Botucatu paleodesert.

Seasonally arid climates in Patagonia and year-round desert conditions in the Paraná Basin (i.e., the Botucatu paleodesert) were coupled with

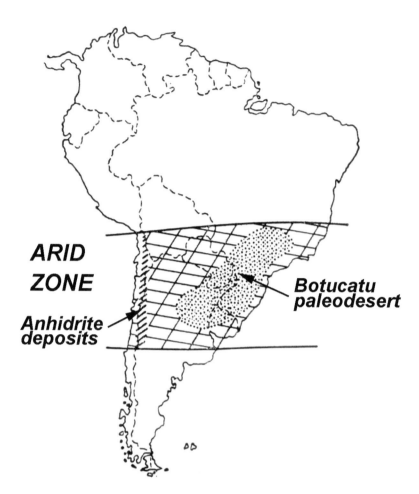

Fig. 3.25. Zone of major aridity in South America during the Late Jurassic and Early Cretaceous. Redrawn from Volkheimer (1969).

intense and widespread volcanic activity, thus creating disturbed environmental conditions that prevailed from Kimmeridgian through Hauterivian times, about 25 million years (Soares 1981; Uliana and Biddle 1988).

The paleodesert of Botucatu as well as the intermittent and recurrent extrusive volcanic events (i.e., lava flows) probably controlled dinosaurian distributions by creating filters for the intracontinental dispersion of both animals and plants. The harsh climatic conditions of this extensive paleodesert may have discouraged the development of forested or vegetated areas, and the intermittent volcanic events may have produced frequent ecological disturbances. These stressed environments may have made it impossible for populations of large plant-eating dinosaurs (particularly sauropods) and their large predators to settle. In contrast, dinosaurs ranging from the size of chickens to the size of ostriches may have been better candidates for tolerating such extreme environments than bigger animals because they may have been capable of surviving in the conditions of a restricted plant supply, high temperatures, sporadic rainfall, and intense solar radiation.

Arid and volcanically unstable environments may have discouraged gregarious behavior in dinosaurs, as suggested by the solitary trackways found in both the La Matilde and Botucatu beds. Also, theropod trackways (presumably corresponding to predatory or omnivorous forms) predominate over those of plant-eating ornithopods, indicating that plant resources were restricted, but a variety of prey (e.g., arthropods, small mammals, tiny ornithopods) was available for small theropods.

The apparent absence of sauropods in the Botucatu desert from the Late Jurassic through the Early Cretaceous contrasts with the notable abundance and diversity of Late Jurassic sauropods in the Morrison Formation in the western United States. Some authors (e.g., Engelmann, Chure, and Fiorillo 2004; Farlow 2007; Farlow, Dodson, and Chinsamy 1995) have argued that sediments of this later formation accumulated under dry climatic conditions. In this context, Engelmann, Chure, and Fiorillo (2004) considered that sauropods, as a group, were adapted to relatively dry environments. Without dismissing this view, however, I would argue that the Botucatu desert may have attained much more extreme environmental conditions, so severe that they made sauropod populations virtually impossible, at least in the numerical abundance that characterizes the Morrison dinosaur fauna.

The extensive eruptions that occurred in vast areas of South America during the Jurassic invite some reflection on the effects of intense volcanic activity on mass extinctions. Cowen has argued that Late Cretaceous volcanic eruptions of the Deccan Traps produced massive levels of aerosols in the stratosphere (Cowen 1995). These traps cover 500,000 square kilometers and may have lasted only about one million years. They probably erupted mainly as lava flows rather than giant explosive eruptions like those of Krakatau (Cowen 1995). In contrast, the magmatism that occurred in Patagonia during most of the Jurassic (Barrio, Panza, and Nullo 1999) is essentially of the Plinian kind of eruption characterized by extreme violence and power and great volumes of ejecta that form columns from 20 to 60 kilometers high. These pyroclastic columns develop later into dense ash falls able to disperse tens of kilometers around the emission center. These important eruptive events occurred in Patagonia and were almost coeval with extensive lava flows in the Paraná, Chaco, Amazonas, and Parnaíba basins. Taken as a whole, these lava flows considerably surpass the volume of erupted lava of the Deccan Traps. Also, the time of volcanic activity in South America was far more prolonged: the Patagonian eruptions lasted around 60 million years and the Serra Geral lava flows lasted around 27 million years, contrasting with the one million years estimated for the Late Cretaceous Indian volcanic event (Cowen 1995). It is expected that ash and aerosols in the form of sulfuric acid droplets (which stay suspended longer than ash and produce long-lasting effects on climate; Cowen 1995) were also introduced into the atmosphere during the Jurassic, although no considerable effects have been detected in the extinction of plants or land vertebrates at the end of that period, at least not with the catastrophic characteristics of the K-T boundary event. In other words, the volcanic activity that occurred during the Jurassic in South America may be used to test how volcanic events affect terrestrial biotas, weakening theories that a less-important volcanic event (i.e., the one that produced the Deccan Traps) could have triggered the mass extinction at the end of the Cretaceous.

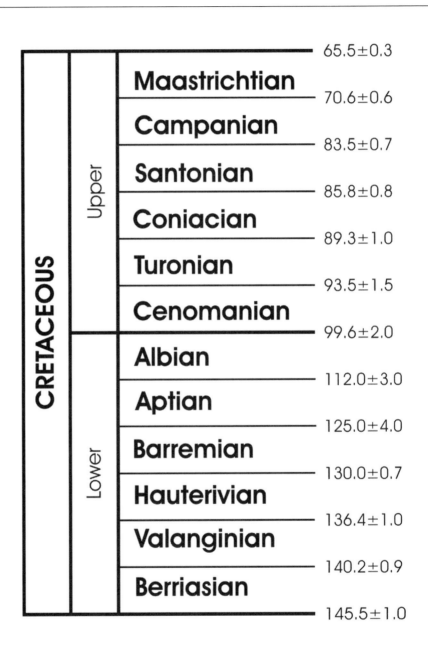

Fig. 4.1. Geological time scale of the Cretaceous Period.

THE FOSSIL RECORD OF CRETACEOUS DINOSAURS IN SOUTH AMERICA

4

The Cretaceous ran from 145 through 65 million years ago (Fig. 4.1). During this interval the supercontinent of Gondwana continued to separate into the continental landmasses of today through the opening of the Atlantic and Indian oceans and the creation of a circum-Antarctic sea. In other words, the continental configuration during the Cretaceous changed from the previous Mesozoic pattern of two supercontinents (Laurasia and Gondwana) straddling an equatorial ocean (Tethys) to a pattern of several continents separated by oceans and subdivided by shallow epicontinental seas that extended into high latitudes (Skelton 2003). The climate was warmer than today (the poles had no glaciers), and thus the latitudinal thermal gradient was less marked than in the present.

At the beginning of the Cretaceous, South America remained united with Africa and Antarctica, a situation that lasted until approximately 130 million years ago. For most of the Cretaceous, the northwestern corner of South America (corresponding to Venezuela, Colombia, Ecuador, and northern Perú) was covered by sea water. The same was true of the southwest corner of South America (southern Argentina and Chile), which was mostly covered by marine waters during most of the Cretaceous Period. The western margins of Argentina, Chile, and southern Perú were flooded by embayments of Pacific waters, a condition that prevailed during the Early Cretaceous but was replaced by continental environments for most of the rest of the period.

During the first half of the Cretaceous, the Botucatu paleodesert in the Paraná Basin was repeatedly affected by lava fissure flows (the basalts of the Serra Geral Formation) resulting from the extensional crust movements that inaugurated the breakup of Gondwana (Fig. 4.2A). Similarly phenomenal extrusive events also occurred (almost simultaneously) in other parts of South America, such as the Parnaíba and Amazonas basins of northern Brazil. It is calculated that these thick volcanic effusions covered around 1.2 million square kilometers in the Paraná Basin and another 100,000 square kilometers in the Parnaíba Basin (Zambrano 1981; Uliana and Biddle 1988). Extrusive volcanism lasted approximately from 147 to 120 million years ago and also affected Namibia in southern Africa (Uliana and Biddle 1988).

The oceanic floor first formed in the South Atlantic around 130 million years ago. However, the continental breakup started in the south and spread northward over the course of the Cretaceous in a process that lasted from 130 to 110 million years ago (Uliana and Biddle 1988; Fig. 4.2).

Thus, connections between South America and Africa remained almost intact north of the Niger Delta (Pitman, Cande, LaBrecque, and Pindell 1993), blocking water circulation between the proto–South Atlantic and the North Atlantic. By Aptian times, around 115 million years ago, South America was more widely separated from Africa, forming a shallow and narrow South Atlantic Ocean that resembled the modern Red Sea (Fig. 4.2B). Increased evaporation of the hypersaline proto-Atlantic triggered the precipitation of salt deposits between Africa and South America (to the north of Río Grande–Walvis Ridge), as also happened almost at the same time along the Pacific margin in the Neuquén Basin with the deposition of the Aptian–Albian Rayoso Group (Zambrano 1981). Although the flow of marine water into the South Atlantic was possible via the Weddell Sea (which appears to have been a substantial ocean basin by that time), no deepwater circulation into the South Atlantic occurred until the Malvinas (Falkland) Plateau separated from Africa about 100 million years ago (Lawver, Gahagan, and Coffin 1992; Fig. 4.2C).

A new tectonic regime was established along the west edge of South America about 115 million years ago, when the subduction rate of the Pacific plates beneath the western margin of the continent accelerated. During Aptian–Albian times, the first incursions of the Atlantic onto the eastern side of South America occurred, flooding the basins of Araripe and Parnaíba in northeastern Brazil, where lake deposits and evaporites are overlain by marine deposits of the Santana Formation. Presumably because of the widening Atlantic aperture, a new system of interior rifts developed in South America, propagating from southern Argentina to southern Brazil and southwest Africa.

These paleogeographic changes occurred along with emplacement of a massive batholith along most of the western margin of the continent and the combination of regional emergence and marine withdrawal, representing the initial stage of the Andean Orogeny (Fig. 4.2C). Marine waters retreated from the Neuquina Basin of western Argentina, where the evaporites of the Aptian–Albian Rayoso Group were deposited, followed by the continental red beds of the Upper Cretaceous Neuquén Group. This rise of the Cordillera changed the slope of the region eastward, which during and after the Late Cretaceous permitted extensive Atlantic marine transgressions across Argentina (Urien, Zambrano, and Martins 1981).

About 100 million years ago the eastern tip of the continental promontory of the Malvinas (Falkland) Plateau finally separated from the apex of South Africa. This permitted an almost uniform extension of the South Atlantic, a change that was associated with magmatic quiescence along the South American continental margin (Uliana and Biddle 1988). Possibly as a result of this transformation, the rate of the spread of the Pacific plate slowed, thus ending the mid-Cretaceous stage of massive batholith emplacement. Another magmatic expression of the new spreading regime was the extrusion of basalt flows in several places of South America (e.g., Patagonia, northwestern Argentina, Bolivia). This tectonic regime persisted until around 80 million years ago, when

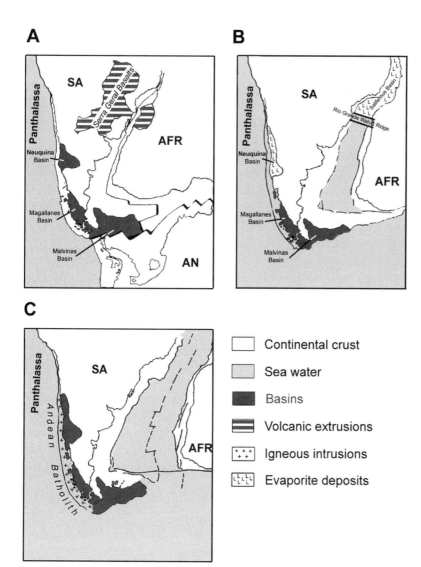

Fig. 4.2. Paleogeographic maps of southern South America (SA) showing the drifting apart from Africa (AFR) and Antarctica (AN) and the opening of the South Atlantic at different stages of the Cretaceous: (A) Hauterivian. (B) Aptian. (C) Cenomanian. *Redrawn from Urien, Zambrano, and Martins (1981).*

North and South America adopted a common rotational pole (Uliana and Biddle 1988). After that time (80 million years before the present), South and North America stopped drifting apart, and the relative position of the Americas has been rather stable ever since (Pitman, Cande, LaBrecque, and Pindell 1993).

South America had a long period, extending from the Early Jurassic to the Early Campanian, of almost continuous separation from North America. Before Campanian times (about 84 million years ago), a Caribbean transform fault system was active, so subaerial connections between the two landmasses would have been possible via an arc of islands similar to that of the Aleutians (Pitman, Cande, LaBrecque, and Pindell 1993). This arc became better developed after about 80 million years ago, when the Caribbean Plate begun to move eastward from the Pacific, inserting itself between the Americas like a great tongue.

Fig. 4.3. Map of South America indicating the Cretaceous sedimentary basin that yielded dinosaur remains. South American countries that yielded dinosaur remains are shown in grey.

The Late Cretaceous (about 70 million years ago) was characterized by marine transgressions (Uliana and Biddle 1988). Late Cretaceous flooding of the Argentine Atlantic margin occurred during a period of tectonic quiescence, when the continental interior was devoid of large topographic barriers (Uliana and Biddle 1988). The absence of collisions between South America and other continents during the Mesozoic means that the mountain chains of the Cretaceous formed because of plate subduction along the Pacific margins. The topographic relief was probably modest compared with the great mountain belts of today (Skelton 2003). The areas under water increased greatly during the Maastrichtian. This was the first great marine transgression from the Atlantic Ocean into the continent, and it flooded vast areas of Patagonia, northwestern Argentina, and Bolivia.

South America completely separated from Africa sometime between 100 and 95 million years ago (Albian–Cenomanian; Pitman, Cande, LaBrecque, and Pindell 1993). By the Early Cretaceous, Africa had become separated from both East and West Antarctica. However, the timing of the separation of South America from West Antarctica is more difficult to determine (Pitman, Cande, LaBrecque, and Pindell 1993). It is not clear whether or when West Antarctica formed a continuous migratory route for land animals between Antarctica and South America. Paleogeographic reconstructions provided by Hay and colleagues (1999) depict the Antarctic Peninsula as connected to the southern Andes throughout the Cretaceous. Some authors (e.g., Pitman, Cande, LaBrecque, and Pindell 1993) think that complete separation between Antarctica and South America did not exist until Eocene times (30 million years ago), when the Drake Passage was formed.

The Cretaceous Dinosaur Record in South America

Discoveries of dinosaur remains (i.e., bones, teeth, eggs, nests, and footprints) have been reported from several Cretaceous formations in South America. The rock units are continental in origin (e.g., lacustrine, fluvial, aeolian) and occur in the following sedimentary basins: the Araripe, São Luis, Sousa, Rio do Peixe, Parnaíba, and Bauru basins in Brazil; Guichón Basin in Uruguay; the Andean and Sub-Andean basins in Perú, Bolivia, Chile, and northwestern Argentina; and the Neuquina, San Jorge–Cañadón Asfalto, and Magallanes basins in Argentina (Fig. 4.3). However, the most abundant and diverse record is that from Argentina. I will briefly review these dinosaur-bearing formations, going from north to south in the continent.

1. Brazil

In general, Lower Cretaceous deposits are located in coastal basins and are accumulations associated with the onset of rifting between South America and Africa. Deposits of Upper Cretaceous age, in contrast, are mainly distributed in the interior of Brazil and are the result of

sedimentary processes in bodies of water in intracratonic basins (Dias-Brito, Musacchio, Castro, Maranhão, Suárez, and Rodrigues 2001). Cretaceous formations of Brazil that have yielded dinosaur remains occur in the Araripe, Sousa, Rio do Peixe, São Luis, Parnaíba, and Bauru basins (Figs. 4.3, 4.4, and 4.5).

1.A. THE ARARIPE BASIN

The Araripe Basin is located in the northeastern region of Brazil, mainly in the states of Ceará, Pernambuco, and Piauí. The most productive locality is Santana do Cariri in the state of Ceará. Unfortunately, no detailed data are available about dinosaur sites because of extensive commercial fossil-collecting by local people. The only site reported in the literature is Ladeira do Berlenga in the state of Piauí, where Price (1959) collected the holotype of the crocodyliform *Araripesuchus gomesii* (Kellner and Campos 2000).

The Santana Formation, thought to be Albian in age, is divided into two members, the Crato (below) and Romualdo (above), both of which are fossiliferous (Fig. 4.4). The Crato Member has yielded isolated flight feathers, semi-plumes, and down feathers revealing birds or feathered dinosaurs closely related to birds (Martins Neto and Kellner 1988; Kellner, Martins Neto, and Maisey 1991; Martill and Figueira 1994; Kellner, Maisey, and Campos 1994; Martill and Frey 1995). By far the most abundant fossils in the concretions of the Romualdo Member are fishes and pterosaurs. Crocodyliforms and turtles also occur (Gaffney and Meylan 1991; Martill 1993; Hirayama 1998), but these are rare (Naish, Martill, and Frey 2004). Reports of dinosaur bones are also rare and presently include the spinosaurid *Irritator challengeri* (Martill, Cruickshank, Frey, Small, and Clarke 1996; Sues, Frey, Martill, and Scott 2002; Kellner and Campos 1996), the coelurosaurians *Santanaraptor placidus* (Kellner 1999) and *Mirischia asymmetrica* (Martill, Frey, Sues, and Cruickshank 2000; Naish, Martill, and Frey 2004), an incomplete sacrum of an indeterminate theropod (Frey and Martill 1995; Kellner 1996b), and a fragmentary bone originally referred to Ornithischia by Leonardi and Borgomanero (1981) but best identified as a thoracic rib of a spinosaurid theropod (Machado and Kellner 2007). Complete dinosaur skeletons have not yet been reported, but partial skeletons in an exceptional state of preservation are known (Kellner 1996a, 1996b; Martill, Frey, Sues, and Cruickshank 2000; Bittencourt and Kellner 2004; Machado, Kellner, and Campos 2005). Moreover, soft tissues were discovered for the theropods *Santanaraptor placidus* (Kellner 1996a, 1996b; 1999) and gut contents were discovered for *Mirischia asymmetrica* (Martill, Frey, Sues, and Cruickshank 2000; Naish, Martill, and Frey 2004).

1.B. THE SOUSA AND RIO DO PEIXE BASINS

These two basins are located in the state of Paraíba. The most abundant dinosaur ichnofauna from northeastern Brazil comes from this region,

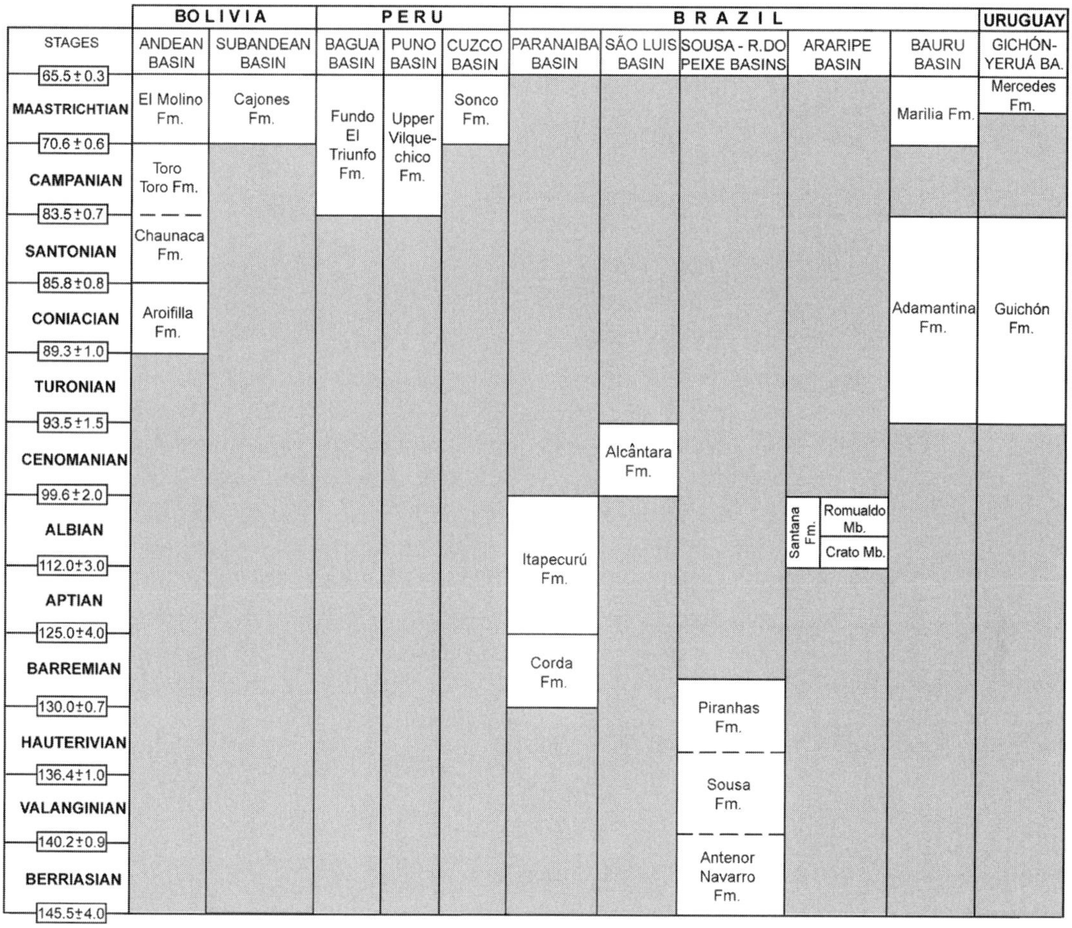

Stages	Bolivia Andean Basin	Bolivia Subandean Basin	Peru Bagua Basin	Peru Puno Basin	Peru Cuzco Basin	Brazil Paranaiba Basin	Brazil São Luis Basin	Brazil Sousa–R. do Peixe Basins	Brazil Araripe Basin	Brazil Bauru Basin	Uruguay Gichón–Yeruá Ba.
Maastrichtian (65.5±0.3 – 70.6±0.6)	El Molino Fm.	Cajones Fm.	Fundo El Triunfo Fm.	Upper Vilquechico Fm.	Sonco Fm.					Marilia Fm.	Mercedes Fm.
Campanian (70.6±0.6 – 83.5±0.7)	Toro Toro Fm.										
Santonian (83.5±0.7 – 85.8±0.8)	Chaunaca Fm.									Adamantina Fm.	Guichón Fm.
Coniacian (85.8±0.8 – 89.3±1.0)	Aroifilla Fm.										
Turonian (89.3±1.0 – 93.5±1.5)											
Cenomanian (93.5±1.5 – 99.6±2.0)							Alcântara Fm.				
Albian (99.6±2.0 – 112.0±3.0)							Itapecurú Fm.		Santana Fm. / Romualdo Mb. / Crato Mb.		
Aptian (112.0±3.0 – 125.0±4.0)											
Barremian (125.0±4.0 – 130.0±0.7)						Corda Fm.					
Hauterivian (130.0±0.7 – 136.4±1.0)								Piranhas Fm.			
Valanginian (136.4±1.0 – 140.2±0.9)								Sousa Fm.			
Berriasian (140.2±0.9 – 145.5±4.0)								Antenor Navarro Fm.			

where 22 ichnofossiliferous sites have been identified (Leonardi 1989; Carvalho 2000, 2004; Fig.4.5). Footprints of theropods, sauropods, and ornithopods (Plate 7) were preserved in sedimentary units of the Berriasian–Lower Barremian Rio do Peixe Group. This group is divided (from bottom to top) into the Antenor Navarro, Sousa, and Piranhas formations (Carvalho 2000, 2004; Fig. 4.4); the Sousa Formation is the most productive. At least 60 track-bearing levels with a total of 276 individual dinosaur trackways were identified in this unit (Carvalho 2000, 2004).

Fig. 4.4. Stratigraphy of dinosaur-bearing Cretaceous formations in Bolivia, Perú, Brazil, and Uruguay.

1.C. THE SÃO LUIS BASIN

The Cenomanian strata of the Alcântara Formation accumulated in the São Luis Basin in the state of Maranhão, located on the Brazilian equatorial margin (Carvalho 2001; Fig. 4.5). In the locality of Laje do Coringa on Cajual Island, the Alcântara Formation contains one of the few bone beds known in South America (Kellner and Campos 2000). Dinosaur specimens collected in the Alcântara Formation are comprised of teeth and bone fragments (Price 1947; Ferreira, Azevedo, Carvalho, Gonçalves, and Vicalvi 1992; Medeiros 2001; Medeiros and Schultz 2001a, 2001b, 2002,

2004; Vilas Boas, Carvalho, Medeiros, and Pontes 1999). Medeiros and co-authors (2007) offered an up-to-date review of the available fossils from this rock unit, describing spinosaurid and carcharodontosaurid teeth, a caudal vertebra resembling these of the iguanodontian *Ouranosaurus*, procoelous caudal vertebrae of titanosaurians, and isolated caudals and teeth probably belonging to rebbachisaurid sauropods. The São Luis area also preserves an extensive megatracksite, with dinosaur trackways occurring over stretch of roughly 50 kilometers. Theropod, ornithopod, and possible ceratopsian footprints have been reported from the Alcântara beds (Carvalho 1994, 2001; Carvalho and Goncalves 1994).

1.D. THE PARNAÍBA BASIN

The pre-Aptian Corda Formation is in this basin, from which Leonardi (1980b, 1989) reported ornithischian and sauropod trackways in outcrops in São Domingos in the state of Goiás (Figs. 4.3 and 4.5). The younger Aptian–Albian Itapecurú Formation, which crops out in the northern region of this basin, has yielded several isolated dinosaur remains, mainly vertebrae and incomplete long bones of the diplodocoid sauropod *Amazonsaurus maranhensis* (Carvalho, Avilla, and Salgado 2003).

1.E. THE BAURU BASIN

The Bauru Group accumulated atop the Lower Cretaceous basalts of the Serra Geral Formation. The group constitutes the most extensive fossiliferous continental sedimentary rock unit in South America (e.g., Fernandes and Coimbra 1996; Kellner 1999; Goldberg and García 2000), with outcrops covering about 370,000 square kilometers in six Brazilian states (São Paulo, Minas Gerais, Mato Grosso, Mato Grosso do Sul, Goiás, and Paraná) and in Paraguay (Fig. 4.3). The group is composed (from bottom to top) of the Caiuá, Santo Anastácio, Adamantina, and Marília formations. Information gathered from ostracods, charophytes, and stratigraphy suggest that the Adamantina Formation was deposited during the Turonian–Santonian interval and that the overlying Marília Formation was deposited during Maastrichtian times (Dias-Brito, Musacchio, Castro, Maranhão, Suárez, and Rodrigues 2001; Fig. 4.4).

Fossils of the Bauru Group occur mainly in the Adamantina and Marília formations. Vertebrate fossils found in the Marília Formation include dinosaur eggs, coprolites, fish scales, frog bones, turtles, lizards, crocodylomorphs, dinosaurs, and mammal teeth (Mezzalira 1980; Bertini, Marshall, Gayet, and Brito 1993; Senra and Silva e Silva 1999; Magalhães Ribeiro and Ribeiro 2000; Magalhães Ribeiro 2002).

Outstanding fossil sites of the Bauru beds occur in the state of Minas Gerais (Fig. 4.5) in a region known as Triangulo Mineiro, which is bounded by the Paranaiba and Grande rivers (Campos and Kellner 1999). Several localities in the Triangulo Mineiro yielded titanosaurid remains, some of which were already known to Friedrich von Huene (1931b). The richest locality is Serra do Veadinho, a hill close to the village of Peirópolis

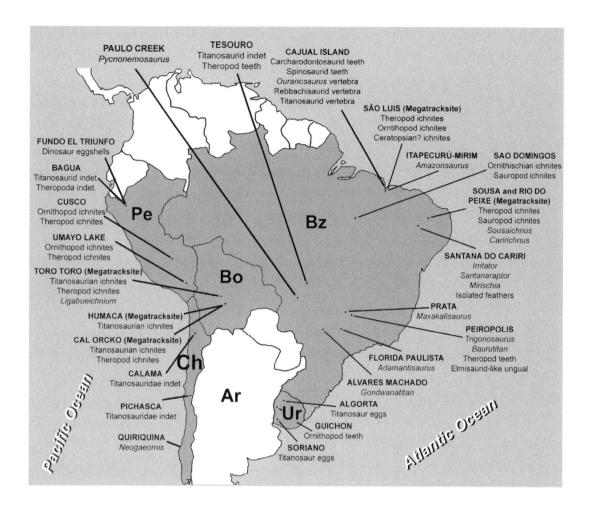

Fig. 4.5. Fossil localities that yielded Cretaceous dinosaur remains in Perú (Pe), Bolivia (Bo), Chile (Ch), Brazil (Bz), and Uruguay (Ur). Fossil sites of Argentina (Ar) are shown in Figs. 5.1, 6.1, and 7.1.

in Minas Gerais, where intense work was carried on by Llewellyn I. Price from 1948 almost continually through 1969 (Campos and Kellner 1999). Peirópolis has furnished a considerable number of bones and teeth of Titanosauridae (Price 1961; Powell 1987a, 2003; Campos and Kellner 1999; Kellner and Campos 2000), including two recently named taxa (*Baurutitan britoi* and *Trigonosaurus pricei*; Kellner, Campos, Azevedo, and Trotta 2005; Campos, Kellner, Bertini, and Santucci 2005), several theropod teeth (Kellner 1995, 1996b), abelisaurid bones (Novas, Carvalho, Borges Ribeiro, and Méndez 2008), and a maniraptoran manual ungual (Novas, Borges Ribeiro, and Carvalho 2005). All these specimens were recovered from beds of the Maastrichtian Marília Formation. Price (1951) reported a titanosaurid egg from the locality of Mangabeira, north of the town of Uberaba. Price's work was so notable that a museum in the village of Peirópolis is named Centro de Pesquisas Paleontológicas Llewellyn Ivor Price.

Other interesting discoveries from the Bauru Group from levels of the underlying Adamantina Formation (Turonian–Santonian) were made in the locality of Álvares Machado in the state of São Paulo, among them several dinosaur specimens including the incomplete skeleton of the titanosaurid *Gondwanatitan faustoi* (Kellner and Azevedo 1999; Fig. 4.5). From the locality of the town of Prata on Minas Gerais, the Adamantina beds yielded *Maxakalisaurus topai* (Kellner, Campos, Azevedo, Trotta, Henriques, Craik, and Silva 2006). From the vicinity of Florida Paulista (São Paulo), Candeiro and colleagues (2004) documented a series of fragmentary theropod and sauropod remains. More recently, Santucci and Bertini (2006) reported the discovery of a new genus and species of titanosaurians, *Adamantisaurus mezzalirai*.

Recently, Kellner and colleagues (2004) reported the discovery of abundant titanosaurid bones and teeth from the Confusão Creek near the town of Tesouro in Mato Grosso (Fig. 4.5). The fossils come from beds of the Bauru Group, but the identification of the formation awaits further research. This discovery is outstanding not only for the geographical provenance but also for the high concentration of well-preserved bones and teeth. This material, still under preparation and study, may belong to a new titanosaurian taxon (Kellner, Azevedo, Carvalho, Henriques, Costa, and Campos 2004).

2. Uruguay

Continental beds of the Guichón and Mercedes Formations crop out on the west side of Uruguay, accumulated into the Gichón-Yeruá Basin (Sprechmann, Bossi, and da Silva 1981; Figs. 4.3, 4.4, and 4.5). The Guichón Formation, presumably Turonian–Santonian in age, may correlate with the Adamantina Formation (Bauru Group, Paraná Basin, mainly in Brazil) and with the Puerto Yeruá Formation (Entre Ríos Province, northeastern Argentina; Fig. 4.3). The Guichón beds have yielded fragmentary ornithopod teeth (von Huene 1934a) and more-complete skeletons of the crocodyliform *Uruguaysuchus*. Overlying Guichón is the Mercedes Formation, which has yielded titanosaurid bones assigned by von Huene (1929b, 1934a, 1934b) to previously described taxa from Patagonia (e.g., *Titanosaurus australis*, *Laplatasaurus araukanicus*, *Antarctosaurus wichmannianus*, and *Argyrosaurus superbus*; von Huene 1929a) as well as titanosaurid eggs and eggshells from outcrops at Soriano and Algorta (Mones 1980; Faccio 1994; Faccio, Ford, and Gancio 1990; Fig. 4.5). The age of the Mercedes Formation is almost probably Maastrichtian, and it is tentatively correlated with the Marília Formation of the Bauru Group.

3. Perú

The pericratonic basin of Bagua is located to the east of the Cordillera Oriental in the Department of Amazonas in northwestern Perú (Fig. 4.3). Cretaceous marine beds and overlying red beds of the Campanian–Maastrichtian Fundo El Triunfo Formation accumulated here and contain fragmentary remains of indeterminate titanosaurids and theropods (Mourier et al. 1986; Mourier et al. 1988; Vildoso Morales and Sciammaro

2005; Fig. 4.5). Discoveries of eggshells (probably corresponding to titanosaurians) were reported from the localities of Fundo El Triunfo and Pongo de Rentema (Mourier et al. 1988; Vianey-Liaud, Hirsch, Sahni, and Sigé 1997).

The Upper Vilquechico Formation (Maastrichtian) is exposed in the Department of Puno, near the Peruvian-Bolivian border. French expeditions to the Vilquechico beds exposed around Umayo Lake (a body of water close to the larger Lake Titicaca) reported the discovery of theropod and ornithopod tracks and dinosaur eggs (Sigé 1968; Kerourio and Sigé 1984; Mourier et al. 1988; Jaillard et al. 1993; Vildoso Morales 1991; Fig. 4.3).

Finally, from the Maastrichtian Sonco Formation, cropping out in the Department of Cusco, Noblet and colleagues (1995) found theropod and ornithopod trackways.

4. Bolivia

Two main geological basins, the Sub-Andean and Andean basins, extend along Bolivia and northwestern Argentina (Figs. 4.3 and 4.4). In the Andean Basin, thick accumulations of the Upper Cretaceous Puca Supergroup are present. The upper units of this group constitute the continental and marine Puca Group, which includes the following Upper Cretaceous units: the Coniacian Aroifilla, Santonian–Campanian Chaunaca, and Toro Toro formations (which are equivalent to the lower and middle Vilquechico Formation of southeast Perú), and the Latest Campanian–Early Paleocene marine El Molino Formation (Gayet, Marshall, and Sempere 1991, Gayet, Marshall, Sempére, Meunier, Cappetta, and Rage 2001). Footprints of an indeterminate quadrupedal dinosaur were reported from the Aroifilla beds (Gayet, Marshall, and Sempere 1991). Close to the villages of Toro Toro and Humaca in the Department of Cochabamba are exposures of the track-bearing Santonian–Campanian formations of Toro Toro and Chaunaca, from which several dinosaur trackways were reported (Leonardi 1981; Suárez Riglos 1995; Meyer, Hippler, and Lockley 2001; Lockley, Schulp, Meyer, Leonardi, and Mamani 2002; McCrea, Lockley, and Meyer 2001; Fig. 4.5).

The overlying El Molino Formation (partially equivalent to the Campanian–Maastrichtian Lecho and Yacoraite beds of northwestern Argentina) has yielded abundant remains of a great diversity of fishes and some turtle and crocodile specimens. Although the El Molino Formation has yielded the richest and best-studied Late Cretaceous ichthyofauna of South America (Gayet, Marshall, and Sempere 1991, Gayet, Marshall, Sempére, Meunier, Cappetta, and Rage 2001), its content of dinosaur body fossils is impoverished: it consists of a single tooth of an indeterminate small theropod that was collected in the lower El Molino Formation at Pajcha Pata (Marshall 1989). In contrast to this scarcity in the fossils of dinosaur bodies, the El Molino Formation is rich in dinosaur footprints.

Cal Orcko in the vicinity of Sucre (Department of Chuquisaca) has extensive limestone exposures with a large number of beautifully preserved dinosaur trackways that correspond to theropods, titanosaurids, and ankylosaurs (Fig. 4.5).

Cretaceous units of the Sub-Andean Basin in Bolivia have produced more bones. In past years, fossils documented in this sedimentary basin were restricted to titanosaurian bones reported by Gutierrez and Marshall in 1994 from the Cajones Formation that yielded little information (presumably Maastrichtian in age; Aguilera, Salas, and Peña 1989). A recent exploration conducted by the present author in the Cajones Formation, which crops out in Amboró National Park about 50 kilometers west of the town of Santa Cruz de la Sierra, resulted in the discovery of bizarre notosuchian crocodyliforms that constitute the most complete Cretaceous vertebrates yet recovered in Bolivia. The red beds of the continental Cajones Formation have been correlated with the El Molino beds (which are exposed at the Altiplano and Cordillera Oriental in Bolivia) as well as with the Yacoraite Formation (in northwestern Argentina) and the Upper Vilquechico Formation (in southeastern Perú; Aguilera, Salas, and Peña 1989; Gutierrez and Marshall 1994).

5. Chile

The dinosaur record in Chile is presently poor (Fig. 4.5). Chilean formations that have yielded dinosaur remains belong to the Andean geological province and include the Lower Cretaceous Chacarilla and Quebrada Monardes formations and the Upper Cretaceous Viñita, Tolar, Quebrada Pajonales, and Quiri Quina formations, which are exposed at different sites in this country. The Chacarilla Formation (Upper Jurassic–Lower Cretaceous), which is exposed in Tarapaca in northern Chile, has yielded large ornithopod footprints (40–50 centimeters in maximum diameter) as well as theropod trackways (Rubilar, Moreno, and Blanco 2000). Also from the Lower Cretaceous are red beds of the Quebrada Monardes Formation, which are exposed at Cerro La Isla, located east of Copiapó in the Atacama region (Bell and Suárez 1989; Rubilar, Moreno, and Blanco 2000). The Quebrada Monardes Formation has yielded footprints of theropod dinosaurs as well as an isolated vertebra of an iguanodontid ornithopod (associated with pterosaur and crocodile bones).

The remaining discoveries come from Upper Cretaceous beds, in which titanosaurid remains (e.g., partial skeletons, teeth, and footprints) predominate over the remains of other kinds of dinosaurs (Rubilar 2003). The most relevant Upper Cretaceous locations are in the La Viñita Formation, which is exposed at Pichasca in northern Chile's Coquimbo Province (Casamiquela, Corvalán, and Franquesa 1969; Chong Díaz and Gasparini 1976; Vargas, Suárez, Rubilar, and Moreno 2000), from which a titanosaurid represented by a scapulocoracoid and some vertebrae and teeth were discovered.

At a promising locality 150 kilometers north of the city of Calama in the Atacama Desert of northern Chile (Fig. 4.5), Chong Díaz and collaborators (2000) communicated the discovery of two incomplete titanosaurid skeletons from the Tolar Formation (Upper Cretaceous). Also in Atacama but from the locality of Cerro Algarrobito, a fragmentary humerus and some ribs of a titanosaurid were found in levels of the Hornitos Formation (Chong Díaz 1985). More recently, Iriarte and colleagues (1999) announced the discovery in the same unit of some forelimb and hindlimb bones and isolated cervical and trunk vertebrae that they referred to an indeterminate titanosaurid.

From the Upper Cretaceous Quebrada Pajonales Formation that is exposed in Sierra de Almeida, II Región (Antofagasta Province), Salinas and colleagues reported the discovery of long dinosaurian bones (Salinas, Sepúlveda, and Marshall 1991).

Finally, from Quiriquina Island and adjacent places in San Vicente Bay (Bío Bío region) Lambrecht found remains of the gaviform bird *Neogaeornis wetzeli* in beds of the Maastrichtian Quiriquina Formation (Lambrecht 1929).

6. Argentina

Most that is known about the evolution of terrestrial Cretaceous reptiles of South America comes from Argentina. Important discoveries have been made since the end of the nineteenth century, especially in the northwestern corner of Patagonia. Discoveries in these highly productive regions continue to be made and published, and a large and growing bibliography exists from this region. I will consider Cretaceous sedimentary basins of Argentina from north to south.

6.A. THE SUB-ANDEAN BASIN

The Sub-Andean Basin extends through the territories of northern Argentina, Bolivia, and Perú (Fig. 4.3). The Sub-Andean Basin of northwestern Argentina accumulated extensive deposits of the Upper Cretaceous Salta Group (Salfity 1982; Gómez, Boll, and Hernández 1989), divided (from bottom to top) into the Las Curtiembres, Los Blanquitos, and Lecho formations (Fig. 4.6A and B). The Las Curtiembres (Turonian–Campanian; Salfity 1999; Salfity and Marquillas 1981; Sabino, Salfity, and Marquillas 1998) has yielded hundreds of specimens of the pipid frog *Saltenia ibanezi* (Báez 1981) and, more recently, tiny enantiornithine birds discovered by the present author and assistants. The Los Blanquitos Formation (Campanian) has yielded remains of the maniraptoran *Unquillosaurus ceibalii* (Powell 1979; Novas and Agnolín 2004) as well as bones of indeterminate titanosaurids. The Lecho Formation (Campanian–Maastrichtian) contains the most productive dinosaur beds of the Salta Group. The quarry of El Brete in the Lecho Formation has been intensively worked by José F. Bonaparte and assistants since 1975 (Fig. 4.6C) and more

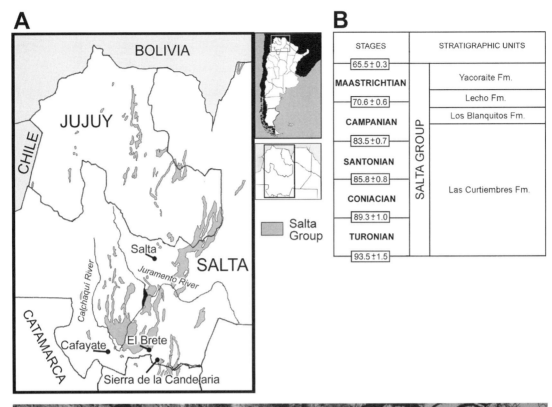

recently by Luis Chiappe, who made a large collection of titanosaurid, noasaurid, and bird bones. Dinosaur taxa (including birds) reported from the Lecho Formation include *Saltasaurus loricatus* (Bonaparte and Powell 1980), *Noasaurus leali* (Bonaparte and Powell 1980), Abelisauroidea indet. (originally described as an Oviraptorosauria indet. by Frankfurt and Chiappe 1999; see Agnolín and Martinelli 2007), *Enantiornis leali* (Walker 1981), *Soroavisaurus australis, Yungavolucris brevipedalis,* and *Lectavis bretincola* (Chiappe 1993). The Salta Group is overlain by the Yacoraite Formation (Maastrichtian), which has yielded some theropod and probably hadrosaurid footprints (Alonso 1989).

6.B. THE GICHÓN-YERUÁ BASIN

The Gichón-Yeruá Basin is shared by northeastern Argentina and western Uruguay (Fig. 4.3). On the Argentine side, the dinosaur-bearing unit is the Puerto Yeruá Formation (Upper Cretaceous), which is equivalent to the Gichón Formation of Uruguay. Titanosaurid remains were found at the Calera Barquín fossil site in Entre Rios Province (northeastern Argentina). The presence of dinosaur remains in Entre Rios Province was mentioned as early as 1912 by Enrique de Carlés, who found an incomplete titanosaurid humerus on the banks of the Uruguay River between the towns of Concordia and Colón. The materials were described by von Huene (1929b) as probably belonging to *Argyrosaurus superbus*, although more recently they have been catalogued as Titanosauridae indet. by Powell (2003). Conical scutes probably belonging to an ankyloraurian as well as theropod teeth were also documented from the Puerto Yeruá beds (De Valais, Apesteguía, and Sauthier 2003).

6.C. THE NEUQUINA BASIN

The dinosaurs of the Neuquina Basin are perhaps the best known of South America (Figs. 4.3 and 4.7). The Cretaceous terrestrial beds accumulated in this basin are some of the most richly fossiliferous and stratigraphically complete in the world (Leanza, Apesteguía, Novas, and de la Fuente 2004; Fig. 4.8). The Neuquina Basin not only constitutes the most productive region of South America for dinosaur discoveries but is also the best place in Argentina for paleontological prospecting for logistical reasons. The main fossil sites are in Neuquén and Río Negro provinces, but some fossil spots occur in Mendoza and La Pampa Provinces as well. In the ravines of the Neuquén River at the city of Neuquén, dinosaur bones have been found since the end of the nineteenth century. Other notable sites are Cinco Saltos, General Roca, and Salitral Moreno (in Río Negro Province), where excavations carried out in 1921 and 1923 by expeditions of the Museo de La Plata that were directed by Santiago Roth and Walter Schiller produced many fossils that Friedrich von Huene described in 1929.

The Neuquina Basin is well developed in west-central Argentina and eastern Chile (Leanza, Apesteguía, Novas, and de la Fuente 2004).

Fig. 4.6. Cretaceous dinosaur-bearing rocks of northwest Argentina. (A) Map of Salta and Jujuy Provinces indicating surface distribution of Upper Cretaceous beds that yielded dinosaur remains. (B) Stratigraphy of the dinosaur-bearing Salta Group. (C) Quarrying works at El Brete fossil site in levels corresponding to the highly productive beds of the Lecho Formation. *(A) Redrawn from Salfity and Marquillas (1999). (C) Courtesy of José Bonaparte.*

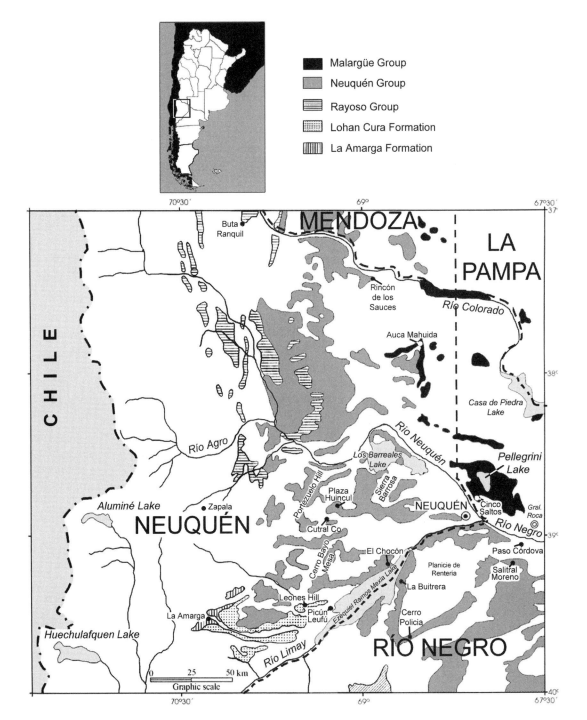

Fig. 4.7. Map depicting geographic distribution of continental strata of the Neuquina Basin. *Redrawn from Leanza, Apesteguía, Novas, and de la Fuente (2004).*

It contains a thick sequence of sediments ranging from the Upper Triassic through the Paleogene (Legarreta and Gulisano 1989). During the Jurassic–Early Cretaceous, the Neuquina Basin accumulated marine sediments from the Pacific Ocean, but during the remainder of the Cretaceous, the basin was filled by thick continental beds. These terrestrial accumulations are stratigraphically subdivided (from bottom to top) into the

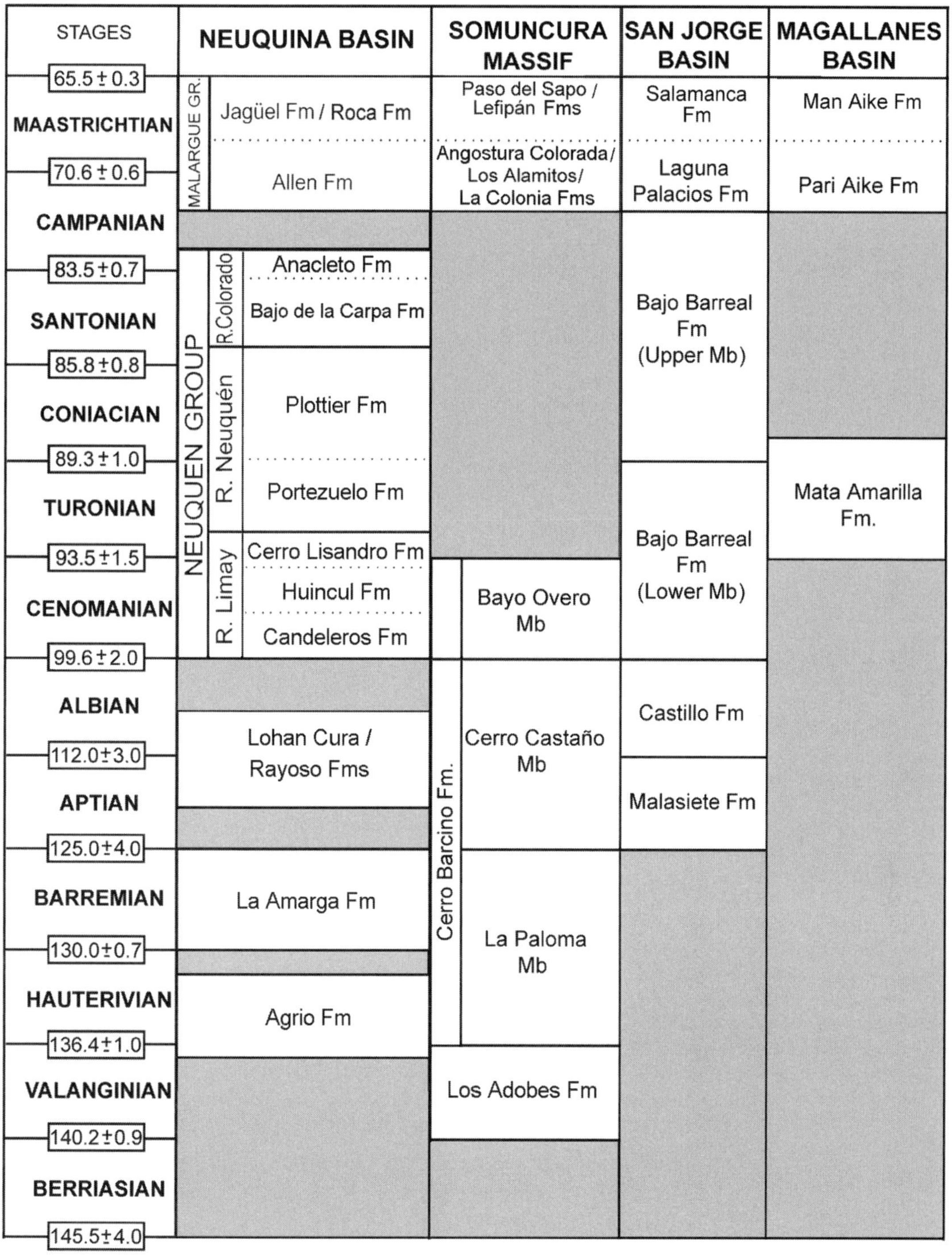

STAGES	NEUQUINA BASIN			SOMUNCURA MASSIF	SAN JORGE BASIN	MAGALLANES BASIN
65.5 ± 0.3 — MAASTRICHTIAN — 70.6 ± 0.6	MALARGUE GR.	Jagüel Fm / Roca Fm		Paso del Sapo / Lefipán Fms	Salamanca Fm	Man Aike Fm
		Allen Fm		Angostura Colorada / Los Alamitos / La Colonia Fms	Laguna Palacios Fm	Pari Aike Fm
CAMPANIAN — 83.5 ± 0.7 — SANTONIAN — 85.8 ± 0.8	NEUQUEN GROUP	R.Colorado	Anacleto Fm		Bajo Barreal Fm (Upper Mb)	
			Bajo de la Carpa Fm			
CONIACIAN — 89.3 ± 1.0 — TURONIAN — 93.5 ± 1.5		R. Neuquén	Plottier Fm			Mata Amarilla Fm.
			Portezuelo Fm		Bajo Barreal Fm (Lower Mb)	
CENOMANIAN — 99.6 ± 2.0		R. Limay	Cerro Lisandro Fm	Bayo Overo Mb		
			Huincul Fm			
			Candeleros Fm			
ALBIAN — 112.0 ± 3.0 — APTIAN — 125.0 ± 4.0			Lohan Cura / Rayoso Fms	Cerro Castaño Mb	Castillo Fm	
					Malasiete Fm	
BARREMIAN — 130.0 ± 0.7			La Amarga Fm	Cerro Barcino Fm.		
HAUTERIVIAN — 136.4 ± 1.0			Agrio Fm	La Paloma Mb		
VALANGINIAN — 140.2 ± 0.9				Los Adobes Fm		
BERRIASIAN — 145.5 ± 4.0						

La Amarga and Lohan Cura formations, the Neuquén Group (which is split into several subgroups and formations), and the Allen Formation.

Fig. 4.8. Stratigraphy of Cretaceous dinosaur-bearing rocks of Patagonia.

6.C.1. THE LA AMARGA FORMATION   This unit may be regarded as Barremian–Early Aptian in age (Leanza and Hugo 1995). Its type locality is La Amarga Creek and the nearby northern slope of the China Muerta Hill (Fig. 4.9). It unconformably overlies the transitional zone of the marine Agrio Formation and is overlain by the continental Lohan Cura Formation. From the La Amarga Formation come the dicraeosaurid sauropod *Amargasaurus cazaui* (Salgado and Bonaparte 1991), the basal diplodocoid *Zapalasaurus bonapartei* (Salgado, Carvalho, and Garrido 2006), the tiny abelisauroid theropod *Ligabueino andesi* (Bonaparte 1996b), and a fragmentary stegosaur (Bonaparte 1996b).

6.C.II. THE LOHAN CURA FORMATION   This unit of continental deposits unconformably overlies the La Amarga Formation, and it is overlain by the Candeleros Formation of the Neuquén Group. The age of the Lohan Cura Formation is interpreted as Late Aptian and Albian, and it is correlated in the central part of the basin with the Rayoso Formation (Leanza 1999, 2003). The probable diplodocimorph *Agustinia ligabuei* (Bonaparte 1999b), the basal titanosaurian *Ligabuesaurus leanzai* (Bonaparte 1999a; Bonaparte, González Riga, and Apesteguía 2006), and a rebbachisaurid sauropod closely related to *Limaysaurus tessonei* (Salgado, Garrido, Cocca, and Cocca 2004) were recovered in Cerro Los Leones near the village of Picún Leufú. Additionally, the rebbachisaurid sauropod *Rayososaurus agrioensis* (Bonaparte 1996b, 1997b) comes from the Rayoso Formation, which is partially equivalent to the Lohan Cura Formation (Leanza, Apesteguía, Novas, and de la Fuente 2004).

6.C.III. THE NEUQUÉN GROUP   The red beds of the Neuquén Group were laid down over a span of nearly 20 million years, from the Cenomanian through the Early Campanian (Leanza, Apesteguía, Novas, and de la Fuente 2004). The group is subdivided into three dinosaur-bearing subgroups (from bottom to top): Río Limay, Río Neuquén, and Río Colorado (Fig. 4.8).

The Río Limay Subgroup includes the Candeleros, Huincul, and Cerro Lisandro formations. The Early Cenomanian Candeleros Formation has a remarkable faunal record, including the basal titanosaur *Andesaurus delgadoi* (Calvo and Bonaparte 1991), the rebbachisaurids *Limaysaurus tessonei* (Calvo and Salgado 1995) and *Cathartesaura anaerobica* (Gallina and Apesteguía 2005), the giant carcharodontosaurid *Giganotosaurus carolinii* (Coria and Salgado 1995), and the abelisaurid *Ekrixinatosaurus novasi* (Calvo, Rubilar-Rogers, and Moreno 2004). The most remarkable locality of this formation is La Buitrera (Río Negro Province), which was discovered by Sebastián Apesteguía in 1999. La Buitrera is distinguished by the abundance of nicely preserved articulated skeletons, including the small dromaeosaurid *Buitreraptor gonzalezorum* (Makovicky, Apesteguía, and Agnolín 2005), large theropods, and rebbachisaurid sauropods as well as many remains of sphenodontid lepidosaurs, araripesuchid crocodyliforms, and basal snakes, collected in an area roughly of 15 square kilometers (Apesteguía, Valais, Gonzáles, Gallina, and Agnolín 2001). Apart from these body fossils, footprints of

Fig. 4.9. Paleontologist Guillermo Rougier excavating the skeleton of the sauropod *Amargasaurus cazaui,* La Amarga Formation. *Courtesy of José Bonaparte.*

titanosauriforms (*Sauropodichnus*), theropods (*Picunichnus, Abelichnus, Bressanichnus, Deferrariischnium*), and large ornithopods (*Sousaichnus, Limayichnus*; Fig. 4.10) have been found in the bedding planes of the Candeleros Formation around Lake Ezequiel Ramos Mexía (Calvo 1991, 1999, 2007).

The Late Cenomanian Huincul Formation conformably overlies the Candeleros beds. This unit has yielded remains of the gigantic basal titanosaurid *Argentinosaurus huinculensis* (Bonaparte and Coria 1993) as well as a variety of medium-sized titanosaurids (Calvo and Salgado 1998; Simón 2001). Rebbachisaurid sauropods were also collected from the Huincul beds (Salgado, Calvo, and Coria 1991; Calvo and Salgado 1998; Calvo 1999), along with the abelisauroid *Ilokelesia aguadagrandensis* (Coria and Salgado 2000) and fragmentary pedal elements of an indeterminate abelisauroid collected in the vicinity of El Chocón (Novas and Bandyopadhyay 2001; De Valais, Novas, and Apesteguía 2002).

Conformably overlying the Huincul beds is the Cerro Lisandro Formation, constituting the youngest unit of the Río Limay Subgroup. The age of the Cerro Lisandro Formation may be regarded as Late Cenomanian–Early Turonian (Hugo and Leanza 2001a). Dinosaur bones found in this unit at Portezuelo Hill (Neuquén Province) include vertebrae and hindlimb bones of tiny ornithischians as well as a coracoid of one of the oldest known neornithine birds (Agnolín, Novas, and Lío 2003, 2006). At Cerro Bayo Mesa, 30 kilometers south of Plaza Huincul, four specimens of the ornithopod *Anabisetia saldiviai* were found in rocks of the Cerro

Fig. 4.10. An ornithopod trackway, *Limayichnus*, in the Candeleros Formation at El Chocón, Neuquén Province. *Courtesy of Jorge Calvo.*

Lisandro Formation (Coria and Calvo 2002). The discovery of a small abelisauroid has also been recently reported (Paulina Carabajal, Coria, and Currie 2003; Coria, Currie, and Carabajal 2007). From El Anfiteatro, located 50 kilometers southwest of Cipolleti in Río Negro Province, Canudo and colleagues (2004) reported the discovery of an isolated tooth referred to Titanosauria, Spinosauridae, Carcharodontosauridae, and Abelisauroidea.

Above the Río Limay Subgroup are the beds of the Río Neuquén Subgroup, which are particularly well exposed in the Sierra Barrosa and Sierra del Portezuelo in the central part of Neuquén Province (Plate 10) as well as around Los Barreales Lake (in Neuquén Province) and in the Planicie de Renteria (Río Negro Province; Fig. 4.7). The Río Neuquén Subgroup includes the Portezuelo Formation below and the Plottier

Formation above (the Portezuelo Formation of northwestern Patagonia should not be confused with the homonymous Portezuelo Formation of La Rioja Province of Triassic age; see chapter 2). According to new geological studies and paleontological information in the region, the age of the Río Neuquén subgroup may be Turonian–Coniacian (Leanza and Hugo 1995; Hugo and Leanza 2001a). The paleontological content of this unit was substantially increased in the last decade, initially by explorations made by the present author (Novas 1996b, 1997d, 1998; Novas and Pol 2005; Novas and Puerta 1997) and more recently by Jorge Calvo and crew, who are carrying out an intensive excavation project along the shore lines of Los Barreales Lake. Dinosaurs documented in the Portezuelo Formation include the basal tetanuran *Megaraptor namunhuaiquii* (Novas 1998; Calvo, Porfiri, Veralli, Novas, and Poblete 2004; Plate 18) and the maniraptorans *Patagonykus puertai*, *Unenlagia comahuensis*, *Unenlagia paynemili*, and *Neuquenraptor argentinus*. Sauropod bones are frequently found in the Portezuelo Formation (Salgado and Calvo 1993), and spectacular specimens of large titanosaurids have been discovered in Los Barreales Lake, such as the recently announced *Futalognkosaurus dukei* (Calvo, Porfiri, Veralli, and Poblete 2001a, 2001b; Calvo, Porfiri, González Riga, and Kellner 2007; Calvo 2002). Ornithopods recently found in the Portezuelo Formation include a beautifully preserved and articulated skeleton of *Macrogryphosaurus gondwanicus*, a basal iguanodontian about 6 meters long (Calvo, Porfiri, and Novas 2005; Calvo, Porfiri, and Novas 2008).

Overlying the Portezuelo beds is the Plottier Formation. Few tetrapod fossils have been recorded from this unit. The most important is a large tetanuran of uncertain relationships (presumably related to *Megaraptor*) that Rodolfo Coria and collaborators excavated at the Barrosa Hill (Coria, Currie, Eberth, Garrido, and Koppelhus 2001; Fig. 4.7). This discovery was originally reported as coming from the top of the Portezuelo Formation, but it was recently reassigned to the Plottier Formation (Coria and Currie 2002b). Bonaparte and Gasparini (1979) indicated that cf. *Antarctosaurus giganteus* also comes from the Plottier Formation at Aguada del Caño near Neuquén City. The titanosaurid *Rinconsaurus caudamirus* (Calvo and González Riga 2003) was excavated from levels of the Río Neuquén Subgroup exposed at Rincón de los Sauces (north of Neuquén Province; Fig. 4.7). The titanosaurid *Mendozasaurus neguyelap* (González Riga 2003, 2005) was found in the upper section of the Rio Neuquén Subgroup in Cerro Guillermo, Mendoza Province.

The uppermost third of the Neuquén Group is the Río Colorado Subgroup, which is widely distributed in the southern half of the Neuquén Basin. As Leanza and colleagues (2004) indicated, excellent outcrops may be seen in the area between Neuquén City and Sierra del Portezuelo (Neuquén Province) and in the region around Planicie de Rentería (Río Negro Province; Fig. 4.7). The Río Colorado Subgroup is divided into the lower Bajo de la Carpa Formation and the upper Anacleto Formation. The age of this subgroup is considered to be Santonian–Early Campanian

(Leanza 1999; Hugo and Leanza 2001a, 2001b; Leanza, Apesteguía, Novas, and de la Fuente 2004).

The Bajo de la Carpa Formation crops out from the Sierra del Portezuelo area in south-central Neuquén to the Bajo de Santa Rosa region in the northern part of Río Negro Province (Hugo and Leanza 2001b). The dinosaur record of this unit includes (Fig. 4.11) the alvarezsaurid *Alvarezsaurus calvoi* (Bonaparte 1991a), the abelisauroid *Velocisaurus unicus* (Bonaparte 1991a), the birds *Neuquenornis volans* (Chiappe and Calvo 1994) and *Patagopteryx deferrariisi* (Alvarenga and Bonaparte 1992), nests containing small eggs that have been ascribed to basal birds (Schweitzer et al. 2002), and the basal ornithopod *Gasparinisaura cincosaltensis* (Coria and Salgado 1996a; Salgado, Coria, and Heredia 1997).

The Anacleto Formation is the highest unit of the Neuquén Group. Its age is considered to be Early Campanian (Leanza 1999; Dingus, Clarke, Scott, Swisher, Chiappe, and Coria 2000; Hugo and Leanza 2001a, 2001b). From west of Paso Córdova (Río Negro Province; Fig. 4.7) in strata belonging to the Anacleto Formation, Wichmann (1916) reported the discovery of the titanosaurid *Antarctosaurus wichmannianus* (von Huene 1929a). Also from the same levels but from the highly productive Cinco Saltos locality (Río Negro Province; Fig. 4.7) comes the abundant saltasaurine titanosaurid *Neuquensaurus australis* (Lydekker 1893; von Huene 1929a; Powell 1986, 2003). Other titanosaurids from the Anacleto Formation are *Laplatasaurus araukanikus* (von Huene 1929a; Heredia and Salgado 1999) and *Pellegrinisaurus powelli* (Salgado 1996). The egg- and nest-producing locality of Auca Mahuida is in beds of the Anacleto Formation (Chiappe and Coria 2000; Chiappe and Dingus 2001; Fig. 4.7). The theropod taxon recorded in the Anacleto beds is *Abelisaurus comahuensis*, from Pellegrini Lake in the vicinity of Cinco Saltos (Río Negro; Fig. 4.7). *Abelisaurus* was originally thought to have come from the overlying Allen Formation (Bonaparte and Novas 1985), but recent stratigraphic review allows this dinosaur to be reassigned to levels of the Anacleto Formation (Heredia and Salgado 1999). Another abelisaurid, *Aucasaurus garridoi*, was recovered from Anacleto beds exposed in Auca Mahuida (Coria and Chiappe 2000; Chiappe and Dingus 2001; Coria, Currie, Eberth, and Garrido 2002). Bird ichnites have been reported from Barrosa Hill (Coria, Currie, Eberth, Garrido, and Koppelhus 2001). Ornithischians from the Anacleto Formation are represented by the basal ornithopod *Gasparinisaura cincosaltensis* (Coria and Salgado 1996a; Salgado, Coria, and Heredia 1997).

The sequence of continental beds in the Neuquina Basin ends with the deposits of the lower part of the Malargüe Group (Fig. 4.8). It is divided into the lower Allen Formation (fluvial and lacustrine beds) and the upper Jagüel and Roca formations (indicating the first Atlantic flooding episode over Patagonia). The Malargüe Group ranges in age from Late Campanian to Danian (earliest Tertiary). It is worth noting that the Cretaceous/Tertiary boundary in the Neuquina Basin occurs in marine sediments of the Jagüel Formation (Leanza, Apesteguía, Novas, and de la Fuente 2004).

The Allen Formation is equivalent to the "Senoniano Lacustre" (Wichmann 1924) of northern Patagonia (Fig. 4.12). The formation displays a varied spectrum of continental sedimentary facies of mostly fluvial and lacustrine environments (Leanza, Apesteguía, Novas, and de la Fuente 2004). It is widely distributed in the eastern region of Neuquén, the northern half of Río Negro, and the southern part of La Pampa Provinces. The Allen Formation is correlated, at least in part, with the Loncoche and Paso del Sapo formations, two units of the Neuquina Basin exposed in the northern half of the basin (corresponding to Mendoza

Fig. 4.11. Some of the reptiles documented in the Bajo de la Carpa Formation. Not to scale. (A) Flightless bird *Patagopteryx deferrariisi* (skull length approximately 7 cm). (B) Enantiornithine bird *Neuquenornis volans* (skull length approximately 4 cm). (C) Basal snake *Dinilysia patagonica* (skull length approximately 10 cm). (D) Crocodyliform *Notosuchus terrestris* (skull length approximately 15 cm). (E) Basal ornithopod *Gasparinisaura cincosaltensis* (skull length approximately 7 cm). (F) Alvarezsaurid theropod *Alvarezsaurus calvoi* (skull length approximately 7 cm). *Illustrations by J. González.*

Province). The Allen Formation also correlates in the south with the Angostura Colorada and Los Alamitos formations, both of which are exposed in the southern half of Río Negro Province (see below), and the base of La Colonia Formation, a sedimentary unit exposed in Chubut Province (see below). The age of the Allen Formation and equivalent units may be Late Campanian–Early Maastrichtian (Hugo and Leanza 2001a, b; Leanza, Apesteguía, Novas, and de la Fuente 2004; Fig. 4.8). Dinosaur fossils recorded in the Allen Formation in the Neuquina Basin include titanosaurid skeletons (*Rocasaurus muniozi, Bonatitan reigii*; Salgado and Azpilicueta 2000; Martinelli and Forasiepi 2004), eggs, and eggshells, which are abundant in those beds, especially at Salitral Moreno and Bajo de Santa Rosa (Powell 1987b; Moratalla and Powell 1994; Faccio 1994; Figs. 4.7 and 4.13). Ornithischians are represented by abundant hadrosaurid remains collected in the Allen Formation at the localities of Salitral Moreno (Río Negro Province; Powell 1987c) and Islas Malvinas (La Pampa Province; González Riga and Casadío 2000) as well as in the equivalent Loncoche Formation at Buta Ranquil (Neuquén Province; Apesteguía and Cambiaso 1999; Fig. 4.7). Ankylosaur remains have been found at Salitral Moreno in beds of the Allen Formation (Coria 1994a; Salgado and Coria 1996; Coria and Salgado 2001). Theropod dinosaurs include the abelisauroid *Quilmesaurus curriei* (Coria 2001; Juárez Valieri, Fiorelli, and Cruz 2007) and the dromaeosaurid *Austroraptor cabazai* (Novas, Pol, Canale, Porfiri, and Calvo 2009), both from the Allen Formation exposed at Bajo de Santa Rosa (Río Negro Province). Carinate birds have been collected from Salitral Moreno (*Limenavis patagonica*; Clarke and Chiappe 2001).

6.D. THE SOMUNCURA MASSIF

At the end of the Mesozoic, the Somuncura Massif was flooded by shallow marine waters from the Atlantic, which are correlated with fluvial and lacustrine dinosaur-bearing beds of the Paso del Sapo and Lefipán formations (Figs. 4.12 and 4.13). In the west and north sections of the basin, continental red beds of the Angostura Colorada and Coli Toro formations were deposited, which have yielded hadrosaurid remains near the town of Ingeniero Jacobacci (Casamiquela 1964b; Bonaparte 1978). Such deposits are roughly equivalent to the Los Alamitos Formation (Upper Campanian–Lower Maastrichtian), which was deposited on the southern corner of the Somuncura Massif and has yielded abundant invertebrate and vertebrate fossil remains, including abundant hadrosaurid bones. Important collections of vertebrates including mammals, dinosaurs (hadrosaurs, titanosaurids, theropods), turtles, lizards, snakes, frogs, and fish were made by José Bonaparte and assistants at the Los Alamitos farm, in the southeastern part of Río Negro Province (Bonaparte 1987). The dinosaurs documented in this locality are "*Kritosaurus*" *australis* (Bonaparte, Franchi, Powell, and Sepúlveda 1984; Bonaparte and Rougier 1987) and *Aeolosaurus rionegrinus* (Powell 1987b; Salgado and Coria 1993).

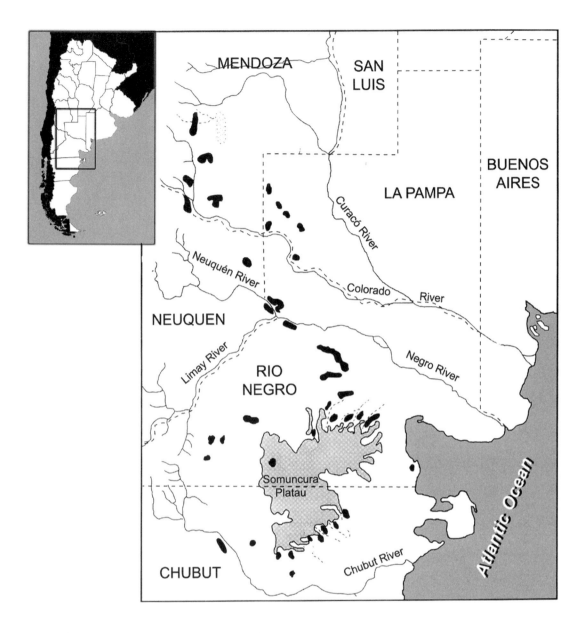

Fig. 4.12. Outcrops of Late Cretaceous beds collectively known as the Senoniano Lacustre (shown in black). These highly fossiliferous rocks are widely distributed in Patagonia; important sites are located around the Somuncura Massif (shown in gray). *Redrawn from Page et al. (1999).*

6.E. THE SAN JORGE AND CAÑADÓN ASFALTO BASINS

Deposits of the Chubut Group cover the Argentine Patagonian provinces of Chubut and the northeastern part of Santa Cruz extensively (Fig. 4.13). The Chubut Group accumulated in two neighboring intracratonic basins, Cañadón Asfalto (bounded to the north by the Somuncura Massif) and the San Jorge Basin (bounded to the south by the Deseado Massif).

In the central region of Chubut Province, the Cañadón Asfalto Basin was filled by deposits of the Chubut Group. In this basin, the group is comprised of (from bottom to top) the Valanginian Los Adobes and the Hauterivian–Albian/Cenomanian Cerro Barcino formations (Page et al. 1999; Rauhut, Cladera, Vickers-Rich, and Rich 2003; Fig. 4.8).

Fossil Record of Cretaceous Dinosaurs

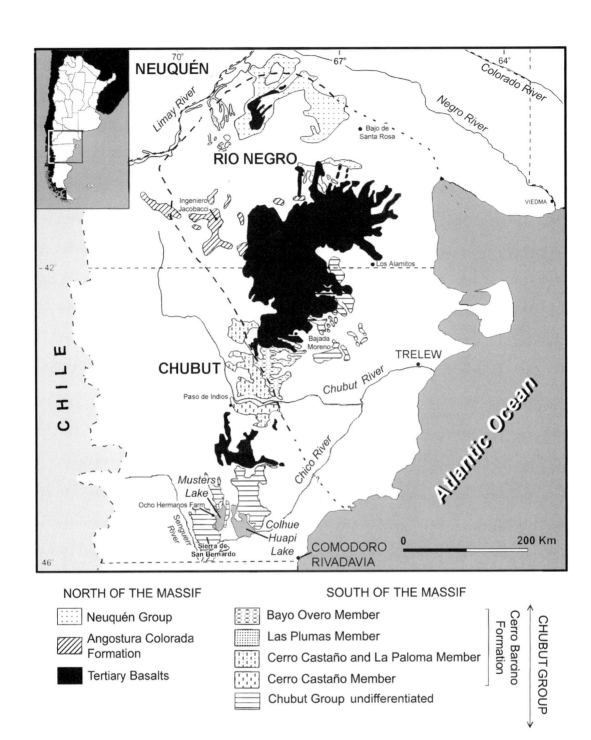

Fig. 4.13. Cretaceous units that crop out north and south of the Somuncura Massif in central Patagonia. *Redrawn from Page et al. (1999).*

The Chubut Group unconformably overlies Jurassic beds (including the dinosaur-bearing Cañadón Asfalto Formation) and is overlain by the Campanian–Maastrichtian La Colonia, Paso de Sapo, and Lefipán Formations.

The levels of the Chubut Group that crop out between the Somuncura Massif and the Chubut River and that have yielded some dinosaur remains correspond to the Cerro Barcino Formation. This unit is comprised of several members (La Paloma, Cerro Castaño, Las Plumas, and Bayo Overo). Dinosaur remains reported from the La Paloma Member (Hauterivian–Barremian) include the oldest Early Cretaceous titanosaur from South America, caudal vertebrae of an abelisaurian theropod (Rauhut, Cladera, Vickers-Rich, and Rich 2003), and fragments of theropod teeth (Vickers-Rich, Rich, Lanus, Rich, and Vacca 1999; Rich, Vickers-Rich, Novas, Cúneo, Puerta, and Vacca 2000).

The overlying Aptian–Albian Cerro Castaño Member has yielded two partial skeletons of the large carcharodontosaurid *Tyrannotitan chubutensis* (Rich, Vickers-Rich, Novas, Cúneo, Puerta, and Vacca 2000; Novas, de Valais, Vickers-Rich, and Rich 2005), a titanosauriform sauropod, and a probable coelurosaurian theropod (Rauhut, Cladera, Vickers-Rich, and Rich 2003), all of which were discovered not far from the town of Paso de Indios (Fig. 4.13). In addition, the fragmentarily known theropod *Genyodectes serus* is thought to have come from these beds (Rauhut 2004).

The Bayo Overo Member has yielded the sauropod *Chubutisaurus insignis* (Del Corro 1975), an undescribed sauropod, and several theropod teeth (Rich, Vickers-Rich, Novas, Cúneo, Puerta, and Vacca 2000). *Chubutisaurus* is thus most probably of Late Cretaceous age (Page et al. 1999; Rich, Vickers-Rich, Novas, Cúneo, Puerta, and Vacca 2000), although it is frequently cited as being Aptian–Albian in age (e.g., Weishampel 1990; Bonaparte 1996a, 1996b; Novas 1997c). The abelisaurid *Carnotaurus sastrei*, originally thought to be from the Albian Cerro Barcino (=Gorro Frigio) Formation (Bonaparte 1985; Bonaparte, Novas, and Coria 1990), is now known to have come from the Maastrichtian La Colonia Formation (Rauhut, Cladera, Vickers-Rich, and Rich 2003; Fig. 4.8).

To the south of Chubut Province, Cretaceous continental beds are more prolific in dinosaur remains than those of the northern half. These beds are exposed around two lakes of southern Chubut: Musters (to the west) and Colhue Huapi (to the east; Fig. 4.13). Pioneering discoveries made in this area include a large sauropod skeleton found by Carlos Ameghino in Pampa Pelada (northeast of Lake Colhué Huapi). An expedition of the La Plata Museum excavated only a forelimb (Ameghino 1921a), which Richard Lydekker described in 1893 under the name *Argyrosaurus superbus*. Also to the east of Lake Colhue Huapi fragmentary material of the ornithischians *Notoceratops bonarelli* and *Secernosaurus koerneri* was collected. Sauropod femora were excavated in 1924 by Elmer Riggs (from the Field Museum of Natural History in Chicago) at Sierra de San Bernardo (west of Lake Musters) that were later described by Friedrich von Huene (1929a). More recently, José Bonaparte and crew and (separately) Rubén D. Martínez and assistants conducted intense field work at sites in Sierra de San Bernardo, producing remarkable dinosaur discoveries.

The following Cretaceous units of the Chubut Group are identified in the area comprised of San Bernardo Hill, Lake Musters, Lake Colhué Huapi, and Chico River (from bottom to top): the Matasiete (Aptian) Formation, the Castillo (Aptian–Albian) Formation, the Lower Member of Bajo Barreal (Cenomanian–Turonian) Formation, the Upper Member of Bajo Barreal (Coniacian–Santonian) Formation, and the Laguna Palacios (Campanian–Maastrichtian) Formation (Page et al. 1999; Uliana and Legarreta 1999; Bridge, Jalfing, and Georgieff 2000; Fig. 4.8).

The Bajo Barreal Formation crops out in southern Chubut and northeastern Santa Cruz. Levels that yielded dinosaur bones correspond to the lower member, predominantly made up of volcanic sandstones and tuffs. The age of the Lower Bajo Barreal Formation is probably Cenomanian–Turonian, based on radiometric dates that yield ages between 91 and 95.8 million years (Bridge, Jalfing, and Georgieff 2000; Martínez, Giménez, Rodríguez, Luna, and Lamanna 2004). Perhaps the most productive locality of the Lower Bajo Barreal Formation is the Ocho Hermanos farm, located at the base of the northwestern slope of San Bernardo Hill. In 1981, José F. Bonaparte and his team recovered from there an articulated series of dorsal and sacral vertebrae of a big sauropod. Intensive work was later done in the area by personnel of the Universidad de la Patagonia, led by Rubén D. Martínez.

Other fossil spots occur on the southern slopes of the San Bernardo Hill, close to the Senguerr River, where José F. Bonaparte and assistants collected a new specimen of *Argyrosaurus superbus* (Bonaparte 1996b; Powell 2003; Fig. 4.13). Dinosaurs documented in the Lower Bajo Barreal Formation include the basal coelurosaur *Aniksosaurus darwini* (Martínez, Lamanna, Smith, Casal, and Luna 1999; Martínez and Novas 1997, 2006), an isolated caudal of an indeterminate diplodocimorph (Sciutto and Martínez 1994), the basal titanosaurid *Epachthosaurus sciuttoi* (Powell 1990, 2003; Martínez, Giménez, Rodríguez, Luna, and Lamanna 2004), the large titanosaurid *Argyrosaurus superbus* (von Huene 1929a; Bonaparte and Gasparini 1979; Bonaparte 1996b; Powell 2003), a marvelously preserved skull of a yet-unnamed titanosaurid (R. D. Martínez 1998a, 1999), the fragmentary maxilla of the sauropod *Campylodon ameghinoi* (von Huene 1929a), the basal iguanodontian *Notohypsilophodon comodoroensis* (R. D. Martínez 1998b), the abelisaurid *Xenotarsosaurus bonapartei* (Martínez, Giménez, Rodríguez, and Bochatey 1986), isolated abelisaurid maxilla (Lamanna, Martínez, and Smith 2002), isolated abelisaurid vertebrae (Martínez, Novas, and Ambrosio 2004), and *Megaraptor*-like unguals (Martínez, Lamanna, Smith, Dodson, and Luna 2000; Lamanna, Martínez, Luna, Casal, Ibiricu, and Ivany 2004).

The Upper Bajo Barreal Formation (Coniacian–Santonian) crops out north of the Chico River to the east of Lake Colhue Huapi. Its age is estimated as Campanian–Maastrichtian. Dinosaurs found in this formation are the possible ceratopsid *Notoceratops bonarelli*, the hadrosaurid *Secernosaurus koerneri*, and the titanosaurid *Aeolosaurus colhuehuapensis* (Casal, Martinez, Luna, Sciutto, and Lamanna 2007).

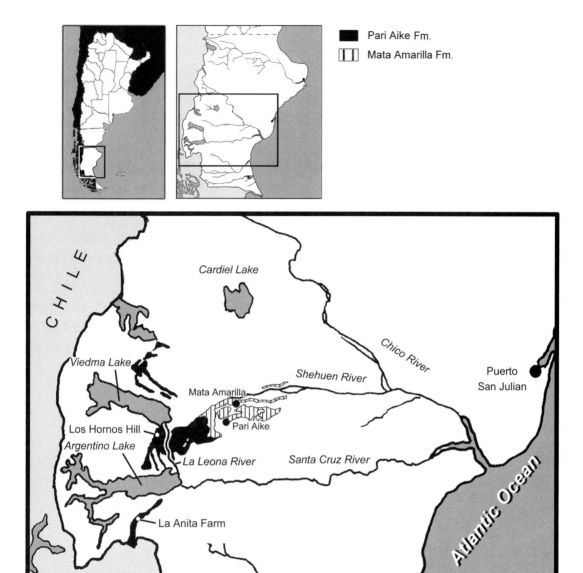

Fig. 4.14. Map of southwestern Santa Cruz Province in southern Patagonia showing geographic distribution of Upper Cretaceous dinosaur-bearing beds.

6.F. THE MAGALLANES BASIN

The Magallanes Basin (also known as the Austral Basin) developed in the southern extremity of Patagonia, more precisely at the southwest corner of Santa Cruz Province (Fig. 4.14). It is predominantly composed of Lower Cretaceous marine sediments overlain by Upper Cretaceous continental deposits of the Turonian Mata Amarilla and the Campanian–Maastrichtian Pari Aike formations (Malumián, Nullo, and Ramos 1983; Novas, Bellosi, and Ambrosio 2002; Fig. 4.8). Until recently, the only available dinosaur materials from this area were isolated and poorly informative bones that were collected at the end of the nineteenth century and the beginning of the twentieth (e.g., Ameghino 1921a; von Huene 1929a). The Mata Amarilla beds have yielded remains of the basal iguanodontian

Loncosaurus argentinus and teeth of the sauropod *Clasmodosaurus spatula* (von Huene 1929a). The dinosaur-bearing Pari Aike Formation is widely exposed on both sides (e.g., east and west) of the La Leona River, and it is bounded by the glacial lakes Viedma and Argentino (Fig. 4.14). The age of the Pari Aike is debated (Novas, Bellosi, and Ambrosio 2002), but a Campanian–Maastrichtian age is widely accepted. At Los Hornos Hill, on the southeastern coast of Lake Viedma, an almost-complete skeleton of the basal iguanodontian *Talenkauen santacrucensis* (Novas, Cambiaso, and Ambrosio 2004; Plate 17), isolated vertebrae of the giant titanosaur *Puertasaurus reuli* (Novas, Salgado, Calvo, and Agnolín 2005), and some teeth and cranial and postcranial bones of the basal coelurosaurian theropod *Orkoraptor burkei* (Novas, Ezcurra, and Lecuona 2008) were recovered. Remains of huge titanosaurid sauropods were also reported by Kenneth Lacovara and colleagues in 2004 that might correspond to *Puertasaurus* or a closely related taxon.

From the south coast of Lake Argentino, José Bonaparte collected some specimens of large titanosaurids from La Anita farm (Bonaparte 1996a; Fig. 4.14).

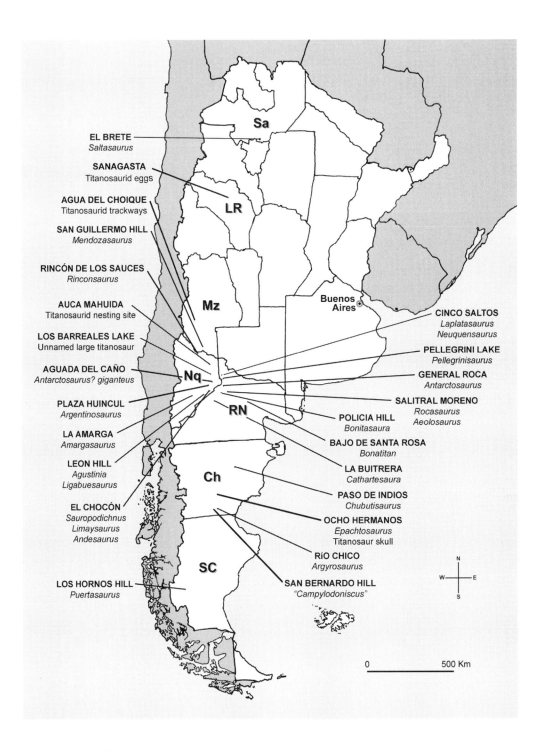

CRETACEOUS SAUROPODS 5

The fossil record of the continental Cretaceous in South America is distinguished by the abundance and diversity of sauropod dinosaurs. This suggests that of the medium to very large herbivores, sauropods were the most common. The number of individuals (and apparently also the species diversity) of sauropods surpasses that of the ornithischian dinosaurs.

The first dinosaurs discovered and named from this continent were Cretaceous sauropods. In 1893, Richard Lydekker named two species of *Titanosaurus* (*T. australis* and *T. nanus*) and the large *Argyrosaurus superbus*. Later, the sample of Upper Cretaceous sauropods increased with the discovery of productive localities at Río Negro Province at the excavations carried out by geologist Ricardo Wichmann in 1912.

The abundance of new material stimulated a large and seminal monograph by Friedrich von Huene (1929a) entitled (translated from the Spanish) "The saurischians and ornithischians from the Cretaceous of Argentina." In this extensive work, von Huene named *Antarctosaurus wichmannianus* (on the basis of a partial skeleton collected by Wichmann) but also offered revisions of the taxa earlier described by Lydekker on the basis of abundant new remains from the rich Cinco Saltos fossil site in Río Negro. In addition, von Huene recognized three new species of Patagonian sauropods (*Laplatasaurus araukanicus*, *Titanosaurus robustus*, and *Campylodon ameghinoi*), and identified *Argyrosaurus*, *Antarctosaurus*, *Laplatasaurus*, and *Titanosaurus* in the Mercedes Formation of Uruguay (von Huene 1929b).

The first notice of titanosaurids in Brazil was published by Friedrich von Huene in 1927 (Huene 1927a, 1927b, 1927c) and consisted of a caudal vertebra collected in the State of São Paulo that he referred to *Titanosaurus australis*.

The Brazilian paleontologist Llewellyn Ivor Price subsequently conducted fruitful excavations in the Bauru Group in the states of São Paulo and Minas Gerais, principally at the productive location of Peirópolis, obtaining a considerable amount of beautifully preserved titanosaurid remains. Although Price opened the Peirópolis quarry in 1948 and worked it almost continually up to 1969 (Campos and Kellner 1999), he published only some brief reports on the fabulous collections he and his assistants made (Price 1951, 1961).

Throughout this interval, the Cretaceous sauropods from Patagonia remained almost unstudied, and it was almost 50 years before the topic was taken up. In 1979, José Bonaparte and Zulma Gasparini offered new interpretations of the taxonomy, stratigraphic provenance, and age of

Fig. 5.1. Map of Argentina indicating Cretaceous fossil localities and a list of sauropod genera discovered in each of them. The provinces of Argentina that yielded sauropod remains are Salta (Sa), La Rioja (LR), Mendoza (Mz), Neuquén (Nq), Río Negro (RN), Chubut (Ch), and Santa Cruz (SC).

the sauropod species previously studied by Lydekker and von Huene. Bonaparte and Gasparini also reviewed a new sauropod, *Chubutisaurus insignis* (Del Corro 1975), thus extending the sauropod record back to the Lower Cretaceous with the inclusion of this Aptian–Albian form, emphasizing that unknown chapters of sauropod evolution remained to be discovered.

Since the mid-1980s, Jaime Powell has added the Patagonian *Epachthosaurus* and *Aeolosaurus* to the list of Cretaceous titanosaurians, offering renewed descriptions and interpretations of the early discoveries by Lydekker and von Huene from Patagonia (Powell 1987b, 1990). He has also expanded our knowledge of the poorly known Brazilian sauropods excavated by Price in Peirópolis. More recent publications were devoted to Price's discoveries since 1999 by Alexander Kellner and Diógenes de Almeida Campos (Campos, Kellner, Bertini, and Santucci 2005; Kellner and Campos 1997; Kellner, Campos, Azevedo, and Trotta 2005).

In the most recent years, explorations and research carried out by José Bonaparte, Leonardo Salgado, Jorge Calvo, and Rodolfo Coria led to the discovery of geologically older and anatomically more primitive members of Titanosauria and, more important, the documentation of sauropod clades other than titanosaurs (e.g., the diplodocoid families Dicraeosauridae and Rebbachisauridae). Studies by these authors of the anatomy, phylogeny, functional anatomy, nesting behavior, and ontogeny of South American sauropods have greatly enhanced our understanding of sauropod evolution as a whole.

The fossil record of sauropod dinosaurs is mostly concentrated in the northwestern corner of Patagonia, although there are some relevant discoveries in Brazil and patchy and less significant Late Cretaceous records from Uruguay, Bolivia, and Chile (Fig. 5.1). No doubt future work in South American countries other than Argentina will produce additional discoveries. Thus far, 33 sauropod species have been documented from different Cretaceous strata in South America belonging to two main neosauropod lineages, Diplodocimorpha and Titanosauriformes, which are phylogenetically related to the Late Jurassic diplodocids and brachiosaurs, respectively (Fig. 5.2).

Diplodocimorphs were very prolific during the Jurassic, when they attained a worldwide distribution, but during the Cretaceous they were geographically restricted mainly to Gondwana. In South America they are represented by the Early Cretaceous *Amargasaurus cazaui* and the Early Late Cretaceous *Limaysaurus tessonei*, both from northwestern Patagonia, as well as by *Amazonsaurus maranhensis* from possibly Aptian beds of Brazil. Basal titanosauriforms (e.g., brachiosaurids and forms more closely related to titanosaurs) attained a worldwide distribution during the Late Jurassic and Early Cretaceous. Titanosaurs are recorded in Cretaceous beds in South America, Africa, Madagascar, India, Australia, Europe, North America, and Asia. After Cenomanian times, titanosaurs became a successful clade in Gondwana and are represented by numerous genera, including forms of relatively small size (e.g., *Saltasaurus*,

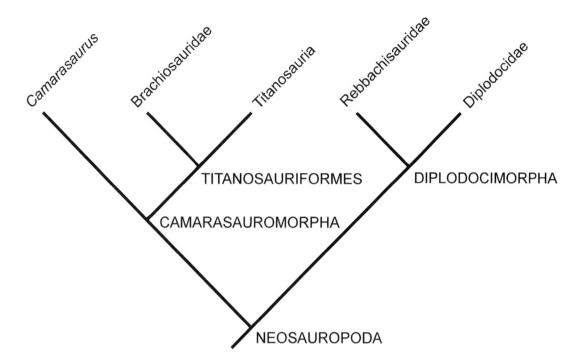

Fig. 5.2. Cladogram of the main groups of Neosauropoda considered in the text.

6 meters long) and perhaps the largest terrestrial animals that ever existed (e.g., *Argentinosaurus*, *Puertasaurus*, *Futalognkosaurus*, and *Ligabuesaurus*, reaching approximately 35 meters in whole length). South American titanosaurs are represented by a diverse array of taxa frequently recorded in Late Cretaceous beds from Argentina, but remains of these dinosaurs have also been recovered in Brazil, Chile, Bolivia, and Perú.

Diplodocimorpha

The Cretaceous fossil record from South America indicates that diplodocimorphs were present and diverse mainly during the Early Cretaceous. As already noted, the end of the Jurassic was the heyday of the diplodocimorphs. After Kimmeridgian times they become rare or absent in the northern hemisphere but remained diverse and abundant in the southern landmasses until their final extinction at the Early Late Cretaceous (Cenomanian). Members of two diplodocimorph lineages have been documented in South America—Dicraeosaurinae and Rebbachisauridae (Figs. 3.15 and 5.3A and B)—as well as basal members of Diplodocimorpha (*Zapalasaurus bonapartei*; Salgado, Carvalho, and Garrido 2006) and bizarre sauropods that probably belong to this clade (*Agustinia ligabuei*; Bonaparte 1999b).

ZAPALASAURUS BONAPARTEI

This new sauropod was collected in the locality of La Picaza, south of Neuquén Province, from the upper levels of the La Amarga Formation. The specimen includes an isolated cervical vertebra, a probably

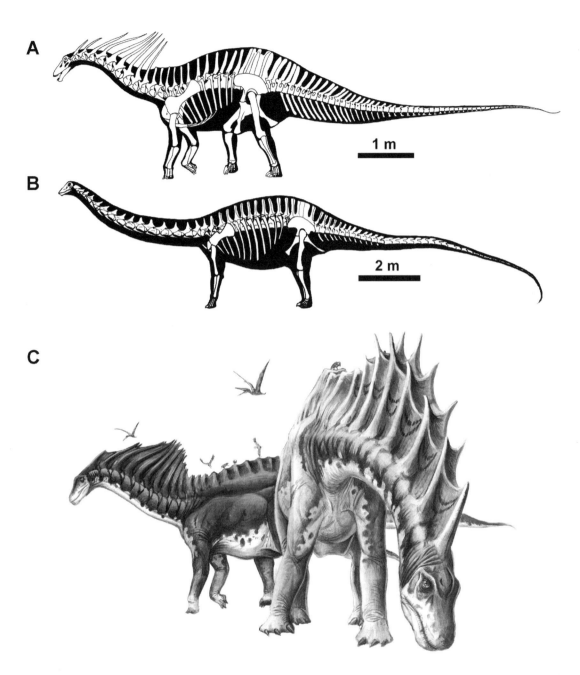

Fig. 5.3. South American diplodocimorphs: (A, B) Skeletal reconstructions of (A) *Amargasaurus cazaui* and (B) *Limaysaurus tessonei*. (C) Life reconstruction of *Amargasaurus cazaui*. Illustrations by J. González.

continuous series of 17 caudals; fragments of the ischium, pubis, and ilium; and an incomplete femur and tibia.

Zapalasaurus bonapartei exhibits some autapomorphic features in the middle to posterior caudals. These include a proximodistally elongate neural spine whose proximodorsal corner is higher than the distal one and caudal centra that progressively increase in length from caudal 1 through caudal 20.

Zapalasaurus has been interpreted by Salgado and colleagues (2006) as a basal diplodocimorph that is more derived than *Haplocanthosaurus*.

Zapalasaurus, Agustinia, Cathartesaura, Limaysaurus, and *Amargasaurus* indicate that during the Early Cretaceous diplodocimorphs were at least as abundant and diverse as the titanosaurs (Salgado, Carvalho, and Garrido 2006).

AGUSTINIA LIGABUEI

This very bizarre sauropod was discovered by José Bonaparte and his assistants in the beds of the Aptian–Albian Cullín Grande Member of the Lohan Cura Formation (Lower Cretaceous; Bonaparte 1999b), which is exposed at Leones Hill in the vicinity of Picún Leufú in Neuquén Province (Figs. 4.7 and 5.1). This sauropod, which is similar in size to *Limaysaurus* (around 17 meters long), is represented by some hindlimb bones, sacral and caudal vertebrae, an isolated dorsal, and what appear to be several dermal plates of curious morphology. The preserved vertebral material is comprised of the neural spines of the posterior dorsals, six fused sacrals, and some proximal caudals that have well-developed lateral crests, a feature present among diplodocimorphs, including rebbachisaurids. The tops of the neural spines of these vertebrae are transversely expanded, a feature that Bonaparte (1999b) interpreted as an accommodation for the large ossifications that were found in association with the specimen (Fig. 5.4). The variety of dermal ossifications in *Agustinia* is distinctive among sauropods. Eight osteoderms were collected, six of them found in natural sequence; they were parallel to one another and followed the line of the posterior dorsal and sacral vertebrae. The remaining two osteoderms were found isolated and had evidently moved from their original position. Four different types of osteoderms are easily distinguished in form and size: 1) a leaf-shaped, unpaired ossification ("o1") that is around 20 centimeters in greatest diameter; 2) a subrectangular, unpaired large ossification ("o2") that is approximately 64 centimeters in transverse diameter and has a pair of laterodorsal projections that presumably represented the core of horny spikes and two robust "feet," possibly for anchoring to the neural spines of the dorsal vertebrae; 3) a third kind of osteoderm ("o3") that resembles the previous one but is divided into two pieces separated along the axial plane; and 4) a fourth kind of ossification ("o4") that is stick-shaped and quite large (80 centimeters long as preserved). These ossifications lack the rugosities, foramina, or vascular grooves that are characteristically present in the dermal ossifications of other dinosaurs (e.g., titanosaurs, stegosaurs, ankylosaurs). Interestingly enough, all of the described osteoderms are composed of a main portion and one or two smaller but thicker pieces that may have connected the larger portion with the top of the neural spine. In the stick-shaped osteoderm ("o4"), the main portion is separated from the proximal and thicker piece. Whether the main portions of these last ossifications articulated movably over the thicker proximal elements is a matter of speculation. Bonaparte (1999b) thinks that the ossifications were located in pairs over the neck, back, and pelvic girdle,

Fig. 5.4. *Agustinia ligabuei:* (A, B) Sequence of preserved neural spines of dorsal, sacral and proximal caudal vertebrae in (A) dorsal and (B) left lateral views. (C) Speculative arrangement of dermal ossifications. Stick-like ossifications were found in articulation with dorsal neural spines and are here tentatively arranged in pairs. The largest, plate-like ossification (o2) coincides fairly well with the transverse width of sacral neural spines, and it is assumed here that this ossification was located above the pelvic girdle. (D) Life restoration of *Agustinia ligabuei.* (A–C) From Bonaparte (1999b). Reprinted with permission of the author and the National Science Museum Monographs, Tokyo. (D) Illustration by J. González.

and it is not unreasonable to think that the small unpaired leaf-shaped ossification occupied the axial plane above the anterior dorsal vertebrae, while the large unpaired subrectangular and spiked kind occupied a more posterior position over the vertebral column. The stick-shaped osteoderms ("o4") were found in articulation with the top of the dorsal vertebrae, but the plate-like osteoderms ("o1" and "o2") were found separated from the rest of the backbone (J. F. Bonaparte, pers. comm.). However, I presume that the largest plate-like osteoderm ("o2") was above the pelvic girdle because the distance between the feet of this ossification matches well with the transverse width of the distal tip of the sacral neural spines, which are wider than in any other of the preserved dorsal and caudal neural spines (Fig. 5.4).

The phylogenetic relationships of *Agustinia* are far from settled, and its referral to Diplodocimorpha needs to be taken with caution. Impressed by the curious morphology of this animal, Bonaparte (1999b) coined a new family name of Sauropoda, Agustiniidae, to emphasize the distinctness of *Agustinia*.

Dicraeosaurinae

The South American record of Cretaceous Dicraeosaurinae is scant at the moment. It includes the well-known *Amargasaurus cazaui* from the Barremian–Early Aptian of northwestern Patagonia and some remains from the Itapecurú Formation (Aptian–Albian, Parnaíba Basin) of northern Brazil (Carvalho, Avilla, and Salgado 2003). South American discoveries are important because *Amargasaurus* constitutes one of the best-known Early Cretaceous sauropods, while the Brazilian specimen, although incomplete, is the latest record for the group, indicating that dicraeosaurines survived until the end of the Early Cretaceous.

AMARGASAURUS CAZAUI

This sauropod was collected from the Early Cretaceous La Amarga Formation, which is exposed at La Amarga creek, approximately 70 kilometers south of Zapala in Neuquén Province (Salgado and Bonaparte 1991). *Amargasaurus* was relatively small by sauropod standards; it was at most only 9 meters long (Fig. 5.3A and C). The skull of *Amargasaurus* (Fig. 5.5A and B) is represented by its posterior half, but nothing of the snout, dentary, and teeth has been recovered (Salgado and Calvo 1992). The supratemporal fenestra is extremely reduced and exposed in lateral view, a bizarre condition that contrasts with that of most other diapsids, in which the upper temporal fenestra is dorsally oriented. The basipterygoid processes are notably elongate, more than three times the width of the basioccipital tubera. As in *Dicraeosaurus hansemanni*, they project anteriorly and slightly ventrally and diverge slightly from each other, a condition that sharply differs from the shorter and divergent basipterygoid processes of other sauropodomorphs (see, for example, the

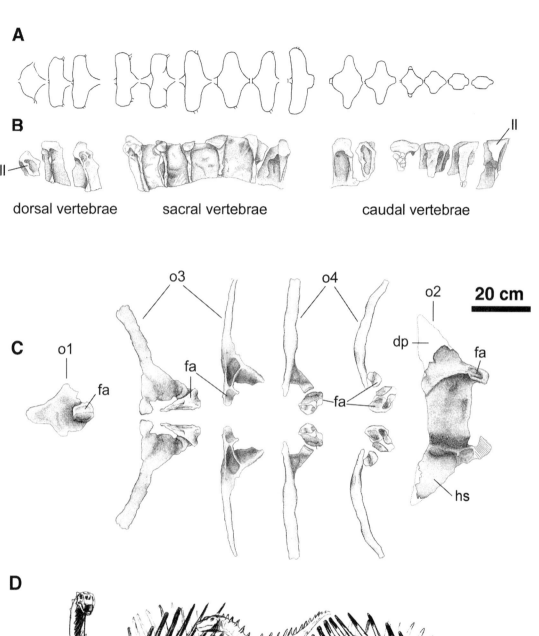

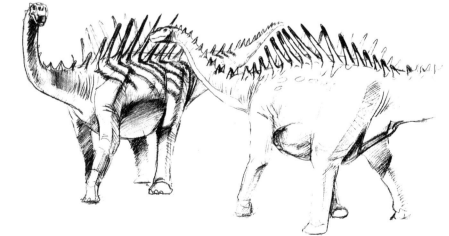

Fig. 5.5. *Amargasaurus cazaui:* (A) Braincase and probable skull shape in lateral view. (B) Braincase in caudal aspect. (C, D) Selected cervical and dorsal vertebrae. Left to right: atlas-axis, cervicals 6 and 8, and dorsals 1, 8, and 9, in (C) lateral and (D) cranial views. *From Salgado and Bonaparte (1991). Reprinted with permission of* Ameghiniana.

rebbachisaurid *Limaysaurus*; Fig. 5.6B). Moreover, the elongation of the basipterygoid processes in *Amargasaurus* is extreme, suggesting that well-developed muscles originated on this part of the basicranium that were functionally significant in movements and postures of the head and jaws of this animal.

The postcranial skeleton of *Amargasaurus* deserves special consideration because of its very bizarre features. The vertebral column is made up of 13 cervicals, 9 dorsals, and possibly 5 sacrals (the number of caudals remains unknown). The cervical and cranial dorsal centra are opisthocoelous, but the remaining dorsals are amphiplatyan. Except for the atlas and the axis, the remaining cervicals are characterized by a deep bifurcation and considerable elongation of the neural spines (Figs. 5.3A and 5.5C and D). Each ramus of the neural spine has a transversely compressed base but becomes subcircular in cross-section over the rest of its shaft and is sharply pointed at the distal tip. Salgado and Bonaparte (1991) suggested that the spines of *Amargasaurus* comprised a colossal defensive weapon. The neural spines of the dorsal vertebrae continue the bifurcate condition of the cervicals except for the last two dorsals, in which the unforked spines acquire a paddle-like aspect. The height of the neural spine reaches 60 centimeters in cervical 8. In the sacrum, the neural spines are tall, unforked, and paddle-shaped and the transverse processes project significantly laterally, indicating that the pelvic girdle was transversely wide and that the ilia were well separated from each other (Salgado and Bonaparte 1991).

The presacral vertebrae of *Amargasaurus* are taller than in *Dicraeosaurus*, and splitting of the neural spines is more pronounced than in its African relative. In *Dicraeosaurus*, bifurcation of the neural spines extends to presacral 17, whereas in *Amargasaurus* it reaches presacral 20. A progressive lengthening of the neural spines in the cervicals and dorsals presumably occurred in the lineage that led to *Amargasaurus*. The elongation of the neural spines reached an extreme in this Patagonian taxon, in which the bifid cervical neural spines are four times longer than the centrum.

The anatomical and functional significance of the spike-like neural spines on cervical vertebrae is far from understood. Did it support a sail of skin or did each spike remain separate from others? We know even less about the possible functions of these bizarre neck devices (e.g., defense, display). In addition to *Amargasaurus*, a number of Early Cretaceous dinosaurs have elongated neural spines on the dorsal, sacral, and proximal caudal vertebrae, including the African iguanodontian *Ouranosaurus* and the theropods *Spinosaurus*, *Suchomimus*, and *Giganotosaurus*. Bailey (1997) came to the interesting conclusion that these elongate neural spines supported a buffalo-like humped back rather than a sail of narrow skin.

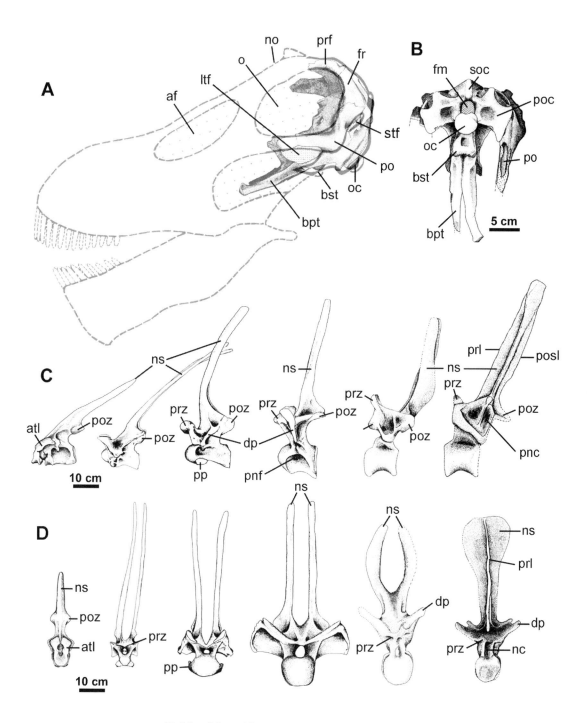

Rebbachisauridae

The Rebbachisauridae is the sister group of Diplodocidae (Fig. 5.2). Although they are anatomically more primitive than diplodocids, rebbachisaurids are the youngest representatives of the clade Diplodocimorpha in the fossil record. However, Rebbachisaurids are among the poorest known sauropods; only five species are currently named and only one of them is represented by a complete skeleton. Rebbachisaurids are a surviving linage of diplodocimorphs that were relatively successful in Gondwana. Recent discoveries reveal that they were some of the most common large herbivores during the Mid-Cretaceous. Their remains are documented in Aptian through Cenomanian beds in Patagonia, Brazil, Africa, and Europe (Lavocat 1954; Sereno et al. 1999; Calvo and Salgado 1995; Medeiros and Schultz 2001a; Pereda-Suberbiola, Torcida Fernández Baldor, Izquierdo, Huerta, Montero, and Pérez 2003; Torcida, Pereda Suberbiola, Huerta Hurtado, Izquierdo, Montero, and Pérez 2003; Canudo and Salgado 2003).

Basal diplodocoids, particularly rebbachisaurids, apparently played a significant role in Patagonian dinosaur communities during the pre-Turonian Cretaceous (Lamanna, Martínez, Luna, Casal, Dodson, and Smith 2001; Gallina, Apesteguía, and Novas 2002). Rebbachisaurid remains were collected from several different sites in South America, including the Lohan Cura, Rayoso, Candeleros, and Bajo Barreal formations in Patagonia and the Aptian–Albian Itapecurú Formation in northeastern Brazil.

The general aspect of rebbachisaurids (Fig. 5.3B) resembles that of the Jurassic *Diplodocus* and *Barosaurus*. The head is small and its buccal apparatus is composed of small, elongate, slender, curved, and peg-shaped teeth. The external nares open at the top of the skull, as is usual among diplodocimorphs. The frontals are rostrocaudally elongate, the ventral margin of the orbit is rounded instead of acuminate, and the supratemporal fenestra is reduced or absent. As far as is known from the African *Nigersaurus* (Sereno et al. 1999; Sereno and Wilson 2005), the rebbachisaurid snout was extremely wide, even wider than in other diplodocimorphs such as *Diplodocus* and *Dicraeosaurus*. The available evidence suggests that similarly shaped snouts, which had right-angled dentaries as seen from above, convergently evolved among titanosaurian sauropods, as shown by *Antarctosaurus wichmannianus* (see below).

The cervical and dorsal vertebrae of rebbachisaurids are remarkably tall, resembling those of dicraeosaurids. Interestingly, rebbachisaurids converged with derived titanosaurid sauropods in losing the hyposphene-hypantrum articulations of the dorsal vertebrae. The tail of rebbachisaurids is extremely elongate, as in other diplodocimorphs. The proximal and mid-caudal centra are quadrangular in cranial view. The scapular blade of rebbachisaurids is characteristically racket-shaped. Of these characteristics, only a racket-shaped scapular blade can be verified in *Limaysaurus*, *Nigersaurus*, and *Rebbachisaurus*, constituting the best-supported synapomorphy of Rebbachisauridae.

The genus that gave its name to the family is *Rebbachisaurus*, reported in 1954 by the French paleontologist René Lavocat from Aptian beds of Morocco. The list of African rebbachisaurids enlarged with the discoveries of *Nigersaurus taqueti* from Niger (Sereno et al. 1999; Sereno and Wilson 2005) and the taxonomic reallocation of the South African *Algoasaurus bauri* to the Rebbachisauridae (Carvalho, Avilla, and Salgado 2003; Canudo and Salgado 2003). Two Patagonian rebbachisaurids have been also found: *Rayososaurus agrioensis* and *Limaysaurus tessonei*. In addition, *Rebbachisaurus* sp. has been reported for the São Luis Basin (Cenomanian, northern Brazil; Medeiros 2001; Medeiros and Schultz 2001a). The recently described basal diplodocoid *Amazonsaurus maranhensis* (Carvalho, Avilla, and Salgado 2003) from the Aptian–Albian of northeastern Brazil may represent another rebbachisaurid.

LIMAYSAURUS TESSONEI

This taxon, originally termed "*Rebbachisaurus*" *tessonei*, was described by Jorge Calvo and Leonardo Salgado (Calvo and Salgado 1991, 1995; Bonaparte 1999a). The specimens were collected in sandstone of the Candeleros Formation, which are widely exposed at Lake Ezequiel Ramos Mexía in Neuquén Province (Fig. 4.7). *Limaysaurus tessonei* comes from stratigraphic levels close to those that yielded the holotype of the huge carcharodontosaurid theropod *Giganotosaurus carolinii*. *Limaysaurus* is represented by several individuals, one of which (the holotype) is 80 percent complete, thus constituting one of the most complete Cretaceous sauropods yet discovered in South America (Calvo 1999). The underlying Lohan Cura Formation yielded remains of a form very similar to *Limaysaurus* (Salgado, Garrido, Cocca, and Cocca 2004).

Limaysaurus tessonei was approximately 17 meters long (Fig. 5.3B). It is distinguished from other rebbachisaurids by an extremely reduced lower temporal fenestra (Fig. 5.6A–C). Its cervicals are strongly opisthocoelous and have double pneumatic pores in both sides of the centrum. The single neural spine is tall and narrow but is formed by four laminae similar to those of *Haplocanthosaurus*. In *Limaysaurus* the cervical neural spines are caudally inclined in the cranial cervicals but become almost vertically oriented in the vertebrae of the base of the neck (Fig. 5.6D). Accessory bone struts on the lateral surface of the neural spines constitute autapomorphic features of this Patagonian rebbachisaurid. The proximal caudal neural spines are distally thickened by the development of "lateral" laminae, terminating in robust bone. Another autapomorphic feature of *Limaysaurus* is the transverse processes of caudal 3 (Fig. 5.7A and B), which are distally split into a dorsal component ("diapophysis") and a ventral component ("parapophysis"), resembling the condition present in the Jurassic *Apatosaurus*. This condition differs from that of *Rebbachisaurus garasbae*, in which the transverse processes are typically wing-like, as in Diplodocidae (Calvo and Salgado 1995).

The scapula of *Limaysaurus* is 87 centimeters long and is characterized by its distal paddle-shaped expansion, as also occurs in *Rebbachisaurus garasbae* (Fig. 5.8C and D). *Limaysaurus* differs from *Rayososaurus* because the scapula lacks the autapomorphy of the latter genus (i.e., an elongate and posteriorly directed acromial process).

Fragmentary vertebrae recovered from beds of the Alcântara Formation (São Luis Basin, Maranhão State, northeastern Brazil; Fig. 4.5) show similarities with those of *Limaysaurus tessonei* (Medeiros and

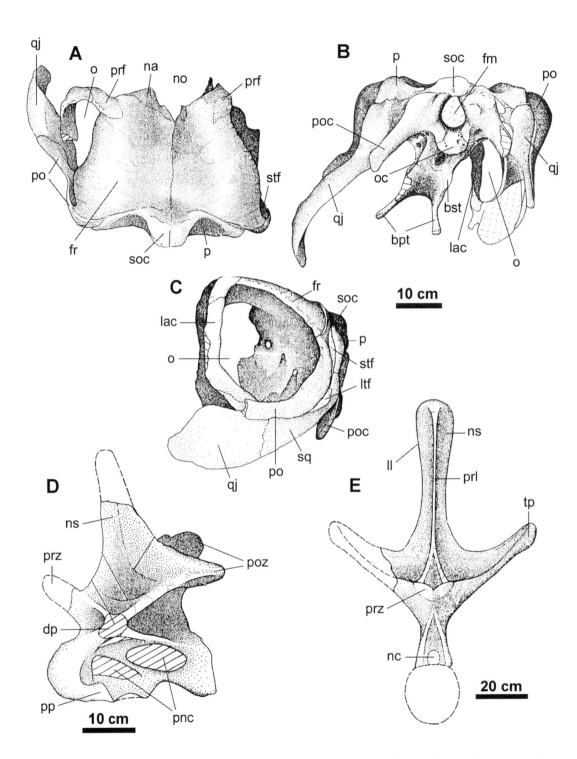

Schultz 2004). The material consists of isolated caudal centra with a subtriangular-shaped articular surface and a dorsally oriented apex and a straight ventral margin. The ventral surface of the centra is almost flat and longitudinally grooved, features that are also present in the Patagonian *Limaysaurus tessonei* (Medeiros and Schultz 2004). An isolated caudal vertebra reported from Lower Member of the Bajo Barreal Formation also has similarities to the proximal caudals of rebbachisaurid sauropods (Sciutto and Martínez 1994; Fig. 5.7E and F).

RAYOSOSAURUS AGRIOENSIS

This rebbachisaurid is known from an isolated scapula that was discovered in beds of the Rayoso Formation (Aptian-Albian) in the Province of Neuquén (Bonaparte 1996b, 1997b). It is slightly older and more primitive than *Limaysaurus tessonei*. The scapula shows the typically rebbachisaurid paddle-shaped distal extremity (Fig. 5.8A). The anteroposteriorly elongate and posteriorly directed acromial process of the scapula of *Rayososaurus* is unique (Bonaparte 1996b, 1997b), in contrast to that of other rebbachisaurids (e.g., *Rebbachisaurus garasbae*, *Limaysaurus tessonei*), in which the acromial process is relatively short and dorsally or posterodorsally oriented.

CATHARTESAURA ANAEROBICA

This new rebbachisaurid, which is similar in size to *Limaysaurus* (approximately 17 meters long), comes from La Buitrera fossil site (Río Negro Province) from levels of the Turonian–Coniacian Huincul Formation (Gallina and Apesteguía 2005). *Cathartesaura* is known from an incomplete skeleton that includes the cervical and some caudal vertebrae, a scapula, the ilium, and a femur. The autapomorphies that diagnose *Cathartesaura* include an accessory lamina on the posterior cervical vertebrae that arises from the mid-length of the prezygodiapophysial lamina and reaches the centrum and thin, wing-like transverse processes on proximal caudal vertebrae, mostly supported by a ventral bony bar that frames a deep triangular fossa. The scapula of *Cathartesaura* is racket-shaped, as is characteristic among rebbachisaurids. However, the morphology of the scapula and femur of *Cathartesaura* are different from those of *Rayososaurus* and *Limaysaurus*, indicating that they are distinct taxa.

AMAZONSAURUS MARANHENSIS

The recovered bones of this sauropod include some dorsal, sacral, and caudal vertebrae, haemal arches and ribs, an ilium, and the pubis. The specimen, which Ismar de Souza Carvalho and his colleagues described in 2003, comes from the Aptian–Albian Itapecurú Formation in the Parnaíba Basin in Itapecurú-Mirim, a fossil location in the state of Maranhão in northeastern Brazil (Fig. 4.5). *Amazonsaurus* (Fig. 5.7C and D) is diagnosed by caudal neural spines that are straight and caudally inclined with lateral laminae that, at least in the most cranial caudal vertebrae, bend cranially in such a way that the cranial surface of the lamina is concave and the caudal one convex. The dorsal vertebral centra are weakly opisthocoelous but are characterized by being very wide and low with a flattened ventral surface. The pleurocoels are deep and well developed, extending from the cranial to the caudal borders of the

Fig. 5.6. *Limaysaurus tessonei:* (A–C) Skull in (A) dorsal, (B) caudal, and (C) lateral views. (D) Cervical vertebra in lateral aspect. (E) Dorsal vertebra in caudal view. *From Calvo and Salgado (1995). Reprinted with permission of* Gaia.

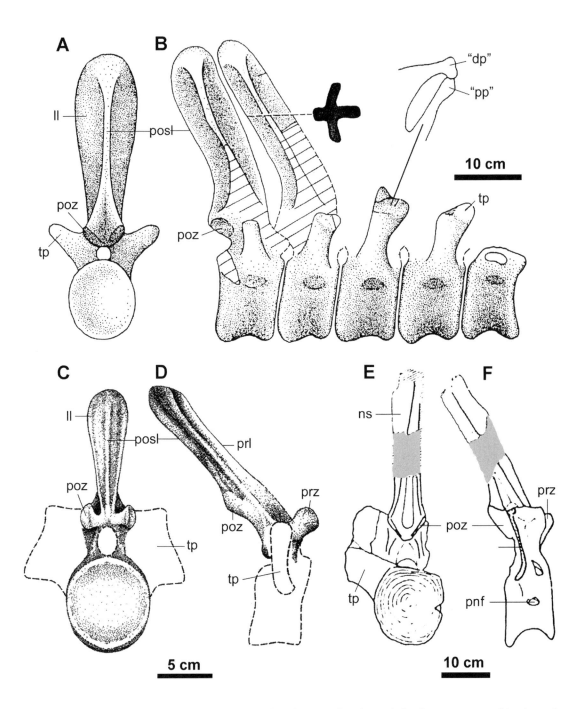

Fig. 5.7. Proximal caudal vertebrae of rebbachisaurid sauropods from South America. (A, B) *Limaysaurus tessonei*: (A) Caudal 5 in caudal view. (B) Caudals 1–5 in right lateral view with a line drawing showing the peculiar transverse

centra. The neural arches are fragile and the laminae resemble that of other diplodocimorphs. The neural arches of the caudal vertebrae are slender and have very high neural spines that are set at an angle of 45 degrees with respect to the axis of the neural canal. The neural spines have a well-developed pair of "lateral" laminae as well as strongly developed pre- and postspinal laminae. The preserved caudal centra are amphiplatyan. The proximally open haemal arches constitute a plesiomorphy the Brazilian form retained, suggesting that its position is basal among diplodocimorphs. Although rebbachisaurid relationships cannot be dismissed, *Amazonsaurus* exhibits some differences from the Patagonian rebbachisaur *Limaysaurus tessonei*, such as arching of the caudal neural spines (and caudally convex and cranially concave "lateral" laminae).

The diplodocimorphs are the primary dinosaurs of the Early Cretaceous record of sauropods in Gondwana. However, in Mid- to Upper Cretaceous beds, the most diverse and abundant sauropods in South America are the titanosaurs. Titanosauria is a prominent dinosaur clade. They were the most diverse and geographically widespread sauropod group and are currently represented by more than 40 genera. The titanosaurs attained worldwide distribution, and, except for Antarctica, their remains have been found in Cretaceous beds of all the continental landmasses. Titanosauria is the only sauropod lineage to successfully survive into the end of the Late Cretaceous and is numerically abundant and taxonomically diverse. The diversity of titanosaurs in South America has been interpreted by Bonaparte (1996b) and Salgado (2003a) as probably linked with the floral variety of Late Cretaceous angiosperms, which offered new ecological opportunities to the herbivores.

The skeletal record of titanosaurs is restricted to Cretaceous beds, but some authors believe that they were already present in the Middle Jurassic on the basis of abundant records of their wide-gauge trackways, a peculiar pattern of footprints produced by limbs that were separated from the body's axial plane during locomotion. In other words, the current fossil record, based on bone remains, supports titanosaurians as a sauropod group that is typically Cretaceous. In South America, the oldest titanosaurian record consists of isolated procoelous caudal vertebrae from the Aptian Rayoso Formation of Neuquén Province (Bonaparte 1996b).

The titanosaurs are stockier than other sauropods (Fig. 5.9). Their neck vertebrae are transversely wide and low and the pectoral and pelvic regions are particularly wide, resulting in a widely spaced posture of the left and right fore- and hindlimbs. Titanosaurian limb bones are robust, and a process of the reduction of ossified carpals and phalanges apparently characterizes the entire group. The tail in derived titanosaurs is composed by caudal vertebrae that are typically procoelous. The cranial and dental morphology is not uniform among titanosaurs: basal forms inherited a *Brachiosaurus*-like skull that has several large and spatulate teeth along most of the buccal margin, but derived titanosaurs developed a *Diplodocus*-like kind of skull that has external nares that are posteriorly placed and slender and cylindrical teeth that are almost completely restricted to the rostral end of the snout. Another unusual feature of advanced titanosaurs was the development of dermal armor composed of small ossicles and large scutes embedded in the skin (Powell 2003; Salgado 2003b). This morphological diversity, which is expressed in variations in body size, head morphology, tooth shape, and vertebral anatomy,

Titanosauria

process of caudal 3 in distal view. (C, D) *Amazonsaurus maranhensis*: Proximal caudal vertebra in (C) caudal and (D) right lateral views. (E, F) Isolated caudal vertebra of a possible rebbachisaurid from the Lower Bajo Barreal Formation in (E) caudal and (F) lateral views. *(A, B) From Calvo and Salgado (1995). Reprinted with permission of* Gaia. *(C, D) Reprinted from* Cretaceous Research *24(6): 697–713. Amazonsaurus maranhensis gen. et sp. Nov. (Sauropoda, Diplodocoidea) from the Lower Cretaceous (Aptian–Albian) of Brazil. Authors: Carvalho, Avilla and Salgado. Page 703 (fig. 8). © Copyright 2003. With permission from Elsevier. (E, F) From Sciutto and Martínez (1994). Reprinted with permission of* Naturalia Patagónica.

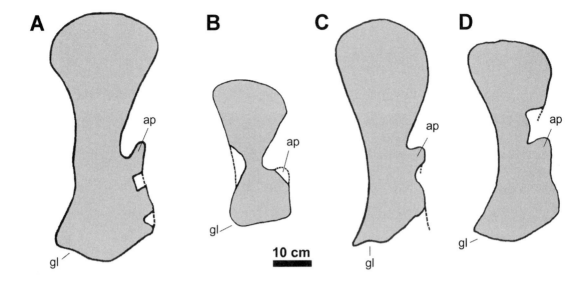

Fig. 5.8. Right rebbachisaurid scapulae compared in lateral view: (A) *Rayososaurus agrioensis*. (B) *Nigersaurus taqueti*. (C) *Rebbachisaurus garasbae*. (D) *Limaysaurus tessonei*. Redrawn from Salgado, Garrido, Cocca, and Cocca (2004).

strongly suggests that Titanosauria were herbivores whose ecological preferences and behaviors differed sharply from each other.

Although the first titanosaur remains were found at the end of the nineteenth century, the origins and interrelationships of titanosaurs remained obscure for many years. One of the reasons was (and still is) the incomplete nature of several holotype specimens. The taxa were founded on the basis of isolated bones that did not always overlap among specimens in a way that would facilitate comparisons. Fortunately, discoveries made and studies carried out in Argentina since 1980 have changed this situation, and now the titanosaur record includes complete skeletons, even embryonic remains. This amount of anatomical evidence sheds light on the phylogenetic relationships of titanosaurians both inside and outside South America.

Titanosaurs, which were particularly successful in Gondwana, figure prominently in discussions of Cretaceous paleobiogeography. The abundance and diversity of titanosaurs in South America has been traditionally understood (e.g., Bonaparte 1986d) as a consequence of the isolation of Gondwana from Laurasia through most of the Cretaceous. Researchers theorize that we find Late Cretaceous titanosaurids in North America and Asia because they emigrated from Gondwanaland. However, the documentation of titanosauriforms and titanosaurians in Early Cretaceous beds in Europe and North America weakens this paleobiogeographical hypothesis, and now most paleontologists agree that titanosaurians attained global distribution after the Early Cretaceous.

To date, 40 titanosaurian genera have been described. Representatives of Titanosauria in the northern hemisphere include *Lirainosaurus astibiae* and *Ampelosaurus atacis* from Europe, *Pleurocoelus nanus* and *Alamosaurus sanjuanensis* from North America, and several representatives from the Upper Cretaceous of Asia, including *Opisthocoelicaudia skarzynskii*, *Nemegtosaurus mongoliensis*, and *Quaesitosaurus orientalis*

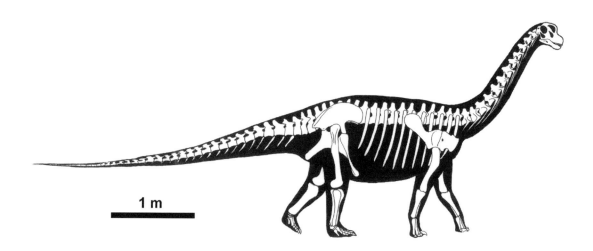

Fig. 5.9. Skeleton of *Neuquensaurus australis*. Redrawn from von Huene (1929a).

from Mongolia and *Phuwiangosaurus sirindhornae* from the Early Cretaceous of Thailand. From Madagascar came the almost-complete skeleton of *Rapetosaurus krausei*. Titanosaurian remains from Africa include the Aptian *Malawisaurus dixeyi* and the Cenomanian *Aegyptosaurus baharijensis* and *Paralititan stromeri*. From India, the first continent in which titanosaurian bones were discovered, the situation is far from optimal, and only one species, *Isisaurus* (=*Titanosaurus*) *colberti*, has been recognized on the basis of diagnosable materials (Jain and Bandyopadhyay 1997; Wilson and Upchurch 2003). Some isolated vertebrae of indeterminate basal titanosaurians are known from the Aptian of Australia (Molnar 2001). Curiously, remains of this prolific group of sauropods have not been reported yet from Antarctica, a situation that may indicate a bias in the fossil record or may reflect some kind of paleobiogeographic barrier for the dispersion of titanosaurians.

In South America (Figs. 4.5 and 5.1), fragmentary titanosaur remains were discovered in Bolivia, Perú, Uruguay, and Chile. More-complete and well-preserved titanosaurs (e.g., *Trigonosaurus*, *Baurutitan*, *Maxakalisaurus*, *Gondwanatitan*, and *Aelosaurus*) were recorded in Brazil, where bones of these animals are commonly found in the Late Cretaceous formations of the Bauru Group (Kellner and Campos 1997; Carvalho, Avilla, and Salgado 2003), particularly from the localities of Peirópolis in the State of Minas Gerais and Presidente Prudente in the State of São Paulo. The most productive location is Peirópolis, where seven sites have yielded most of the Brazilian titanosaurid remains. The specimens have been found in conglomeratic sandstone that in some places form a bone bed. Titanosaur remains have been also reported from the Confusão Creek in Brazil's State of Mato Grosso from levels that are referred to the Bauru Group (Kellner, Campos, de Azevedo, Silva, and Carvalho 1995) and in Mid Cretaceous beds of the São Luis Basin in northeastern Brazil (Medeiros and Schultz 2001b, 2002).

Argentina has yielded the most complete record of titanosaurs, most of which were discovered in Patagonia (Fig. 5.4). Titanosaurs are the

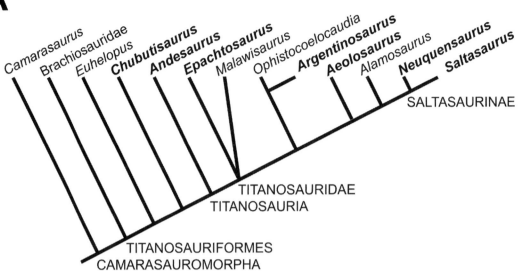
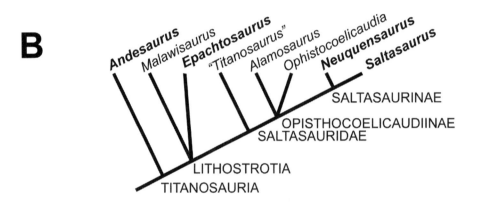
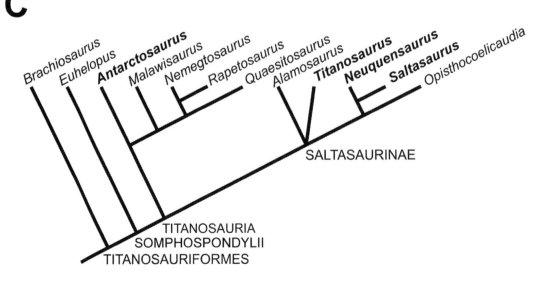

best-represented Cretaceous dinosaurs in Argentina, both in the number of species and their relative abundance. They provide remarkable anatomical details about osteology, skin impressions, embryos, eggs, nests, and footprints. The Argentine fossil record of titanosaurs constitutes a superb biostratigraphic sequence, including taxa from Aptian through Maastrichtian times, thus recording approximately 50 million years of the evolutionary history of this sauropod clade.

Phylogenetic Relationships of Titanosauria

After Friedrich von Huene's monograph on Cretaceous dinosaurs from Argentina was published (von Huene 1929a), most authors grouped titanosaurs with diplodocid sauropods, primarily on the basis of narrow tooth crowns and the *Diplodocus*-like aspect of the reconstructed skull of *Antarctosaurus wichmannianus*. But now sauropod specialists are convinced that titanosaurs were more closely related to Jurassic brachiosaurids than to the diplodocids. In addition, important progress has been made in recent years regarding the phylogenetic relationships among known species of titanosaurs themselves. Leonardo Salgado and colleagues were the first to argue for the existence of a clade named Titanosauriformes that includes *Brachiosaurus*, *Chubutisaurus* and Titanosauria and all of their descendants (Salgado, Coria, and Calvo 1997). These authors also recognized *Camarasaurus* as a more remote outgroup of Titanosauriformes and coined the name Camarasauromorpha to include these two taxa. More recently, Wilson and Sereno (1998) recognized *Euhelopus* as a taxon that is more closely related to Titanosauriformes than to Camarasauridae and gave the name Somphospondylii to the clade that includes Titanosauria and *Euhelopus* (Fig. 5.10).

Titanosauriforms are distinguished from other sauropods by morphological traits in the vertebral column, such as the development in the posterior trunk vertebrae of a medial prespinal lamina (Fig. 5.11A) to which the interspinal ligaments attached. The titanosauriform manus is modified from the ancestral sauropod pattern by a progressive reduction of the ungual phalanx of digit I (Fig. 5.11B). The pelvic girdle manifests transformations presumably related to the ability to acquire a bipedal pose, including a broadly expanded and upwardly directed preacetabular lobe of the ilium and an anterodorsally inclined sacrum that forms a right angle with the long axis of the pubic pedicle of the ilium (Fig. 5.11C). The femur is distinguished by a prominent lateral bulge below the greater trochanter, a feature interpreted as indicative of a widely spaced posture of the hindlimb (Fig. 5.11D and E). Finally, in the tail vertebrae, the neural arches are positioned anteriorly in the mid- and distal caudal vertebrae (Fig. 5.12).

Titanosauriforms more derived than *Brachiosaurus* and *Euhelopus* evolved a distal end of the tibia that is broader transversely than anteroposteriorly, thus reverting to the primitive saurischian condition. Examples

Fig. 5.10. Cladograms depicting alternate hypotheses about the phylogenetic relationships of titanosaurian sauropods. (A) Hypothesis defended by Salgado, Coria, and Calvo (1997). (B) Hypothesis defended by Upchurch, Barrett, and Dodson (2004). (C) Hypothesis defended by Curry Rogers and Forster (2004).

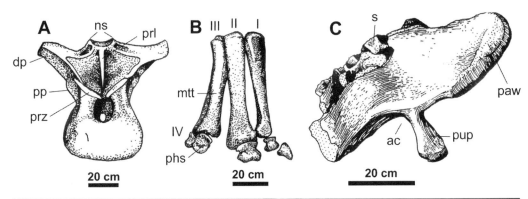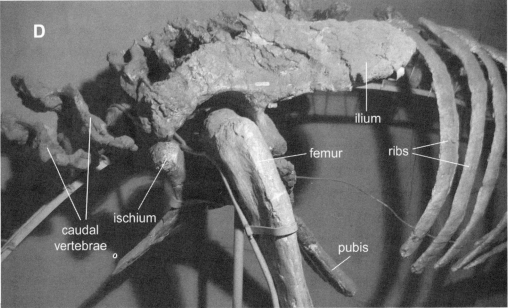

186   The Age of Dinosaurs in South America

of basal titanosauriforms from South America are the Early Cretaceous *Chubutisaurus insignis* and *Ligabuesaurus leanzai*.

Titanosauria is the clade that includes Titanosauridae and the most recent common ancestor of *Andesaurus delgadoi* and all of its descendants (Salgado, Coria, and Calvo 1997). Most of the derived characteristics that diagnose Titanosauria (Fig. 5.13) refer to the vertebral column, including proportionally small, eye-shaped, and caudally acuminate pneumatic cavities on the sides of trunk vertebrae and a two or more deep depressions with angled margins on the ventral portion of the neural arches, especially on the posterior trunk vertebrae. These depressions are delimited by new struts of bone, including a centro-parapophyseal lamina and a crest that unites the parapophysis with the widened base of the infradiapophyseal lamina. Also, the infradiapophyseal lamina is ventrally widened and slightly forked (Fig. 5.13A). One remarkable feature of titanosaurids is a pubis that is considerably longer than the ischium.

Titanosaurs that are more derived than *Andesaurus* are gathered within the Titanosauridae (Fig. 5.10). They are characterized by procoelous caudal vertebrae with a ball-and-socket articulation consisting of prominent and strongly convex posterior articular surfaces that articulate with well-excavated anterior articular cups (Fig. 5.12). The procoelous condition is one of the most remarkable traits of titanosaurids, with the exception of *Opisthocoelicaudia skarzynskii*, whose caudal vertebrae exhibit the opposite opisthocoelous condition. This condition is interpreted as an autapomorphic trait of this Mongolian taxon. Distinctive titanosaurids include *Epachthosaurus*, *Malawisaurus*, *Argentinosaurus*, *Opisthocoelicaudia*, *Aeolosaurus*, *Alamosaurus*, *Rapetosaurus*, *Saltasaurus*, and *Neuquensaurus*.

Researchers agree about recognizing a derived clade of titanosaurids, the Saltasaurinae, which are represented by *Saltasaurus*, *Neuquensaurus*, *Rocasaurus*, and *Bonatitan*. They also agree that the remaining titanosaurs form a series of sister taxa, although the arrangement of these taxa is far from settled (Fig. 5.10). However, increasing evidence supports the recognition of another clade of derived titanosaurids, the Aeolosaurinae, to gather *Aeolosaurus*, *Gondwanatitan*, and *Rinconsaurus* (Costa Franco-Rosas, Salgado, Carvalho, and Rosas 2004; Casal, Martinez, Luna, Sciutto, and Lamanna 2007). Aeolosaurines, saltasaurines, *Alamosaurus*, *Baurutitan*, and probably also *Antarctosaurus*, appear to be more closely related among titanosaurids because they share a biconvex first caudal that contrasts with the procoelous condition of the first caudal in more-primitive titanosaurs. The presence of a dorsal prominence on the inner face of the scapula may also be a derived characteristic diagnostic of a clade composed of Aeolosaurinae, *Alamosaurus*, and Saltasaurinae.

Although no consensus exists about the phylogenetic arrangement of more-primitive titanosaurids, the phylogenetic position of *Epachthosaurus* at least is generally accepted as basal within Titanosauridae. Some authors think that *Argentinosaurus*, *Opisthocoelicaudia*, and *Alamosaurus* share a common ancestor with saltasaurines and aeolosaurines because

Fig. 5.11. Selected bones that indicate synapomorphic features of Titanosauriformes. (A) Posterior trunk vertebra in anterior view indicating the development of a medial prespinal lamina. (B) Manus of *Brachiosaurus* showing the reduction in the size and number of phalanges. (C) Right lateral view of the ilium and sacrum of *Saltasaurus*. (D, E) Hindquarters of *Saltasaurus* in lateral (D) and caudodorsal (E) views, showing the broadly expanded preacetabular lobe of the ilium and the craniodorsally inclined sacrum. *(A, B) Redrawn from Salgado (2003a). (C) Redrawn from Powell (2003). (D, E) Photograph by Fernando Novas.*

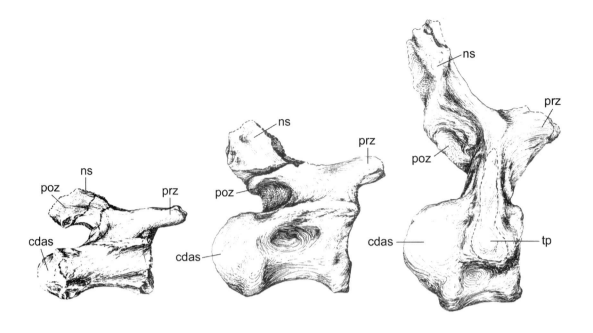

Fig. 5.12. *Neuquensaurus australis:* Caudal vertebrae in right lateral view, showing a characteristically procoelous ball-and-socket articulation. Proximal caudal on the right, mid-caudal on the center, and distal caudal on the left. Not to scale. *From von Huene (1929a).*

they share quadrangular coracoids and prespinal lamina that extend down to the base of neural spine in the posterior trunk vertebrae. Jorge Calvo and colleagues (2007) have recently communicated an interesting novelty regarding the diversity of basal titanosaurids and their systematic arrangement. They have gathered *Mendozasaurus* and *Futalognkosaurus* within Lognkosauria, a clade of basal titanosaurids that is less derived than *Epachthosaurus*.

Aside from the issues of titanosaurian phylogeny that have not been settled yet, serious problems of nomenclature affect the validity of the family name Titanosauridae. As Wilson and Upchurch (2003) noted, the name *Titanosaurus* was erected by Richard Lydekker in 1877 for some isolated caudal procoelous vertebrae, a condition that at that time was known to occur only in this sauropod. Over time, the distribution of procoelous caudal vertebrae has broadened to encompass most titanosaurs. Consequently, this characteristic must be abandoned as diagnostic of *Titanosaurus*. Of the 13 species that have been referred to *Titanosaurus*, most of them, including the type species *T. indicus*, are non-diagnostic, rendering the genus invalid. Only two of the species of *Titanosaurus* are recognized as diagnostic, *T. colberti* and *T. araukanicus*, which consequently had to be renamed: *Titanosaurus colberti* was renamed as *Isisaurus colberti*, and *Titanosaurus araukanicus* has returned to its original denomination, *Laplatasaurus araukanicus*. Because *Titanosaurus* is not a valid taxon, the International Code of Zoological Nomenclature recommends that familiar names deriving from it (e.g., Titanosauridae, Titanosauroidea) should also be considered invalid. Although the name Titanosauridae is deeply entrenched in paleontological literature, Wilson and Upchurch (2003) strongly suggested that it be abandoned. However, because the term Titanosauridae has a long historical use and because it

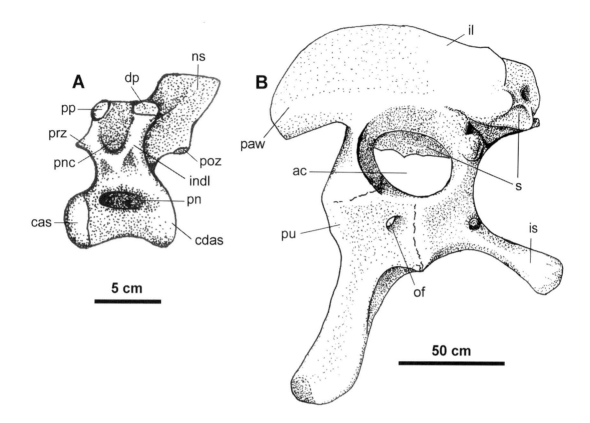

Fig. 5.13. Selected bones that indicate synapomorphic features of Titanosauria: (A) Posterior dorsal vertebra (in left lateral view) of an indeterminate titanosaurid from Brazil showing eye-shaped and caudally acuminate pleurocoels and modifications of depressions and bony struts on the ventral side of transverse processes. (B) Pelvic girdle of *Opisthocoelicaudia* in left lateral view. (A) Redrawn from Salgado (2003a). (B) Redrawn from Borsuk-Bialynicka (1977).

was applied to a clade that was defined on the basis of cladistic phylogenetics (Salgado, Coria, and Calvo 1997), I will keep this familiar name but will follow the interpretations proposed by Salgado (2003c) to stabilize this nomenclatural problem.

Recently, the name "Lithostrotia" was coined by Paul Upchurch and colleagues (2004) to refer to the clade containing derived titanosaurs that bear dermal armor (Fig.5.10B). However, the presence of scutes in the basal titanosaurs *Malawisaurus* and *Mendozasaurus* (considered to be the sister groups of the clade named Lithostrotia) weaken support for the latter name. Furthermore, a dermal skeleton is lacking in the largely complete and articulated skeleton of *Epachthosaurus sciuttoi*, which is interpreted as one of the most basal members of the clade named either Titanosauridae or Lithostrotia.

The Main Anatomical Features of Titanosauria

SKULL MORPHOLOGY

Aside from the partially preserved skull of *Antarctosaurus wichmannianus* and the isolated maxilla of "*Campylodoniscus ameghinoi*" (von Huene 1929a; Bonaparte and Gasparini 1979; Powell 2003), little was known about the cranial anatomy of titanosaurs until recently. From the time of the original description of *Antarctosaurus wichmannianus*,

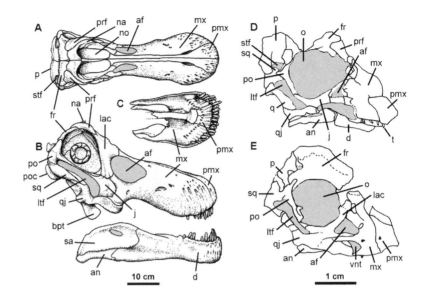

Fig. 5.14. Titanosaurian skull and jaws: (A–C) *Nemegtosaurus mongoliensis* in (A) dorsal and (B) lateral views. (C) Detail of snout with teeth in ventral view. (D, E) Embryonic titanosaurids discovered in Auca Mahuevo in right lateral view. *(A-C) Redrawn from Nowinski (1971). (D, E) Redrawn from Chiappe, Salgado, and Coria (2001).*

the titanosaurian skull and teeth were taken to be diplodocid-like, and on this evidence several authors considered Titanosauria to be the sister group of Diplodocidae. More recently, the skull became partially known in a variety of titanosaurian genera, including *Saltasaurus* (Powell 2003), *Rapetosaurus* (Curry Rogers and Forster 2004), *Bonitasaura* (Apesteguía 2004), *Quaesitosaurus* and *Nemegtosaurus* (Salgado and Calvo 1997; Nowinski 1971; Wilson 2005; Fig. 5.14A–C), a well-preserved skull of a primitive titanosauriform from the Bajo Barreal Formation (Martínez 1998a), and an undescribed titanosaurid skull from Rincón de los Sauces (Coria and Salgado 1999). Exquisitely preserved skulls of titanosaurian dinosaurs are also known in embryonic remains from the Late Cretaceous nesting site at Auca Mahuevo (Neuquén Province) in beds of the Anacleto Formation (Chiappe, Salgado, and Coria 2001; Fig. 5.14D and E). This information on titanosaurian cranial anatomy is supplemented by isolated premaxillae discovered at different fossil localities in Argentina (Powell 1979, 2003; Bonaparte 1996b; Coria and Chiappe 2001). Available information indicates that the titanosaurian skull was *Brachiosaurus*-like, albeit more elongate and depressed. Distinctions among titanosaurs appear to correspond to tooth shape: some forms retained coarse leaf-shaped teeth (e.g., the Bajo Barreal basal titanosaur "*Campylodoniscus*") while others (such as *Rapetosaurus, Antarctosaurus, Saltasaurus, Nemegtosaurus*) developed slender and pencil-like teeth resembling those of *Diplodocus* (Fig. 5.15).

The embryonic skull from the nesting site of Auca Mahuevo shows a low rostral portion of the mandible, a derived feature known for the titanosaurs *Rapetosaurus, Antarctosaurus*, and an undescribed adult titanosaurid from Rincón de los Sauces (Río Colorado Formation). The extreme width of the skull and a large mandibular fenestra are derived characteristics that were also reported for the Rincón de los Sauces

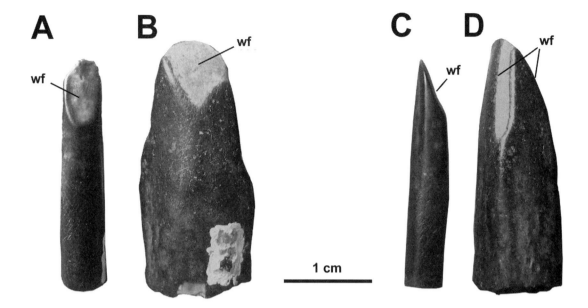

Fig. 5.15. Titanosaurian tooth. (A, C) Teeth found in the Portezuelo Formation (Turonian, Neuquén Province) that exemplifies the slender, pencil-like type of tooth that is similar to those of *Diplodocus*. (B, D) Tooth found in the Mata Amarilla Formation (Turonian, Santa Cruz Province) that exemplifies the coarse, leaf-shaped kind of tooth that is similar to those of *Brachiosaurus*. *Photographs by Fernando Novas.*

titanosaur. A distinct ventral notch caudal to the maxillary dentigerous margin (Fig. 5.14D and E) is shared with *Rapetosaurus*. Anatomical information available in titanosaurid embryos sheds light on sauropod skull evolution, such as a retraction of the external nares, forward rotation of the braincase, and the shortening of the infraorbital region. Previous interpretations of the evolution of the sauropod skull (Upchurch 1999) proposed that narial retraction was coupled with rostral rotation of both the quadrate and the braincase. But the new embryonic evidence shows that narial retraction and braincase rotation likely evolved independently (Chiappe, Salgado, and Coria 2001; Salgado, Coria, and Chiappe 2001). It is possible that the position of the external nares migrated backward during ontogeny, accompanied by lengthening of the snout, a change in the shape of the orbit from circular to elliptical, and elongation of the jugal (Chiappe, Salgado, and Coria 2001; Salgado, Coria, and Chiappe 2001).

VERTEBRAL COLUMN AND THORAX

Some authors (Powell 2003) estimate that titanosaurids had 13 cervical, 12 dorsal, 6 sacral, and more than 30 caudal vertebrae (at least 11 dorsals have been reported for *Opisthocoelicaudia* and *Rapetosaurus*; Borsuk-Bialynicka 1977; Curry Rogers and Forster 2001). The cervical vertebrae of titanosaurs (Fig. 5.16) are characterized by a dorsoventrally depressed centrum and an unforked neural spine; some forms have an unusually inflated distal tip of the spine. Interestingly, Rubén D. Martínez (1999) reported three rods of ossified tendons along both sides of the neck of a titanosaurid specimen. Two of these rods are 1 centimeter in diameter, but the third is only 3 millimeters wide. The tendons along the neck probably helped it support its head and control movement.

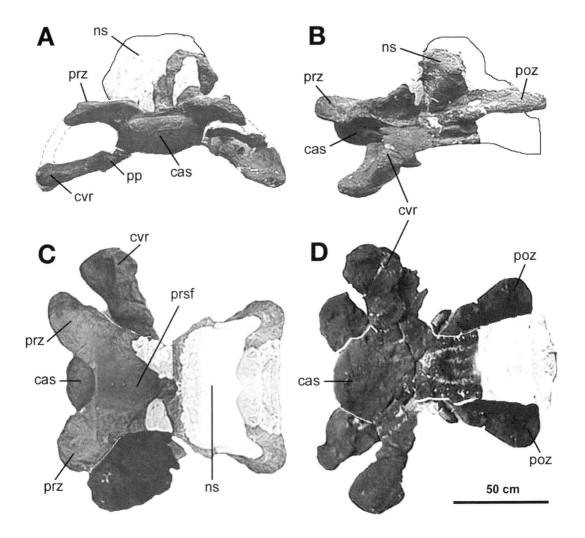

Fig. 5.16. Cervical 9? of the giant titanosaurid *Puertasaurus reuili* in (A) cranial, (B) left lateral, (C) dorsal, and (D) ventral views. *From Novas, Salgado, Calvo, and Agnolín (2005). Reprinted with permission of* Revista del Museo Argentino de Ciencias Naturales.

Most titanosaurians exhibit low and elongate cervicals that have axially elongate neural spines. But thanks to new discoveries, it has become evident that the morphology of neck vertebrae within Titanosauria was much more diverse than expected. Some forms have dorsally inflated neural spines (e.g., *Puertasaurus*), while others have craniocaudally short centra that are associated with fan-shaped (i.e., transversely expanded) neural spines (e.g., *Mendozasaurus*). The documented diversity in the shapes and proportions of titanosaurian cervical vertebrae has been preliminarily explored (e.g., González Riga 2005; Novas, Salgado, Calvo, and Agnolín 2005), but such studies reveal that titanosaurian sauropods evolved long-necked as well as short-necked forms, probably reflecting contrasting feeding strategies.

The trunk vertebrae of most titanosaurs are characterized by dorsoventrally depressed and anteroposteriorly elongate dorsals that have well-developed convex cranial articulations (opisthocoely, Fig. 5.13A). In addition, most dorsal neural spines are craniocaudally short, transversely narrow at their tips, and steeply inclined toward the rear, such that the

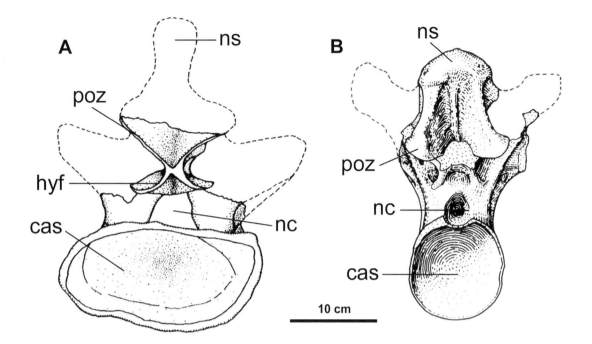

Fig. 5.17. Examples of dorsal vertebrae in Titanosauria. (A) Mid-dorsal of the basal titanosaurid *Epachthosaurus sciuttoi* in caudal view. (B) Posterior dorsal vertebra of the derived titanosaurid *Saltasaurus loricatus* in caudal view. Restored parts of the vertebrae are indicated in dotted lines. (A) Redrawn from Martínez, Giménez, Rodriguez, Luna, and Lamanna 2004. (B) Redrawn from Powell (2003).

tip of the spine clearly extends beyond the caudal edge of the centrum (Fig. 5.13A). The lateral surface of the neural spine has sharp laminae that extend from both sides of the prespinal lamina toward the transverse processes. This system of laminae delimits depressions on the neural spines. The lateral surface of the transverse processes is anteroposteriorly wide and has a well-developed U-shaped lamina below both the diapophysis and parapophysis that delimits a deep pneumatic excavation (Figs. 5.13A). The transverse processes are anterolaterally directed in the posterior dorsals (especially in D10), thus giving the impression that they "embrace" the preceding vertebra (D9) in dorsal view.

Basal titanosaurs inherited the ancestral saurischian condition of the articulation of the hyposphene and the hypantrum in trunk vertebrae. In some basal titanosaurs (e.g., *Epachthosaurus*, *Argentinosaurus*), these accessory intervertebral articulations are remarkably well developed (Fig. 5.17A). However, the hyposphene-hypantrum articulations apomorphically disappeared in more-derived titanosaurs (e.g., *Argyrosaurus*, *Aeolosaurus*, *Pellegrinisaurus*, saltasaurines; Fig. 5.17B). Some authors believe that the loss of hyposphene-hypantrum articulations combined with the pronounced opisthocoely of the dorsal centra suggest an increased range of trunk motion compared with the range of motion of more primitive members of the clade (Wilson and Carrano 1999).

Titanosaurs increased the number of sacral vertebrae to six by incorporating a new vertebral segment. Interestingly, an ossified ligament or tendon dorsally uniting the neural spines of the sacral vertebrae (even reaching the last dorsals; Fig. 5.18) has been reported for some titanosaurids (e.g., *Epachthosaurus* and an unnamed titanosaurian from the Late Cretaceous of Brazil, currently known as "Series C"; Powell 1987a;

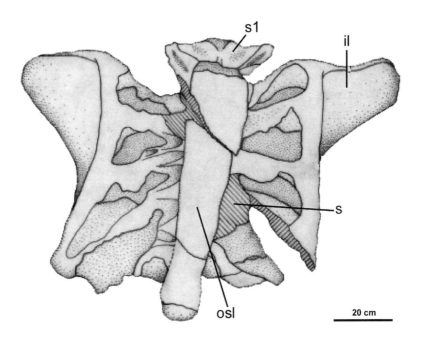

Fig. 5.18. Sacrum of *Epachthosaurus sciuttoi* in dorsal aspect, indicating ossified ligament or tendon above neural spines (osl). *From Martínez, Giménez, Rodriguez, Luna, and Lamanna (2004).* © Copyright 2004 the Society of Vertebrate Paleontology. Reprinted and distributed with permission of the Society of Vertebrate Paleontology.

Campos and Kellner 1999; Martínez, Giménez, Rodríguez, Luna, and Lamanna 2004). This wide band of connective tissue may have increased the strength of the pelvic girdle as a whole.

The characteristically procoelous caudal vertebrae of titanosaurs, which have ball-and-socket articulations along the entire length of the tail, evolved from a less-derived version present in basal titanosaurs, in which slightly procoelous centra were limited to the proximal portion of the tail. Apparently, procoely advanced progressively from the proximal to the distal segments of the tail (Bonaparte 1996b).

GIRDLES AND LIMBS

The anatomy of the forelimbs and hindlimb as well as characteristics of both the pectoral and pelvic girdles indicate that titanosaurs had a wide separation between the left and right limbs. Also, the shape of the pelvis suggest that they may have been able to acquire a bipedal posture. The thorax of titanosaurs, for example *Saltasaurus*, *Epachthosaurus*, and *Opisthocoelicaudia*, was very broad and almost cylindrical at the back, thus contrasting with the narrower and dorsoventrally deep thorax present in most sauropods (Powell 2003).

Derived features of the titanosaurid forequarters (Giménez 1992; Fig. 5.19) include extremely wide sternal plates, a strongly developed deltopectoral crest of the humerus, a proportionally short and wide radius, a distally symmetrical metacarpal I, metacarpals with an almost flat distal surface, and the loss of manual phalanges. The large crescent-shaped sternal plates of titanosaurids, as documented in *Epachthosaurus*, *Opisthocoelicaudia*, *Neuquensaurus* and *Saltasaurus*, suggest a broader shoulder carriage and an expanded surface for attaching the neck and pectoral

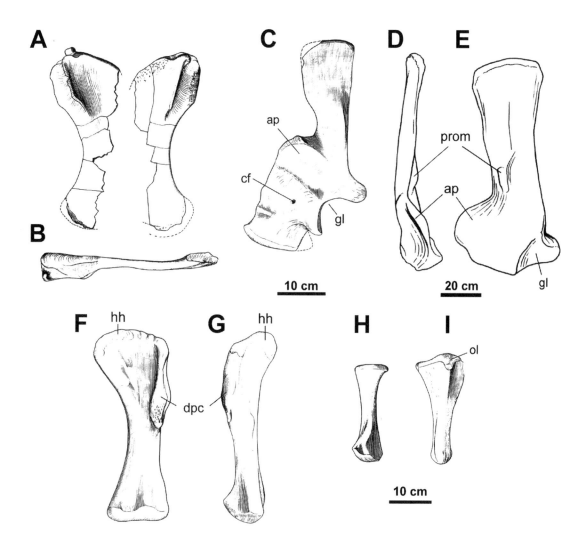

Fig. 5.19. Scapular girdle and forelimb bones of titanosaurids. (A–C, F–I) *Neuquensaurus australis:* (A) Paired sternal plates in ventral view. (B) Left sternal plate in lateral view. (C) Left scapulocoracoid in lateral view. (F, G) Right humerus (reversed) in (F) cranial and (G) lateral views. (H) Right radius in medial view. (I) Right ulna in lateral view. (D, E) *Saltasaurus loricatus*: Left scapula in (D) cranial and (E) lateral views. (A–C, F–I) From von Huene (1929a). (D, E) Redrawn from Powell (2003).

musculature (Wilson and Carrano 1999). The presence of large sternal plates is consistent with a broad-chested animal whose scapular glenoid was widely separated from the midline. This peculiar forelimb carriage is congruent with the wide separation of left and right hindlimbs suggested by the shape of the femur, which angled outward from the pelvis to articulate with an almost-vertical tibia. Such a repertoire of locomotor adaptations suggest that titanosaurids were wide-gauge trackmakers (Wilson and Carrano 1999).

It has already been mentioned that some vertebral features (e.g., lack of the hyposphene-hypantrum, opisthocoely) may indicate an increased range of motion. This may also be related to pelvic features that suggest that titanosaurs were facultative bipeds. Titanosaurid ilia are characterized by an enlarged preacetabular lobe that is nearly horizontal and outwardly projected. This could have created space for the iliofemoral muscle, which is involved in femoral protraction. The enlarged sacrum, which is strengthened in some titanosaurs by the presence of an ossified dorsal tendon or ligament, suggests that the pelvic girdle acted as a fulcrum for pivoting the vertebral column over the hindlimbs. This peculiar morphology of the titanosaurid ilium sharply differs from that seen in more-basal sauropods (e.g., *Haplocanthosaurus, Diplodocus*), and Powell (2003) believes it is evidence of occasional bipedalism in titanosaurids (Fig. 5.20). As Powell notes, the lateral expansion of the iliac blades superficially resembles the pelvic girdle of the large Pleistocene ground sloths such as *Paramylodon* and *Megatherium* (Fig. 5.20E and F). Perhaps the flexible tail (as indicated by the strongly developed ball-and-socket articulations of the centra) was advantageous during a shift from a quadrupedal to a bipedal posture (Wilson and Carrano 1999). Moreover, development of ball-and-socket articulations in the caudal vertebrae may have had important benefits in force transmission if titanosaurids used the tail against the ground to help support the body during a bipedal stance.

Derived titanosaurs were also peculiar in having a flexed posture of the forelimb. Instead of being columnar, saltasaurine forelimbs were flexed to some degree, as inferred from the cranial facing of the distal condyles of the humerus (Fig. 5.19F and G) and the hypertrophied development of ulnar olecranon process (Fig. 5.19I) (Wilson and Carrano 1999). Also notable are transformations of the manus that elongated the metacarpals. When articulated, they formed a vertically oriented cylinder that allowed them to efficiently support the great weight they carried (Martínez, Giménez, Rodríguez, Luna, and Lamanna 2004).

As mentioned earlier, Titanosauriformes showed a trend toward reduction of the claw of manual digit I, to the point that in titanosaurids the ungual phalanx was lost. The complete set of manual phalanges is absent in Titanosauridae, in sharp contrast with the conspicuous manual phalanges of other sauropods. In addition, titanosaurid metacarpals do not have the typical convex phalangeal articular facets; rather, the facets are flat, broad, and rugose surfaces that are not exposed on the anterior sides of metacarpals (a condition that would be consistent with the absence of phalanges), suggesting that these surfaces contacted the ground through soft tissue (e.g., cartilage pads or the dermis). The absence of manual phalanges would explain the absence of a pollex claw impression in titanosaur trackways (Wilson and Carrano 1999).

DERMAL OSSIFICATIONS

An outstanding feature of many titanosaurids is their well-developed and morphologically diverse dermal ossifications, a feature that is unique among sauropods (Fig. 5.21). Although French paleontologist Charles Depéret suggested (1896) that osteoderms developed in titanosaurids, it was not until the discovery of *Saltasaurus loricatus* that this development was confirmed (Bonaparte and Powell 1980; Powell 1980, 1986, 1992, 2003). Now we know that several titanosaurids, including *Mendozasaurus, Aeolosaurus, Ampelosaurus, Laplatasaurus, Lirainosaurus, Malawisaurus, Neuquensaurus,* and *Saltasaurus* had osteoderms, some

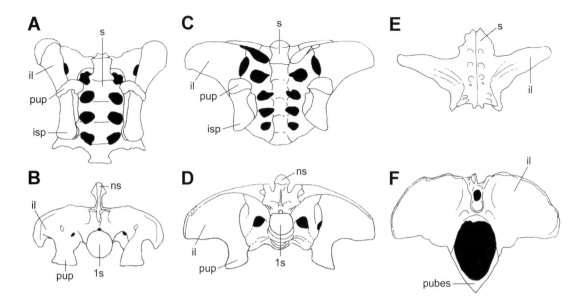

Fig. 5.20. Pelvic girdles of sauropods and mammals compared. (A, B) *Haplocanthosaurus*. (C, D) indeterminate Titanosauridae from Brazil. (E, F) Giant ground sloth *Paramylodon*. (A, C) Ventral, (E) dorsal, and (B, D, F) cranial views. Not to scale. *Redrawn from Powell (2003).*

of which were very large and thick (Fig. 5.21A and B). Titanosaurid osteoderms have been reported from Late Cretaceous beds in Argentina, India, Malawi, France, Spain, and Madagascar (Dodson, Krause, Forster, Sampson, and Ravoavy 1998). Interestingly, the extreme development of hyposphene-hypantrum articulations and the absence of dermal elements in *Epachthosaurus* is consistent with the hypothesis of Le Loeuff and colleagues (1994) that osteoderms may have served to limit dorsal mobility in titanosaurians, most of which lack these articulations (Martínez, Giménez, Rodríguez, Luna, and Lamanna 2004). This interpretation, however, counters another hypothesis (Wilson and Carrano 1999) that the strongly opisthocoelous trunk vertebrae conferred flexibility to the back.

There are good indications that in saltasaurines, at least, two different kinds of dermal ossifications were present: large scutes that were presumably dispersed over the surface of the body (Fig. 5.21C and D) and smaller ones that formed a compact paviment that was buried in the skin (Fig. 5.21E). The large osteoderms are dorsoventrally flattened and oval in shape. The texture of the bone surface is coarse and it has several grooves, pits, and tiny projections. Researchers believe that some large ossifications with a depression on the external surface supported a spike of keratin or a blunted horn. These larger scutes presumably were distributed over the backs of these animals, a hypothesis that cannot be tested at the moment because the dermal armor was not found in a natural position with the skeletons (Azevedo and Kellner 1998; Powell 2003; Salgado 2003b; Marinho and Candeiro 2005).

South American Titanosaurs

Most of the sauropods documented in Upper Cretaceous beds in South America belong to Titanosauria. The great majority of titanosaurs have

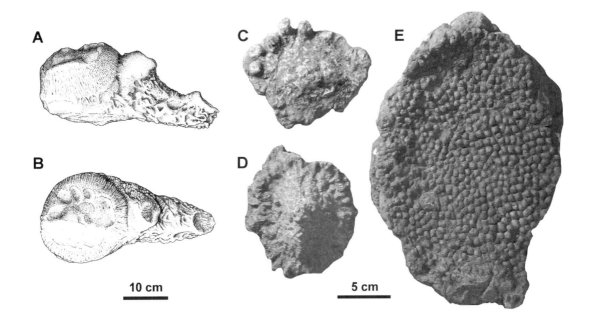

Fig. 5.21. Examples of dermal ossifications in titanosaurians: (A, B) Large dermal ossification of an indeterminate titanosaurid from Salitral Moreno (Río Negro Province, Argentina) in (A) side and (B) dorsal views. (C–E) Dermal armor of *Saltasaurus loricatus* including (C, D) scutes and (E) a set of bean-like ossicles. *(A, B) From Powell (2003). Reprinted with permission of the author. (C–E) Photographs by Fernando Novas.*

been collected from Santonian through Maastrichtian beds of northwestern Patagonia, but more recent discoveries include older forms that reach back to Aptian times (e.g., *Chubutisaurus, Ligabuesaurus*). I will first describe basal representatives of Titanosauria and then consider progressively more-derived members of the clade, following the phylogenetic arrangements depicted in Figure 5.10.

CHUBUTISAURUS INSIGNIS

This dinosaur was briefly described by Guillermo Del Corro in 1975, but a more recent review of this taxon was published by Leonardo Salgado in 1993. *Chubutisaurus* is known from isolated and incomplete vertebrae (mostly from the caudal region) and some forelimb and hindlimb bones (Fig. 5.22A–D). The specimen comes from the upper levels of the Cerro Barcino Formation (i.e., the Bayo Overo Member, which is presumably Albian–Cenomanian) at a fossil site near Paso de Indios, located in the center of Chubut Province. The distal end of tibia of *Chubutisaurus* is transversely expanded, constituting a synapomorphy of Titanosauria (Salgado and Bonaparte 2007). The amphicoelous condition of the posterior dorsal and slightly procoelous caudal vertebrae (Fig. 5.22C and D) demonstrates the primitive condition of *Chubutisaurus*, thus distinguishing it from more-derived titanosaurs, in which the dorsals are strongly opisthocoelous and the caudals exhibit a pronounced procoely. The humerus-to-femur proportions of *Chubutisaurus* that McIntosh (1990) used to support the brachiosaurid affiliation of this taxon must be dismissed, because these proportions are not different from the humerus-to-femur ratio seen in primitive titanosaurs (Bonaparte 1996b; Bonaparte, González Riga, and Apesteguía 2006). *Chubutisaurus* retained metacarpals with distal articular surfaces, which suggests that its phalanges were ossified (Salgado 2003a). This condition would contrast with that of titanosaurids, in which manual phalanges became lost.

LIGABUESAURUS LEANZAI

This new basal titanosaur was recovered from the Aptian–Albian Cullín Grande Member of the Lohan Cura Formation (Bonaparte, González Riga, and Apesteguía 2006) at the same levels that yielded the possible diplodocimorph *Agustinia ligabuei* (Bonaparte 1999b). The holotype specimen of *Ligabuesaurus leanzai* was found at Leones Hill, located 10 kilometers west of Picún Leufú in Neuquén Province. It is represented by an incomplete maxilla with teeth, some cervical and dorsal vertebrae (Fig. 5.22E and F), several bones of the pectoral girdle, and limb elements. The maxillary teeth of *Ligabuesaurus* resemble those of *Brachiosaurus* in that they are inserted perpendicular to the alveolar margin of the maxilla and have sharply inclined wear facets, a feature that is widely distributed among Titanosauriformes (Bonaparte, González Riga, and Apesteguía 2006). *Ligabuesaurus* is peculiar among sauropods in the morphology of cervical and cranial dorsal vertebrae; the neural spines are craniocaudally compressed and transversely expanded, showing a rhomboid-shaped contour in cranial view. The scapula has a widely expanded acromial process, resembling that of other titanosauriforms. The forelimbs of *Ligabuesaurus* are proportionately elongate, with a humerus-to-femur ratio of 0.9 (149 centimeters and 166 centimeters long, respectively), thus indicating that this condition is not diagnostic of Brachiosauridae (Bonaparte, González Riga, and Apesteguía 2006). The humerus of *Ligabuesaurus* is particularly elongate and slender, which is different from the robust version of more-derived titanosaurids. *Ligabuesaurus* enlarges the meager record of Aptian–Albian South American sauropods, which presently includes the diplodocimorphs *Rayososaurus*, *Limaysaurus*, *Amazonsaurus*, and *Agustinia* and the basal titanosaur *Chubutisaurus*.

ANDESAURUS DELGADOI

This is one of the most primitive members of Titanosauria. It comes from the Candeleros Formation of the Río Limay Subgroup (Calvo and Bonaparte 1991) from levels close to those in which the skeletons of *Argentinosaurus* and *Limaysaurus* were excavated and in beds containing a rich association of dinosaur footprints (Calvo 1991, 1999, 2007). *Andesaurus* is considered a basal titanosaur because it has all the requisite titanosaur synamorphies (for example, eye-shaped pleurocoels in the dorsal vertebrae and a pubis that is considerably longer than the ischium; Salgado and Bonaparte 2007), but its caudal vertebrae are only slightly procoelous. This contrasts with the strongly procoelous caudals that are diagnostic of more-derived members of Titanosauria (Fig. 5.23). *Andesaurus* exhibits axially expanded neural spines on the caudal vertebrae, a condition also reported in other basal titanosaurians.

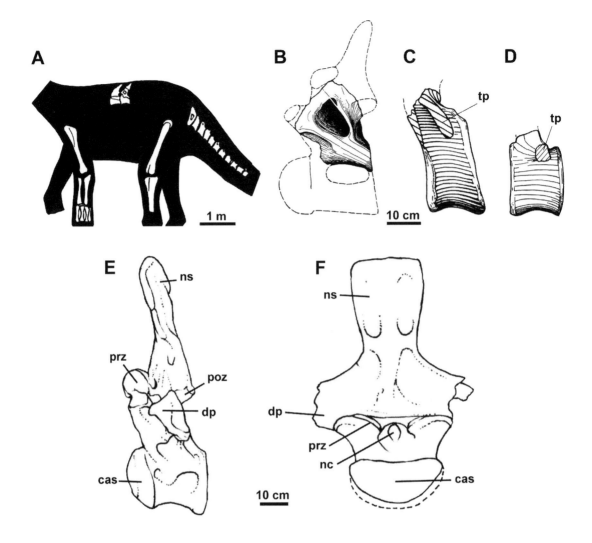

Fig. 5.22. Basal titanosaurs from South America. (A–D) *Chubutisaurus insignis:* (A) Silhouette indicating recovered bones. (B) Neural arch of mid-dorsal vertebra in left lateral view. (C) Proximal caudal centrum in left lateral view. (D) Mid-caudal centrum in left lateral view. (E, F) *Ligabuesaurus leanzai:* Anterior dorsal vertebra in (E) left lateral and (F) cranial views. *(A–D) From Salgado (1993). Reprinted with permission of* Ameghiniana. *(E, F) Redrawn from Bonaparte, González Riga, and Apesteguía (2006).*

MENDOZASAURUS NEGUYELAP

This titanosaur, which was described by Bernardo González Riga in 2003, comes from the upper portion of the Rio Neuquén Subgroup from a fossil site at Guillermo Hill in western Argentina's Mendoza Province. *Mendozasaurus* was a large titanosaurian (18–25 meters long) characterized by fan-shaped neural spines in the cervical vertebrae (e.g., craniocaudally compressed and transversely expanded; Fig. 5.24A and B) that are similar to those of *Ligabuesaurus*. *Mendozasaurus* resembles *Malawisaurus* and *Andesaurus* in its laminated and anteroposteriorly elongate mid-caudal neural spines. *Mendozasaurus* has large subconical-spherical osteoderms that differ from those of *Saltasaurus loricatus*; it lacks the peripheral ring of tubercles and the longitudinal ventral crest present in *Saltasaurus loricatus*. In *Mendozasaurus* the proximal caudals are strongly procoelous but the middle ones are platycoelous. Thus, the caudal series of this taxon illustrates the first phylogenetic stage in acquiring cranially concave caudal centra. For this reason, *Mendozasaurus* is interpreted as the

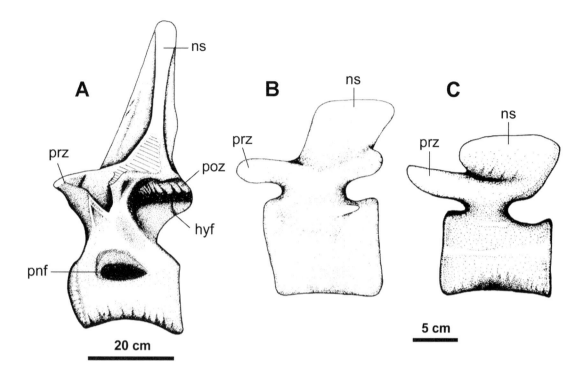

Fig. 5.23. *Andesaurus delgadoi* vertebrae in left lateral view: (A) Posterior dorsal. (B) Proximal caudal. (C) Mid-caudal. *From Calvo and Bonaparte (1991). Reprinted with permission of* Ameghiniana.

sister group of Titanosauridae, in which the middle caudal vertebrae are strongly procoelous.

Jorge Calvo and his colleagues (2007) have recently coined the name Lognkosauria for a clade of primitive titanosaurids including *Mendozasaurus* and *Futalognkosaurus*. They recognized the following features in support of the monophyly of Lognkosauria: craniocaudally short and high posterior cervical vertebrae; transversely expanded posterior cervical neural spines that are wider than the centrum; and transversely wide neural spines on the proximal caudal vertebrae.

FUTALOGNKOSAURUS DUKEI

A beautifully preserved skeleton of this titanosaurian was excavated by Jorge Calvo and crew on the coast of Lake Barreales, an artificial lake located 90 kilometers northwest of Neuquén City (Fig. 4.7). The specimen comes from a richly fossiliferous site of the Portezuelo Formation that also produced abundant plant material, yet-undescribed sauropod remains, new specimens of the theropods *Megaraptor* and *Unenlagia*, the bones of the new ornithischian dinosaur *Macrogryphosaurus*, and crocodylomorphs and fish (Calvo, Porfiri, González Riga, and Kellner 2007).

Futalognkosaurus (Fig. 5.25) is represented by ten cervicals, one dorsal, a sacrum, two proximal caudal vertebrae, an ilia, a right pubis, and an ischium. The available material belongs to a robust and giant titanosaurid whose mid-cervicals are 113 centimeters high and 102 centimeters long. The pubis is 135 centimeters long and the maximum width of the pelvis is 280 centimeters, a measurement that was taken at the level

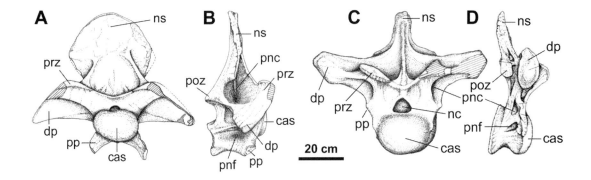

Fig. 5.24. *Mendozasaurus neguyelap:* (A, B) Posterior cervical in (A) cranial and (B) right lateral views. (C, D) Dorsal 1? in (A) cranial and (B) right lateral views. *From González Riga (2003, 2005). Reprinted with permission of Ameghiniana.*

of the broadly expanded preacetabular lobe of the ilia (Fig. 5.25D). The following derived characteristics allow us to include this animal within Titanosauridae: procoelous cranial caudals, a sacrum formed by six fused vertebrae, dorsals that lack the hyposphene-hypantrum articulation, caudals with a cranially placed neural arch, and a pubis that is longer than the ischium. *Futalognkosaurus* is unusual in that its cervical vertebrae have high, sail-like, neural spines; its proximal caudals exhibit neural spines that are laterally expanded at their distal ends; and the top of the spine has a strong prespinal lamina (Fig.5.25B).

Futalognkosaurus dukei is the most complete giant sauropod ever found. Its estimated length is between 32 to 34 meters (Calvo, Porfiri, González Riga, and Kellner 2007), a size comparable to that of *Argentinosaurus huinculensis* (Bonaparte and Coria 1993) and *Puertasaurus reuili* (Novas, Salgado, Calvo, and Agnolín 2005).

The discovery of *Futalognkosaurus* and *Mendozasaurus* indicates a new lineage of basal titanosaurids that have strong and huge necks. This characteristic sets them apart from the remaining members of this group and increases the diversity in neck anatomy of these large dinosaurs.

EPACHTHOSAURUS SCIUTTOI

This dinosaur comes from the Lower Bajo Barreal Formation (Cenomanian). The first specimen collected consisted of a series of articulated dorsal vertebrae that were excavated by José Bonaparte and his crew in 1981 and were later studied by Jaime Powell in 1990, who coined the scientific name for this titanosaur (see Powell 2003). However, the most complete specimen of *Epachthosaurus* was excavated by Rubén D. Martínez and his crew at San Bernardo Hill west of Lake Musters in Chubut Province and consists of an articulated skeleton lacking a skull, a neck, and some dorsal vertebrae (Martínez, Giménez, Rodríguez, Luna, and Lamanna 2004). It was preserved resting on its ventral surface (e.g., thorax, belly, and pelvic girdle) with the forelimbs widely extended, both hindlimbs flexed, and the tail extended distally and curved (Fig. 5.26A–C). *Epachthosaurus* was a medium-sized titanosaur that was an estimated 8–9 meters long. Although this animal is more derived than *Andesaurus* in that its caudal vertebrae are strongly procoelous, it retained some

Fig. 5.25. *Futalognkosaurus dukei:* (A, B) Posterior cervical vertebra in (A) right lateral and (B) cranial views. (C) Caudal vertebrae in distal view. (D) Ilia

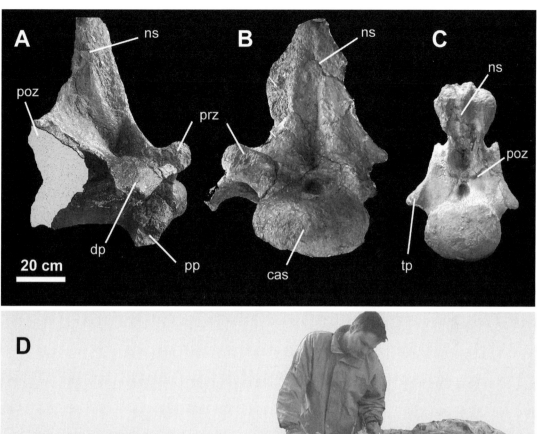

primitive features such as strongly developed hyposphene-hypantrum extra-articulations in the dorsal vertebrae (Figs. 5.17A and 5.26D and E). In addition, *Epachthosaurus* is more primitive than the remaining titanosaurids because its prespinal lamina is bifurcated at its base instead of being unforked, as in more-derived titanosaurs. The dorsal centra of *Epachthosaurus* are wider than they are high and have rather flat ventral articulated with sacrum in cranioventral view, paleontologist Juan Porfiri standing behind. *Photographs courtesy of Jorge Calvo, head of Proyecto Dino (Los Barreales Lake).*

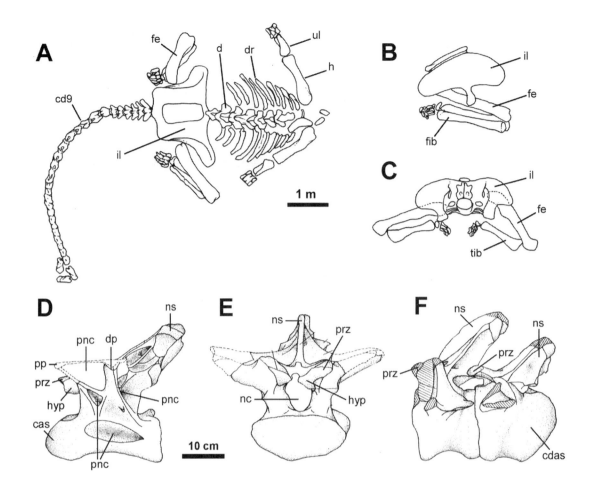

Fig. 5.26. *Epachthosaurus sciuttoi:* (A–C) Skeleton of the most complete specimen of *Epachthosaurus* as it was found in the field. (A) Skeleton in dorsal view. (B) Right ilium and hindlimb in lateral view. (C) Pelvis and hindlimbs in cranial view. (D, E) Mid-dorsal vertebra in (D) left lateral and (E) cranial views. (F) Caudals 1 and 2 in left lateral view. *From Martínez, Giménez, Rodriguez, Luna, and Lamanna (2004). © Copyright 2004 the Society of Vertebrate Paleontology. Reprinted and distributed with permission of the Society of Vertebrate Paleontology.*

surfaces. *Epachthosaurus* is also distinguished from other titanosaurs by well-developed pleurocoels in the dorsal vertebrae, which are larger than in any other known titanosaur (Bonaparte 1996b). Notably, the hyposphene-hypantrum extra-articulations are also present in caudals 1–14. These features may be diagnostic of this taxon, because the hyposphene-hypantrum articulations are limited to the dorsal vertebrae in *Andesaurus* and these structures are absent in the remaining titanosaurs (with the exception of *Argentinosaurus*).

ARGENTINOSAURUS HUINCULENSIS

This basal titanosaurid is characterized by its formidable size. The only available specimen of *Argentinosaurus huinculensis* was recovered not far from Plaza Huincul in the center of Neuquén Province (Fig. 4.7) from conglomeratic bedrock of the Huincul Formation of the Río Limay Subgroup. The animal was originally described by Bonaparte and Coria in 1993 on the basis of several dorsal vertebrae, an ilium, a fibula, and some ribs. *Argentinosaurus* is unquestionably one of the largest dinosaurs ever discovered, rivaling the titanosaurids *Puertasaurus reuili* and

Antarctosaurus? giganteus in size. For example, the length of the fibula of *Argentinosaurus* is 160 centimeters, and one of the posterior dorsals is 160 centimeters tall and 130 centimeters wide. The centra are extremely large, reaching 57 centimeters in transverse diameter (Fig. 5.27). As is usual in many titanosaurs, in which the internal tissue of most of the vertebrae is cancellous, the internal structure of the dorsal and sacral vertebrae of *Argentinosaurus* shows enormous cavities (around 4 to 6 centimeters in diameter) bordered by very thin walls (Bonaparte 1996b).

The phylogenetic relationships of *Argentinosaurus* within Titanosauridae remain obscure, mainly due to the fragmentary nature of the skeleton and the difficulty of examining (in the articulated vertebrae) the morphology of the intervertebral articulations below both the pre- and postzygapophyses. Consequently, some authors (Bonaparte and Coria 1993) concluded that *Argentinosaurus* has hypertrophied hyposphenal-hypantral articulations resembling those of the basal titanosaurid *Epachthosaurus* (compare Figs. 5.17 A with 5.27 J). Conversely, Powell (2003), Salgado and Martínez (1993), and Sanz and colleagues (1999) thought that hyposphenal-hypantral articulations are absent in the dorsal vertebrae of *Argentinosaurus* and consider the complex system of articular structures to be derived modifications of reinforced laminae below the postzygapophyses. However, I concur with Bonaparte and Coria (1993) in thinking that below the prezygapophyses the hypantrum projects lateroventrally, forming a considerably widened surface for articulation with a similarly well-developed hyposphene. The latter articular structure is dorsally placed with respect to the hypantrum, making an extended secondary articulation between contiguous vertebrae (Bonaparte 1996b).

A review of the dorsal vertebrae of *Argentinosaurus* (Novas and Ezcurra 2006) indicate that the cranial dorsals of *Argentinosaurus* are morphologically intermediate between those of basal camarasauromorphs (which have vertical neural spines) and advanced titanosaurids (which have neural spines that are strongly inclined backward). Also, in derived titanosaurids (e.g., *Trigonosaurus*; Powell 2003; Campos, Kellner, Bertini, and Santucci 2005), the transverse processes of the cranial dorsals are laterodorsally oriented, whereas in *Argentinosaurus* they are horizontally projected, thus retaining the primitive camarasauromorph condition. In sum, the morphology of dorsal vertebrae of *Argentinosaurus* supports the hypothesis that this taxon constitutes a basal member of Titanosauria.

PUERTASAURUS REUILI

This taxon was described by the present author and colleagues (Novas, Salgado, Calvo, and Agnolín 2005) on the basis of some isolated but well-preserved vertebrae (Figs. 5.16 and 5.28). *Puertasaurus reuili* is one of the largest known dinosaurs and sheds light on the morphology of the neck vertebrae of a giant titanosaur. The specimen was recovered at Los Hornos Hill, near La Leona River in Santa Cruz Province, Argentina (Fig. 4.14). It is from the Pari Aike Formation, which is thought to be

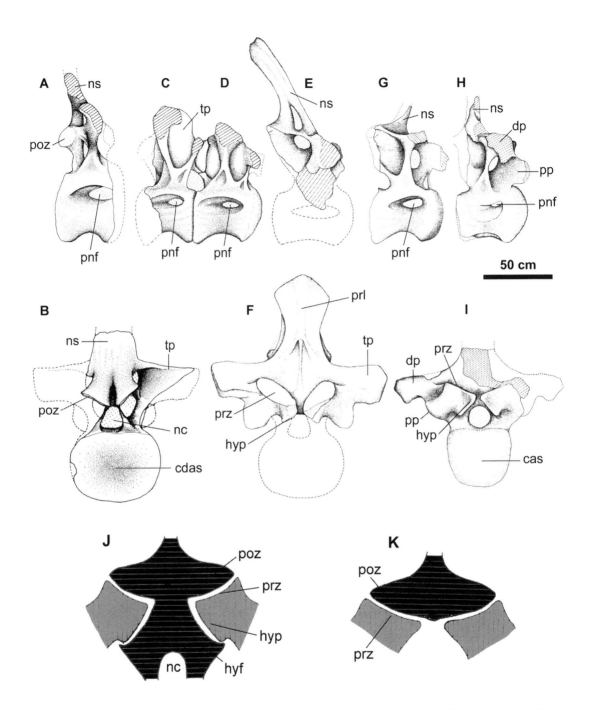

Early Maastrichtian in age (Kraemer and Riccardi 1997; Novas, Cambiaso, and Ambrosio 2004). *Puertasaurus reuili* is diagnosed on the basis of the following combination of characteristics: gigantic size; a cervical neural spine that is considerably inflated, is transversely wider than the vertebral centrum, and has strong dorsolateral ridges; and cranial dorsal vertebrae that are extremely short, more so than in other sauropods. The neck vertebra of *Puertasaurus* (presumably cervical 9; Fig. 5.16) is 118 centimeters long (measured from the pre- to the postzygapophyses) and is exceptionally wide; its transverse width is 140 centimeters (including fused ribs). The neural spine has deep and wide pre- and postspinal fossae for cradling well-developed interspinal ligaments as well as a considerably inflated distal end, suggesting a powerful neck ligament and cervical muscles. The dorsal enlargement of the neural spine is more derived than the transversely narrower neural spine of all other titanosauriforms (e.g., *Brachiosaurus*, *Euhelopus*, *Neuquensaurus*) or the transversely expanded but craniocaudally compressed neural spine of *Mendozasaurus* and *Ligabuesaurus*. The cervical of *Puertasaurus* exhibits features that are seen in other titanosaurians, such as laterally projecting diapophyses and parapophyses and a low neural arch combined with a high neural spine. In the new specimen, the zygapophyseal articulations are positioned low on the neural arch and the centrum is even more depressed than in other titanosaurids (e.g., *Saltasaurus*). Consequently, the system of bony struts on the sides of the vertebra (and the pneumatic fossae they define) are dorsoventrally flattened. The cervical vertebra is poorly pneumatized and lacks pleurocoels.

The cranial dorsal vertebra (probably dorsal 2) of *Puertasaurus* is craniocaudally short, in sharp contrast with the cervical (Fig. 5.28). The centrum is strongly opisthocoelous and proportionally shorter than in other Titanosauridae. Dorsal 2 of *Puertasaurus* is 106 centimeters high but 168 centimeters across the ends of the wing-like transverse processes, thus exceeding by nearly 45 centimeters the width of dorsal 3 of *Argentinosaurus huinculensis*. Dorsal 2 of *Puertasaurus* is considerably wider (in absolute terms) than in other known sauropods. The transverse processes are dorsoventrally deep at their bases, resulting in a wing-like appearance in cranial view. In *Puertasaurus* the transverse processes of dorsal 2 are perpendicular to the axial plane, as is the case in *Argentinosaurus* (Fig. 5.27E) and the basal titanosauriform *Euhelopus*, instead of being laterocaudally oriented as in more-derived titanosaurids (e.g., the saltaurine *Saltasaurus* and the aeolosaurine *Trigonosaurus*; Powell 2003). The neural spine is dorsoventrally low, transversely expanded, and vertically oriented and is perpendicular to the anteroposterior axis of the centrum, thus resembling *Argentinosaurus*, for example. The pre - and postspinal fossae of the dorsal vertebra are wider and deeper than in more-derived titanosaurids (e.g., *Saltasaurus*, *Opisthocoelicaudia*; Powell 2003; Borsuk-Bialynicka 1977), resembling basal titanosauriforms such as *Euhelopus* and *Brachiosaurus* (Bonaparte 1999c). Two mid-caudal centra are preserved. They are procoelous, a common feature among Titanosauridae (Salgado, Coria, and Calvo 1997).

The morphology of the caudal region indicates that *Puertasaurus* was more derived than *Andesaurus*, but some features of the dorsal vertebra suggest that *Puertasaurus* may constitute a basal member of Titanosauridae. These features include the plesiomorphically retained vertical neural spines, transverse processes that are perpendicularly oriented, and

Fig. 5.27. *Argentinosaurus huinculensis*: (A, B) Dorsal 10 in (A) right lateral and (B) caudal views. (C, D) Articulated dorsals 7 and 6 in right lateral view. (E, F) Dorsal 5 in (E) right lateral and (F) cranial views. (G) Dorsal 4 in right lateral view. (H, I) Dorsal 3 in (H) right lateral and (I) cranial views. (J, K) Schematic shape of zygapophyses and hyposphene-hypantrum in transverse cross-section of a mid-dorsal vertebra of (J) *Argentinosaurus* and (K) *Saltasaurus* (not to scale). *From Bonaparte and Coria (1993). Reprinted with permission of Ameghiniana. Note that the arrangement of the available dorsals depicted here differs from the original interpretation offered by these authors.*

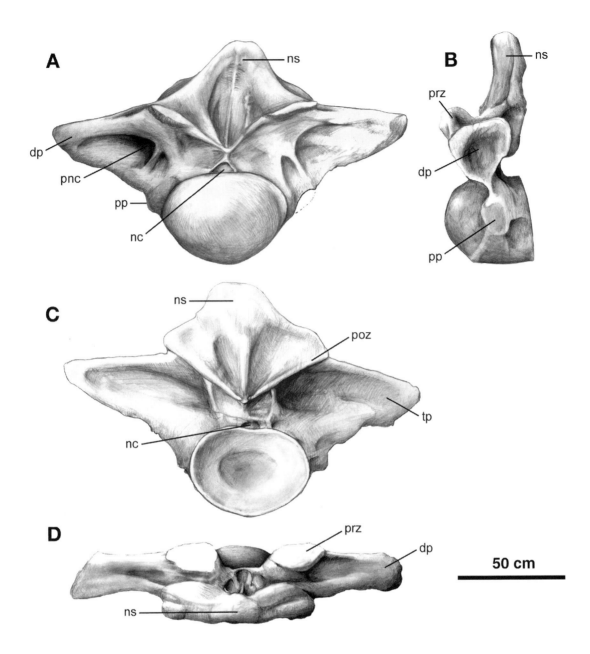

Fig. 5.28. *Puertasaurus reuili:* Dorsal 2 in (A) cranial, (B) left lateral, (C) caudal, and (D) dorsal views. *From Novas, Salgado, Calvo, and Agnolín (2005). Reprinted with permission of* Revista del Museo Argentino de Ciencias Naturales.

neural arches that lack the complex system of laminae and pneumatic depressions characteristic of derived titanosaurids.

The discovery of *Puertasaurus* demonstrates that the disparity in neck anatomy among sauropod dinosaurs is greater than was suspected. The low and wide titanosaurian cervicals differ from the deep cervicals of the remaining neosauropods (e.g., diplodocids, *Brachiosaurus,* and *Euhelopus*) in that the latter have rounded articular surfaces of the centra, ribs that are mostly ventrally oriented, and zygapophyses that are elevated compared to the centrum. How these anatomical distinctions affect posture and the mechanics of movement have not been explored, and they may have important consequences for functional studies of sauropod necks (Wedel, Cifelli, and Sanders 2000; Stevens and Parrish 1999).

A LARGE SAUROPOD CERVICAL FROM THE BAURU GROUP

A large cervical vertebra was collected near the city of Presidente Prudente in the state of São Paulo in Brazil (Simbras, Oliveira, Campos, and Kellner 2007; Fig. 4.5). The material, which came from beds of the Bauru Group, consists of a nearly complete posterior cervical vertebra (F. Simbras, pers. comm.). This specimen measures approximately 57 centimeters long and is larger than any cervical found in Brazil so far.

The caudal articular surface of the vertebra is 32.7 centimeters wide and is transversely wider than it is high. The preserved portion of the neural arch shows that it was proportionally higher than in other Brazilian titanosaurs (e.g., *Trigonosaurus*, *Maxakalisaurus*; Simbras, Oliveira, Campos, and Kellner 2007). The vertebra of this large sauropod differs from that of *Puertasaurus*, for example; it has a pleurocoel on the side of the centrum while *Puertasaurus* does not.

This new material, which at present has been reported only preliminarily, is the largest dinosaur yet recovered from Brazil, indicating that large titanosaurs were present there (Simbras, Oliveira, Campos, and Kellner 2007).

LAPLATASAURUS ARAUKANICUS

This animal was discovered at the Cinco Saltos and Rancho de Avila fossil sites in Río Negro Province (Fig. 4.7). Although Powell (2003) theorized that the fossils of *Laplatasaurus* come from the lower portion of the Allen Formation (Maastrichtian), other authors (Heredia and Salgado 1999; Salgado and Bonaparte 2007) believe that specimens assigned to *Laplatasaurus* were collected from the older Anacleto Formation (Campanian).

Several postcranial elements of *Laplatasaurus araukanicus* have been collected, but a holotype of this species was not designated in the original description (von Huene 1929a; Fig. 5.29). Bonaparte and Gasparini (1979) designated a lectotype consisting of a tibia and a fibula that presumably belonged to a single individual. Because of their slenderness, the hindlimb bones of *Laplatasaurus* sharply differ from those of *Saltasaurus*, *Neuquensaurus*, and *Aeolosaurus*. Other specimens that have been referred to this taxon are distinguished by their slender construction; a humerus assigned to *Laplatasaurus* is more slender than the robust humeri of *Saltasaurus* and *Argyrosaurus*, for example. The fibula has a prominent and double lateral tuberosity for the M. biceps femoris, a feature Powell interprets as autapomorphic of *Laplatasaurus* (Powell 2003), although this characteristic is also present in the fibula of *Epachthosaurus* (Martínez, Giménez, Rodríguez, Luna, and Lamanna 2004), thus indicating

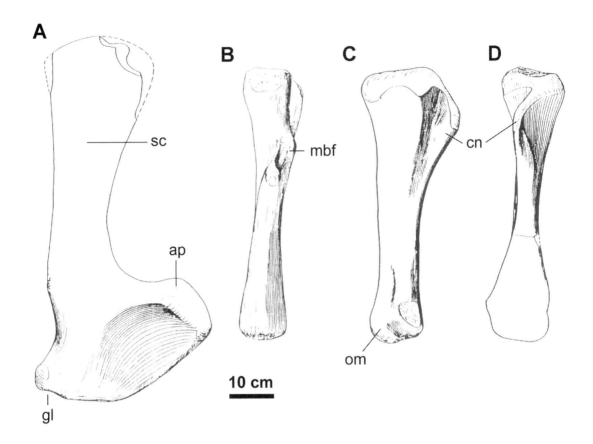

Fig. 5.29. *Laplatasaurus araukanicus:* (A) Right scapula in lateral view. (B) Right fibula in lateral view. (C, D) Right tibia in (C) lateral and (D) cranial views. *From von Huene (1929a).*

a wider distribution among titanosaurs. The scapula of *Laplatasaurus* resembles that of other titanosaurids (e.g., *Argyrosaurus, Neuquensaurus*); its distal end is expanded and the angle formed by the scapular blade and the acromial process is acute. However, the scapula is devoid of the process for muscular attachment that is present on the medial surface of this bone seen in *Aeolosaurus, Saltasaurus,* and *Neuquensaurus* (Powell 2003). The pubis of *Laplatasaurus* is also long and narrow compared with the stouter one of *Argyrosaurus* and *Saltasaurus,* for example. A posterior cervical vertebra (probably cervical 9) assigned to *Laplatasaurus* is much longer than the equivalent one of *Neuquensaurus australis.* The caudal centra are relatively short and high and have a narrow ventral moiety. The lateral walls are slightly concave and lateroventrally oriented. Moreover, the caudal vertebrae of *Laplatasaurus* lack the cancellous bone tissue that is characteristic of *Saltasaurus* and related forms (Powell 2003). The phylogenetic relationships of *Laplatasaurus* with other titanosaurids are far from being clear, and this taxon is here tentatively depicted as one of the sister taxa of Saltasaurinae.

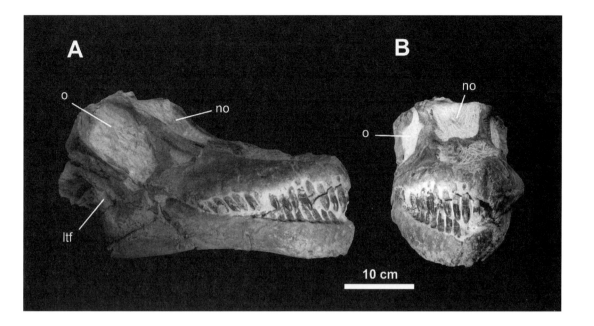

Fig. 5.30. Skull of a presently unnamed titanosaur from the Lower Bajo Barreal Formation in (A) lateral and (B) rostral views. *Courtesy of Rubén Martínez.*

AN UNNAMED BASAL TITANOSAUR
FROM THE BAJO BARREAL FORMATION

Rubén D. Martínez (1998a) announced the discovery of a well-preserved skull with jaws and teeth in beds of the Lower Bajo Barreal Formation, which crops out at the Ocho Hermanos Farm, located at the foot of San Bernardo Hill in Chubut Province (Fig. 4.13). The overall aspect of the skull is reminiscent of that of *Brachiosaurus* (Fig. 5.30); it has a prenasal depression in the snout, as does the Jurassic African genus. The teeth, which also resemble those of *Brachiosaurus*, are thick and form a long tooth row (15 teeth on the maxillary and premaxillary arcade and 11 teeth on the dentary). The facets of the teeth are sharply inclined, as in *Nemegtosaurus* and *Brachiosaurus*. The skull looks primitive in the shape of the supratemporal fenestra, which is rostrocaudally elongate and dorsally exposed, and the jugal, which does not participate in the margin of the preorbital fenestra.

"*CAMPYLODONISCUS AMEGHINOI*"

This taxon was described by Friedrich von Huene in 1929. It consists of a left maxilla with some teeth that was recovered from San Bernardo Hill west of Lake Musters in Chubut Province. The stratigraphical provenance of the specimen is unclear because two units of the Chubut Group crop out in the area: the Castillo and Bajo Barreal formations. The maxilla is incomplete and the shape of the teeth is "intermediate between the spatulate-shaped one of *Camarasaurus* and the peg-like one of *Diplodocus*" (von Huene 1929a, 83). The morphology of the teeth resembles that of *Antarctosaurus wichmannianus*; it has the same pattern of rugosities (Powell 2003). The fragmentary nature of the specimen indicates that "*Campylodoniscus ameghinoi*" should be considered a nomen vanum (Bonaparte and Gasparini 1979).

BONITASAURA SALGADOI

This is a new titanosaurian represented by a partially articulated skeleton recovered from beds of Bajo de la Carpa Formation, which is exposed close to the town of Cerro Policía in Río Negro Province. The skeleton is approximately 9 meters long. *Bonitasaura* is distinguished by its unusual rectangular lower jaw, which has narrow anteriorly restricted teeth and (most notably) an edentulous caudal region that forms a sharp dorsal edge and has a profusely vascularized lateral side that probably was covered by a sharp keratinous sheath (Apesteguía 2004; Fig. 5.31). This discovery demonstrates that some titanosaurs acquired a highly derived mandibular configuration that worked to cut plant material.

Bonitasaura has a rostrocaudally straight mandibular ramus that turns medially at almost a right angle to meet the opposite ramus in a transverse symphysis, resembling the ramus of *Antarctosaurus* and other titanosaurs (Powell 1979; Coria and Chiappe 2001). In contrast with the broad-crowned, compressed teeth of primitive titanosaurs (R. D. Martínez 1998a; Figs. 5.15, 5.30), *Bonitasaura* has narrow, pencil-like teeth, as in the derived titanosaurs *Rapetosaurus*, *Antarctosaurus*, and *Nemegtosaurus*. The teeth of *Bonitasaura* have well-marked, non-denticulated carinae on their mesial and lateral edges that separate the labial and lingual sides, as also occurs in *Antarctosaurus* and *Rinconsaurus*.

The available information suggests that *Bonitasaura* is closely related to the Late Cretaceous Malagasy titanosaur *Rapetosaurus*, the Mongolian taxa *Nemegtosaurus* and *Quaesitosaurus*, and possibly to the Patagonian *Antarctosaurus*. These sauropods share a dentary symphysis that is almost perpendicular to the mandibular rami, and narrow, pencil-like teeth that are cylindrical in cross-section and are mostly restricted to the anteriormost portion of the lower jaw. However, *Bonitasaura* differs from *Antarctosaurus* in having the guillotine crest, a less-straight angle of the symphysis, and a rather flat (instead of sinuous) caudal surface of the parietal.

Bonitasaura and *Antarctosaurus* demonstrate that a square-shaped muzzle evolved among advanced titanosaurs convergently with diplodocimorphs, as was originally proposed by Friedrich von Huene (1929a) and later by Jorge Calvo (1994) and Leonardo Salgado (2003a).

The horny sheath constitutes a curious adaptation that has not been recorded before among Sauropoda. Also intriguing is the fact that a similar keratinous covering developed on the sides of the jaws instead on the front of the snout, as usually occurs among beaked tetrapods (e.g., turtles, birds).

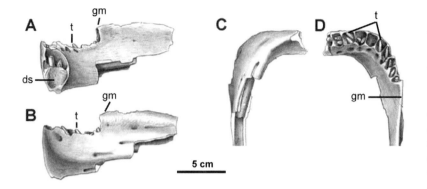

Fig. 5.31. *Bonitasaura salgadoi:* (A–D) Right dentary in (A) medial, (B) lateral, (C) ventral, and (D) dorsal views. *From Apesteguía (2004). Reprinted from* Naturwissenschaften *91: 493–497.* Bonitasaura salgadoi *gen. et sp. nov.: a beaked sauropod from the Late Cretaceous of Patagonia. Author: Apesteguía, S., p.495 (fig. E). © Copyright 2004. With kind permission of Springer Science and Business Media.*

ANTARCTOSAURUS WICHMANNIANUS

This dinosaur is known from a single specimen recovered in 1912 by geologist Ricardo Wichmann on the south coast of the Negro River that is currently known as Paso Córdova, 15 kilometers southwest of the city of General Roca in Río Negro Province (Wichmann 1916; Fig. 4.7). Friedrich von Huene, who coined the name for this dinosaur, published its description in 1929. *Antarctosaurus* comes from levels of the Campanian Rio Colorado Subgroup, probably from the Anacleto Formation (Bonaparte and Gasparini 1979; Powell 2003; Leanza, Apesteguía, Novas, and de la Fuente 2004; but see Salgado and Bonaparte 2007 for a different view on the stratigraphical provenance of the specimen). Several other fragmentary specimens recovered in Brazil, India, and Argentina have been referred to the genus *Antarctosaurus*, but the taxonomic status of such discoveries is dubious.

The holotype specimen of *Antarctosaurus wichmannianus* includes a partially preserved skull that has jaws with teeth and several bones of the postcranial skeleton (Fig. 5.32). However, the peculiarities observed in the jaw (which is square-shaped) and teeth (which are numerous, proportionally small and slender, and restricted to the rostral end of the jaw) have led to a variety of opinions about the nature of the holotype specimen of *Antarctosaurus wichmannianus* as well as the phylogenetic relationships of this taxon. Some authors (e.g., Jacobs, Winkler, and Gomani 1993; Sereno et al. 1999; Upchurch 1999) considered this peculiar morphology of the jaw and teeth as enough evidence to reinterpret *Antarctosaurus wichmannianus* as a Late Cretaceous diplodocimorph, more precisely the youngest record of a rebbachisaurid sauropod. However, two lines of evidence may be used to support the interpretation that *Antarctosaurus* is a titanosaurid, not a rebbachisaurid. First, the morphology of the skull is congruent with that of other titanosaurids; and second, the quarry in which *Antarctosaurus* was excavated did not yield any other postcranial bones but those belonging to Titanosauridae. Regarding the first point, Chiappe, Salgado, and Coria (2001) found evidence of unquestionable titanosaur affiliations for *Antarctosaurus*, such as the downwardly curved paraoccipital process, a sharp angle between the facets of wear and the

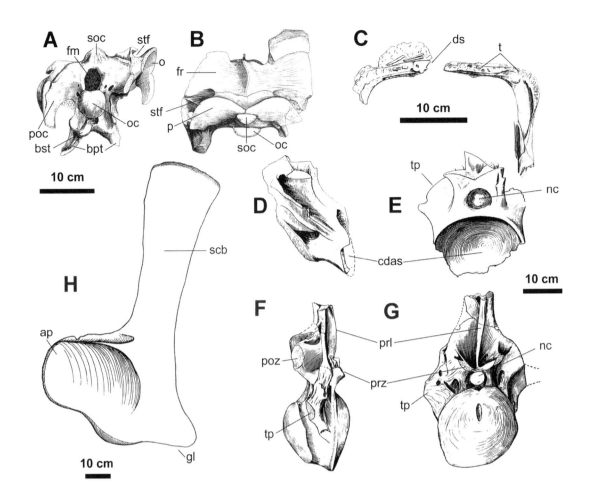

Fig. 5.32. *Antarctosaurus wichmannianus:* (A, B) Braincase in (A) caudal and (B) dorsal views. (C) Rostral ends of both dentaries as seen from above. (D, E) Posterior cervical vertebra in (D) left lateral and (E) caudal views. (F, G) Caudal 1 originally referred by von Huene (1929a) to *Laplatasaurus araukanikus* in (F) right lateral and (G) cranial views. (H) Left scapula in lateral aspect. *From von Huene (1929a).*

longitudinal axes of teeth, and an angle of roughly 90 degrees between the main axes of the symphysis and the mandible. Further supporting the interpretation that *Antarctosaurus* is deeply nested among titanosaurs is the cladistic analysis presented by Curry Rogers and Forster (2004), who depicted *Antarctosaurus* as a member of a titanosaurid subclade that also includes *Rapetosaurus, Nemegtosaurus,* and *Quaesitosaurus*. The right angling of both dentaries and the fact that the numerous slender teeth are restricted to the rostral end of the jaw look highly derived compared to *Malawisaurus, Rapetosaurus, Bonitasaura,* and the Bajo Barreal titanosaur, indicating that *Antarctosaurus* was a large titanosaur with a specialized feeding mechanism (Fig. 5.33).

A first caudal was recovered close to the skull of *Antarctosaurus*, but von Huene (1929a) referred it to *Laplatasaurus araukanicus* because it seemed (to von Huene) proportionally small compared to the preserved bones of *Antarctosaurus*. However, the size of this caudal vertebra matches the size of the holotypic posterior cervical of *Antarctosaurus wichmannianus,* keeping the relative proportions with respect to the femur seen in other titanosaurs (e.g., *Saltasaurus*). If the caudal vertebra indeed belongs to *Antarctosaurus wichmannianus,* then a biconvex first

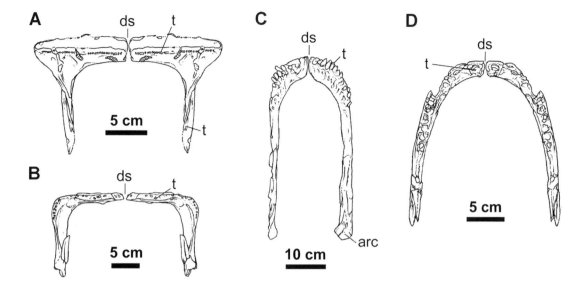

Fig. 5.33. Sauropod lower jaws in dorsal view. (A) Nigersaurus. (B) Antarctosaurus. (C) Nemegtosaurus. (D) Rapetosaurus. (A) From von Huene (1929a). (B) Redrawn from Sereno et al. (1999). (C, D) Redrawn from Nowinski (1971) and Curry Rogers and Forster (2004), respectively.

caudal constitutes an important characteristic that would indicate that this taxon is located within the titanosaurid subclade composed of *Opisthocoelicaudia*, *Aeolosaurus*, *Pellegrinisaurus*, and *Alamosaurus*, in which a biconvex first sacral is present.

Unfortunately, the forelimb bones of *Antarctosaurus* are very incomplete for reliable comparisons with other titanosaurs, but the humerus and femur look elongate and slender and thus different from the more-robust proportions of other derived titanosaurs. The scapula of *Antarctosaurus* is narrow and its cranial margin forms a right angle with the upper border of the acromial process, differing from the wider and cranially notched scapula present in *Argyrosaurus* and saltasaurines. The morphology of the scapula may indicate that *Antarctosaurus* is located outside the remaining titanosaurids that have a biconvex first sacral vertebra. Powell (2003), who agrees with this phylogenetic interpretation, observed that the cranial anatomy of *Antarctosaurus* looks more primitive than that of *Saltasaurus*.

ANTARCTOSAURUS? GIGANTEUS

This animal is one of the biggest dinosaurs known (others include *Argentinosaurus* and *Puertasaurus*; Van Valen 1969). The species was named by von Huene (1929a), who interpreted it as presumably belonging to the genus *Antarctosaurus*. Although likely a valid species, it is highly probable that it belongs to a new genus (Powell 2003), but confirming this belief will require new and more complete specimens. The holotype of *Antarctosaurus? giganteus* consists of both femora (Fig. 5.34), both incomplete pubes, the distal end of a tibia, rib fragments, and two distal caudal vertebrae. It comes from Aguada del Caño, located approximately 22 kilometers west of the city of Neuquén, from levels that probably belong to the Plottier Formation of the Río Neuquén Subgroup (Bonaparte and

Gasparini 1979). Despite its colossal size, the construction of the femoral shaft of *Antarctosaurus? giganteus* is relatively gracile: the left femur is 2.31 meters long and its transverse width at mid-shaft is 31 centimeters. Although the phylogenetic relationships of this beast are far from being understood, it seems evident that it is not a member of the saltasaurines, which are characterized by femora that are more robust.

Fig. 5.34. *Antarctosaurus giganteus*: Femora exhibited at Museo de La Plata. In the photograph, Francisco Novas.

ARGYROSAURUS SUPERBUS

This is probably the best known of the Late Cretaceous giant titanosaurids from Argentina (Bonaparte 1996b). The holotype specimen of this titanosaur includes an almost-complete forelimb (a humerus, a radius, an ulna, some carpal bones, and five metacarpals) that was recovered from Río Chico, northeast of Lake Colhué-Huapi in Chubut Province, probably from beds of the Lower Bajo Barreal Formation. Recognition of this taxon was originally based on forelimb anatomy. However, a new and more complete specimen that probably belongs to the same species (*Argyrosaurus superbus?*; Fig. 5.35) was recovered by José Bonaparte and his assistants from the right bank of the Senguerr River on the southern extremity of the San Bernardo Hill in Chubut Province (Fig. 4.13). This partial skeleton includes forelimb bones that are similar in morphology to *Argyrosaurus superbus* and some dorsal and caudal vertebrae, a scapula, a pubis, and fragments of hindlimb bones that offer a fairly complete view of the anatomy of this animal. Among the specimens excavated from other Patagonian localities are some femora of considerable size recovered by Elmer Riggs in 1924 on the same hill (San Bernardo) and probably from close to or even the same stratigraphic unit. Jaime Powell (2003) has cautiously referred these robust bones to the same taxon. As recommended by Salgado and Bonaparte (2007), a comprehensive review of all the specimens referred to *Argyrosaurus superbus* is needed, especially through checking the stratigraphical provenance of each of the available specimens referred to this taxon.

Argyrosaurus was a large sauropod; the humerus of one of its specimens attaining a length close to 130 centimeters and its pubis was around 115 centimeters long. Other specimens have a femur reaching 2 meters in length. The heavy construction of the humerus (Fig. 5.35E) makes *Argyrosaurus* different from *Antarctosaurus* and *Laplatasaurus*, which have limbs that are more slender. Instead, *Argyrosaurus* resembles the remaining titanosaurids (e.g., *Epachthosaurus*, *Paralititan*, saltasaurines). The mid-dorsals of *Argyrosaurus* have a characteristically titanosaurid pattern, with strongly opisthocoelous centra that are depressed and elongate and neural arches with well-developed lateral surfaces below the transverse processes (Fig. 5.35A). However, a marked vertebral zonation is evident in the posterior half of the trunk: the posterior dorsal vertebrae (Fig. 5.35B) are short and deep with centra that are cranially slightly convex. This construction resembles the proportions of the basal titanosaur *Andesaurus*, thus contrasting with the strongly opisthocoelous and elongate centra of

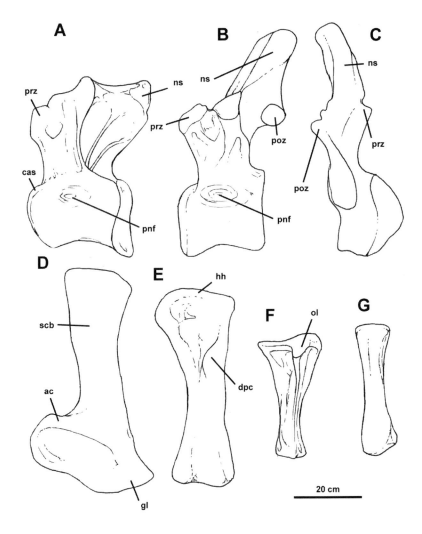

Fig. 5.35. *Argyrosaurus superbus:* (A–C) Vertebrae in left lateral view: (A) Mid-dorsal. (B) Posterior dorsal. (C) Proximal caudal. (D–G) Bones of left scapular girdle and forelimb: (D) scapula in lateral view; (E) humerus in cranial view; (F) ulna in lateral view; (G) radius in lateral view. *(A, B) Redrawn from Bonaparte (1999c). (C–G) Redrawn from Powell (2003).*

other titanosaurids (e.g., *Epachthosaurus*, saltasaurines). The dorsal vertebrae of *Argyrosaurus* lack hyposphene-hypantrum articulations (Powell 2003), so it is assumed that this taxon is more closely related to saltasaurines than to *Epachthosaurus* and *Argentinosaurus*. In this context, the apparently primitive proportions of the dorsal centra are tentatively interpreted as an autapomorphic reversion that is diagnostic of *Argyrosaurus*. Consistent with this, the proximal caudals of this taxon are very short and deep (Bonaparte 1996b, 1999c), differing from the more elongate ones of the remaining titanosaurs. Although caudal 2 of *Argyrosaurus* is strongly procoelous (Powell 2003), caudal 1 is so poorly preserved that it is impossible to determine whether the centrum was biconvex or procoelous. Like those of aeolosaurine titanosaurs, the neural spines of the proximal caudals of *Argyrosaurus* incline forward (Fig. 5.35C), thus contrasting with the caudally inclined neural spines present in other titanosaurs (e.g., *Epachthosaurus, Saltasaurus, Opisthocoelicaudia, Baurutitan*).

Plate 1. Landscape of Ischigualasto (also known as Valley of the Moon), San Juan Province, northwestern Argentina. At the front, grayish sandstone and mudstone of the Carnian Ischigualasto Formation, with some large geoforms labrated by erosion. At the back, great ravines made up of reddish sandstone of the Los Colorados Formation.

Plate 2. *Herrerasaurus ischigualastensis* feeding on a carcass of a rhynchosaur. *Illustration © by Emilio López Rolandi.*

Plate 3. Reconstruction of the basal saurischian *Eoraptor lunensis*. Illustration © by Carlos Papolio.

Plate 4. Reconstruction of the early sauropodomorph *Unaysaurus tolentinoi*. Natural-size sculpture (2.75 m long) mounted at the Museu Nacional, Universidade Federal do Rio de Janeiro. Model © by Orlando Grillo.

Plate 5. Reconstruction of the predatory dinosaur *Herrerasaurus ischigualastensis*. Illustration © by Carlos Papolio.

Plate 6. Skull of coelophysoid theropod *Zupaysaurus rougieri* in right lateral view.

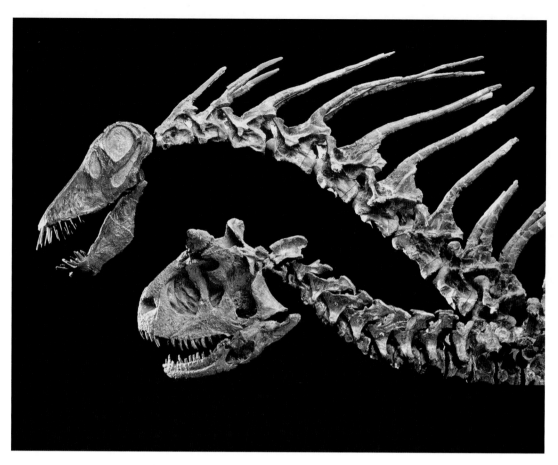

Plate 7. Ornithopod trackway left on a bedplane of the Sousa Formation, Sousa Basin, northeastern Brazil. *Photograph by Ismar de Sousa Carvalho.*

Plate 8. Skull and neck of the dicraeosaurid sauropod *Amargasaurus* (above) and the abelisaurid theropod *Carnotaurus* (below). These two bizarre dinosaurs are among the most remarkable discoveries made in southern continents. *Photograph courtesy of José Bonaparte.*

Plate 9. Reconstruction of the abelisaurid theropod *Ekrixinatosaurus novasi*. *Illustration © by Carlos Papolio.*

Plate 10. Sierra del Portezuelo, Neuquén Province, northwestern Patagonia. This beautiful hill yielded the remains of the theropods *Patagonykus, Unenlagia, Neuquenraptor,* and *Megaraptor.*

Plate 11. Reconstruction of the alvarezsaurid theropod *Patagonykus puertai.* Illustration © by Carlos Papolio.

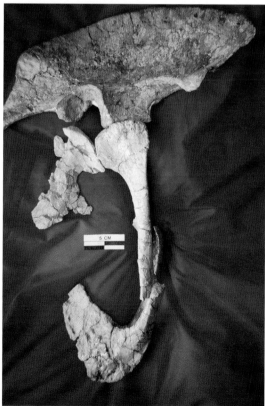

Plate 12. Reconstruction of the dromaeosaurid theropod *Unenlagia comahuensis*. Illustration © by Carlos Papolio.

Plate 13. Pelvic girdle of *Unenlagia comahuensis* in right lateral view.

Plate 14. Reconstruction of the dinosaur fauna at the Santana do Cariri locality, northeastern Brazil, 110 million years ago. The large spinosaurid *Irritator* frightens the compsognathid *Mirischia*, which is robbing a hatchling *Santanaraptor* from its nest. A large pterosaur, *Tropeognathus mesembrinus*, flies behind them. *Illustration © by Maurilio Oliveira.*

Plate 15. Reconstruction of the dinosaur fauna at La Juanita locality, Chubut Province, Patagonia, 110 million years ago. A couple of carcharodontosaurids *Tyrannotitan chubutensis* attack a herd of large titanosaurs. *Illustration © by Maurilio Oliveira.*

Plate 16. Reconstruction of the titanosaurid *Maxakalisaurus* being harassed by a group of abelisaurids. The partial skeleton of this sauropod was collected in beds of the Turonian–Santonian Adamantina Formation, State of São Paulo, Brazil. *Illustration © by Maurilio Oliveira.*

Plate 17. Reconstruction of the basal iguanodontian *Talenkauen santacrucensis*, which lived in southern Patagonia at the end of the Cretaceous, approximately 70 million years ago. *Illustration © by Gabriel Lio.*

Plate 18. Reconstruction of the basal tetanurine theropod *Megaraptor namunhuaiquii* approaching an ornithopod dinosaur. *Illustration © by Gabriel Lio.*

Plate 19. Left hand of *Megaraptor* as it was discovered during an excavation at Los Barreales Lake. *Photograph courtesy of Jorge Calvo.*

Plate 20. Block preserving the negative skin impressions of *Carnotaurus* that correspond to the proximal region of the tail. In the image above, impressions of successive awl-like transverse processes are preserved along the top margin of the block (cranial is to the left). A series of elongate ventrally inclined haemal arches are also visible. The bottom image is a detail of the impression, showing one of the haemal arches and a row of impressions that corresponds to conical dermal scutes.

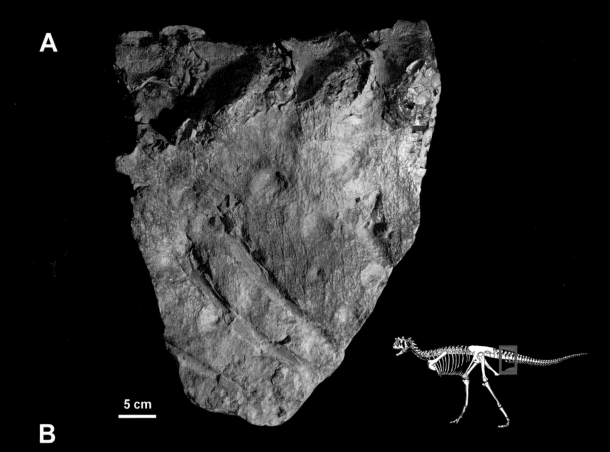

Aeolosaurinae

The tribe Aeolosaurini was originally recognized by Costa Franco-Rosas and colleagues (2004) to include *Aeolosaurus*, *Rinconsaurus*, and *Gondwanatitan*, although the list may be enlarged with the incorporation of the newly described Brazilian taxa *Trigonosaurus* and *Adamantisaurus*, I prefer to use the term Aeolosaurinae for a subfamily of the Titanosauridae. These sauropods share cranially inclined proximal and middle caudal centra; neural arches of the middle caudals placed at the level of the cranial half of the centrum; a craniodorsally projected neural spine (at least in middle caudals); and elongated prezygapophyses.

The available fossil record indicates that aeolosaurine titanosaurids were widely distributed in South America during the Late Cretaceous.

AEOLOSAURUS RIONEGRINUS

This titanosaurid has been recorded in the Maastrichtian Allen Formation, which is widely exposed in Salitral Moreno, 20 kilometers southwest of General Roca in Río Negro Province (Salgado and Coria 1993; Fig. 4.7). Specimens that have been referred to *Aeolosaurus* were also reported from the Los Alamitos and Angostura Colorada formations. Recently, Casal and colleagues (2007) documented a new species of *Aeolosaurus* (*A. colhuehuapensis*) from levels of the Upper Bajo Barreal Formation (Coniacian–Santonian).

Aeolosaurus is approximately 10 meters long. It is distinguished from other titanosaurs by caudal centra that are transversely compressed, resulting in deep lateral surfaces and a narrow ventral margin (Fig. 5.36). In addition, the neural arches of the caudal vertebrae are steeply inclined forward and the neural spine extends beyond the cranial edge of the vertebral centrum. The prezygapophyses accompany this inclination and are longer than in other titanosaurids (Fig. 5.36C). The articular facets of the prezygapophyses are wide, particularly in the proximal caudals. In the more distal caudals the entire neural arch is positioned on the cranial half of the centrum.

On its internal side, the scapular blade of *Aelosaurus* has a peculiar prominence for attaching muscles that is also present in *Saltasaurus* and *Neuquensaurus* among titanosaurs (Fig. 5.19D and E; prom). As in the latter taxa, the humerus and metacarpals are robust. The dermal plates of *Aeolosaurus* are thick and are distinguished from those of *Saltasaurus* by a depression on the ventral surface and the absence of a central projection on the dorsal surface.

Aeolosaurus, *Alamosaurus*, *Opisthocoelicaudia*, *Pellegrinisaurus*, saltasaurines, and probably *Antarctosaurus* share a biconvex first caudal, an unusual condition that is absent in other sauropods. This feature may constitute a synapomorphy that unites these taxa, but this notable feature is absent in *Saltasaurus* as well as in *Gondwanatitan*, a taxon that is interpreted as closely allied to *Aeolosaurus*.

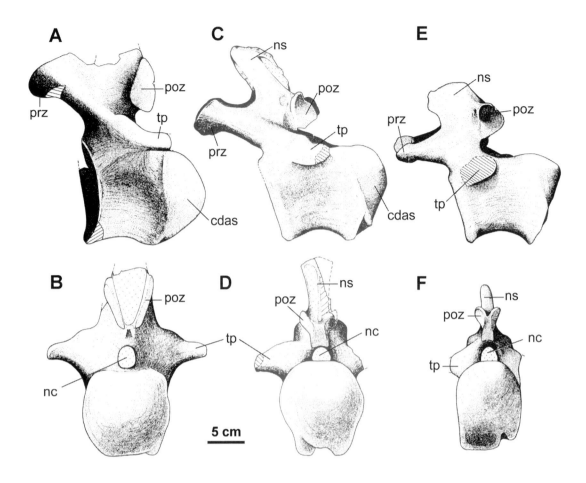

Fig. 5.36. *Aeolosaurus rionegrinus:* Caudal vertebrae in (A, C, E) left lateral and (B, D, F) caudal views. (A, B) Proximal caudal. (C, D) Mid-caudal. (E, F) Distal caudal. *From Salgado and Coria (1993). Reprinted with permission of* Ameghiniana.

GONDWANATITAN FAUSTOI

This sauropod, which constitutes the most complete sauropod yet found in Brazil, was discovered not far from Álvares Machado in the state of São Paulo in beds of the Adamantina Formation (Kellner and Campos 1997; Kellner and Azevedo 1999; Fig. 4.5). *Gondwanatitan* was a relatively small titanosaur that was 6–7 meters long. As occurs in most titanosaurids, the dorsal vertebrae does not have a hyposphene-hypantrum articulation (Fig. 5.37). The sacrum is biconvex (i.e., sacral 1 is opisthocoelous and the last sacral is procoelous), a condition that distinguishes *Gondwanatitan* from *Opisthocoelicaudia, Aeolosaurus, Pellegrinisaurus, Alamosaurus,* and presumably also *Antarctosaurus*, in which the last sacral is opisthocoelous in order to articulate with the biconvex first caudal. The last sacral and first caudal centra of *Gondwanatitan* thus resemble those of *Saltasaurus*.

As seen in caudal view, the proximal and mid-caudal vertebrae of *Gondwanatitan* are heart-shaped in outline. Kellner and Azevedo (1999) interpret this as a main feature that distinguishes *Gondwanatitan* from all other described titanosaurids (e.g., *Alamosaurus, Saltasaurus, Neuquensaurus*), in which the centrum is dorsoventrally depressed and has convex

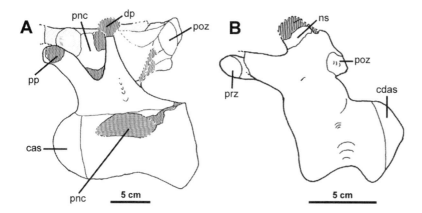

Fig. 5.37. *Gondwanatitan faustoi:* (A) Dorsal 6 in left lateral view. (B) Caudal 5 in lateral view. *From Kellner and Azevedo (1999). Reprinted with permission of the authors and the National Science Museum Monographs, Tokyo.*

lateral faces. Nevertheless, Salgado and García (2002) dismissed the diagnostic value of this feature, because the heart-shaped contour of the caudal centra varies with the position of the vertebra along the caudal series and comparison of non-homologous caudal segments may result in equivocal distribution of this feature.

Because *Gondwanatitan* has a cranially directed neural spine on the proximal caudals, it bears the greatest similarities to *Aeolosaurus rionegrinus* (Kellner and Azevedo 1999; Fig. 5.37B). Bertini and colleagues (2000) interpreted this characteristic, along with other appendicular features, as strong evidence for reconsideration of *Gondwanatitan* as a junior synonym of *Aeolosaurus*. If this taxonomic referral proves to be correct, then *Aeolosaurus* may constitute a form of titanosaurid that attained a wide geographic distribution in South America.

RINCONSAURUS CAUDAMIRUS

This new titanosaur, which was an estimated 11 meters long, was described by Jorge Calvo and Bernardo González Riga in 2003. The holotype specimen of *Rinconsaurus caudamirus* comes from the Turonian–Coniacian Río Neuquén Subgroup at Rincón de los Sauces, north of Neuquén Province. *Riconsaurus* is known by an articulated series of 13 proximal caudals that were associated in "life position" with two ilia. Also, in the same quarry, teeth, skull bones, cervical and dorsal vertebrae, and appendicular bones were collected that were referred to the same taxon (Calvo and González Riga 2003). *Rinconsaurus* is characterized by steeply inclined neural spines in the cranial and mid-dorsal vertebrae (which are more caudally inclined than in any other known titanosaur; Fig. 5.38A). However, the most remarkable feature is the variable condition of the distal caudal centra, as seen in two different articulated series of vertebrae (Fig. 5.38B and C). In one of the series, a procoelous centrum articulates distally with an amphicoelian centrum and the latter is distally articulated with a biconvex centrum. In the other series, an opisthocoelous centrum distally articulates with a biconvex vertebra that distally fits into a procoelous centrum.

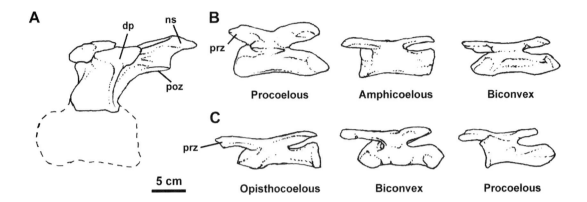

Fig. 5.38. *Rinconsaurus caudamirus:* Vertebrae in left lateral view: (A) Neural arch of an anterior dorsal (reconstructed centrum indicated by dotted line). (B) Series of originally articulated distal caudals with procoelous, amphicoelous, and biconvex centra. (C) Series of originally articulated distal caudals of another specimen of *Rinconsaurus caudamirus* with opisthocoelous, biconvex, and procoelous centra. Redrawn from Calvo and González Riga (2003).

The teeth of *Rinconsaurus* are elongate and thin. The crowns are D-shaped in cross-section due to longitudinal carinae that separate both the labial and the lingual surfaces. This kind of tooth fills a morphological gap between the broad spatulate teeth of basal titanosauriforms and the completely cylindrical teeth of advanced titanosaurs (Apesteguía 2004).

Calvo and González Riga (2003) concluded that *Rinconsaurus* is more closely related to *Aeolosaurus* than to other titanosaurids because they both have elongate prezygapophyses on the caudal vertebrae. Nevertheless, Wilson and colleagues (1999) have interpreted its biconvex distal caudal centra (which Calvo and González Riga theorize is an autapomorphy of *Rinconsaurus*) as diagnostic of a clade that includes *Opisthocoelicaudia* and saltasaurids.

TRIGONOSAURUS PRICEI

The holotype specimen of *Trigonosaurus pricei* is comprised of a continuous sequence of vertebrae that includes cervicals 9 to 13 (the last one of which is preserved in articulation with dorsal 1), 10 dorsals, a sacrum, and an ilium. Another specimen referred to the same taxon includes an incomplete sequence of 10 proximal caudals. Both specimens were excavated at Veadinho Hill, close to Periópolis in the Brazilian state of Minas Gerais, from sandstone of the Serra da Galga Member of the Marília Formation (Campos, Kellner, Bertini, and Santucci 2005; Fig. 4.5).

The cervicals of *Trigonosaurus* (Fig. 5.39) are proportionally longer than in *Saltasaurus* and *Neuquensaurus* and the prezygapophyses are not curved upward, as occurs in the latter two Patagonian taxa. Furthermore, the caudals are craniodorsally inclined, the prezygapophyses are proximally projected, and the neural spines are vertical or even cranially inclined (Fig. 5.40). This condition contrasts with that of *Saltasaurus* and *Neuquensaurus* (Fig. 5.12), in which the caudal vertebrae have a strongly caudally directed neural spine (Salgado, Apesteguía, and Heredia 2005) and are dorsoventrally more depressed (Campos, Kellner, Bertini, and Santucci 2005).

The caudals of *Trigonosaurus* lack the hyposphene-hypantrum articulation observed in more-basal titanosaurids (e.g., *Epachthosaurus*). The

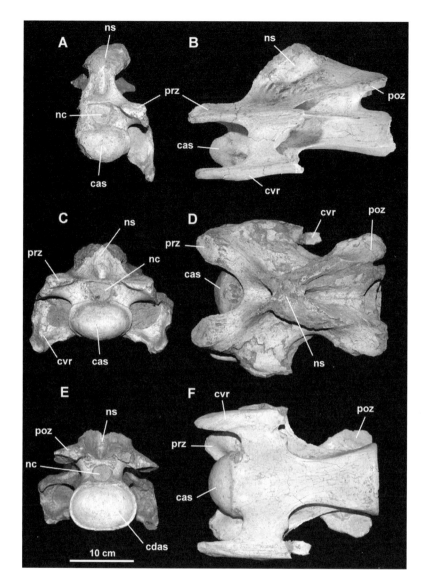

Fig. 5.39. *Trigonosaurus pricei*: (A, B) Cervical 9 in (A) cranial and (B) lateral views. (C–F) cervical 11 in (C) cranial, (D) dorsal, (E) caudal, and (F) ventral views. From Campos, Kellner, Bertini, and Santucci (2005). Reprinted with permission of Arquivos do Museu Nacional do Río de Janeiro.

anteriorly displaced neural arch and the anterodorsally projected neural spine of *Trigonosaurus pricei* suggest some affinities with *Aeolosaurus rionegrinus* and *Gondwanatitan faustoi*. However, none of the preserved elements of the new taxon show a constricted ventral part of the centrum, which would give the posterior surface of the centrum the heart-shaped appearance that is characteristic of these aeolosaurines (Campos, Kellner, Bertini, and Santucci 2005).

The anterior caudals of *Trigonosaurus* differ from those of *Baurutitan* (which was collected from the same fossil spot) in that the neural spine is further directed anterodorsally and is placed on the anterior half of the centrum close to the anterior margin. Further, *Trigonosaurus* lacks the dorsal tuberosity in the contact region between the transverse process and the neural arch that is present in *Baurutitan* (Campos, Kellner, Bertini, and Santucci 2005; Kellner, Campos, Azevedo, and Trotta 2005).

Fig. 5.40. *Trigonosaurus pricei*: Sequence of originally articulated proximal caudal series in (A) right lateral and (B) distal views. *From Campos, Kellner, Bertini, and Santucci (2005). Reprinted with permission of Arquivos do Museu Nacional do Río de Janeiro.*

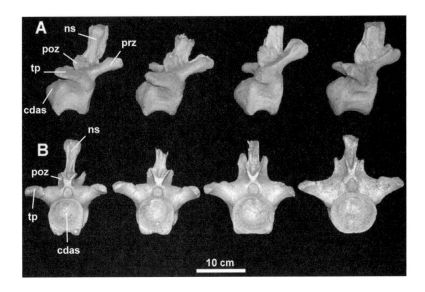

The morphological pattern of the pelvic girdle of *Trigonosaurus* is similar to that described for other titanosaurids (e.g., *Epachthosaurus*, *Saltasaurus*); the preacetabular wings of the ilia are flared out. The degree of expansion of the ilia in *Trigonosaurus* seems greater than in the basal titanosaurid *Epachthosaurus* (Fig. 5.18). The Veadinho Hill fossil site yielded three beautifully preserved sacra in articulation with at least one ilia (Campos and Kellner 1999). One of these has been referred to *Trigonosaurus*, but the others are currently assigned as Titanosauridae indet. (Fig. 5.41).

Some features of *Trigonosaurus* indicate that it is not a member of Saltasaurinae (at least), but some similarities it shares with *Gondwanatitan* and *Aeolosaurus*, including a biconvex sacrum and proximodorsally inclined caudal vertebrae, suggest that *Trigonosaurus* may constitute another member of the Aeolosaurinae.

ADAMANTISAURUS MEZZALIRAI

Santucci and Bertini (2006) recognized this taxon based on a series of six articulated proximal caudal vertebrae (probably caudals 2 to 8) that were excavated not far from the city of Florida Paulista (São Paulo State). The vertebrae have straight or slightly backward-projecting neural spines whose distal ends are strongly expanded laterally. The postzygapophyses of *Adamantisaurus*, which have concave articular facets, resemble those of *Trigonosaurus* and *Aeolosaurus*, and the laterally expanded neural spines and stout prespinal lamina of *Adamantisaurus* resemble those of *Trigonosaurus*.

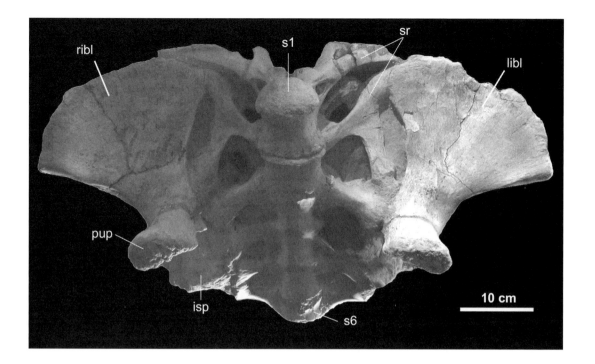

Fig. 5.41. Ilia and sacrum (in cranioventral view) of a titanosaurid indet. Collected at the Peirópolis fossil site, Brazil. *Photographs by Fernando Novas.*

Titanosaurids Probably Related to Saltasaurinae

The following three titanosaurians *(Baurutitan britoi, Maxakalisaurus topai,* and *Pellegrinisaurus cincosaltensis)* show some similarities with the Saltasaurinae. However, their systematic allocation must await a rigorous phylogenetic analysis of the South American titanosaurs.

BAURUTITAN BRITOI

Llewellyn Ivor Price conducted productive field seasons at Veadinho Hill. From beds of the Maastrichtian Marília Formation (Serra da Galga Member), he recovered an almost-complete and partially articulated series of cervical, dorsal, sacral, and caudal vertebrae pertaining to different titanosaurian specimens; the sacral vertebrae were found in articulation with one or two of the ilia. Argentine paleontologist Jaime Powell published preliminary descriptions of these beautifully preserved materials (1986, 1987a, 2003). He informally named these specimens (each part of a separate individual) Series A (a continuous sequence of articulated vertebrae from cervicals 3 to 12 and dorsal vertebrae 1 to 3), Series B (cervical vertebrae 9 to 11, dorsal vertebrae 1 to 9, a sacrum with the left ilium articulated, and 15 nearly articulated caudal vertebrae), and Series C (the last sacral and 18 caudals). However, Powell (2003) referred Series C to cf. *Titanosaurus* sp. and Series A and B to Titanosauridae indet.

Recently, Alexander Kellner and colleagues (2005) and Diogenes de Almeida Campos and colleagues (2005) offered a detailed reappraisal of the titanosaurid remains found in Peirópolis that recognized two different

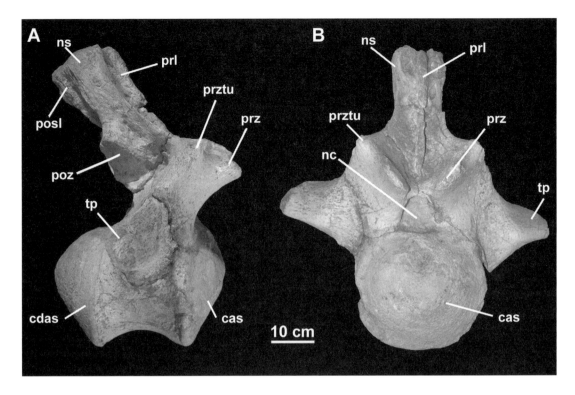

Fig. 5.42. *Baurutitan britoi:* Caudal 1 in (A) right lateral and (B) cranial views. From Kellner, Campos, Azevedo, and Trotta (2005). Reprinted with permission of Arquivos do Museu Nacional do Río de Janeiro.

taxa: *Baurutitan britoi* (based on the specimen originally named Series C) and *Trigonosaurus pricei* (based on the specimen originally named Series B).

Baurutitan britoi is represented by the last sacral and 18 caudal vertebrae and thus is one of the most complete articulated caudal sequences for a titanosaurian. This sample has made it much easier to place isolated caudal elements in a tail series (Kellner, Campos, Azevedo, and Trotta 2005). Based on comparisons with other titanosaurian specimens, the length of *Baurutitan britoi* is estimated at between 12 and 14 meters. The last sacral of *Baurutitan* is not fused to the previous sacral and its posterior articular surface is strongly concave. The first caudal vertebra is biconvex (Fig. 5.42A), resembling *Aeolosaurus*, *Pellegrinisaurus*, and *Alamosaurus*, for example. The proximal caudals have a prominent tuberosity on the neural arch that turns into a lateral ridge in the middle caudals (Fig. 5.42B; prztu). The anterior caudals of *Baurutitan* have short prezygapophyses and neural spines that are posteriorly inclined, thus distinguishing it from aeolosaurines (e.g., *Aeolosaurus*, *Gondwanatitan*), in which the neural arch of the proximal caudals is cranially inclined and the prezygapophyses are notably elongate.

Baurutitan britoi lacks several caudal traits found in the saltasaurines. These include a strongly dorsoventrally flattened centrum of the anterior caudals, a centrum that is significantly wider than it is high, and a neural spine whose anterodorsal edge is positioned behind the anterior border of the postzygapophyses.

MAXAKALISAURUS TOPAI

This taxon is known by an incomplete maxilla (with teeth; Fig. 5.43A–C), 12 cervical vertebrae (including several cervical ribs; Fig. 5.43D), parts of seven dorsal vertebrae (and ribs), one sacral centrum, six caudals, several haemal arches, parts of both scapulae, both sternal plates, a fragmentary ischium, both humeri, some metacarpals, an incomplete fibula, and one osteoderm. The specimen was recovered about 45 kilometers west of the town of Prata in a region of Brazil's Minas Gerais State called Serra da Boa Vista (Fig. 4.5). The materials come from beds regarded as part of the Turonian–Santonian Adamantina Formation of the Bauru Group.

The estimated length of this dinosaur is roughly 13 meters. The preserved portion of the maxilla curves gently inward, suggesting that the anterior part of the skull of *Maxakalisaurus* was rounded, as has been observed in other titanosaurs whose skulls are known (Kellner et al. 2006). The teeth are pencil-like and have anterior and posterior carinae, which are not normally found in other titanosaurid teeth (other titanosaurs with carinated teeth are *Rinconsaurus* and *Rapetosaurus*; Calvo and González Riga 2003; Curry Rogers and Forster 2004; Kellner et al. 2006).

The cervical vertebrae of *Maxakalisaurus* are opisthocoelous and have no pleurocoels (Fig. 5.43D). The tip of the neural spine is blunt and is transversally thickened. As Kellner and colleagues (2006) remark, *Maxakalisaurus* falls in the category of elongated cervical vertebrae, therefore differing from the condition observed in *Mendozasaurus*, for example, which has shorter centra and taller neural spines (González Riga 2005). The cervical vertebrae of the Saltasaurinae *Saltasaurus* also display a morphology that is quite distinct from those preserved in *Maxakalisaurus*. It is shorter because of short prezygapophyses that are dorsally inclined, and the articular facets are near the level of the diapophyses (a sinapomorphy of the Saltasaurinae), the postzygapophyses are posteriorly extended, the neural spine is low, and deep pleurocoels perforate the centrum (Kellner et al. 2006).

The preserved caudals of *Maxakalisaurus* are strongly procoelous. However, the presence of a biconvex mid-caudal vertebra indicates that this Brazilian sauropod had a variation in the morphology of its centrum that is probably similar to that described for *Rinconsaurus* (Calvo and González Riga 2003).

The humerus of *Maxakalisaurus* is slender and elongate, differing from the robust humeri of *Opisthocoelicaudia*, *Argyrosaurus*, *Neuquensaurus*, and *Saltasaurus*.

PELLEGRINISAURUS POWELLI

This titanosaurid was found around Pellegrini Lake in northwestern Río Negro Province. The specimen was recovered from the Anacleto Formation, which is considered to be Early Campanian in age (Leanza, Apesteguía, Novas, and de la Fuente 2004). Powell (1986) originally felt

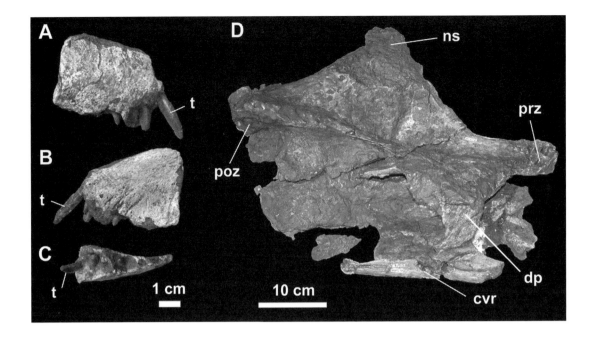

Fig. 5.43. *Maxakalisaurus topai:* (A–C) Right maxilla with teeth in (A) lateral, (B) medial, and (C) ventral views. (D) Cervical vertebra in right lateral view. From Kellner et al. (2006). Reprinted with permission of Boletim do Museu Nacional, Geológia.

that the holotype specimen pertained to the basal titanosaurid *Epachthosaurus* because of details of the dorsal vertebrae. Later, Salgado (1996) reviewed this specimen and recognized derived features that are absent in *Epachthosaurus* and the remaining titanosaurs, thus indicating that the specimen is representative of a new taxon, *Pellegrinisaurus powelli* (Salgado 1996; Fig. 5.44). *Pellegrinisaurus* is a large titanosaurid (its total length is estimated to be 20–25 meters) and is distinguished from other titanosaurids in that its mid- to distal caudals are craniocaudally elongate and have dorsoventrally depressed neural spines whose anterior corners are at a higher position than the posterior ones (Fig. 5.44F and G). The autapomorphic characteristics of *Pellegrinisaurus powelli* include dorsal centra that are strongly depressed, especially the posterior dorsal ones, in which the transverse width of the centrum is approximately twice the maximum dorsoventral depth (Fig. 5.44B).

Two different morphologies of caudal centra are recognized within Titanosauridae. In the basal members of Titanosauridae (e.g., *Andesaurus, Epachthosaurus, Aeolosaurus*), the caudal centra are craniocaudally short and dorsoventrally deep and have lateral surfaces that are slightly concave craniocaudally. In Saltasaurinae, in contrast, the caudal centra are proportionally low (e.g., they are wider than they are high) and have dorsoventrally convex lateral sides and a broad ventral surface. *Pellegrinisaurus* exhibits the apomorphic condition, although it is restricted to the mid- to posterior caudal vertebrae. Thus, this titanosaurid is interpreted as the sister taxon of Saltasaurinae (e.g., the group composed of *Neuquensaurus australis, Saltasaurus loricatus,* and *Rocasaurus muniozi*), in which the entire tail series shows the derived condition (Salgado 1996).

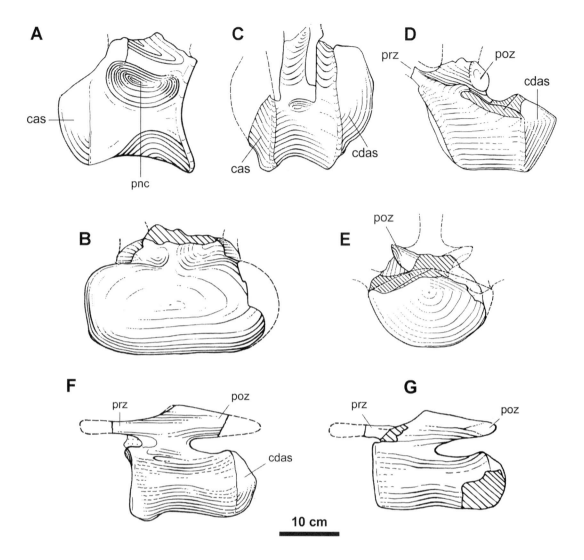

Fig. 5.44. *Pellegrinisaurus powelli:* (A, B) Posterior dorsal vertebra in (A) left lateral and (B) cranial views. (C) Caudal 1 in left lateral view. (D, E) Proximal caudal vertebra in (D) left lateral and (E) distal views. *From Salgado (1996). Reprinted with permission of* Ameghiniana.

Saltasaurinae

The name Saltasaurinae was coined by Jaime Powell in 1986 to designate the clade that includes *Saltasaurus loricatus*, from Salta Province in northwestern Argentina, and *Neuquensaurus australis*, from northwestern Patagonia. More recently, the Patagonian taxa *Rocasaurus muniozi* and *Bonatitan reigii* have been added to the list of saltasaurines (Salgado and Azpelicueta 2000; Martinelli and Forasiepi 2004). Saltasaurinae includes titanosaurs (Fig. 5.9) around 6 meters long; these (along with dicraeosaurid diplodocoids) are the smallest known sauropods (Salgado 1999). Saltasaurines are diagnosed by the following modifications of the vertebral column (Fig. 5.45A and B): short prezygapophyses in the cervical vertebrae with articular facets located near level of diapophyses; postzygapophyses of the cervical vertebrae that project outward and backward, extending beyond the caudal margin of the vertebral centrum;

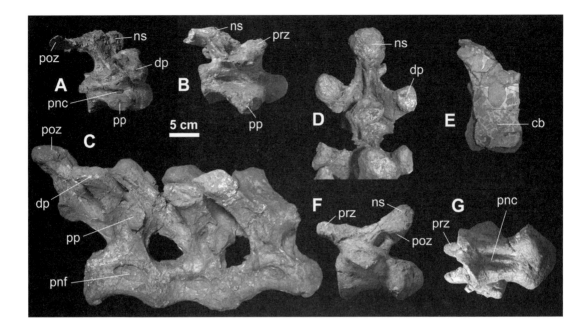

Fig. 5.45. *Saltasaurus loricatus:* (A) Cervical vertebra in right lateral view. (B) A more posterior cervical vertebra in right lateral view. (C, D) Articulated sequence of three mid-dorsal vertebrae in (C) right lateral aspect, and (D) detail of neural arch of the last of these vertebrae in dorsal view. (E) Cross-section of a mid-dorsal vertebra showing cancellous bone tissue. (F, G) Distal caudal vertebra in (F) left lateral and (G) ventral views. *Photographs by Fernando Novas.*

dorsoventrally depressed caudal vertebrae; a deeply excavated ventral surface of the caudal centra; mid-caudals with neural spines posteriorly placed with respect to the anterior base of postzygapophyses (a reversal to the probable ancestral condition for Sauropodomorpha); and cancellous bone in the caudal centra. Cancellous bone (Fig. 5.45E) is also documented among basal titanosaurids, for whom it is restricted to the dorsal vertebrae. But Saltasaurines developed this kind of bony tissue in the caudal centra as well. Leonardo Salgado (2003b) suggested that this extensive development of cancellous tissue in the caudal vertebrae may be a side effect of calcium loss during the formation of the eggshell of these dinosaurs rather than an adaptive device for lightening the axial skeleton. Variations in the degree of development of cancellous bone is verified among saltasaurines specimens (L. Salgado, pers. comm.). Following Salgado's hypothesis, we expect that specimens with a greater development of cancellous bone would be the females (J. Farlow, pers. comm.). This interesting but still poorly explored topic requires testing. It may be a useful tool for distinguishing the sex of samples of titanosaurian sauropods.

Titanosaur diversity increased in South America toward the end of the Cretaceous, when the saltasaurines *Saltasaurus*, *Neuquensaurus*, *Rocasaurus*, and *Bonatitan* appeared. The fossil record of saltasaurine titanosaurs is confined to Argentina. Because saltasaurines appear to be absent in other parts of the world, Salgado, Coria, and Calvo (1997) thought that saltasaurines evolved after the fragmentation of Gondwana, principally after the complete opening of the South Atlantic Ocean.

NEUQUENSAURUS AUSTRALIS

The holotype of this species consists of six caudal vertebrae that were discovered at the end of the nineteenth century in the ravines of the Neuquén River. Richard Lydekker, who studied these remains in 1893, coined the name *Titanosaurus australis* for this taxon. Later, more specimens were discovered at the productive location of Cinco Saltos in the Río Negro Province, and von Huene (1929a) referred the later specimens to this species (Figs. 5.9, 5.12, 5.19 and 5.46). Currently, paleontologists believe that these remains come from the Anacleto Formation, Early Campanian. More recently, Powell (2003) interpreted the anatomy of *Titanosaurus australis* as different enough from that of *Titanosaurus indicus* to merit a new genus, *Neuquensaurus australis*. As Bonaparte (1996b) pointed out, the most significant differences between *Titanosaurus* and *Neuquensaurus* are in the morphology of the caudal vertebrae and in the bones of the forelimb. *Neuquensaurus* is, in turn, similar to *Saltasaurus*.

The evidence at hand demonstrates that *Neuquensaurus* was armored (Salgado, Apesteguía, and Heredia 2005), and it is highly possible that the scutes von Huene (1929a) described from Cinco Saltos as belonging to the ankylosaur "*Loricosaurus scutatus*" correspond in fact to *Neuquensaurus* (Powell 1980; Bonaparte 1996b).

The diagnostic features of *Neuquensaurus* (Powell 1986; Salgado, Apesteguía, and Heredia 2005) include a fibula with a strong lateral tuberosity and a bent shaft, seven sacral vertebrae, transversely narrow sacral centra 3 to 5, mid- to posterior caudal vertebrae with a transversely wide non-keeled ventral depression that is bounded by lateral rounded ridges culminating in articular facets for the hemapophyses, and lateral walls of caudal centra that are little exposed in ventral view. *Neuquensaurus* is distinguished from *Laplatasaurus*, *Argyrosaurus*, *Antarctosaurus*, and *Saltasaurus* in having a short and paddle-like scapular blade (Fig. 5.19C). The prominent acromial process forms a strong U-shaped notch with the scapular blade, which is different from the wider angle present in that portion of the cranial margin of the scapulae of *Antarctosaurus* and *Saltasaurus*. The coracoid is deep and rectangular-shaped in side view, as in other titanosaurids in which this bone is known (e.g., *Saltasaurus* and *Opisthocoelicaudia*). The humerus is short and robust, similar to that of *Saltasaurus* and *Argyrosaurus*, for example.

Neuquensaurus differs from its close relative *Saltasaurus* in details of the proximal caudal vertebrae, which have small but well-defined oval pleurocoel-like openings. This feature developed convergently among advanced diplodocids (Upchurch, Barrett, and Dodson 2004). The most significant difference in the morphology of the dorsal vertebrae is in the height of the neural spines, which are higher than in *Saltasaurus* (Salgado, Apesteguía, and Heredia 2005). Unlike other titanosaurs, in which the number of sacral vertebrae is invariably six, the sacrum of *Neuquensaurus* is composed of seven vertebrae (Salgado, Apesteguía, and Heredia

2005). The centra of the first six vertebrae are solidly co-ossified, but the seventh centrum is unfused. The last sacral is apparently biconvex, so the first caudal is strongly procoelous.

Differences are also seen in the appendicular skeleton, mostly in the ilium and ischium. As Salgado and colleagues (2005) noted, the ilia of *Neuquensaurus* are slightly more expanded than in more basal titanosaurids (e.g., *Epachthosaurus*, *Trigonosaurus*). The ischium of *Neuquensaurus* is similar to that of *Saltasaurus* in its relatively narrow caudoventral shaft, although it differs in that the iliac peduncle is large relative to the whole bone (Powell 1986; Salgado, Apesteguía, and Heredia 2005).

Salgado and colleagues (2005) have noted that available *Neuquensaurus* specimens can be sorted into two different morphs—slender and robust. This situation also applies to the specimens of *Saltasaurus* found in the El Brete fossil quarry (Powell 1986). Von Huene (1929a) interpreted these two classes of bones as reflecting two species of *Neuquensaurus* (i.e., the stout *N. robustus* and the more slender *N. australis*). Powell (1986) dismissed this specific separation, hypothesizing that such dissimilarities could be due to individual or sexual variation.

SALTASAURUS LORICATUS

Most of what we know about the anatomy and paleobiology of saltasaurines in particular and titanosaurids in general is based on this taxon.

Saltasaurus was found by José Bonaparte and crew in beds of the Early Maastrichtian Lecho Formation exposed at the El Brete fossil location in Salta Province in northwestern Argentina. *Saltasaurus loricatus* is represented by several specimens, including a partial skull and many well-preserved postcranial bones (Figs. 5.11C–E, 5.17B, 5.19D and E, 5.21C–E, and 5.45). As is *Neuquensaurus*, *Saltasaurus* is distinguished from more-primitive titanosaurs by its smaller body size, which is around 6 meters long, and its short and robust appendicular bones. *Saltasaurus* teeth are cylindrical and slightly spatulate at the tip. The neck is composed of 13 cervical vertebrae that are craniocaudally shorter than in other titanosaurs (Fig. 5.45A and B). As mentioned earlier, *Aeolosaurus*, *Alamosaurus*, and *Opisthocoelicaudia* have a biconvex caudal 1. The procoelous first caudal of *Saltasaurus* may be a secondary reversal to a more-primitive condition; this was recently reported for its close relative *Neuquensaurus* (Salgado, Coria, and Calvo 1997), thus constituting another synapomorphic feature of saltasaurines. As in other titanosaurids, the iliac blades of *Saltasaurus* flare craniolaterally (Fig. 5.11E). The unique features of *Saltasaurus* are (among others) fused basal tubera, proximal caudal vertebrae that are twice as wide as they are high, and proximal caudal ribs that are robust and expanded dorsoventrally at their distal ends (Upchurch, Barrett, and Dodson 2004).

Saltasaurus has circular dermal scutes approximately 10 centimeters in diameter and small ossicles the size of beans that are closely packed together (Fig. 5.21C–E). It is possible that these small ossicles were

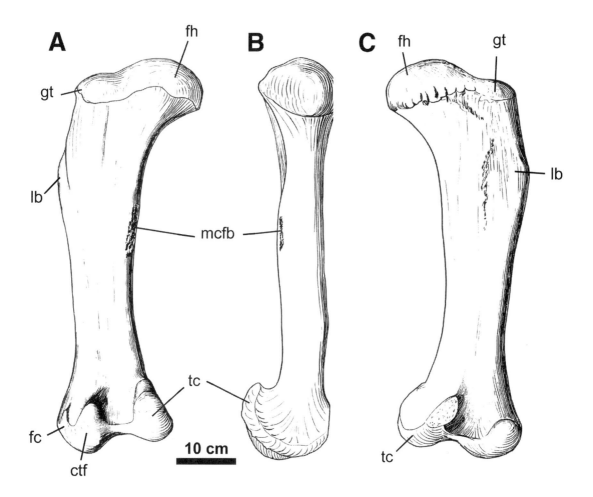

Fig. 5.46. *Neuquensaurus australis:* Left femur of robust morph of *N. australis* in (A) caudal, (B) medial, and (C) cranial views. *From von Huene (1929a).*

embedded in the skin, as they are in mylodontid edentates (Cenozoic ground sloths).

ROCASAURUS MUNIOZI

The holotype specimen of this species was described by Salgado and Azpelicueta in 2000. It comes from the lower member of the Allen Formation (Powell 1986), from the Salitral Moreno fossil site, which is approximately 25 kilometers south of the city of General Roca in Río Negro Province. *Rocasaurus* (Fig. 5.47) is distinguished in having caudal centra that are much more dorsoventrally depressed than in other titanosaurs (including saltasaurines). Also, *Rocasaurus* has a plate-like ischium that is notably more expanded than in *Saltasaurus* and *Neuquensaurus*. In *Rocasaurus* the system of internal cavities of the caudal vertebrae is more developed than in any other titanosaur, and this system communicates with the exterior through small pneumatic pores on the ventral and lateral surfaces of the centrum (Fig. 5.47E). *Rocasaurus* is closely related to *Saltasaurus*, as indicated by the shared presence of proximal and mid-caudal centra with a ventral cavity that is separated by a longitudinal septum (compare

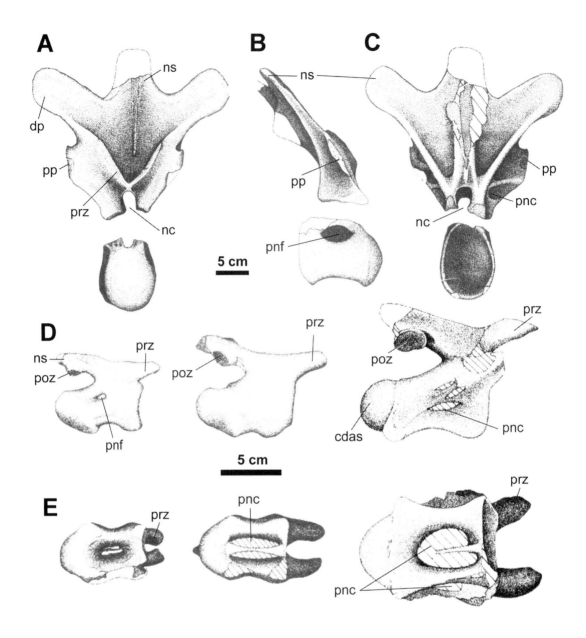

Fig. 5.47. *Rocasaurus muniozi:* (A–C) Dorsal vertebra in (A) cranial, (B) right lateral, and (C) caudal views. (D, E) Series of distal caudal vertebrae in (D) right lateral and (E) ventral views. From Salgado and Azpelicueta (2000). Reprinted with permission of *Ameghiniana*.

Figs. 5.45G and 5.47E). However, Martinelli and Forasiepi noted (2004) that the limb bones of *Rocasaurus* are elongate and slender, as they are in more-basal titanosaurids, and thus they differ from the robust bone structure of *Saltasaurus* and *Neuquensaurus*.

BONATITAN REIGII

This titanosaurid was recently named by Agustín Martinelli and Analía Forasiepi (2004) in honor of its discoverer, José Bonaparte, and the late Osvaldo Reig. *Bonatitan* is a probable saltasaurine titanosaurid represented by two incomplete but beautifully preserved specimens of different size that were found mixed at the same fossil location (Bajo de Santa

Rosa, Río Negro Province; Fig. 4.13) in beds of the Allen Formation. The basicranium preserves interesting details and resembles that of other titanosaurids (Fig. 5.48A and B). The frontals are rostrocaudally short and faced dorsally. They form the rostral margin of the supratemporal fenestra, as occurs in other titanosaurids (e.g., *Antarctosaurus, Rapetosaurus, Saltasaurus*). The frontals failed to fuse together, but they fuse caudally with the parietals.

The parietals are wider than in *Antarctosaurus*, but they are similar in proportion to the parietals of *Rapetosaurus*. Where both parietals contact, a tuberosity continues and enlarges backward over the supraoccipital. This tuberosity, which presumably served for the attachment of neck musculature, defines a longitudinal groove that extends from the parietal to the foramen magnum. Although a similar parietal groove is present in other titanosaurids (e.g., *Rapetosaurus*), that of *Bonatitan* extends rostrally. In *Antarctosaurus* the supraoccipital forms a distinctive process, but a longitudinal groove is not evident. The basipterygoid processes of basisphenoid are elongate and slender and are rostroventrally projected. They are less divergent than those of *Antarctosaurus* (in which they form an angle of nearly 90 degrees), but they are more divergent than in *Rapetosaurus*. The basisphenoidal tubera are elongated and narrow and are not fused to each other; they delimit the lower surfaces of the pituitary fossa. The basisphenoid tubera of *Bonatitan* differ from those of *Antarctosaurus* and *Rapetosaurus*, in which they are dorsoventrally shorter and transversely wider.

The caudal vertebrae of *Bonatitan* are strongly procoelous (Fig.5.48D). The centrum is wide, short, and high and has a prominent axial ventral keel. This feature is absent in other currently known titanosaurs and is considered an autapomorphy of *Bonatitan*. It contrasts with the structure of *Saltasaurus, Neuquensaurus*, and *Rocasaurus*, in which the proximal caudal vertebrae are dorsoventrally depressed and have an axial ventral depression. In *Rocasaurus* this depression is relatively deeper and has an axial septum (Figs. 5.47E).

The craniodorsal corner of the neural spine is at the same level as the caudal edge of postzygapophyses, a feature considered to be a synapomorphy of the Saltasaurinae (Salgado, Coria, and Calvo 1997) because it occurs in *Saltasaurus, Neuquensaurus*, and *Rocasaurus*.

Although some vertebral features suggest that *Bonatitan* is a member of Saltasaurinae, some characteristics of the forelimb bones are not consistent with expectations for this subfamily. For example, the humerus of *Bonatitan* is more slender than in *Neuquensaurus* and *Saltasaurus*. The humeral head and deltopectoral crest are poorly developed; this is similar to the condition present in less-derived titanosaurids (e.g., *Laplatasaurus, Lirainosaurus* and *Rapetosaurus*). In contrast, *Neuquensaurus* and *Saltasaurus* have a more prominent head and a more robust deltopectoral crest. The ulna of *Bonatitan* is more slender than in *Neuquensaurus* and *Saltasaurus*. A similar situation occurs with the hindlimb bones, which are more elongate and gracile than in saltasaurines. The proximal end of

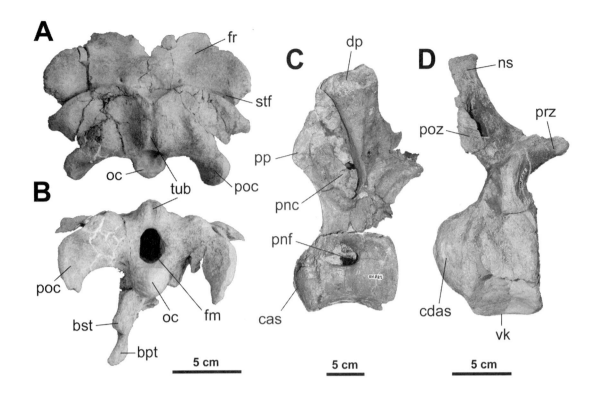

the tibia in *Bonatitan* is craniocaudally narrower than in *Neuquensaurus* and *Saltasaurus*, and the cnemial crest is transversely thin and poorly developed; in *Neuquensaurus* and *Saltasaurus* it is robust and prominent.

Cretaceous Sauropod Eggs and Nests

South America has yielded a significant record of eggs, nests, and embryos of sauropod dinosaurs that almost surely correspond to Titanosauridae. These discoveries currently are restricted to the Campanian–Maastrichtian time interval from localities in Perú (in the Vilquechico Formation in Laguna Umayo near the Bolivian border and in the Bagua Formation in northern Perú; Sigé 1968; Mourier et al. 1988; Kerourio and Sigé 1984; Vildoso Morales 1991), in Brazil (in the Marília Formation in Uberaba; Price 1951; Magalhães Ribeiro 2002; Magalhães Ribeiro and Ribeiro 2000), in Uruguay (in the Mercedes Formation in Soriano and Algorta; Mones 1980; Faccio 1994; Faccio, Ford, and Gancio 1990; Fig. 4.5), and in Argentina. The Argentine sites are the most productive as well as the most informative paleobiologically. Except for some isolated reports of eggshells from Sanagasta in La Rioja Province in northwestern Argentina (Hünicken, Tauber, and Leguizamón 2001), abundant eggshells and eggs are concentrated in Patagonia, with an astonishing abundance of material in beds of the Anacleto Formation in Auca Mahuevo in Neuquén Province (Chiappe, Coria, Dingus, Jackson, Chinsamy, and Fox 1998) and the Allen Formation (and equivalent units) in Río Negro and La Pampa provinces (Frenguelli 1951; Powell 1985, 1987b; Manera de Bianco 1996, 2000a; Manera de Bianco and Bolognani 2003; Manera de Bianco, Montalvo, Casadío, and Parras 1999; Manera de Bianco, Parras, Montalvo, and Casadío 2000; Casadío, Manera, Parras, Montalvo, and Cornachione 2000; Casadío, Manera, Parras, and Montalvo 2002; Martinelli and Forasiepi 2004; Fig. 5.1).

Most of these South American discoveries consist of fragmentary eggshells as well as eggs in different states of preservation, and they frequently occur in beds where titanosaurid bones are also abundant. The eggs attributed to titanosaurid sauropods are large, with diameters that reach 18 centimeters and volumes of up to 5,500 milliliters, and are spherical or subspherical. The shells are usually very thick (5 millimeters on average) and have a nodular surface and a multispherulitic structure consisting of very long, thin, crystalline units (Faccio 1994). Eggs of this kind are usually called *Megaloolithus*, but they were previously named *Sphaerouvum erbeni* by Alvaro Mones in 1980, and technically this is the proper generic name for large spherical megaloolithid eggs (Carpenter 1999).

Powell (1985) and Moratalla and Powell (1994) reported the recovery and mapping of several groups of eggs (thought to be nests) at Salitral Moreno in Río Negro Province in Argentina. The nests, containing up to 12 eggs, are close together. Faccio (1994) presented some interesting and reasonable speculations about nesting behavior based on discoveries in Uruguay (Algorta and Soriano fossil sites) and Salitral Moreno. He suggested that the proximity of different clutches to each other plus the large size of the adult might mean that adults did not directly care for eggs and hatchlings. The absence of the skeletal remains of babies and juveniles in the nest-bearing layer combined with the presence of many hatched eggs led Faccio to suggest a precocial behavior for the titanosaurid species that laid the eggs. The superposition of two levels of eggs suggests that the parent returned to the same nesting area each year (Faccio 1994).

Auca Mahuevo, located on the southeastern slope of the extinct volcano Auca Mahuida in Neuquén, Argentina, is by far the most outstanding site for titanosaurid eggs and nests. This wonderful location, which Luis Chiappe and his colleagues discovered in 1998, has yielded hundreds of well-preserved eggs (Fig. 5.49A) buried in rocks of the Campanian Anacleto Formation. Auca Mahuevo provides the only reliable evidence of sauropod embryos contained inside eggs with a megaloolithid microstructure. Auca Mahuevo is also remarkable because of its enormous size (the egg layers extend laterally for several kilometers) and for its several egg layers (Chiappe, Coria, Dingus, Jackson, Chinsamy, and Fox 1998; Chiappe, Jackson, Dingus, Grellet-Tinner, and Coria 1999; Chiappe et al. 2000; Chiappe, Salgado, and Coria 2001).

The distribution of the eggs as well as the characteristics of the egg-bearing sediments rule out the possibility that the eggs were relocated after they were laid. Consequently, Auca Mahuevo is an important site that clarifies aspects of the reproductive biology of sauropod dinosaurs and provides clear evidence about their nesting behavior (Chiappe, Coria, Dingus, Jackson, Chinsamy, and Fox 1998; Chiappe, Coria,

Fig. 5.48. *Bonatitan reigi*: (A, B) Basicranium in (A) dorsal and (B) caudal views. (C) Mid-dorsal vertebra in left lateral view. (D) Proximal caudal in right lateral view. *From Martinelli and Forasiepi (2004). Reprinted with permission of* Revista del Museo Argentino de Ciencias Naturales.

Dingus, Salgado, and Jackson 2001; Chiappe, Jackson, Dingus, Grellet-Tinner, and Coria 1999; Chiappe et al. 2000; Chiappe, Salgado, and Coria 2001).

The concentration of eggs at Auca Mahuevo is so remarkable that the egg layers of the formation can be better characterized as egg beds. At one quarry, more than 200 whole eggs were mapped in a 25-square-meter area. Spatial analysis showed that the eggs in each level were clustered rather than randomly distributed, suggesting that the egg distribution represents individual egg clutches and not fluvial transport. The egg clutches at Auca Mahuevo typically contain many eggs. Of the nine or more clutches identified in the quarry, seven have between 15 and 34 eggs. This number is significantly larger than the 10 eggs or fewer commonly reported for clutches of megaloolithid eggs.

The fact that the egg clutches are densely grouped suggests a gregarious nesting behavior. Sauropods returned to this site at least five separate times to lay their eggs. However, it is not yet possible to estimate the time intervals between these nesting occurrences. The short distance that separates clutches of whole eggs at the quarry in egg bed 3 and the large adult size of all known sauropods suggests that brooding or any other kind of parental care involving specific egg clutches was unlikely among the Auca Mahuevo sauropods.

Several nest structures preserved in sandstone indicate that the sauropods laid their eggs in excavated nests and that the eggs were not buried in the substrate. The features of the soil of the egg-bearing strata suggest that semi-arid conditions prevailed during the period of occupation.

The stratigraphic and geographic distribution of these sauropod egg clutches as well as their taphonomic framework support several inferences about behavior: (1) gregarious nesting; (2) fidelity to a site; (3) a preference for environments that were far away from active rivers; (4) nest building; and (5) no prolonged brooding of neonates by adults.

Perhaps the most remarkable feature of Auca Mahuevo is the abundance of embryonic remains (at least ⅓ of the in situ eggs examined contain bones). In addition, dozens of egg fragments produced casts of embryonic skin.

Cretaceous Sauropod Footprints

In contrast with the notable abundance of bone remains of Sauropoda that characterize the Cretaceous rocks in South America, the footprint record of this dinosaur clade is scarce. This was noted by Leonardi (1989), who observed that ornithopod trackways are proportionally more common than those of sauropods and hypothesized that the former were more active walkers than sauropods.

The Lower Cretaceous record of sauropod footprints is restricted mainly to the Antenor Navarro and Corda formations in Brazil (Leonardi 1989). The Upper Cretaceous is much more informative and is represented at two main South American localities: Sucre in Bolivia and El Chocón in northwestern Patagonia (Fig. 5.50).

Fig. 5.49. (A) Group of titanosaurid eggs from Auca Mahuevo, Neuquén Province. (B) Life restoration of baby titanosaurs. *Illustration by J. González.*

The first reports of sauropod dinosaur tracksites in Bolivia were made by Giuseppe Leonardi in 1984 and 1994, and a thorough review was published by Martin Lockley and colleagues in 2002. The most famous, largest, and best-documented sites near Sucre are Toro Toro, Cal Orko, and Humaca, all in the Department of Chuquisaca (Figs. 4.5 and 5.50A–C). These localities demonstrate the rich ichnological potential of the track-bearing El Molino Formation, which also includes trackways that have been attributed to theropods and ankylosaurs (Leonardi 1994; Meyer, Hippler, and Lockley 2001; McCrea, Lockley, and Meyer 2001). At the Humaca site titanosaurid tracks predominate, and study of the site has revealed 11 parallel trackways of subadult sauropods traveling as a group (Lockley, Schulp, Meyer, Leonardi, and Mamani 2002). The Humaca trackways are interpreted as evidence of social behavior among small titanosaurids, and together with the Toro Toro site they represent some of the scant Late Cretaceous evidence of herding behavior among

Fig. 5.50. Titanosaurid trackways from (A) Toro Toro, (B) Cal Orcko, (C) Humaca, and (D) El Chocón. Dotted lines indicate mud rims. Solid lines indicate the width of the inner trackway in trackways from Cal Orcko and Toro Toro. (E) A recently discovered titanosaurid trackway on the shoreline of the Lake Ezequiel Ramos Mexía, Río Negro Province. *(A–C) Redrawn from Lockley, Schulp, Meyer, Leonardi, and Mamani (2002). (D) Redrawn from Calvo (1991). (E) Photograph by Fernando Novas.*

sauropods (Lockley, Schulp, Meyer, Leonardi, and Mamani 2002). The nearby Cal Orcko site preserves more variable trackways produced by larger individuals.

In Patagonia, sauropod trackways have been found at the El Chocón fossil site (Figs. 4.7 and 5.50D and E), from which Jorge Calvo (1991, 1999, 2007) described sauropod footprints left in bedding planes of the Candeleros Formation. Each footprint is large (90 centimeters in diameter) and circular. They were named *Sauropodichnus giganteus*, and because they fit within the wide-gauge pattern, it is possible that the producer of these trackways was a titanosaurian sauropod (Calvo 1999, 2007; Fig. 5.50D).

A new trackway site in Argentina has been recently studied by Bernardo González Riga and Jorge Calvo (2007). The site is Agua del Choique, close to the town of Malargüe, in southern Mendoza Province (Fig. 5.1). More than 150 sauropod ichnites were reported from the bedding planes of the Campanian–Maastrichtian Loncoche Formation. All the trackways have a wide-gauge pattern, with left and right footprints separated from the midline, thus resembling *Brontopodus birdi* from the Lower Cretaceous of the United States (Farlow, Pittman, and Hawthorne 1989; Farlow 1992). The manus print is reniform, as is usual among sauropods (Farlow, Pittman, and Hawthorne 1989), and lacks impressions of the manual unguals, which is congruent with the manual anatomy of titanosaurian sauropods (González Riga and Calvo 2007). The body fossils collected in the Loncoche Formation include pencil-like teeth that are characteristic of derived titanosaurids (González Riga and Calvo 2007). One of the trackways from Agua del Choique measures 46 meters long.

Recent exploration conducted by myself and my assistants in the Candeleros Formation, which crops out at the shoreline of Lake Ezequiel Ramos Mexía (Río Negro Province), resulted in the discovery of a well-preserved trackway of a sauropod (Fig. 5.50E). The trackway shows details of the manus and pes that were not documented before for the sauropod ichnites (i.e., *Sauropodichnus*) that had previously been described for these beds (Calvo 1991; Calvo and Mazzetta 2004).

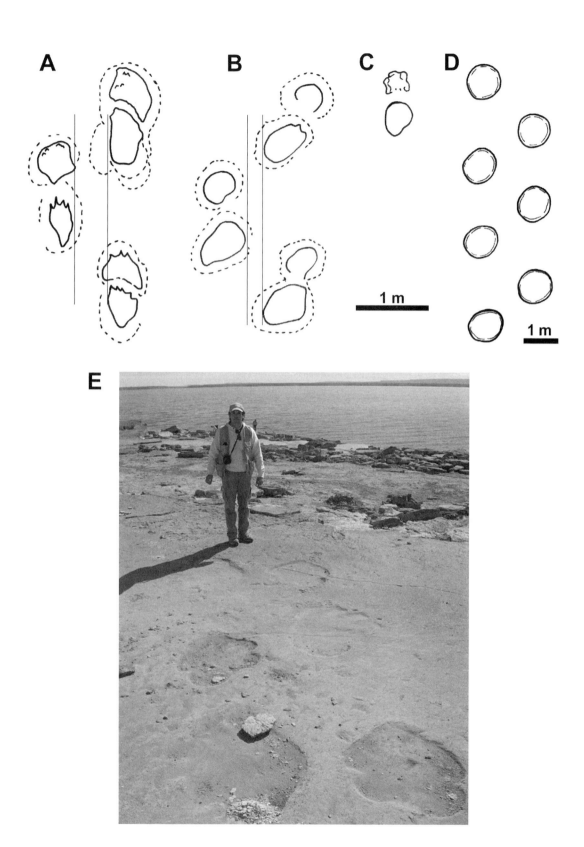

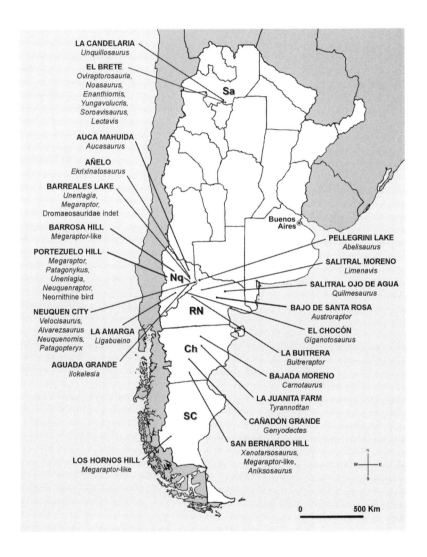

Fig. 6.1. Map of Argentina indicating Cretaceous fossil localities and a list of theropod genera discovered in each of them. Provinces of Argentina that yielded theropod remains are: Salta (Sa), Neuquén (Nq), Río Negro (RN), Chubut (Ch), and Santa Cruz (SC).

CRETACEOUS THEROPODS   6

South America has yielded the most impressive and comprehensive record of Cretaceous theropods from the southern hemisphere. The record is rich in the number of species and the respective clades they represent, and some examples are known on the basis of almost-complete skeletons. Most important, theropods recorded in Argentina offer interesting insights about the evolution and paleobiogeography of meat-eating dinosaurs as a whole (Fig. 6.1).

This amount of information is mainly the result of continued explorations carried out over the last 20 years, mainly in Argentina and Brazil. In contrast, pioneering discoveries made at the end of the nineteenth and beginning of the twentieth centuries are both anatomically and phylogenetically less informative. In 1901, the British paleontologist Arthur Smith Woodward described *Genyodectes serus* on the basis of a portion of a badly preserved snout with teeth discovered by Santiago Roth in Cretaceous strata from central Patagonia. The fragmentary nature of the specimen severely restricted elucidation of the phylogenetic relationships of *Genyodectes* until recently (Rauhut 2004).

Most of the early discoveries that purportedly belonged to Theropoda are, in fact, bones and teeth that belong to other dinosaur groups. Among the first South American dinosaurs originally thought to be theropods is *Loncosaurus argentinus*, described by Florentino Ameghino in 1899 on the basis of a proximal fragment of femur (Ameghino 1921b). He—and later Friedrich von Huene in 1929—considered *Loncosaurus* to be a member of Coelurosauria, but more recent investigations revealed that *Loncosaurus* is an ornithopod ornithischian, as demonstrated by Molnar (1980b) and Coria and Salgado (1996b).

The fertile fossil site of Cinco Saltos in Argentina's Río Negro Province, so productive in titanosaurid bones, yielded some elements that Friedrich von Huene (1929a) interpreted as belonging to Theropoda. They consisted of three isolated distal caudals (which von Huene found similar to the distal caudals of *Ornithomimus*), an isolated metacarpal III, and a curved and trenchant ungual phalanx. Aside from this last element, which unambiguously belongs to a small theropod, the remaining bones correspond better to young titanosaurids. Another of the early discoveries presumed to belong to Theropoda is *Clasmodosaurus spatula*, a taxon von Huene (1929a) founded on the basis of three isolated teeth discovered in southern Patagonia. However, their morphology (which is not matched in any other known theropod dinosaur) is suggestive of a large titanosaurian sauropod.

An unfortunate stasis in theropod discoveries followed the 1929 publication of von Huene's monograph, and for most of the twentieth century understanding of theropod evolution in South America relied upon mostly fragmentary evidence. Fifty years elapsed before new discoveries of theropod remains were described. In 1979, Jaime Powell published the finding of the large theropod *Unquillosaurus ceibalii* (originally interpreted as a carnosaurian). A year later, José Bonaparte and Jaime Powell announced the discovery of the small predatory dinosaur *Noasaurus leali*, which they described as a member of Coelurosauria. However, the discoveries that really changed previous misconceptions about predatory dinosaur evolution in Gondwana were *Abelisaurus comahuensis*, represented by an almost-complete skull José Bonaparte and Fernando Novas described in 1985, and *Carnotaurus sastrei*, an outstanding specimen José Bonaparte published in 1985. These discoveries offered valuable cranial and postcranial information that was not previously available, permitting substantial progress in interpretation of the evolutionary history of theropod dinosaurs in the southern hemisphere. The discovery of *Abelisaurus* and *Carnotaurus* made it possible to theorize that they represented a novel group of theropods that deserved its own family, the Abelisauridae, that is phylogenetically unrelated to the Cretaceous tyrannosaurids from Laurasia but instead is related to the Jurassic *Ceratosaurus*. The large abelisaurids were soon associated with the smaller *Noasaurus*, forming a more comprehensive assemblage of Gondwanan theropods, the Abelisauroidea. The rapid discovery of more abelisauroids led to the interpretation that in Gondwana the role of large predators was mainly played by abelisaurids, in sharp contrast with North America and East Asia, where the large-bodied theropods were the Tyrannosauridae. Discoveries announced by Bonaparte of the chicken-sized *Velocisaurus* (described in 1991) and the even smaller *Ligabueino andesi* (published in 1996) demonstrated that the evolutionary radiation of abelisauroids and their kin also included tiny predators, another distinction from the dinosaurian assemblages from Laurasia, where abelisauroids seem to have been absent and the role of small carnivorous was played by coelurosaurian maniraptorans (Bonaparte and Kielan-Jawarowska 1987).

During the 1990s, however, our understanding of theropod evolution on the southern continents dramatically changed again, due to documentation of the most complete carcharodontosaurid skeletons to date, a highly informative spinosaurid skull, and, most important, an unexpected series of coelurosaurian theropods that came to light from different Cretaceous localities in Patagonia and Brazil.

Among coelurosaurians, José Bonaparte described *Alvarezsaurus calvoi* in 1991 as a theropod convergent in many respect with Laurasian ornithomimosaurians. But with the discovery of *Patagonykus puertai* in 1996 it became clear that these two gracile animals were closely related to the Mongolian *Mononykus olecranus*, forming part of a distinct branch of maniraptoran theropods, the Alvarezsauridae (Novas 1996b, 1997d). The list of basal coelurosaurians increased with the descriptions

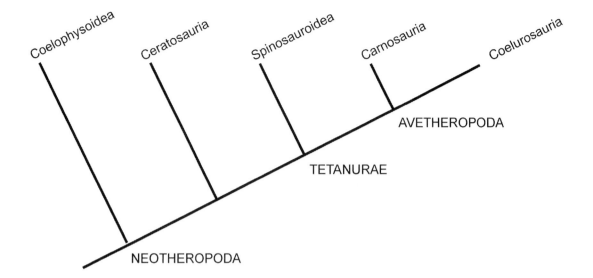

Fig. 6.2. Simplified cladogram of Neotheropoda.

of *Santanaraptor placidus* (Kellner 1999) and *Mirischia asymmetrica* (Naish, Martill, and Frey 2004), both from the Aptian Santana Formation, and *Aniksosaurus darwini*, from the Cenomanian Bajo Barreal Formation (Central Patagonia; Martínez and Novas 2006). Also, an increasing variety of deinonychosaurian theropods has come to light from Late Cretaceous beds of Patagonia, including *Unenlagia comahuensis* (Novas and Puerta 1997), *Unenlagia paynemili* (Calvo, Porfiri, and Kellner 2004), *Neuquenraptor argentinus* (Novas and Pol 2005), *Austroraptor cabazai* (Novas, Pol, Canale, Porfiri, and Calvo 2009), and *Buitreraptor gonzalezorum* (Makovicky, Apesteguía, and Agnolín 2005). Finally, the clade Aves is represented by several disparate members, including Enantiornithes, *Patagopteryx deferrariisi*, basal ornithuromorphs, and early neornithines.

Thus far, 28 Cretaceous non-avian theropod genera have been named on the basis of body fossils. They belong to two of the main lineages into which Theropoda is divided (Fig. 6.2), Ceratosauria and Tetanurae. Ceratosauria is phylogenetically defined as the group containing the common ancestor of *Ceratosaurus*, Abelisauridae, and all its descendants (Sereno, Wilson, and Conrad 2004) and is represented in South America by the Cretaceous *Genyodectes*, Noasauridae, and Abelisauridae. Tetanurae is defined as the neotheropods more closely related to Neornithes than to *Ceratosaurus* (Holtz, Molnar, and Currie 2004) and in South America is represented by Spinosauridae, *Megaraptor namunhuaiquii*, the allosauroid Carcharodontosauridae, and a wide range of coelurosaurians. The coelurosaurians constitute a diverse clade of derived theropods, represented in South America by compsognathids, oviraptorosaurs, alvarezsaurids, dromaeosaurids, and birds. The disparity of ecological types recorded in the Cretaceous of this continent strongly indicates that an important evolutionary diversification of meat-eating dinosaurs occurred in the southern landmasses.

Fig. 6.3. Phylogenetic relationships of ceratosaurian theropods: (A) Cladogram of Ceratosauria and Abelisauroidea. (B) Cladogram of Abelisauridae. Genera recorded in South America are indicated in bold.

Geographically, the fossil record of Cretaceous theropods comes from Argentina and Brazil. Stratigraphically, most discoveries were made in beds of Upper Cretaceous age of Argentina. In contrast, theropod material from Brazil is presently rare. It is restricted to the Late Cretaceous abelisaurid *Pycnonemosaurus nevesi* (Kellner and Campos 2002), an isolated elmisaurid-like manual ungual (Novas, Borges Ribeiro, and Carvalho 2005), and three taxa from Aptian strata of the Santana Formation of the Araripe Basin, including the spinosaurid *Irritator challengeri* (Martill, Cruickshank, Frey, Small, and Clarke 1996; Sues, Frey, Martill, and Scott 2002; Kellner and Campos 1996) and the coelurosaurians *Santanaraptor placidus* (Kellner 1999) and *Mirischia asymmetrica* (Naish, Martill, and Frey 2004). Even though it is restricted to a few species, the fossil record from Brazil is the most informative from this continent for the Lower Cretaceous.

Isolated theropod teeth are known from Uruguay, Bolivia, Brazil, and Argentina, but a comprehensive study of this material is still lacking. Cretaceous theropod footprints are known mainly from Argentina and Brazil (Leonardi 1989; Calvo 1991, 1999, 2007).

A review of the main theropod discoveries made in South America follows.

Ceratosauria

Ceratosaurians (Fig. 6.3A) originated in the Jurassic, but few forms of this initial stage have so far been discovered. Jurassic representatives include *Ceratosaurus nasicornis*, from the Morrison Formation in the United States (Gilmore 1920; Madsen and Welles 2000); the graceful and bizarre *Elaphrosaurus bambergi*, from the Tendaguru beds of Tanzania in central Africa (Janensch 1920, 1925; Holtz 2000); and *Berberosaurus liassicus*, a new theropod recently reported from Early Jurassic beds of the Moroccan High Atlas Mountains and interpreted as a basal abelisauroid (Allain et al. 2007).

More-derived ceratosaurians are documented in Early Cretaceous rocks from Africa, such as *Spinostropheus gautieri*, from the Berriasian–Hauterivian Tiouraren Formation in Niger, which Sereno and colleagues (2004) interpreted as a taxon that was intermediate in both age and phylogenetic position between the Late Jurassic *Elaphrosaurus* and the Late Cretaceous abelisauroids. *Spinostropheus* shares with abelisauroids a prominent ridge on the cervical neural arches that connects the prezygapophyses with the base of the epipophyses. It is interesting to note, however, that Rauhut (2005b) has recently interpreted *Ozraptor subotaii*, which is known on the basis of a single bone fragment from the Middle Jurassic Colalura Sandstone of western Australia (Long and Molnar 1998), as a possible Jurassic abelisauroid.

In South America, the record of Cretaceous basal ceratosaurians is restricted to the possible case of *Genyodectes serus*.

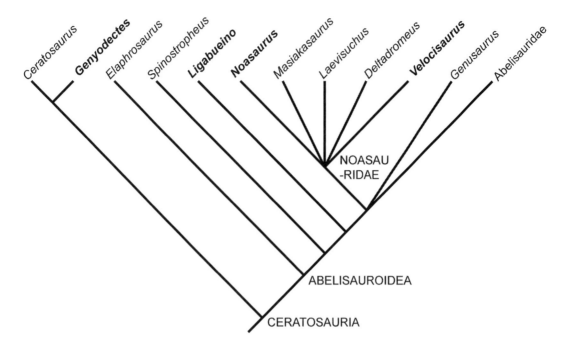
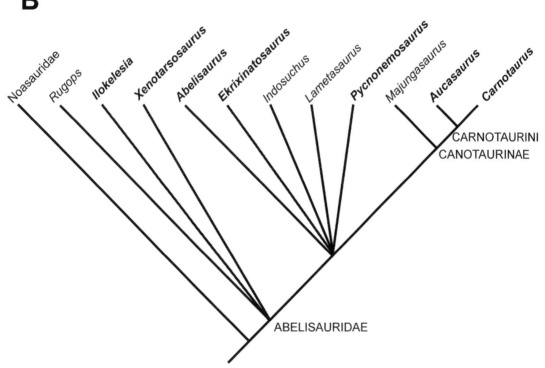

Fig. 6.4. *Genyodectes serus:* Left premaxilla, maxilla, and dentary in lateral view. *Photograph by Fernando Novas.*

GENYODECTES SERUS

The only (and badly preserved) specimen of this taxon comes from Cañadón Grande, almost in the center of Chubut Province in Argentina, from beds presumably corresponding to the Aptian–Albian Cerro Castaño Member of the Cerro Barcino Formation (Fig. 6.4). This massive theropod was originally thought to be a megalosaurid (von Huene 1929a) and later was thought to be a probable member of Tyrannosauridae (e.g., von Huene 1932), but Molnar (1990) dismissed such preliminary interpretations. Paul (1988) noted that *Genyodectes* probably represented an abelisaurid, an idea also expressed by Bonaparte (1996b), who noted some similarities between the snout of *Genyodectes* and abelisaurids. More recently, Olivier Rauhut (2004) has re-prepared and reviewed the type specimen of this theropod and has concluded that *Genyodectes serus* is a valid taxon diagnosed on the basis of some derived dental features (e.g., premaxillary teeth arranged in an overlapping en echelon pattern, crowns of the longest maxillary teeth that are longer apicobasally than the minimal dorsoventral depth of the mandible). Moreover, Rauhut noted that *Genyodectes* lacks several abelisaurid synapomorphies, such as a strong external sculpturing of the external skull bones and longitudinal striations on the interdental plates. One derived character that *Genyodectes* shares with abelisaurids is a longitudinal groove on the dentary that is deeper ventrally than it is dorsally.

The maxillary teeth crowns of *Genyodectes* are symmetrical and very strongly compressed labiolingually, more so than the premaxillary teeth or the maxillary teeth of most theropods. The crown of one of the maxillary teeth is around 95 millimeters long. *Genyodectes* shows premaxillary tooth crowns that are considerably smaller than the maxillary crowns, a feature that is also found in *Ceratosaurus*. The strongly transversely compressed maxillary crowns of *Genyodectes* contrast with the teeth of abelisaurids, which are rather short and not as flattened transversely as in *Genyodectes*. Likewise, although the premaxillary body below the nares in *Genyodectes* is high, it is almost as long as it is high and thus differs from the premaxillae in abelisaurids, which are higher than they are long.

Rauhut interpreted *Genyodectes* as a basal neoceratosaur and argued that it was less probably a member of Ceratosauridae. It shares with *Ceratosaurus* completely fused interdental plates (also found in abelisauroids), but the extreme transverse flattening of the lateral maxillary crowns is found both in *Ceratosaurus* and in advanced carcharodontosaurids. However, the teeth of carcharodontosaurids differ from those of *Genyodectes* in down-pointing grooves at the bases of the marginal denticles and pronounced enamel wrinkles on the crown. A characteristic uniquely shared by *Genyodectes* and *Ceratosaurus* is the extreme length of the maxillary tooth crowns, which exceed the minimal height of the dentary.

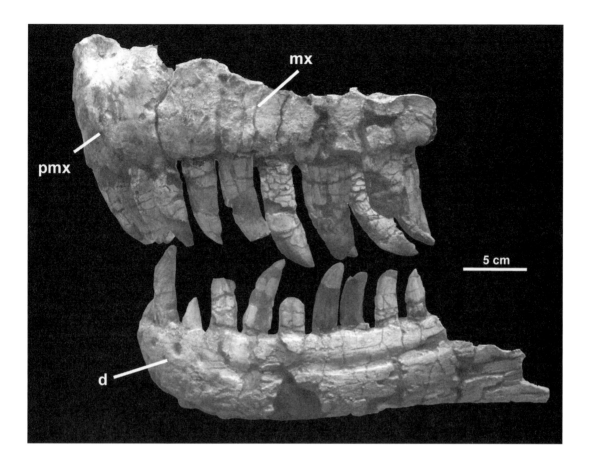

Abelisauroidea

Abelisauroidea is a ceratosaurian clade that is includes *Carnotaurus sastrei* and *Noasaurus leali* (Sereno, Wilson, and Conrad 2004; Fig. 6.3A), their most recent ancestor, and all descendents. Abelisauroids underwent a significant evolutionary radiation in Gondwana during the Cretaceous. They attained a considerable morphological disparity, species diversification, and numerical abundance in the southern continents. Their bone remains were left in South America, Africa, Madagascar, and India, and some reports from France (i.e., *Genusaurus sisteronis*) demonstrate that abelisauroids also inhabited Europe. These ceratosaurian dinosaurs have not, however, been recorded in the highly productive Late Cretaceous dinosaur beds from North America and Asia, thus supporting the hypothesis that abelisauroids mainly evolved in the southern landmasses. Abelisauroids include predatory forms ranging from the size of a hen (for example, *Velocisaurus unicus*) to large and bulky animals around 10 meters long such as *Abelisaurus comahuensis* and *Lametasaurus indicus*.

Synapomorphies of Abelisauroidea (Fig. 6.5) include a maxilla with a subvertical ascending ramus, cervical vertebrae with hypertrophied epipophyses and reduced neural spines, a pubic pedicle of the ilium oriented approximately 60° from the horizontal, a robust ischiac pedicle of ilium that articulates in a cup-shaped proximal process of ischium, and a cranially invaginated and crescentic fibular fossa (Bonaparte 1991b;

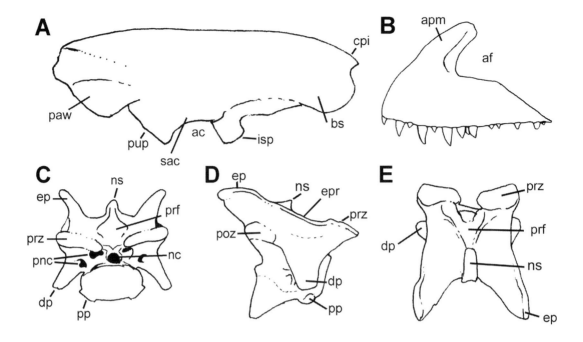

Fig. 6.5. Diagnostic features of Abelisauroidea as illustrated by some bones of *Majungasaurus atopus*. (A) Left ilium in lateral view. (B) Left maxilla in lateral view. (C–E) Cervical vertebra in (C) cranial, (D) right lateral, and (E) dorsal views. Not to scale. *Redrawn from Sampson et al. (1998).*

Bonaparte, Novas, and Coria 1990; Novas 1991, 1992a). Abelisauroids differ from other theropods, including less-derived ceratosaurians (e.g., *Spinostropheus, Ligabueino*), in having square-shaped rather than laminar (e.g., craniocaudally extended) neural spines on the cervical vertebrae.

The large abelisauroids are the best anatomically documented. They are grouped within the family Abelisauridae and include basal forms such as *Rugops primus* and *Ilokelesia aguadagrandensis* as well as the highly derived and bizarre *Carnotaurus sastrei*. Another lineage of smaller abelisauroids, the Noasauridae, includes *Noasaurus leali* from northwestern Argentina and possibly also *Velocisaurus unicus* from Patagonia, *Laevisuchus indicus* from India, *Masiakasaurus knopfleri* from Madagascar, and *Deltadromeus agilis* from Africa. A diminutive Patagonian taxon, *Ligabueino andesi*, may be a poorly known branch of the early radiation of Abelisauroidea.

The Geographic and Stratigraphic Distribution of Abelisauroids

Abelisauroid discoveries in Argentina are perhaps the most complete and informative for providing a comprehensive view of the anatomy and phylogeny of the entire group. The abelisauroid record of Argentina provided a firm basis for the taxonomical identification of isolated bones recorded in India (Novas and Bandyopadhyay 1999, 2001; Novas, Agnolín, and Bandyopadhyay 2004), the phylogenetic relationships of which had been obscure since their original description more than 70 years ago (von Huene and Matley 1933). The Argentine record of abelisauroids is important not only for the number of named species (currently ten) but also for the state of preservation of the specimens, including complete skulls and large parts of the postcranium, including even skin impressions for *Carnotaurus*. Moreover, biostratigraphic documentation of South American abelisauroids includes Barremian–Early Aptian representatives (e.g., *Ligabueino andesi*) through Maastrichtian forms (e.g., *Carnotaurus sastrei*, *Noasaurus leali*, *Quilmesaurus curriei*). Fragmentary abelisaurid remains are known from the Hauterivian–Barremian La Paloma Member of the Cerro Barcino Formation (Rauhut, Cladera, Vickers-Rich, and Rich 2003). Some abelisauroids are Cenomanian (e.g., *Ilokelesia aguadagrandensis*, *Xenotarsosaurus bonapartei*, *Ekrixinatosaurus novasi*), and others correspond to the Campanian (e.g., *Abelisaurus comahuensis*, *Aucasaurus garridoi*, *Velocisaurus unicus*). Specimens of these theropods come from different Cretaceous units of the provinces of Neuquén, Río Negro, Chubut, and Santa Cruz (Fig. 6.1). Abelisauroid remains from Brazil include the ponderous *Pycnonemosaurus nevesi* (Kellner and Campos 2002), from beds of the Bauru Group of Mato Grosso, as well as some isolated bones and teeth from the Bauru Group (Bertini 1996; Bittencourt and Kellner 2002; Candeiro et al. 2004; and Novas, Carvalho, Borges Ribeiro, and Méndez 2008).

Dale Russell (1996) presented the first remains of possible abelisaurids from the Late Cretaceous of Africa, but more recently Paul Sereno and colleagues (2004) offered unambiguous evidence that basal abelisaurids (i.e., *Rugops primus*) were present during the Cenomanian on this continent. In congruence with these discoveries, Novas, Dalla Vechia, and Pais (2005) reported the discovery of abelisauroid pedal unguals from Cretaceous beds of Morocco, and Mahler (2005) referred an isolated maxilla to the same family. Also, *Deltadromeus agilis* (from the Cenomanian Kem Kem beds from Morocco) was originally considered to be a basal coelurosaurian, but a recent review interpreted this taxon as a member of Noasauridae (Sereno, Wilson, and Conrad 2004). The abelisauroid *Genusaurus sisteronis* (Accarie, Beaudoin, Dejax, Friès, Michard, and Taquet 1995; Buffetaut and le Louff 1995) has also been reported from Albian marine beds of southeastern France, constituting firm evidence that supports the theory that there were paleobiogeographical connections between Europe and the Gondwanan continents (i.e., Africa) during the Cretaceous.

Beautifully preserved and very informative specimens have been collected in the last few years in Madagascar, thanks to explorations in the Maevarano Formation by Dave Krause, Cathy Forster, Scott Sampson, and Matthew Carrano. This work resulted in the discovery of marvelous specimens of the abelisaurid *Majungasaurus atopus* (Sampson et al. 1998; Sampson and Krause 2007) and the presumed noasaurid *Masiakasaurus knopfleri* (Sampson, Carrano, and Forster 2001; Carrano, Sampson, and Forster 2002).

Abundant abelisauroid remains have been excavated from several fossil sites in central India from beds of the Late Cretaceous (Maastrichtian) Lameta Formation (von Huene and Matley 1933; Chatterjee and Rudra

1996; Novas and Bandyopadhyay 1999, 2001; Novas, Agnolín, and Bandyopadhyay 2004; Novas, Chatterjee, Rudra, and Datta in press). Named theropods from this formation include the small *Laevisuchus indicus* and the abelisaurids *Indosuchus raptorius, Indosaurus matleyi, Lametasaurus indicus,* and *Rajasaurus narmadensis* (Wilson, Sereno, Srivastava, Bhatt, Khosla, and Sahni 2003), although the taxonomic validity of these four abelisaurids is far from settled (Novas, Agnolín, and Bandyopadhyay 2004). A new Indian abelisaurid, *Rahiolisaurus gujaratensis,* has recently been described from the Lameta beds (Novas, Chatterjee, Rudra, and Datta in press). To this point, no records of abelisauroids have been reported in Antarctica and Australia (but see Agnolín, Ezcurra, and Pais 2005).

Interestingly, several abelisaurid specimens are known by almost completely preserved skeletons that include the skull (e.g., *Carnotaurus, Aucasaurus,* and *Skorpiovenator bustingorry*; Canale, Novas, and Scanferla 2006; Canale, Scanferla, Agnolín, and Novas 2009). Some other abelisaurids are known by disarticulated but abundant bone remains that belong to different individuals (e.g., *Rahiolisaurus, Majungasaurus*; Chatterjee and Rudra 1996; Sampson et al. 1998; Novas, Chatterjee, Rudra, and Datta in press). Whether these recurrent taphonomic modes of almost undisturbed corpses and abundant individuals reflect the environmental preferences of abelisaurids remains unknown.

The Phylogenetic Relationships of Abelisauroids

Since *Abelisaurus* was first described, there has been agreement among authors that this Patagonian taxon is closely related to *Ceratosaurus nasicornis* from the Upper Jurassic of North America. Furthermore, *Ceratosaurus* was thought to be tied phylogenetically with the Triassic and Early Jurassic coelophysoids (e.g., *Coelophysis, Syntarsus, Dilophosaurus*), for which the name Ceratosauria was coined (Gauthier 1986; Rowe and Gauthier 1990). In this context, abelisauroids were almost unanimously considered to be Cretaceous ceratosaurians, and the name Neoceratosauria (Novas 1991) was erected for the clade comprised of *Ceratosaurus* plus Abelisauroidea. However, more-recent studies (Carrano and Sampson 1999; Forster 1999; Carrano, Sampson, and Forster 2002; Rauhut 1998, 2003a) have demonstrated that coelophysoids retained the pelvic girdles and hindlimbs of the primitive basal theropod pattern and are currently considered to be representatives of an early radiation of theropods that is not closely linked with the evolution of true ceratosaurians (e.g., *Ceratosaurus* plus abelisauroids). In other words, *Ceratosaurus* plus Abelisauroidea share a common ancestor with Tetanurae rather than with Coelophysoidea. Within this new phylogenetic arrangement, the name Neoceratosauria became a junior synonym of Ceratosauria (Wilson, Sereno, Srivastava, Bhatt, Khosla, and Sahni 2003).

Currently, no consensus exists about the phylogenetic relationships within Abelisauridae, on the one hand, and Noasauridae, on the other, so confusion about nomenclature has increased in the last years. There is no agreement about the taxa constituting Noasauridae, and the proposed unique features joining them within this clade remain controversial. I have reservations about these groupings, mostly because the involved taxa are fragmentarily known and the characteristics in the hindfoot and cervical vertebrae that diagnose Nosauridae and Abelisauridae may be, in fact, primitive states already present in the common abelisauroid ancestor.

The Main Anatomical Features of Abelisauroids

SKULLS, JAWS, AND TEETH

The skull of abelisauroids (Fig. 6.6) is well known in several genera, so certain cranial characteristics can be recognized as synapomorphic for Abelisauridae, which allows us to postulate some diagnostic features of Abelisauroidea. Among the features that probably emerged in the abelisaurid common ancestor is a proportionally short and deep skull that has a rugose pattern of ornamentation on the external surface of the dermal bones. In these theropods, the nasals are thick and are decorated with conspicuous rugosities consisting of numerous pits and spikes that are suggestive of a thick and rugose epidermal cover.

In some abelisaurids (e.g., *Abelisaurus*, *Majungasaurus*) the nasals are strongly fused, but in all known members of the group, the frontals are fused and dorsally inflated, a condition that is present in an initial stage in *Abelisaurus*. In contrast, a pair of bull-like frontal horns is present in *Carnotaurus* and a single large frontal eminence reminiscent of the bony dome of pachycephalosaurs occurs in *Majungasaurus* (Sues and Taquet 1979; Sampson and Witmer 2007; Fig. 6.7A and B). This bone development was likely covered by keratinous cones, as occurs in the crest of the ratite bird cassowary (*Cassuarius cassuarius*) and horns of the three-horned chameleon (*Chamaeleo jacksoni*). In addition, a remarkable feature that is present in *Ilokelesia*, *Rugops*, and the remaining abelisaurids is a wide articulation between the lacrimal and the postorbital that defined a thick brow above the orbits. This indicates that the top of the skull formed a rigid structure above the orbits. It is usual among abelisaurids for bone to nearly encircle the orbit due to the deeply notched orbital margins of both the lacrimal and the postorbital, as is seen in *Abelisaurus*, *Majungasaurus*, and *Carnotaurus*. Additionally, in these abelisaurids the postorbital has a rostroventrally projecting fan-shaped expansion that almost contacts the lacrimal. A less-derived condition in the encircling of the orbit is documented in *Ilokelesia*, in which the jugal ramus of the postorbital is almost vertically oriented and a precursor of the fan-shaped expansion is represented by a small laminar expansion. The functional significance of structures that encircled the orbit and the development of a sub-orbital flange was probably protection of the eye. Because the greatest development of the flange (and the encircling of the eye) occurs in the horned abelisaurids *Majungasaurus* and *Carnotaurus*, I suspect

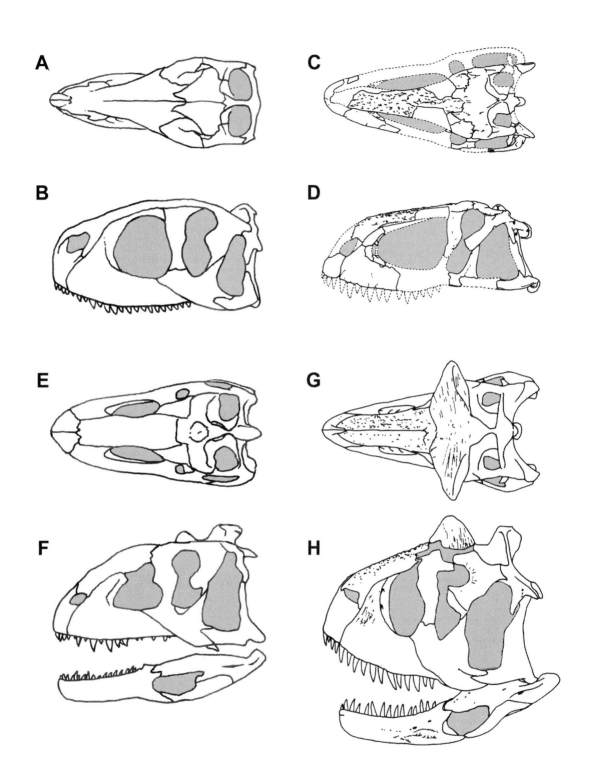

that such modification of the postorbital reduced the risk of wounds to the eyes from sudden blows to the head.

Abelisaurid skulls are also characterized by an interorbital wall (presumably constituted by the parasphenoid bone) that hangs vertically below the mid-frontal suture. This interorbital wall is seen in *Indosaurus*, *Abelisaurus*, *Carnotaurus*, and *Majungasaurus*. Moreover, the cranial half of the parasphenoid ends in a diamond-shaped structure that has a double foramen for the passage of the olfactory nerve. Ossified parasphenoids that are tightly fused to the skull roof and have a double exit for the olfactory nerve (nerve I) are features not exclusive to Abelisauridae, since they are also present in *Ceratosaurus*, *Acrocanthosaurus*, and some tyrannosaurids.

Cranial kinesis in abelisaurids remains unexplored, and thus only general considerations can be offered. The maxillary ramus of the lacrimal is strongly reduced in abelisaurids and the nasal-frontal suture is transversely oriented, thus giving the impression that the snout pivoted against the caudal half of the skull. In *Carnotaurus* and *Majungasaurus* (Sampson et al. 1998) the nasal bone overlaps the lacrimal and has small peg-like caudal projections that fit into a rostrolateral notch of the lacrimal bone, suggesting a rather movable articulation between the two bones.

In *Carnotaurus*, *Majungasaurus*, and *Abelisaurus*, the frontals form a single structure with the parietals, which are strongly fused with the supraoccipital and are dorsally projected, but in *Carnotaurus* and *Majungasaurus* the supraoccipital extends above the level of the dorsal surface of the skull. The occipital crest is high and transversely wide, and this expansion of the occipital surface suggests a strong development of several nuchal muscles, such as M. complexus, M. biventer cervicis, and M. splenius capitis. Also, the strong development of the supraoccipital indicates that a powerful nuchal ligament was present.

In the lower jaw, abelisauroids are characterized by a large external mandibular fenestra. This developed at the expense of the bones around it, which have reduced articular contacts. This kind of reduction in the articulation of the dentary-postdentary bones is peculiar among theropods and suggests important intramandibular movements.

The teeth of abelisaurids look tiny compared to the size of the skull. They are transversely compressed and rostrocaudally wide and have serrated rostral and caudal margins (Sampson, Krause, Dodson, and Forster 1996; Bittencourt and Kellner 2002). The number of premaxillary teeth is four, but in the maxilla the number of teeth varies from 19 in *Rugops* to 12 in *Carnotaurus*. Among abelisauroids, *Masiakasaurus* shows a bizarre condition having in procumbent teeth at the rostral end of the dentary (Sampson, Carrano, and Forster 2001).

Notably, abelisaurids evolved some convergent resemblances in the skull to the contemporaneous carcharodontosaurid carnosaurs (e.g., *Giganotosaurus carolinii*, *Carcharodontosaurus saharicus*, *Tyrannotitan chubutensis*), including a reduced antorbital fossa; a similar pattern of rugosities decorating the nasals, maxillae, and dentaries; a thick brow

Fig. 6.6. Skulls of abelisaurids in (A, C, E, G) dorsal and (B, D, F, H) lateral views. (A, B) *Rugops primus*. (C, D) *Abelisaurus comahuensis*. (E, F) *Majungasaurus atopus*. (G, H) *Carnotaurus sastrei*. Not to scale.

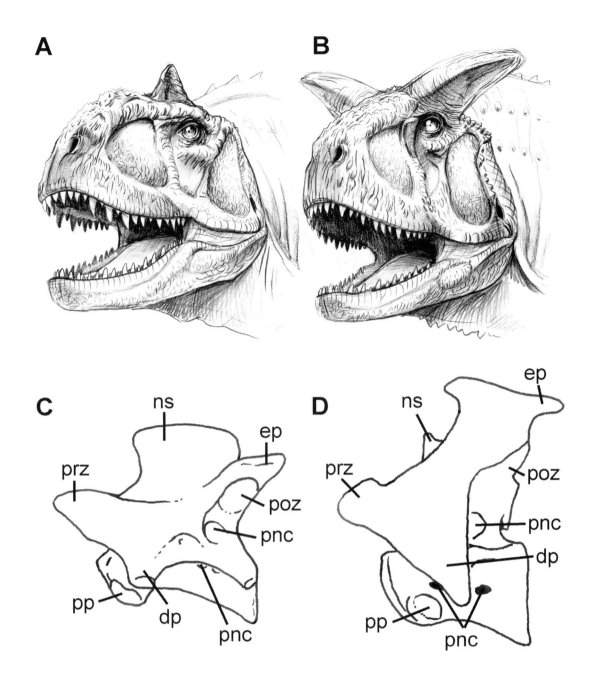

Fig. 6.7. Reconstructed heads of abelisaurids: (A) *Majungasaurus* and (B) *Carnotaurus*. (C, D) Cervical vertebrae of abelisauroids in left lateral view: (C) *Spinostropheus* compared with (D) the derived abelisaurid *Carnotaurus*. Not to scale. *(A, B)* Illustration by J. González.

formed by the lacrimal and the postorbital above the orbit; and sinuous orbital margins of both the lacrimal and postorbital.

VERTEBRAL COLUMNS

Neck vertebrae are known for a wide variety of abelisauroids. However, the total number of cervicals remains unknown for most members of the group. The exceptions are *Carnotaurus* and *Majungasaurus*, in which 10 cervicals are counted. Abelisauroids have a pair of pneumatic foramina (pleurocoels) on the cervical centra, a feature that is apparently ancestral

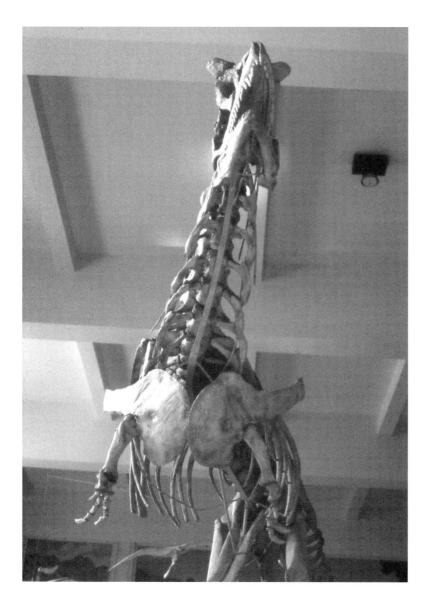

Fig. 6.8. Skull and neck of *Carnotaurus sastrei* in ventral view. Skeleton mounted in the Museo Argentino de Ciencias Naturales Bernardino Rivadavia, Buenos Aires.

for Ceratosauria (e.g., *Ceratosaurus*; Novas 1991). Most abelisauroids are distinguished from other theropods in that the neural arch is transversely wide (Fig. 6.5C and E). Basal members of Abelisauroidea (e.g., *Spinostropheus*, *Ligabueino*) have retained laminar and craniocaudally extended neural spines on the cervical vertebrae (Fig. 6.7C). However, in more-derived abelisauroids, the neural spines became craniocaudally short and square-shaped in cross-section and are bounded by deep and broad pre- and postspinal fossae (Fig. 6.5C and E and 6.7D). The epipophyses are notably enlarged, particularly in derived members of the clade (e.g., *Carnotaurus*, *Noasaurus*), in which the epipophyses are dorsoventrally deep and expanded forward and backward, suggesting that the muscles involved in side-to-side movements of the neck were strongly developed. Another feature that is widely distributed among Abelisauroidea is a

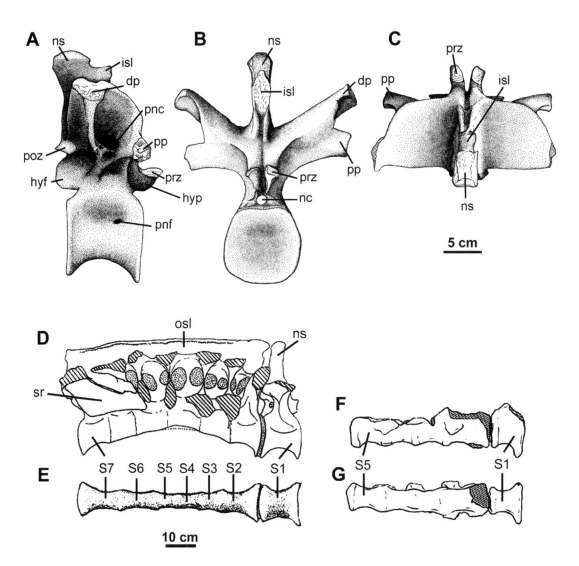

ridge connecting the epipophyses with the prezygapophyses, thus defining a table-like dorsal surface distinct from the lateral surfaces of both diapophyses.

The evidence at hand indicates that basal abelisauroids (*Elaphrosaurus*, *Spinostropheus*) and probably also noasaurids were long-necked theropods resembling the ornithomimid coelurosaurs, in contrast with abelisaurids, which were short and robust-necked theropods (Agnolín and Martinelli 2007).

The abelisaurid neck looks robust (Fig. 6.8), particularly when the cervical vertebrae are observed in articulation with the ribs, which are characterized by wing-like processes and caudally elongate rods, all of which indicate a remarkably strong and muscular neck. This set of derived cervical features is consistent with the morphology of the occipital surface of skull described above and reflects the development of a bulky epaxial musculature.

Majungasaurus (O'Connor 2007) has 13 dorsal vertebrae but the more-derived *Carnotaurus* has 11 (Fig. 6.9A–C; Bonaparte, Novas, and Coria 1990). The parapophyses are prominent and located high on the neural arch, as is usual among ceratosaurs. The pneumatic features are well developed ventral to the transverse processes and have wide and deep pneumatic cavities. The dorsal vertebrae show strong articular interlocking, suggesting that this region of the vertebral column was fairly rigid. In this regard, the extra-articulations of the hyposphene and the hypantrum are characterized by their complexity: the hypantrum is delimited by pendant processes ventral to the prezygapophyses, and the hyposphene is cone-shaped.

The sacrum is composed by five vertebrae in *Majungasaurus* (O'Connor 2007) and seven vertebrae in *Carnotaurus*, two more vertebrae than is usual among non-avian Theropoda (Fig. 6.9D–G). Some distinctions in the shape of the sacral centra are verified among abelisauroids. For example, in *Lametasaurus*, *Rajasaurus*, and *Masiakasaurus*, a conservative kind of sacral centra is present, consisting of transversely broad sacral elements. In *Carnotaurus*, in contrast, the sacral centra are strongly reduced in transverse diameter, a condition that is also documented among the abelisauroid fossil assemblage from the Lameta Formation of India. The phylogenetic significance of this feature among abelisauroids remains uncertain (Novas, Agnolín, and Bandyopadhyay 2004).

The caudal vertebrae of Abelisauridae (Fig. 6.10) are curious in showing transverse processes that are distally fan-shaped, and in most members of this group, they have a slender anterior projection on the transverse processes that contacts with the transverse process of the preceding caudal (Coria and Salgado 2000; Coria, Chiappe, and Dingus 2002). Moreover, in *Aucasaurus* the cranial projections of the transverse processes of caudal 1 contact with the caudal end of both ilia, a feature as yet unreported in non-abelisaurid theropods (Coria, Chiappe, and Dingus 2002). In the holotype of *Carnotaurus*, molds of the transverse processes of the caudal vertebrae that are associated with skin patches indicate that the condition described for *Aucasaurus* is also present in *Carnotaurus*. In *Aucasaurus*, the awl-like projections of its transverse processes laterally embrace the lateral ends of the ilia. Although these projections are not preserved in the first caudals of *Carnotaurus*, notches for their reception are visible on the lateral surface of the dorsocaudal corner of the ilia.

In *Spinostropheus*, ossified tendons of the epaxial musculature run along the side of the neural spines of the sacral vertebrae (Sereno, Wilson, and Conrad 2004). This condition, which resembles that in ornithischian dinosaurs, suggests that abelisauroids strengthened the proximal end of the tail, probably restricting movements at its base, in contrast with what occurred in derived maniraptorans (e.g., dromaeosaurids and birds), in which the base of the tail was highly movable and the distal two-thirds of the tail were stiffened by longitudinal tendons (Gatesy 2001).

The peculiar morphology described above for the skull and vertebral column prompts speculation about the functional morphology

Fig. 6.9. Dorsal and sacral vertebrae of abelisaurids. (A–C) Dorsal 8 of *Carnotaurus* in (A) lateral, (B) cranial, and (C) dorsal views. (D, E) Sacrum of *Carnotaurus* in (D) lateral and (E) ventral views. (F, G) Sacrum of *Rajasaurus* in (F) lateral and (G) ventral views. *(A–E) From Bonaparte, Novas, and Coria 1990. Reprinted from* Contributions in Science *416: 1–42 with permission of the Natural History Museum of Los Angeles County. © Copyright 1990. (F, G) Redrawn from Wilson et al. (2003).*

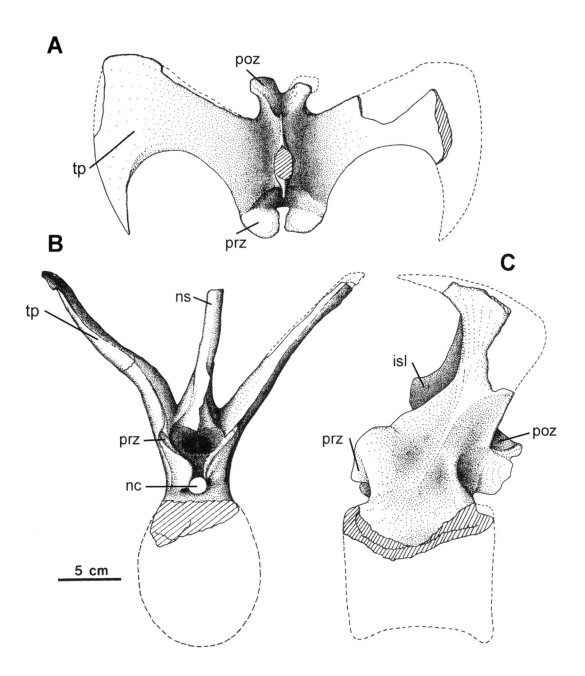

and behavior of abelisaurids. Several skull and cervical features, such as prominent horns, an orbit surrounded by bone, and the inferred development of powerful nuchal muscles, ligaments, and tendons, may constitute adaptations that served for delivering blows. The shortening of the skull (with respect to the length of the presacral vertebral column) is notable in *Carnotaurus*, *Aucasaurus*, and *Majungasaurus*, and this may also reflect improved control of movements through reducing the moment of inertia of the head. The neck is very robust; it is transversely as wide as the head. (In other theropods, for example in *Tyrannosaurus rex*, the neck is notably narrow in comparison with the transverse width of the head.)

The strength of the neck was also conferred by the peculiar cervical ribs, which interlock through cranial and caudal projections. This condition, along with the peculiar shape of neural spines and epipophyses that are indicative of a novel muscle arrangement, suggest that the unusually robust neck of abelisaurids was a formidable device involved in strong and quick head movements and in absorbing shocks.

In contrast with the neck, most of the abelisaurid backbone seems constructed to reduce lateral movements, presumably to counter violent shocks experienced by the head and neck system. The presence of well-developed hyposphene-hypantrum articulations in the dorsal vertebrae, an enlarged synsacrum made by seven sacrals, and the peculiar overlapping of the transverse processes of successive proximal caudals are indicative that most of the post-cervical column was rigid and may have acted as a firm point of pivot for the (inferred) mobility of the head and neck.

The movement of the robust and muscled neck of abelisaurids against a rigid post-cervical vertebral column probably allowed the head to deliver blows against competitors or prey. It is not illogical to suppose that horned abelisaurids were prone to engage in combat with members of their own species, using their horns to intimidate rivals and assure access to mates, food, or territories. Whether the small size of the teeth was related to this alternative means of subduing rivals or food needs to be explored in more detail.

Fig. 6.10. Caudal 6 of *Carnotaurus* in (A) dorsal, (B) cranial, and (C) left lateral views. Reconstructed centrum expressed in dotted lines. *From Bonaparte, Novas, and Coria (1990). Reprinted from* Contributions in Science *416: 1–42 with permission of the Natural History Museum of Los Angeles County.* © Copyright 1990.

SHOULDER GIRDLES AND FORELIMBS

The abelisauroid shoulder girdle is characterized by a strap-like scapular blade and an unusually wide elliptical-shaped coracoid, a condition verified in *Carnotaurus, Majungasaurus, Aucasaurus, Indosuchus,* and the basal abelisauroid *Deltadromeus* (Fig. 6.11). The enlarged coracoid seems incongruent with the pronounced reduction in size of the entire forelimbs. The wide coracoidal surface may instead reflect an enlargement of the area for attaching muscles of the neck rather those muscles that inserted on the atrophied forearms. Thus, the peculiar morphology of the scapular girdle may correlate well with the form and function of the head and neck system hypothesized above for abelisaurids.

Among abelisauroids, the forelimb (or parts of it) is known in *Carnotaurus, Aucasaurus, Masiakasaurus, Majungasaurus,* and *Deltadromeus* (Carrano 2007). They share an almost straight humeral shaft in both cranial and lateral views that is associated with a globous proximal articular head and a reduced deltopectoral crest, features that are also present in the Jurassic ceratosaur *Elaphrosaurus*. However, in contrast with the slender and elongate humeral shafts of *Elaphrosaurus* and *Masiakasaurus*, the humerus in Abelisauridae is remarkably short and stout, as is seen in *Carnotaurus* and *Aucasaurus*, in which the bone is approximately 30 percent of the femur length. The Late Jurassic *Elaphrosaurus bambergi* from Africa (Janensch 1920, 1925; Galton 1982) has a humerus resembling that of abelisauroids in the shape of the shaft and reduction of the deltopectoral crest. It is highly possible that abelisauroids inherited

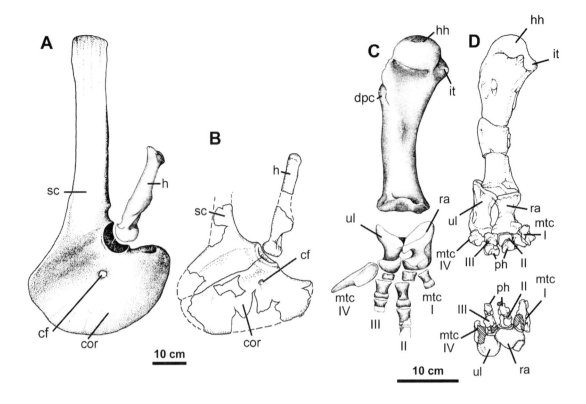

Fig. 6.11. Shoulder girdles and forelimb bones in abelisauroids. (A, B) Left scapulocoracoid and articulated humerus in lateral view: (A) *Carnotaurus*. (B) *Deltadromeus*. (C, D) Right forelimbs of abelisauroids: (C) *Carnotaurus* in cranial view and (D) right forelimb (reversed) of *Aucasaurus* in caudal view. *(A, C) From Bonaparte, Novas, and Coria (1990). Reprinted from* Contributions in Science *416: 1–42 with permission of the Natural History Museum of Los Angeles County. © Copyright 1990. (B, D) Redrawn from Sereno et al. (1996) and Coria, Chiappe, and Dingus (2002), respectively.*

this peculiar humeral morphology from their Jurassic ancestors (Novas, Ezcurra, and Agnolín 2006).

Almost-complete forearms and hands of abelisauroids are currently known in *Carnotaurus* and *Aucasaurus* (Bonaparte, Novas, and Coria 1990; Coria, Chiappe, and Dingus 2002). The ulna and radius are strongly abbreviated and the metacarpals are also extremely shortened. The distal ends of both the ulna and the radius form large and convex articular surfaces. The hands of *Aucasaurus* and *Carnotaurus* are formed by four metacarpals, of which metacarpals I and IV are roughly conical-shaped structures that appear to have carried no phalanges. Metacarpals II and III, in contrast, have typical distal articulations and articulate with one and two phalanges, respectively. It is unclear whether these digits had claws (Fig. 6.11).

THE PELVIC GIRDLE AND HINDLIMBS

This part of the skeleton is documented in several abelisauroid taxa (Fig. 6.12). The ilium is craniocaudally elongate in abelisaurids, covering a large number of sacral vertebrae. Rodolfo Coria and colleagues (2002) have documented a curious feature in *Aucasaurus* that consists of an iliac postacetabular wing with a prominent caudal process (Fig. 6.12; cpi) that articulates with a prong-shaped cranial projection of the first caudal transverse process. This feature of the ilium is also present in other abelisauroids (e.g., *Ligabueino, Carnotaurus, Indosuchus, Majungasaurus*;

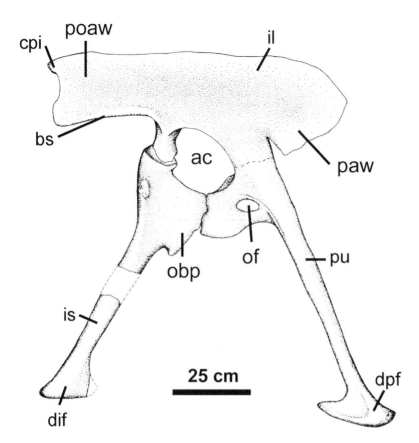

Fig. 6.12. Pelvic girdle of *Carnotaurus* in right lateral view. From Bonaparte, Novas, and Coria (1990). Reprinted from Contributions in Science 416: 1–42 with permission of the Natural History Museum of Los Angeles County. © Copyright 1990.

Fig. 6.5A), so it is probable that this peculiar iliac-caudal articulation was widely present within the group. Another outstanding feature of the ilium is the morphology of the pubic pedicel, which is dorsoventrally deep and has an almost cranially oriented articular surface. The proximal end of the pubis is proximodistally deep and retains a wide fenestration (i.e., the obturator foramen). Moreover, a strong iliopubic fusion is documented in *Carnotaurus* and *Indosuchus*. The abelisauroid pubis is a long and slender bone that is transversely narrow in cranial aspect and is distally expanded into a modest pubic boot. The ischium is a stout bone with a well-developed obturator process and a prominent distal foot.

The hindlimbs of abelisauroids are rather conservative and closely resemble those of the Jurassic *Ceratosaurus* in having a large, rounded, and craniomedially oriented femoral head; a low anterior trochanter; and a prominent and rugose trochanteric shelf (Fig. 6.13A–D). However, abelisauroids seem more derived than basal tetanurans in that the 4th trochanter is slightly developed and the mediodistal crest of the femur is proximodistally extended and sharp. The abelisauroid tibia is distinguished by a large cranioproximally projected hatchet-shaped cnemial crest, as exemplified in *Quilmesaurus* (see below). In the derived abelisaurid *Lametasaurus indicus* from the Lameta Formation of central India (Matley 1923), the tibia is notably short and stout and has a prominent cnemial crest. This massive kind of tibia may be associated with short and

Fig. 6.13. Hindlimb bones of abelisaurids: (A–D) Right femur of *Xenotarsosaurus* in (A) lateral, (B) cranial, (C) caudal, and (D) medial views. (E) Left metatarsals II–IV of *Rahiolisaurus* in cranial view. (F, G) Pedal unguals of an indeterminate abelisaurid from Patagonia. (H, I) Pedal unguals of an indeterminate abelisaurid from India. *(F–I) From Novas and Bandyopadhyay (2001). Reprinted with permission of* Ameghiniana.

stout femora that were recovered from the same fossil site of the Lameta Formation (von Huene and Matley 1933; Novas, Agnolín, and Bandyopadhyay 2004) and may be a peculiar adaptive type of large Cretaceous theropods with bulky and short hindlimbs.

It is common among abelisauroids that the tibia, fibula, and proximal tarsals form a fused tibiotarsus that is shorter than the femur. In abelisauroid theropods, the astragalus and calcaneum are tightly fused and the astragalar ascending process is dorsoventrally low. The abelisauroid foot (Figs. 6.13E) is symmetrical. The side metatarsals (II and IV) are elongate and narrow and have asymmetrically arranged distal ginglymoids, but metatarsal III is stout and transversely wide(De Valais, Novas, and Apesteguía 2002; Carrano 2007). This kind of metatarsus, which is widely documented among abelisauroids (e.g., *Deltadromeus*, *Velocisaurus*, *Masiakasaurus*, *Aucasaurus*, *Majungasaurus*, and *Indosuchus*), may be termed "eurymetatarsal" (from Greek: *eury*, wide), in contrast to the "arctometatarsalian" condition recorded in many coelurosaurian theropods (e.g., Tyrannosauridae, Ornithomimidae, derived Alvarezsauridae, Troodontidae, and basal Dromaeosauridae), which have a proximally thin metatarsal III (Holtz 1994, 1995).

The thinning of metatarsal II implies a reduction of metatarsal I and its correspondent digit. This digit is documented only in *Aucasaurus* among abelisauroids, and it is, in fact, a strongly reduced pedal element.

In contrast with other theropods, abelisauroids have non-ungual pedal phalanges with a distal ginglymoid that forms a cylindrical articular surface that is smooth and uniformly convex and has a rectangular outline in distal view. Abelisauroid pedal unguals differ from those of other theropods because they have proximally bifurcated grooves (which sometimes delimits a rounded bump) and a narrow and deep furrow on the ventral surface. This kind of ungual is documented in *Velocisaurus*, *Masiakasaurus*, and the abelisaurids, and isolated specimens have been documented in Cretaceous beds in Patagonia, India, and Africa (Novas and Bandyopadhyay 2001; Novas, Agnolín, and Bandyopadhyay 2004; Novas, Dalla Vecchia, and Pais, 2005).

SKIN IMPRESSIONS

Patches of skin impressions of *Carnotaurus sastrei* were recovered while excavating the holotype specimen (Bonaparte, Novas, and Coria 1990; see Plate 16). The patches belong to the anterior cervical region, the scapular area near the glenoid, and portions of the thoracic region, but the largest available skin impression corresponds to the proximoventral region of the tail. The impressions vary little among different regions of the body. The surface of the skin is made up of low and conical scutes around 4–5 centimeters in diameter that form a row in which the scutes are separated from each other by 8–10 centimeters. The rest of the skin surface is made up by small mosaics about 5 millimeters in diameter. Narrow parallel wrinkles are also present (Fig. 6.14). The importance of

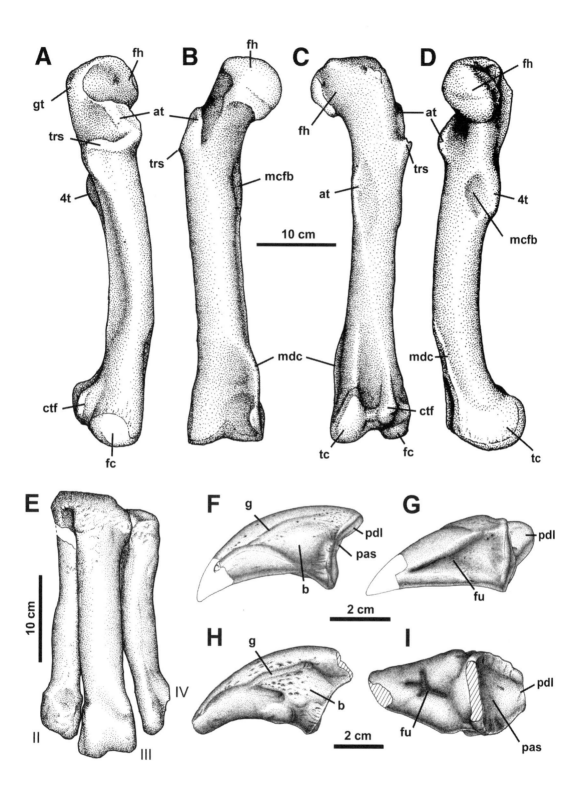

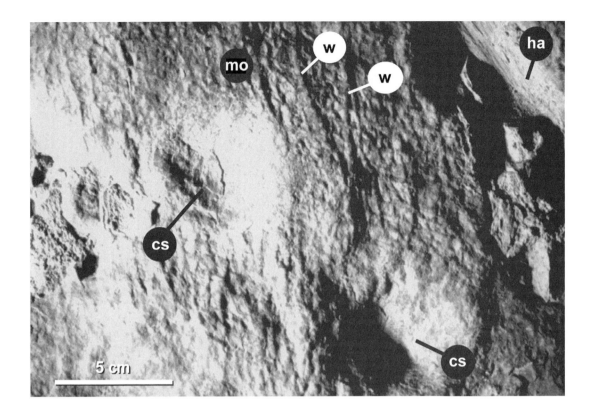

Fig. 6.14. Skin impressions of *Carnotaurus*.

this discovery relates to the phylogenetic development of feathers: in this ceratosaurian subclade, at least, this epidermal feature was absent, thus giving support to the possibility that feathers evolved later in theropod phylogeny (e.g., Coelurosauria).

Abelisauroids Documented in South America

BASAL ABELISAUROIDS

Under this informal category, I describe a set of small (e.g., pigeon- to rhea-sized) abelisauroids that include *Ligabueino*, *Velocisaurus*, and *Noasaurus*. However, the phylogenetic relationships among these poorly known taxa remain uncertain.

LIGABUEINO ANDESI José Bonaparte described this ceratosaurian in 1994 and 1996 on the basis of a single incomplete specimen found in the Lower Cretaceous (Barremian–Early Aptian) La Amarga Formation at a fossil site located south of La Amarga Creek, Neuquén Province. *Ligabueino* is the oldest known record of Abelisauroidea in South America. The holotype of *Ligabueino andesi* (Fig. 6.15) corresponds to one of the smallest theropod specimens currently known; its femur is 6.2 centimeters long and its cervical neural arch is 0.6 centimeters wide. Although the possibility cannot be dismissed that the specimen is a juvenile, the neural arch and centra available distal caudal vertebra are fused

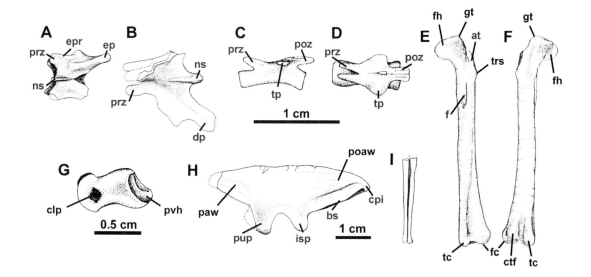

Fig. 6.15. *Ligabueino andesi:* (A) Cervical vertebra in dorsal view. (B) Dorsal vertebra in dorsal aspect. (C, D) Caudal vertebra in (C) left lateral and (D) dorsal views. (E, F) Left femur in (E) cranial and (F) caudal views. (G) Pedal phalanx in lateral view. (H) Left ilium in lateral view. (I) Paired ischial shafts in cranial view. *From Bonaparte (1996b). Reprinted with permission from* Münchner Geowissenschaftliche Abhandlungen, *Series A, vol. 30.*

(Bonaparte 1996b). Its small size supports the idea that *Ligabueino* ate insects or small vertebrates.

This taxon has abelisauroid features on the cervical vertebra, such as a wide dorsal surface that is separated from the lateral surface of the diapophyses by a sharp epipophyseal-prezygapophyseal ridge (Fig. 6.15A) and well-developed cervical epipophyses (Bonaparte 1996b). The neural spine of *Ligabueino* is dorsoventrally low but craniocaudally extended, in contrast with that of more-derived abelisauroids (e.g., *Masiakasaurus*, *Laevisuchus*, *Noasaurus*, *Carnotaurus*), in which the neural spine is craniocaudally short. The neural arches, which were originally interpreted as belonging to a dorsal vertebrae (Bonaparte 1996b), may correspond to proximal caudals, as suggested by the laminar condition of the neural spine and especially the strongly laterocaudally projected transverse processes. The only available distal caudal vertebra (Fig. 6.15C and D) is long and low and also has extensive transverse processes resembling those of the larger Indian abelisauroids (Novas, Agnolín, and Bandyopadhyay 2004). The ilium of *Ligabueino* (Fig. 6.15H) is long and low, and its pubic peduncle is oriented almost cranially, as in other abelisauroids. The brevis shelf is well developed, as is usual among basal theropods, including abelisauroids. A sharply pointed caudal process extends backward from the postacetabular blade of the ilium, as also occurs in *Carnotaurus*, *Aucasaurus*, *Indosuchus*, and *Majungasaurus*. Among the pelvic elements are two slender bones that Bonaparte (1996b) originally described as corresponding to the distal halves of the pubes (Fig. 6.15I). However, their delicacy, coupled with the total absence of a pubic apron or a distal pubic foot, raises doubts about this identification. These tightly appressed bones may correspond better to a pair of ischial shafts.

The femur of *Ligabueino* is long and slender (Fig. 6.15E and F). It has a large foramen on the cranial surface of the proximal shaft, the anatomical significance of which is unknown. Also notable is the morphology of a non-ungual pedal phalanx, which has a well-developed proximoventral "heel" that superficially resembles that of dromaeosaurid theropods (Fig. 6.15G). Whether or not this phalanx supported a trenchant ungual as in the latter theropods is unknown pending further discoveries. However, as Bonaparte (1996b) pointed out, the peculiar foot morphology of *Ligabueino* suggests phylogenetic relationships with *Noasaurus*, which also shows bizarre pedal phalanges. Even if this interpretation is incorrect (the available pedal phalanges of *Ligabueino* and *Noasaurus* are, in fact, quite different), it is interesting to note that at least two members of Abelisauroidea evolved specialized adaptations on their hindfeet.

NOASAURUS LEALI   This taxon was documented in beds of the Lecho Formation (Maastrichtian) exposed at El Brete in Salta Province (Bonaparte and Powell 1980; Bonaparte 1991b). *Noasaurus* was a relatively small predator approximately 1.50 meters long that has derived features on the pedal ungual of digit II. The holotype specimen is represented by a maxilla with teeth, a quadrate, the neural arch of a cervical vertebra, the centrum of a dorsal vertebra, two cervical ribs, a metatarsal II, a non-ungual phalanx (possibly belonging to pedal digit IV), and two unguals, originally described as corresponding to pedal digit II (Fig. 6.16). In the original description, Bonaparte and Powell (1980) described a squamosal, but this bone was later identified as part of a posterior cervical rib (Novas 1989b).

The maxilla is proportionally short and deep, resembling that of abelisaurids, especially in the abbreviated ascending process, which almost certainly failed to contact the lacrimal, a condition also present in other abelisauroids (Fig. 6.16A and B). A sharp ridge delimits the preorbital depression, which is wider than in abelisaurids. The interdental plates are fully fused. Eleven alveoli are present in the maxilla, but some teeth of small size are preserved. They are slightly curved and serrated along both the cranial and caudal margins. The quadrate is not fused with the quadratojugal, as occurs in other basal abelisauroids (i.e., *Ilokelesia*) but differs from some derived abelisaurids (e.g., *Carnotaurus*, *Abelisaurus*), in which both bones are tightly fused.

The neural arch of the cervical vertebra (presumably a 6th cervical) is beautifully preserved (Fig. 6.16C–F). It is dorsoventrally depressed, transversely wide, and craniocaudally elongate. The dorsal surface is almost flat and is clearly separated from the diapophyses by a sharp ridge. The pre- and post-spinal fossae are wide and deep, as is usual among abelisauroids. The region corresponding to the neural spine is broken. However, if the neural spine existed, it was extremely reduced, perhaps more so than in *Carnotaurus* and *Masiakasaurus*. The neural arch is deeply pneumatized on the cranial and caudal surfaces and below the diapophyses. The epipophyses look extremely elongated craniocaudally in *Noasaurus*, more so than in any other abelisauroid, so this is interpreted as an autapomorphic trait of this taxon.

It must be emphasized that the discontinuous distribution among abelisauroids of cranially pointed epipophyses (e.g., they present in

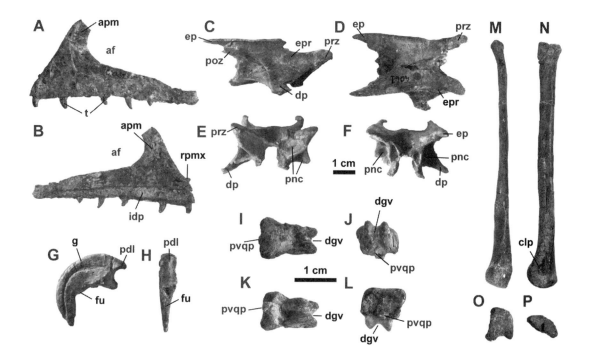

Noasaurus, Carnotaurus and Aucasaurus but are absent in the remaining abelisauroids) constitutes a confusing phylogenetic signal. However, the ambiguous distribution of this feature may reflect the degree of ossification of tendons attached to the cranial end of the epipophyses, an attribute that may be due to ontogeny or individual variation.

Two cervical ribs of Noasaurus were recovered. One of the ribs (originally described by Bonaparte and Powell as a squamosal) closely resembles the posterior cervical ribs of Carnotaurus because of a sharp cranial process. Another rib is more laminar and has a flat and wide external surface that resembles the cranial cervical ribs of other abelisauroids and kin (e.g., Spinostropheus, Carnotaurus, Ilokelesia). Pneumatic excavations are present between the capitulum and tuberculum. However, the state of preservation of the holotype specimen makes it impossible to determine if the shaft was distally rod-like, as in other members of Abelisauroidea.

The hindfoot of Noasaurus is represented by a metatarsal II, a non-ungual phalanx, and two ungual phalanges (Fig. 6.16G–P). Metatarsal II is proximally narrow, as in the remaining abelisauroids, but the available phalanges have interesting features that are absent in other abelisauroids (Fig. 6.16M and N). The preserved non-ungual phalanx (Fig. 6.16I–L) is proportionally short, dorsoventrally compressed, and robust (proportions that may indicate that it belongs to pedal digit II) but has a proximally expanded (both ventrally and dorsally) distal ginglymoid as well as a peculiar quadrangular-shaped process on the ventral margin of the proximal articular end, features that do not match pedal phalanges of abelisaurids. Interestingly, this phalanx articulates with the available trenchant ungual.

Fig. 6.16. *Noasaurus leali:* (A, B) Left maxilla in (A) lateral and (B) medial views. (C–F) Neural arch of cervical vertebra in (C) right lateral, (D) dorsal, (E) cranial, and (F) caudal views. (G, H) Pedal? ungual in (G) side and (H) proximal views. (I–L) Non-ungual pedal phalanx in (I) dorsal, (J) distal, (K) ventral, and (L) proximal views. (M–P) Right metatarsal II in (M) plantar, (N) medial, (O) distal, and (P) proximal views. *Photographs by Fernando Novas.*

However, the most outstanding phalangeal feature of *Noasaurus* is a strongly curved and trenchant ungual that superficially resembles that of pedal digit II of deinonychosaurian theropods (Bonaparte and Powell 1980; Fig. 6.16G and H). It was recently proposed that the available unguals of *Noasaurus* may belong to the manus rather than to the pes (Agnolín, Apesteguía, and Chiarelli 2004), as suggested by the strong lateral compression, the strongly curved blade, the symmetrical condition of the collateral grooves, and the sharply defined median keel on the proximal articular surface. However, this ungual of *Noasaurus* has a well-excavated pocket for tendon attachment on its ventral surface, a feature that resembles the pedal unguals of abelisaurids (Novas and Bandyopadhyay 2001; Fig. 6.13G and I). Moreover, the ungual in question is devoid of proximoventral tuberosity, a feature that is present (and is usually prominent) in the manual unguals of most theropods.

Whether the trenchant unguals of *Noasaurus* belong to the foot or to the hand, they are clearly different from both the pedal and manual unguals recorded in other abelisauroids (e.g., *Masiakasaurus*, *Aucasaurus*). Interestingly, neither the pedal nor the manual unguals of *Masiakasaurus* (Carrano, Sampson, and Forster 2002) have the strongly curved condition of *Noasaurus*. The curious morphology of non-ungual and ungual phalanges plus the highly derived *Carnotaurus*-like cervical epipophyses clearly show *Noasaurus* to be distinct in comparison with the remaining abelisauroids.

AN EL BRETE ABELISAUROID ORIGINALLY DESCRIBED AS AN OVIRAPTOROSAUR   Frankfurt and Chiappe (1999) described a virtually complete neck vertebra (Fig. 6.17) as belonging to an oviraptorosaurian coelurosaur. The specimen was recovered from the Lecho Formation (Early Maastrichtian) in the El Brete fossil locality of Salta Province in northwestern Argentina from the same fossil spot where the holotype specimen of *Noasaurus leali* was excavated.

Frankfurt and Chiappe (1999) noted several features resembling those of oviraptosaurs (e.g., a longitudinal ventral groove that is shallow and broad and is flanked by lateral ridges that merge cranially with the parapophyses; a U-shaped space between the prezygapophyses; and well-developed diapophyses that extend laterally well beyond the plane of the centrum, where the vertebra is widest).

A review of the specimen recently published by Federico Agnolín and Agustín Martinelli (2007) indicates that features that have been recognized to support oviraptorosaurian relationships are also present in abelisauroid theropods. The neural spine is not preserved in the El Brete cervical, but it seems to be reduced (Fig. 6.17B). This feature was originally interpreted to support the oviraptorosaurian affinities of the El Brete theropod (Frankfurt and Chiappe 1999), but this condition is also present in abelisauroids (e.g., *Ligabueino*, *Noasaurus*; Agnolín and Martinelli 2007; Figs. 6.15A and 6.16D). The centrum is low, narrow, and elongate and has a ventral longitudinal groove (Fig. 6.17C), which is usual among oviraptorosaurs but is also present in the basal abelisauroid *Elaphrosaurus*

(Janensch 1920; Agnolín and Martinelli 2007). The prezygapophyses are small and stout and their articular facets are oval (Fig. 6.17B). In dorsal view, the space between the prezygapophyses is U-shaped, a condition resembling that of oviraptorosaurians (Frankfurt and Chiappe 1999), but it also occurs among abelisauroids, such as *Elaphrosaurus* (Janensch 1920), *Ligabueino* (Bonaparte 1996b), and *Carnotaurus* (Bonaparte, Novas, and Coria 1990).

The cervical vertebra from El Brete also has a well-developed prespinal fossa (Fig. 6.17B; prsf), a feature that also occurs in neoceratosaurs (i.e., *Ceratosaurus* and abelisauroids; Agnolín and Martinelli 2007). In anterior view, the cervical vertebra has two peduncular foramina (i.e., pneumatic cavity) on the neural arch, which are located below the level of the prezygapophyses, as in the abelisauroids *Carnotaurus* and *Ilokelesia* (Agnolín and Martinelli 2007).

Following Agnolín and Martinelli (2007), the specimen from El Brete could be assigned to Neoceratosauria on the basis of the following features: an anteroposteriorly reduced neural spine of the cervical vertebrae (despite the lack of neural spine in this specimen, the surrounding area indicates that it was very reduced) and laterally displaced zygapophyses. Also, the cervical shares with Abelisauroidea a cervical centrum that is more than 20 percent broader than tall in anterior view (Holtz 2000), neural arches whose dorsal surfaces are clearly delimited from the lateral surface of the diapophyses (Bonaparte 1991a; Coria and Salgado 2000), and a deep prespinal depression (Novas 1992a).

Agnolín and Martinelli (2007) argue that because the cervical vertebra was found associated with the holotype of *Noasaurus leali* and the material resembles this taxon both in morphology (e.g., the similar position of peduncular foramina) and size, it belongs to an anteriormost cervical vertebra of *Noasaurus leali* (Bonaparte and Powell 1980). However, the neural arches of the cervical of both *Noasaurus* and El Brete look quite different, and I am not aware of any abelisauroids (e.g., *Spinostropheus*, *Elaphrosaurus*, *Carnotaurus*, *Majungasaurus*) that have strong changes in morphology in cervical vertebrae.

In sum, the cervical of El Brete (Frankfurt and Chiappe 1999; Agnolín and Martinelli 2007) is referred as belonging to Abelisauroidea.

VELOCISAURUS UNICUS The only available specimen of this chicken-sized theropod was found in beds of the Bajo de la Carpa Formation in the city of Neuquén. *Velocisaurus* is represented by an incomplete right hindlimb, originally described by Bonaparte (1991a, 1996b) as a possible member of Ceratosauria on the basis of the metatarsal construction (Fig. 6.18A). The metatarsals are elongate and slender, resembling those of *Masiakasaurus*. Metatarsal III is the most robust bone of the foot and is exposed along the entire cranial surface. Interestingly, the phalanges of pedal digit IV are transversely compressed, a condition that is not uniformly present among abelisauroids (Novas, Agnolín, and Bandyopadhyay 2004).

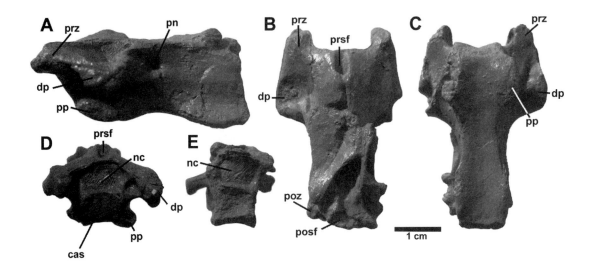

Fig. 6.17. Abelisauroid cervical vertebra from El Brete in (A) left lateral, (B) dorsal, (C) ventral, (D) cranial, and (E) caudal views. *Photographs by Fernando Novas.*

The unguals of *Velocisaurus* are not trenchant, being similar to these of *Masiakasaurus* and abelisaurids, thus differing from the trenchant ungual presumably belonging to the foot of *Noasaurus*. Because of the non-trenchant unguals, Bonaparte (1991a) inferred omnivorous habits for this creature. But because almost the same kind of ungual is present in *Masiakasaurus* and in abelisaurids (whose dental morphology suggests carnivorous habits), that assumption can be dismissed.

The distal end of the tibia of *Velocisaurus* has a faint longitudinal ridge running vertically along the cranial surface of the bone (Fig. 6.18B). This trait has been also recognized in other abelisauroid tibiae (Rauhut 2005b).

Abelisauridae

The South American members of this family are *Carnotaurus*, *Abelisaurus*, *Aucasaurus*, *Ekrixinatosaurus*, *Skorpiovenator*, *Xenotarsosaurus*, *Pycnonemosaurus*, and *Ilokelesia*. Together with their African (e.g., *Rugops*), Indian (e.g., *Rahiolisaurus*, *Indosuchus*, *Lametasaurus*, *Rajasaurus*, and *Indosaurus*), and Malagasy (e.g., *Majungasaurus*) cousins they share several synapomorphies absent in the remaining abelisauroids. Distinctive characteristics of Abelisauridae include a craniocaudally short and deep premaxilla, a dorsoventrally deep snout at the level of the narial openings, dorsoventrally reduced antorbital fossae, dorsoventrally thickened frontals, a wide and smooth caudal surface of the basioccipital, a short dentary with a convex ventral margin, and loose contacts among the dentary and splenial and postdentary bones.

The African *Rugops* is one of the most basal abelisaurids, lacking many of the synapomorphic features of more-derived abelisaurids (for example, *Rugops* lacks a thickened skull roof and horns, and the orbital brow, albeit present, is poorly developed and retains a fenestra between the brow formed by the lacrimal plus the postorbital and the original orbital margin). *Ilokelesia* has a primitive kind of postorbital, suggesting its

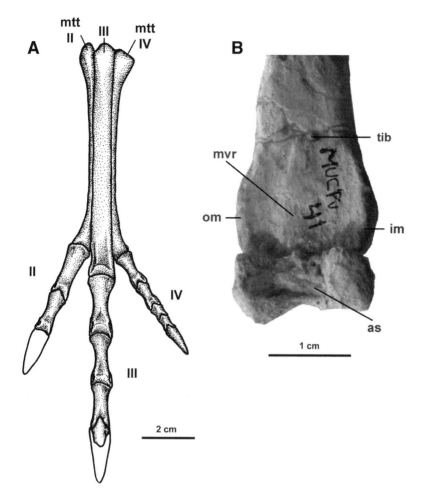

Fig. 6.18. *Velocisaurus unicus:* (A) Left foot in dorsal view. (B) Right tibia and astragalus in cranial view, showing the cranial articular surface for the ascending process of the astragalus. *(A) Redrawn from Bonaparte (1991a). (B) Photograph by Fernando Novas.*

basal position within the family. *Abelisaurus* is more closely related to the remaining members of the Abelisauridae, sharing with them a ventral ramus of the lacrimal that is cranially curved, a reduced dorsal exposure of the lacrimal, a derived postorbital, and the absence of a dorsally projected supraoccipital prominence. Some of the remaining abelisaurids (i.e., *Carnotaurus, Majungasaurus*) may be joined in an abelisaurid subclade, Carnotaurinae (Sereno 1999). However, the phylogenetic relationships of most abelisaurid taxa (e.g., *Aucasaurus, Indosaurus, Indosuchus, Rahiolisaurus, Rajasaurus, Lametasaurus, Ekrixinatosaurus*) remain unclear.

ABELISAURUS COMAHUENSIS

This was the first described member of this theropod branch, constituting the type genus from which the family name Abelisauridae was coined (Bonaparte and Novas 1985). The holotype of *Abelisaurus comahuensis* comes from the upper levels of the Campanian Anacleto Formation (Heredia and Salgado 1999), which is exposed along the shores of Pellegrini Lake in Río Negro Province. The elongate but deep snout seems

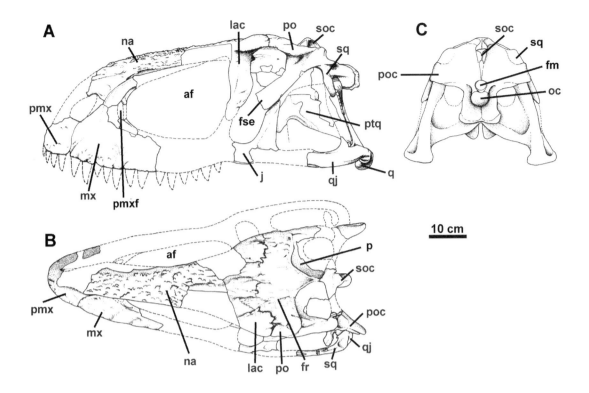

Fig. 6.19. *Abelisaurus comahuensis:* (A–C) Skull in (A) lateral, (B) dorsal, and (C) caudal views. *(A, B) From Bonaparte and Novas (1985). Reprinted with permission of Ameghiniana. (C) Redrawn from Bonaparte (1991b).*

to be a unique condition of this taxon, as does the triangular-shaped and expanded lateral temporal fenestra (Fig. 6.19). Moreover, both premaxillae of *Abelisaurus* are tightly fused, in contrast with the paired bones of *Carnotaurus, Indosuchus,* and *Majungasaurus.* The preorbital opening is rostrocaudally expanded, in clear contrast with that of the remaining abelisaurids (and probably abelisauroids, as is suggested by the rostrocaudally short maxillary proportions of *Noasaurus*).

The general shape of the skull of *Abelisaurus* resembles that of carcharodontosaurid tetanurans (Novas 1997a; Sampson et al. 1998; Lamanna, Martínez, and Smith 2002). Some features in common are a reduced antorbital fossa, rugose nasals and maxilla, an anteroposteriorly expanded preorbital opening, wide contact between the lacrimal and the postorbital forming a thick brow above the orbit, an orbital fenestra invaded by the lacrimal and the postorbital, and strongly caudoventrally inclined quadrates resulting in a mandibular articulation that is far caudal to the occiput. Lamanna and colleagues (2002) suggested the possibility that *Abelisaurus* constitutes a member of Carcharodontosauridae rather than Abelisauridae. However, *Abelisaurus* lacks several distinctive features of carcharodontosaurids, including a rostrodorsally inclined occiput and a transversely wide sagittal crest of the parietal bones. Moreover, *Abelisaurus* shares with *Carnotaurus* and *Majungasaurus* several synapomorphies (e.g., dorsoventrally deep nasals, especially above the nasal openings; a densely ornamented dorsal surface of the skull roof; a cup-shaped parietal crest in dorsal view; short and deep premaxilla;

dorsoventrally thickened frontals resulting in a dorsal massiveness; a wide and smooth caudal surface of the basioccipital), thus supporting close relationships among these theropods. In this phylogenetic context, resemblances that *Abelisaurus* and carcharodontosaurids share are better interpreted as homoplasies (Sereno, Wilson, and Conrad 2004; Novas, De Valais, Vickers-Rich, and Rich 2005).

The lacrimals of *Abelisaurus* are widely exposed in dorsal view, as also occurs in other abelisaurids (e.g., *Rugops*, *Indosuchus*, *Majungasaurus*). In *Abelisaurus*, the frontolacrimal suture is subquadrangular in dorsal view and forms a wide groove behind the lacrimal; also, a sharp peg of the lacrimal projects medially into a frontal socket (Fig. 6.19B). This distinctive articulation between these two bones of the skull roof is also present in *Indosuchus* (Novas, Agnolín, and Bandyopadhyay 2004). In this context, it is important to emphasize the notable resemblance in the skull roof and basicranial morphology between *Abelisaurus* and the Indian *Indosuchus*, thus suggesting close relationships between these two taxa. The skull roof of *Abelisaurus* is similar to that of *Indosuchus* in that it is dorsoventrally thick but lacks prominences above the skull roof. These two taxa seem more conservative than *Carnotaurus* and *Majungasaurus*, for example, and for this reason it is probable that *Indosuchus* and *Abelisaurus* are basal members of the group. Another plesiomorphic feature supporting a basal position of *Abelisaurus* within the Abelisauridae is the absence of a dorsally projected supraoccipital prominence, a structure that is well developed in the more advanced *Carnotaurus* and *Majungasaurus*. In addition, the jugal of *Abelisaurus* is dorsoventrally low, in contrast with the dorsoventrally deep jugals of *Carnotaurus* and *Majungasaurus*.

ILOKELESIA AGUADAGRANDENSIS

The remains of this abelisauroid were excavated in Aguada Grande, Neuquén Province, from beds of the Turonian Huincul Formation. The holotype specimen (Fig. 6.20) was described by Rodolfo Coria and Leonardo Salgado in 2000. *Ilokelesia* is a basal stage in the evolution of abelisaurids as revealed by the morphology of the postorbital, which has a descending ramus that is almost perpendicular to the articular process for the lacrimal and a modestly developed fan-shaped projection that is different from the wider one of more-derived abelisaurids (e.g., *Carnotaurus*, *Majungasaurus*). However, the postorbital has some features that are shared with other members of the family (e.g., *Abelisaurus*, *Majungasaurus*), such as a thick process for articulation with the lacrimal that forms a prominent edge above the orbit. The cervical vertebrae have abelisauroid features such as the shape of the centrum (e.g., it is opisthocoelous and has double pleurocoels) and the neural arch (e.g., a craniocaudally short neural spine, deep pre- and postspinal fossae, a broad dorsal surface of the neural arch that is laterally bounded by sharp epipophyseal-prezygapophyseal ridges, enlarged epipophyses, and a well-developed system of

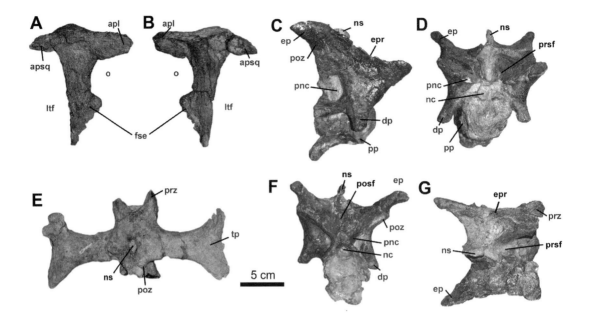

Fig. 6.20. *Ilokelesia aguadagrandensis*: (A, B) Right postorbital in (A) lateral and (B) medial views. (C, D, F, G) Cervical vertebra in (C) right lateral, (D) cranial, (F) caudal, and (G) dorsal views. (E) Caudal vertebra in dorsal view. *Photographs by Fernando Novas.*

pneumatic cavities). The epiphyses lack cranial projections of the kind found in a variety of abelisauroids such as *Noasaurus*, *Carnotaurus*, and *Aucasaurus*. However, the cervical epiphyses in *Majungasaurus* lack cranial projections.

The transverse processes of the mid-caudal vertebrae are distally fan-shaped, bearing cranial and caudal projections. The development of a strong caudal projection was once thought to be an autapomorphic condition of *Ilokelesia* (Coria and Salgado 2000), but it is also seen in *Aucasaurus* and *Ekrixinatosaurus*. The only preserved pedal ungual of *Ilokelesia* is badly damaged but has a patch with rugosities and grooves on one side like those in pedal unguals of other abelisauroids.

CARNOTAURUS SASTREI

Because of its unusual combination of bull-like horns and extremely reduced forelimbs, *Carnotaurus sastrei* constitutes one of the most spectacular Cretaceous dinosaurs yet discovered. It is also one of the best-known abelisaurids, represented by a fairly complete specimen collected in 1984 by José Bonaparte and assistants. The holotype was found near Bajada Moreno north of Chubut Province in Argentina (Fig. 4.13) in beds of the Early Maastrichtian La Colonia Formation.

Carnotaurus is the most derived abelisaurid yet recorded, judging by the anatomy of skull, cervical vertebrae, and forelimbs. This taxon is distinguished from other abelisaurids in a pair of prominent cone-shaped horns above the orbits, constituting a bizarre feature unknown in other meat-eating dinosaurs (Fig. 6.21). This, plus the very short and deep snout, seem to constitute cranial autapomorphies of this abelisaurid (Fig. 6.22). Additional autapomorphic features of *Carnotaurus* are

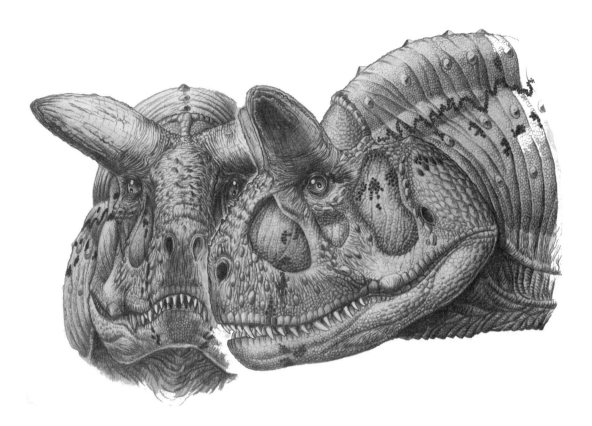

Fig. 6.21. Reconstruction of the head of *Carnotaurus sastrei* (size of the skull 57 cm). *Illustration by J. González.*

identified in the forelimbs, which have humeri that are more robust than in other abelisauroids and that end proximally in an articular head that is hemispherical (rather than elliptical in proximal view, as in less-derived abelisauroids) and a distal articular end that is rather flat instead of having rounded condyles (Bonaparte 1991b). The ulna and radius are proportionally wider and shorter than in other members of the family (e.g., *Aucasaurus*; Fig. 6.11D).

Carnotaurus appears to be closely related to *Majungasaurus* (based on cranial features; Sereno 1999) and *Aucasaurus* (based on available information about the postcranial skeleton; Coria, Chiappe, and Dingus 2002), but the available information is not enough to decide to which of these two later taxa it is more closely allied. The main reason for this ambiguity is that the skull of *Aucasaurus* is poorly known (Coria, Chiappe, and Dingus 2002). *Carnotaurus* shares with *Majungasaurus* a supraoccipital nuchal wedge that is well above the frontoparietal skull table (Sereno, Wilson, and Conrad 2004). Also, the postorbitals are distally fan-shaped and medially set off (the anatomical meaning of this feature is unknown), and the articulation with the ascending ramus of jugal consists of a sharp cone-shaped projection. These features of the postorbital bone contrast with the more conservative postorbitals of the basal abelisaurids *Abelisaurus* and *Ilokelesia*. It is important to emphasize here that not all of the features listed by previous authors to support phylogenetic relationships between *Carnotaurus* and *Majungasaurus* seem

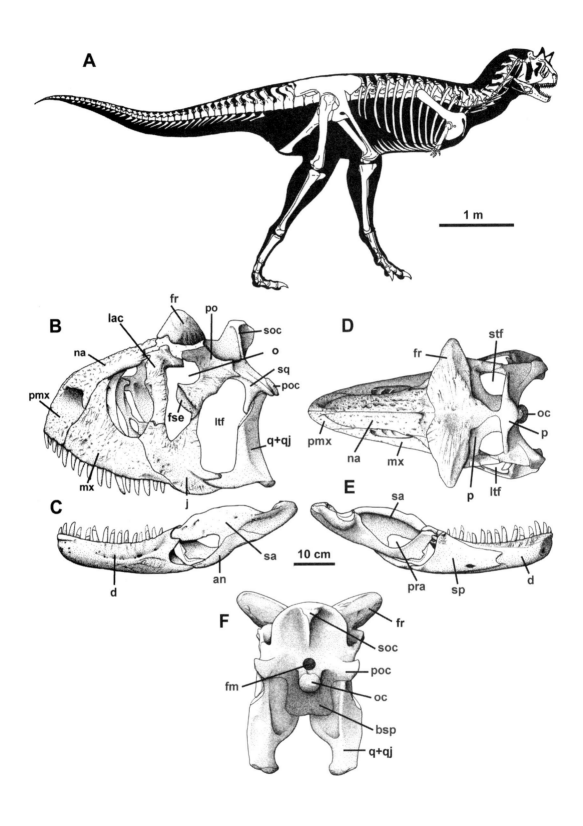

correct. For example, Sereno and colleagues (2004) proposed horns as a synapomorphic feature uniting *Carnotaurus* and *Majungasaurus*, but such projections are not homologous. In *Carnotaurus* the pair of "horns" project laterally, while in *Majungasaurus* a single horn is located on top of the frontals (Fig. 6.7A and B).

Postcranial characteristics support the interpretation that *Carnotaurus* and *Aucasaurus* are closely related to each other. But some of the apomorphic resemblances these two Patagonian taxa share may constitute synapomorphies within the Abelisauridae, some of which are bizarre. These include proportionally short forelimbs, particularly the distal elements (i.e., epipodials and metapodials) that have strongly reduced manual phalanges. Moreover, it is not improbable that the ungual phalanges of digits I to IV of *Carnotaurus* and *Aucasaurus* were completely lost, thus distinguishing these abelisaurids from more basal abelisauroids (e.g., *Masiakasaurus*), in which normally shaped manual ungual phalanges were present (Carrano, Sampson, and Forster 2002). Bonaparte and colleagues (1990) interpreted the largest element of the reconstructed hand of *Carnotaurus* as a highly modified metacarpal IV (Fig. 6.11C). If the hand of *Carnotaurus* has been correctly interpreted, then the claw-shaped external metacarpal may constitute another bizarre peculiarity of this taxon.

AUCASAURUS GARRIDOI

This abelisaurid is represented by an almost complete and articulated skeleton discovered at the Auca Mahuevo fossil site at the top of the Anacleto Formation (Río Colorado Subgroup, Campanian) from levels above those that yielded hundreds of titanosaurid nests (see the section on Cretaceous sauropod eggs and nests in chapter 5). The holotype specimen of *Aucasaurus* was found articulated in an unusual laminated unit composed of varves of unknown periodicity (Coria, Chiappe, and Dingus 2002).

Aucasaurus is still only briefly described (Coria, Chiappe, and Dingus 2002; Coria 2007). The nearly complete caudal series of *Aucasaurus* is comprised of the first 13 elements of the tail, which show strong morphological resemblances with *Carnotaurus*. The transverse processes are dorsolaterally pointed, and the distal ends form cranially projected awl-like process. The latter are 40–50 percent of the length of the distal end of the transverse processes, and their tips are laterally overlapped by the transverse processes of the contiguous posterior caudal vertebra (Fig. 6.23A).

The complete articulated forelimbs of *Aucasaurus* match the unusual morphology of *Carnotaurus*, although they are proportionally longer. The forelimbs of *Aucasaurus* are less derived than in *Carnotaurus*, as indicated by the humeral shaft, which is more gracile and craniocaudally more compressed than the stout subcylindrical shaft of *Carnotaurus*. The humeral distal condyles of *Aucasaurus* are more rounded, thus resembling

Fig. 6.22. *Carnotaurus sastrei:* (A) Skeletal reconstruction. (B, D, F) Skull in (B) lateral, (D) dorsal, and (F) caudal views. (C, E) Left lower jaw in (C) lateral and (E) medial aspects. From Bonaparte, Novas, and Coria (1990). Reprinted from Contributions in Science 416: 1–42 with permission of the Natural History Museum of Los Angeles County. © Copyright 1990.

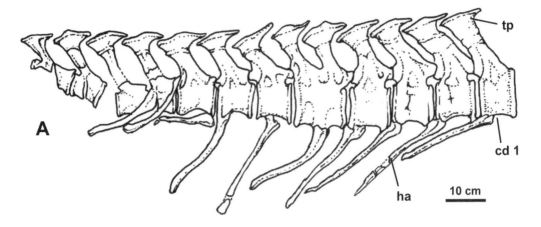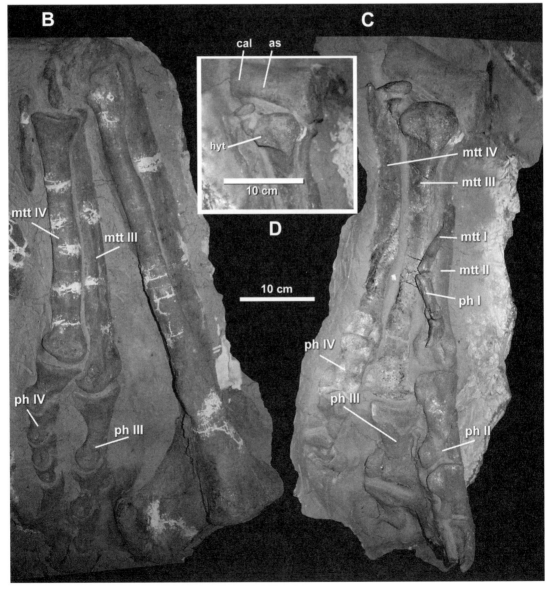

the usual theropod condition, in contrast with the flattened distal end of the humerus described for *Carnotaurus*. The relative lengths of the ulna and radius are greater (the ulna is close to ⅓ the length of the humerus instead of ¼, as in *Carnotaurus*), and the radius lacks the hooked proximal process for contacting the ulna. The shaft of the radius has the same craniolateral "osseous process" present in *Carnotaurus* (Bonaparte, Novas, and Coria 1990). The hand of *Aucasaurus* was found in full articulation and almost probably in natural position, showing that the palmar surface was caudally oriented, thus refuting a previous interpretation (Bonaparte 1991b), which was based on the partially disarticulated hand of *Carnotaurus*, that the palmar surface was craniomedially faced.

Aucasaurus also offers, for the first time, valuable information about foot construction in abelisaurids (Fig. 6.23B–D). Metatarsal III is significantly broader than metatarsals II and IV, and metatarsal I is notably reduced (Coria, Chiappe, and Dingus 2002). In *Aucasaurus* there is a robust caudal buttress on the proximal end of metatarsal III that expands to metatarsal IV; this structure approaches the development of some avian hypotarsi (Coria, Chiappe, and Dingus 2002).

As already noted, the skull of *Aucasaurus* is very damaged, and only a few characteristics are discernible in the available specimen. Coria and colleagues (2002) indicated that in *Aucasaurus* the frontals form low buttresses and the rostrum is proportionally longer than in *Carnotaurus*. In addition, *Aucasaurus* lacks the distinctly convex outline of the dentigerous margin of the maxilla of *Carnotaurus*.

Remarkably, Coria and colleagues (2002) overlooked comparisons with *Abelisaurus comahuensis* (Bonaparte and Novas 1985), an abelisaurid recovered from the same sedimentary unit (Heredia and Salgado 1999). *Aucasaurus* resembles *Abelisaurus* in having a rostrum and antorbital fenestra that are longer and lower than in *Carnotaurus*, a ventral margin of the antorbital fenestra that is horizontal (instead of strongly curved as in *Carnotaurus*), a complete lateral exposure of the maxillary fenestra, and frontal swells instead of horns (Coria, Chiappe, and Dingus 2002). Future studies comparing the skull of *Aucasaurus* and *Abelisaurus* in detail will reveal if these two taxa are synonyms.

EKRIXINATOSAURUS NOVASI

This new abelisaurid, represented by an incomplete skeleton, was recently discovered by Jorge Calvo and assistants in beds of the Cenomanian Candeleros Formation, which is exposed near Añelo in Neuquén Province. *Ekrixinatosaurus* is known from some skull elements (e.g., a postorbital, a maxilla, the dentary, the skull roof, and part of the basicranium) as well as isolated cervical, dorsal, sacral, and caudal vertebrae and parts of the pelvic girdle and hindlimbs (Fig. 6.24). *Ekrixinatosaurus* is a large abelisaurid around 7 to 8 meters long characterized by a fenestra between the postorbital and the anterior border of the frontal and craniocaudally compressed cervicals with flattened ventral sides.

Fig. 6.23. *Aucasaurus garridoi:* (A) Articulated series of proximal caudals in right lateral view. (B) Articulated right tibia, fibula and pes in lateral view. (C) Left foot in plantar view. (D) Detail of figure (C). *(A) Redrawn from Coria, Chiappe, and Dingus 2002. (B–D) Photographs by Fernando Novas.*

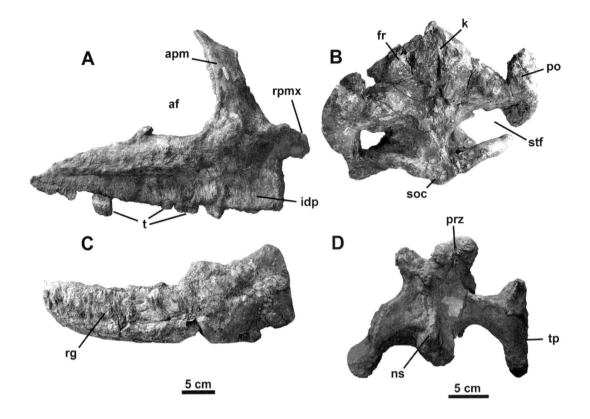

Fig. 6.24. *Ekrixinatosaurus novasi:* (A) Left maxilla in medial view. (B) Frontoparietal region of skull in dorsal view. (C) Left dentary in lateral view. (D) Caudal vertebra in dorsal view. *Photographs by Fernando Novas.*

Although the maxilla is proportionally short and deep (Fig. 6.24A), its size (42 centimeters long and 28 centimeters deep) makes *Ekrixinatosaurus* one of the largest known abelisaurids (in comparison, the maxilla of *Carnotaurus* is 30 centimeters long and 24 centimeters deep). The maxilla of *Ekrixinatosaurus* is distinguished from that of other abelisaurids by of the wide development of the antorbital fossa. Based on the articulation with the jugal, the antorbital fenestra appears to have been rounded and rostrocaudally very short, resembling the condition of *Carnotaurus*. As in *Carnotaurus*, the ascending ramus of the maxilla is almost vertical, lacking the backward inclination and caudal projection present in *Majungasaurus*, for example. The dentary is also short, robust, and curved (Fig. 6.24C). Calvo, Rubilar-Rogers, and Moreno (2004) estimate that the length of the skull of *Ekrixinatosaurus* is around 80 centimeters, slightly longer than the femur. This proportion contrasts with that of *Carnotaurus*, in which the skull length (59 centimeters) is nearly half the length of femur (103 centimeters). This means that although the shape of the skull of *Ekrixinatosaurus* (which is inferred on the basis of the maxilla and dentary) resembles that of *Carnotaurus*, the size of the skull of *Ekrixinatosaurus* was not reduced, as in *Carnotaurus*.

The frontal of *Ekrixinatosaurus* is a wedge-shaped block with an almost flat dorsal surface, although a small central knob is present (Fig. 6.24B). The contact between the frontal and the postorbital is fused, but a fenestra is present between these two bones, a feature also reported for the basal abelisaurid *Rugops* (Sereno, Wilson, and Conrad 2004). The parietals are fused to the frontals, and the supraoccipital is caudally projected and slightly dorsally offset from the frontal surface. The cranial evidence suggests that *Ekrixinatosaurus* retained some primitive features (compared to *Carnotaurus*), such as a proportionally large head, a non-prominent supraoccipital, and a skull roof that failed to ossify in the region surrounded by frontals and postorbitals.

The posterior cervicals of *Ekrixinatosaurus* look proportionally shorter than in *Carnotaurus* and have well-developed epipophyses that are caudally (but not cranially) projected. The size of the available posterior cervical of *Ekrixinatosaurus* is similar to cervical 10 of *Carnotaurus*. The mid-caudal vertebrae have transverse processes with distal extremities that are projected both cranially and distally (Fig. 6.24D), as is also documented in *Ilokelesia* and *Aucasaurus*.

SKORPIOVENATOR BUSTINGORRY

A nearly complete and articulated skeleton has been recently found by Juan Canale and assistants (2006, 2008) in beds of the Turonian–Santonian Huincul Formation, close to the town of El Chocón in Neuquén Province. *Skorpiovenator* and *Carnotaurus* are the most informative abelisaurids yet recovered in this continent, providing valuable details about the anatomy of skull and neck. The specimen, measuring roughly 6 meters long, has hypertrophied cervical epipophyses that are devoid of the cranial projections that characterize *Noasaurus*, *Carnotaurus*, and *Aucasaurus*. Furthermore, the jaw of the new abelisaurid has wide contacts between the dentary and postdentary bones, a condition that is interpreted as less derived than that of *Carnotaurus* and *Majungasaurus* (Canale, Novas, and Scanferla 2006). The new abelisaurid lacks frontal horns, retaining simple (albeit thickened) supraorbital margins. The postorbital bone is distally fan-shaped, thus differing from the modestly expanded condition present in the basal abelisaurid *Ilokelesia* (Coria and Salgado 2000), a taxon also recorded from beds of the Huincul Formation.

PYCNONEMOSAURUS NEVESI

This ponderous abelisaurid was described by Alexander Kellner and Diógenes de Almeida Campos in 2002 and is the best-known abelisaurid from Brazil. *Pycnonemosaurus* is represented by incomplete teeth, caudal vertebrae, rib fragments, partial pubis, tibia, and fragment of fibula. The specimen comes from a conglomeratic sandstone of an unidentified formation of the Bauru Group that is exposed close to Paulo Creek in Mato Grosso State. The caudal vertebrae of *Pycnonemosaurus* have abelisaurid features such as distally fan-shaped transverse processes and a cranial projection, although in the more distal caudals the expansion diminishes. This contrasts with other abelisaurids (e.g., *Aucasaurus*), for whom the awl-like projections of the transverse processes are present in most of the caudal vertebrae. The pubis of *Pycnonemosaurus* is elongate and slender,

Fig. 6.25. *Pycnonemosaurus nevesi:* Right pubis in (A) lateral and (B) cranial views. *Photograph by Fernando Novas.*

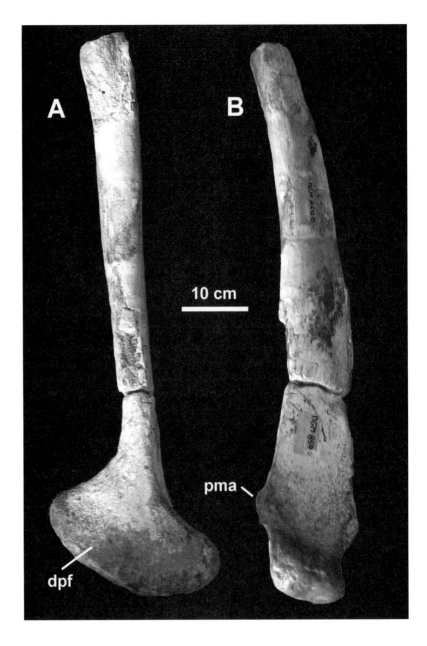

almost straight in side view (Fig. 6.25). The pubic foot has a protuberance for muscle attachment on its anterolateral surface, a feature not reported yet in other members of the family. The tibia is 84.5 centimeters long. It is a stout and robust bone, resembling that of the Indian *Lametasaurus* but differing from tibiae of the Patagonian *Carnotaurus, Aucasaurus, Xenotarsosaurus,* and *Quilmesaurus*. The tibia of *Pycnonemosaurus* is characterized by a cranially extended hatchet-shaped cnemial crest, but it is not proximodistally expanded to the degree present in *Quilmesaurus*. Distally, the tibia is transversely expanded, and its articular surface for the ascending process of astragalus has a triangular depression for roughly 19 percent of the total length of the bone.

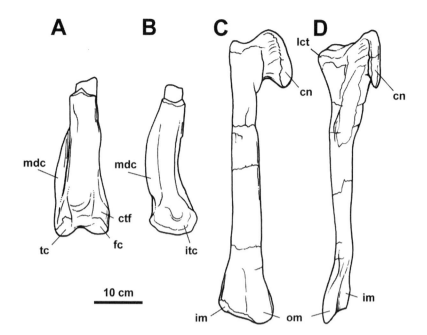

Fig. 6.26. *Quilmesaurus curriei:* (A, B) Distal end of right femur in (A) caudal and (B) medial views. (C, D) Right tibia in (C) caudal and (D) lateral views. Redrawn from Coria (2001).

QUILMESAURUS CURRIEI

This taxon was originally interpreted by Rodolfo Coria (2001) as a theropod of uncertain relationships, but its membership in Abelisauridae has been recently supported by Kellner and Campos (2002) and Juárez Valieri and colleagues (2007). *Quilmesaurus curriei* is represented by a distal half of the femur and a complete tibia recovered in Salitral Ojo de Agua, located 40 kilometers south of the city of General Roca in Río Negro Province, from outcrops of the Allen Formation (Campanian–Maastrichtian). The characteristics of *Quilmesaurus* that support abelisaurid affinities include a rostrally expanded cnemial crest and a well-developed mediodistal crest on the distal end of the femur. Both the femur and the tibia are long and slender, differing from the more robust bones of other abelisaurids (e.g., *Aucasaurus, Pycnonemosaurus, Lametasaurus, Ekrixinatosaurus*). Also curious for *Quilmesaurus* is the shape of the cnemial crest, which is hatched-shaped in side view (Coria 2001; Fig. 6.26). *Quilmesaurus* and probably also *Carnotaurus* are the youngest records of abelisaurids in South America.

XENOTARSOSAURUS BONAPARTEI

The beds in which *Xenotarsosaurus* was found constitute the Lower Member of the Bajo Barreal Formation, accumulated during the Cenomanian–Turonian, a time of abelisaurid proliferation (Martínez, Giménez, Rodríguez, and Bochatey 1986; Martínez, Novas, and Ambrosio 2004; Lamanna, Martínez, Luna, Casal, Ibiricu, and Ivany 2002). *Xenotarsosaurus* is represented by the femur (Fig. 6.13A–D), tibia, fibula, and tarsus from the right hindlimb and a badly preserved cranial dorsal vertebra.

Unfortunately, no derived characteristics (e.g., autapomorphies) have yet been recognized in the preserved bones of *Xenotarsosaurus bonapartei* to support the validity of this taxon (Novas 1989b; Coria and Salgado 2000). Although originally referred to the Abelisauridae, *Xenotarsosaurus* has been more recently interpreted as a neoceratosaurian of uncertain relationships (Coria and Salgado 2000; Coria and Rodriguez 1993). However, I list this genus within Abelisauridae because *Xenotarsosaurus* has a femur, tibia, and tarsus that are virtually identical to these of *Carnotaurus*, *Majungasaurus*, *Indosuchus*, *Aucasaurus*. and *Ekrixinatosaurus*. The large size and robust proportions of these bones are also congruent with the bones of these abelisauroids and not with the slender proportions of the smaller abelisauroids *Velocisaurus* and *Masiakasaurus*.

AN INDETERMINATE ABELISAURID MAXILLA

This specimen was found in beds of the Bajo Barreal Formation (Late Cenomanian–Early Turonian) at the Ocho Hermanos farm, located at the foot of the Sierra de San Bernardo in Chubut Province. The maxilla was described by Matt Lamanna and colleagues (2002), who remarked that the importance of this isolated bone is that it constituted the first definitive abelisaurid theropod recorded from pre-Senonian deposits. Moreover, they defended the hypothesis that the specimen is the oldest-known member of a derived subgroup of abelisaurids, the Carnotaurinae (sensu Sereno 1998), a clade that gathers all known abelisaurids excepting *Abelisaurus* (Lamanna, Martínez, and Smith 2002).

The maxilla (Fig. 6.27) has a densely rugose ornamentation on its external surface, a reduced rostral ramus that resembles more the condition present in *Carnotaurus* than that in *Noasaurus* and *Majungasaurus* (in which the rostral ramus of the maxilla is longer), a promaxillary fenestra that is hidden in lateral view, a maxillary body with subparallel dorsal and ventral margins, and an elongate maxilla-jugal contact. All these features are widely distributed among abelisaurids and probably do not help resolve the phylogenetic relationships of the specimen within Abelisauridae. Moreover, Sereno and colleagues (2004) emphasized the close resemblance between this isolated maxilla from Patagonia with that of the African basal abelisaurid *Rugops*, citing the similar pattern of rugosities on the external surface of the maxilla. In sum, the noted similarities between the maxillae of the Bajo Barreal and the basal abelisaurid *Rugops* and the uneven distribution of maxillary features among abelisaurids necessitate caution about proposing the monophyly of Carnotaurinae; the morphology of the maxilla described above is diagnostic of a group wider than Carnotaurinae.

Tetanurae

Tetanurae is defined to include birds and all theropod taxa sharing a more recent common ancestor with birds than with Ceratosauria (Holtz, Molnar, and Currie 2004). Although tetanurine phylogeny is under continuous review, three main subgroups are frequently recognized (Fig. 6.28):

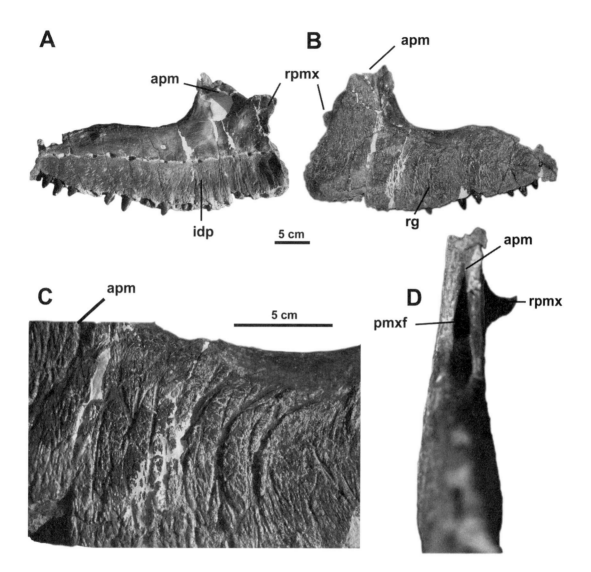

Fig. 6.27. Right maxilla of an indeterminate abelisaurid from the Bajo Barreal Formation in (A) medial and (B) lateral views. (C) Detail of the decoration pattern on the external surface of the bone. (D) Ascending ramus of the maxilla in caudal view showing the promaxillary fenestra. *Photographs by Fernando Novas.*

Spinosauroidea (which includes the Cretaceous spinosaurids and the Jurassic "megalosaurs"), Carnosauria (which includes allosaurids and carcharondontosaurids), and Coelurosauria (which includes a diverse set of derived theropods more closely related to birds than to Carnosauria). Fossil representatives of these three lineages have been discovered in Cretaceous beds of South America.

Spinosauridae

Spinosauridae is a distinctive Cretaceous theropod family that may have phylogenetic ties with some Jurassic forms (e.g., *Torvosaurus*; Galton and Jensen 1979). Spinosaurids and their kin form the superfamilial group Spinosauroidea. Spinosaurids were large theropods; some of them probably measured 16–18 meters long, thus constituting the largest known carnivorous dinosaurs. (Dal Sasso, Maganuco, Buffetaut, and Mendez 2005).

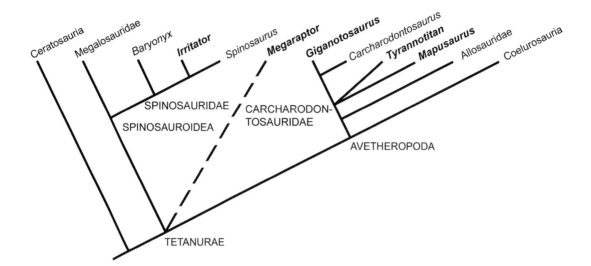

Fig. 6.28. Cladogram depicting phylogenetic relationships among main clades of Tetanurae.

They were characterized by a bizarre combination of crocodile-like skulls and bison-like humped backs. Spinosaurids had elongate and depressed but narrow snouts that had numerous conical teeth. The rostral extremity of the upper and lower jaws forms a well-developed spoon-shaped expansion, or rosette, and the symphysial end of the dentary is upturned and has a constricted region immediately posterior to it (Milner 1997, 2001). Seven teeth are present on the premaxillary, thus surpassing the tooth count of other theropods. Spinosaurid teeth differ from those of other theropods in being almost subcircular in cross-section with crowns that are slightly recurved or straight and finely serrated or unserrated.

However, spinosaurids are unusual in the degree of elongation of the neural spines in most dorsal vertebrae. This is especially the case for *Spinosaurus aegyptiacus*, in which the neural spines reached 1.65 meters tall (Stromer 1915).

Spinosaurids are documented from Barremian to Cenomanian beds in Europe, northeastern Brazil, and (especially) northern Africa, where isolated teeth and bones referable to Spinosauridae are fairly common (Sues, Frey, Martill, and Scott 2002). In South America, remains of these animals are documented in Albian beds of the Santana Formation, in the Araripe Basin (e.g., Kellner 1996b; Martill, Cruickshank, Frey, Small, and Clarke 1996; Sues, Frey, Martill, and Scott 2002), and in Cenomanian beds of the Alcântara Formation in the state of Maranhão in northeastern Brazil (Medeiros 2001; Vilas Boas, Carvalho, Medeiros, and Pontes 1999). The only South American spinosaurid named so far is *Irritator challengeri*, which constitutes a senior synonym of *Angaturama limai* that Kellner and Campos (1996) recovered from the same locality and horizon.

Notably, no bone remains belonging to Spinosauridae have yet been recovered in the productive Cretaceous beds of Patagonia, where abelisaurid and carcharodontosaurid remains are frequently discovered. However, it is probable that the bizarre basal tetanuran *Megaraptor namunhuaiquii* (Novas 1998; Calvo, Porfiri, Veralli, Novas, and Poblete

2004) and another as-yet-unnamed theropod (Coria and Currie 2002b) may have phylogenetic ties with Spinosauridae.

IRRITATOR CHALLENGERI

This is the only spinosaurid so far discovered in South America. The holotype specimen of this animal almost surely comes from the highly fossiliferous Romualdo Member (possibly Mid- to Late Albian) of the Santana Formation, which crops out at a site near Santana do Cariri, Ceará State, Brazil (Figs. 4.3 and 4.5). *Irritator challengeri* is represented by a partial skull nearly 60 centimeters deep that constitutes the most complete skull of a spinosaurid known to date (Sues, Frey, Martill, and Scott 2002; Fig. 6.29). *Irritator* most closely resembles *Spinosaurus* because of several apomorphic features in the dentition, including maxillary tooth crowns that are straight or slightly recurved and conical rather than labiolingually compressed, carinae of tooth crowns that are devoid of serrations, a thin layer of enamel layer on the teeth, and vertical ridges ("fluting") on both the labial and lingual surfaces of the crown. Sues and colleagues (2002) thought that *Irritator* might well prove to be congeneric with *Spinosaurus*, but the current lack of cranial material for the latter makes detailed comparisons between the two taxa impossible.

The skull of *Irritator* is elongate and low, especially in its rostral region, a condition shared with the remaining spinosaurids. However, *Irritator* is distinguished from other members of this family in having nasals with a prominent median bony crest (a continuation of the crest present on the premaxillae; Kellner and Campos 1996) that terminates posteriorly in knob-like, somewhat dorsoventrally flattened projection (Fig. 6.29; knp). Also, the dorsal surface of the parietals faces posterodorsally and the vertical axis of the braincase is anteroventrally inclined. Seven premaxillary and at least 11 maxillary teeth are present in *Irritator*. The teeth are deeply implanted and vertically oriented and are well separated from each other. The tallest teeth are located rostrally, probably in accordance with the piscivorous habits inferred for these dinosaurs (Charig and Milner 1986). The jaws of *Irritator* are notably deep, and the surangular has a broad lateral shelf.

POSTCRANIAL SPINOSAURID REMAINS FROM THE SANTANA FORMATION

The same formation that yielded the specimens of *Irritator* also produced two sets of bones that may belong to Spinosauridae. One set of bones consists of an articulated series of three posterior sacrals and six anterior caudal vertebrae (Bittencourt and Kellner 2004), and the other set, which was not found in association with the anterior vertebrae, consists of articulated dorsal and sacral vertebrae, ilia, pubes, ischia, a femur, and manual elements. The sacral and caudal vertebrae were presumably collected in the surroundings of the city of Santana do Cariri in southern Ceará State, where most of the nodules from the Santana Formation were collected

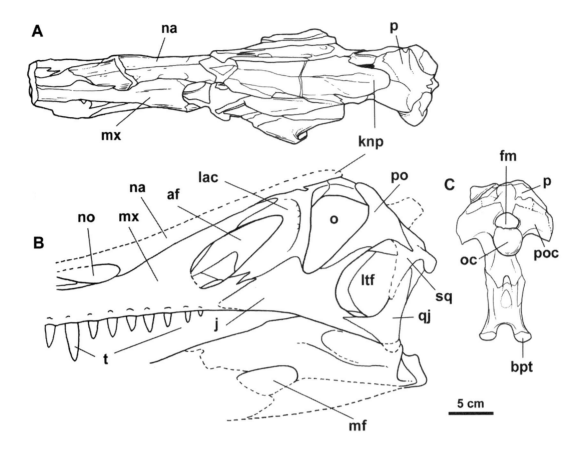

Fig. 6.29. *Irritator challengeri:* Skull in (A) dorsal, (B) lateral, and (C) occipital views. *Redrawn from Sues, Frey, Martill, and Scott (2002).*

(Bittencourt and Kellner 2004). As Bittencourt and Kellner have noted, the most notable trait of these proximal caudals is two buttresses below the transverse process delimiting three fossae, a condition that they interpreted as a potential synapomorphy of Spinosauroidea because it is also shared by the spinosaurid *Baryonyx* and the torvosaurid *Torvosaurus* from the Late Jurassic Morrison Formation.

The second specimen referred to Spinosauridae was illustrated by Campos and Kellner (1991) and Kellner (2001) and briefly described by Machado, Kellner, and Campos (2005). Bittencourt and Kellner (2004) tentatively referred it to Spinosauridae because of the very tall neural spines in the sacral vertebrae and the peculiar configuration of the manual ungual.

A SPINOSAURID TOOTH FROM THE CANDELEROS FORMATION

Canudo and colleagues (2004) reported the discovery of an isolated spinosaurid tooth from El Anfiteatro, not far from the town of General Roca in Río Negro Province. The specimen comes from levels of the Cerro Lisandro Formation (Cenomanian). It is conical-shaped, elongate, and slightly compressed lateromedially. Both the mesial and distal margins are finely serrated, and, most important, the crown has longitudinal ridges, as typically occurs in spinosaurid theropods (Charig and Milner 1997; Canudo, Salgado, Barco, Bolatti, and Ruiz-Omeñaca 2004). This is the first evidence to support the possible presence in Patagonia of this clade of bizarre theropod dinosaurs.

MEGARAPTOR NAMUNHUAIQUII

A group of large-bodied South American theropods is represented by *Megaraptor*, a basal tetanuran with enlarged ungual phalanges on the hands (Novas 1998; Calvo, Porfiri, Veralli, Novas, and Poblete 2004; Fig.6.30A; Plates 18 and 19). The phylogenetic relationships of this animal are obscure, but some features suggest that *Megaraptor* may have phylogenetic relationships with spinosaurids (Calvo, Porfiri, Veralli, Novas, and Poblete 2004), although other authors (Smith, Makovicky, Hammer, Currie 2007) have recently interpreted this animal as a derived member of Carcharodontosauridae. Two incomplete specimens of this animal were collected in Neuquén Province in northwestern Patagonia, and fragmentary but complementary remains of *Megaraptor*-like creatures were more recently discovered in Chubut Province (Lamanna, Martínez, Luna, Casal, Ibiricu, and Ivany 2004), thus indicating that these dinosaurs were more numerous than expected. Initially, *Megaraptor* was interpreted as a gigantic basal coelurosaurian with a raptorial ungual on pedal digit II (Novas 1998), but recent discoveries clearly demonstrate that the enlarged and trenchant ungual corresponds to manual digit I (Calvo, Porfiri, Veralli, Novas, and Poblete 2004). *Megaraptor namunhuaiquii* is a startling example of a large and bulky theropod (probably 8 meters long, based on comparisons with *Allosaurus*) with specialized raptorial hands. Although the phylogenetic relationships of *Megaraptor* are still poorly understood, it is clear that this creature does not belong to the theropod clades frequently documented in South America (e.g., Carcharodontosauridae, Spinosauridae, Abelisauridae, Deinonychosauria).

The cervicals of *Megaraptor* are proportionally short and deep, indicating a thick and powerful neck (Fig. 6.30B). The available caudal vertebrae of *Megaraptor* (Calvo, Porfiri, Veralli, Novas, and Poblete 2004; Lamanna, Martínez, Luna, Casal, Ibiricu, and Ivany 2004) are characterized by pleurocoels, a feature that is also documented among Carcharodontosaurids (Fig. 6.30C). Moreover, the transverse processes are supported from below by a pair of robust oblique buttresses, which bound presumed pneumatic depressions, similar to those reported for the possible spinosaurid from Santana Formation as well as for *Baryonyx* and *Torvosaurus* (Bittencourt and Kellner 2004).

The pectoral girdle is quite distinctive in that the scapular blade is craniocaudally expanded and the coracoid is almost flat and elliptical (Fig. 6.30D), similar to the coracoid of the basal spinosaurid *Baryonyx*, from the Barremian of Europe (Charig and Milner 1997). The forearms of *Megaraptor* are well developed and robust, and although the humerus is currently unknown it is likely that it was an elongate and

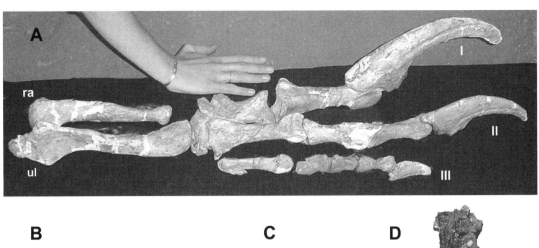
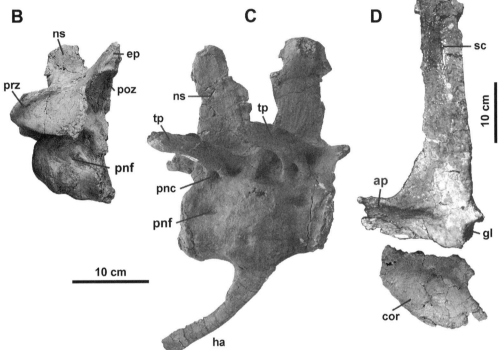

Fig. 6.30. *Megaraptor namunhuaiquii:* (A) Right forearm in lateral view. (B) Cervical vertebra in left lateral view. (C) Articulated proximal caudal vertebrae and haemal arch in right lateral view. (D) Left scapula and coracoid in lateral view. *Photographs courtesy of Jorge Calvo.*

robust bone. The ulna is stout and has a prominent and transversely compressed olecranon process that is triangular-shaped in proximal view. It superficially resembles that of the spinosaurids *Baryonyx* (Charig and Milner 1997) and *Suchomimus* (casts housed at the Natural History Museum, London) from the Aptian of Africa. The manus of *Megaraptor*, which reached 70 centimeters in maximum length, is impressive for the enormous trenchant unguals on digits I and II. Digit III also ends in a laterally compressed and sharply curved ungual, albeit considerably reduced in size. Four metacarpals are present, of which metacarpal IV is rudimentary and lacks phalanges. The ungual of digit I is the sharpest of the manual unguals, especially because its cutting edge is strongly compressed transversely. In the holotype specimen, the ungual of digit I has both lateral and medial grooves, with the internal furrow occupying a higher position than the external one. Notably, this trenchant ungual resembles the sickle-like ungual of pedal digit II of dromaeosaurids, especially because it is transversely compressed and the collateral grooves are asymmetrically placed.

The pubis of *Megaraptor* has a stout appearance, especially because of its proximal end, which is dorsoventrally deep and craniocaudally expanded, which suggests that the proximal end of the ischium was also high. This condition resembles that present in basal tetanurans (e.g., *Torvosaurus, Piatnitzkysaurus*), thus contrasting with that of more-derived tetanurans (e.g., *Allosaurus, Giganotosaurus, Bahariasaurus*, coelurosaurs), in which the pubis is proximodistally low. The pubic apron developed at mid-length and a pubic foot was present, although its size and shape remain unknown.

The phylogenetic relationships of *Megaraptor* are far from clear. Since *Megaraptor* retained metacarpal IV in the manus and its pubis had a notably expanded proximal end, it is likely that this taxon originated prior to the differentiation of Avetheropoda (e.g., the clade including allosauroids plus coelurosaurians; Holtz, Molnar, and Currie 2004). The scapula of *Megaraptor* retained some primitive features (e.g., its scapular blade is craniocaudally wide compared to its proximodistal length and a third of the blade is fan-shaped) present in *Ceratosaurus* and basal tetanurines (e.g., *Baryonyx*). On the other hand, the slender scapulae characteristic of the more restricted group Avetheropoda are not present in the Patagonian taxon.

As noted by Calvo, Porfiri, Veralli, Novas, and Poblete (2004) and later by Smith and colleagues (2007), some features of the cervical vertebrae of *Megaraptor* resemble those of carcharodontosaurid allosauroids (e.g., parapophyses located at the midlength of the centrum; a prominent ridge that connects the prezygapophysis with the epipophysis; a posterodorsally pointed epipophyses; a low neural spine that is craniocaudally wide at its base rather than on its upper end). Such features of the neck vertebrae of *Megaraptor* may lead to the interpretation that it is a derived member of Carcharodontosauridae, as Smith and colleagues (2007) hypothesize. Although I do not dismiss this possibility, it is important to note that the rest of the cervical vertebral anatomy of *Megaraptor* does not correspond to the kind of cervical vertebrae present in carcharodontosaurids (e.g., *Giganotosaurus, Tyrannotitan*). The caudal vertebrae resemble these of *Carcharodontosaurus* (Stromer 1931) in having a proportionally large pleurocoel, but it must be emphasized that this characteristic is not unique to carcharodontosaurids. It is not uniformly distributed among this theropod family (Novas, de Valais, Vickers-Rich, and Rich 2005). Finally, other parts of the skeleton of *Megaraptor* do not fit well with the currently known carcharodontosaurid pattern. For example, the forelimbs of *Megaraptor* are remarkably elongate and robust, contrasting with the reduced humerus documented in *Giganotosaurus* and *Mapusaurus*.

Megaraptor, whether it is related to spinosaurids or carcharodontosaurids, is representative of a theropod clade not previously identified.

This clade has a particularly interesting combination of a bulky construction of the body (as suggested by the cervical vertebrae and the proportions of the scapulocoracoid, hands, pubis, and metatarsal IV) and exceptional raptorial abilities in the hands. In this regard, spinosaurids may compare with the Patagonian taxon in powerful arms and developed ungual phalanges, but even in these theropods the hands are not enlarged to the degree seen in *Megaraptor*.

LARGE THEROPODS POSSIBLY RELATED TO *MEGARAPTOR*

Two new theropods of unknown phylogenetic relationships were collected in recent years from different Late Cretaceous fossil sites in Argentina. These large animals are estimated to be about 6 to 9 meters long. One of the specimens was recovered in Barrosa Hill in Neuquén Province from the Portezuelo Formation, the same sedimentary unit that produced *Megaraptor*. The discovery of this new dinosaur was briefly announced by Rodolfo Coria and Philip Currie (2002b), who interpreted it as a coelurosaurian theropod. It is represented by a beautifully preserved basicranium and several cranial and postcranial bones.

A second theropod, *Aerosteon riocoloradensis* (Sereno et al. 2008), was found in outcrops of the Campanian Anacleto Formation in southern Mendoza Province. *Aerosteon* is represented by some cranial and postcranial bones. It shares with the Sierra Barrosa theropod similarities in the shape of the postorbital, the ilium, and particularly the caudal vertebrae, thus suggesting that they may be closely related.

The bones and teeth available from these two creatures clearly demonstrate that they are not members of Abelisauridae, Carcharodontosauridae, or Deinonychosauria, theropod groups that have been documented in the Cretaceous of Patagonia. On the contrary, both *Aerosteon* and the Barrosa Hill theropod have a pleurocoel on each side of the caudal centrum as well as a pair of buttresses bounding three fossae on the ventral surface of the base of the transverse processes of the tail vertebrae, conditions resembling those of the basal tetanuran *Megaraptor*, *Baryonyx*, and the possible spinosaurid from the Santana Formation described above (Bittencourt and Kellner 2004).

Aerosteon shares with *Megaraptor* (Calvo, Porfiri, Veralli, Novas, and Poblete 2004) outstanding similarities in the shape of cervical and caudal vertebrae, the scapula, and the pubis.

Studies of these animals may shed light about their phylogenetic relationships within Tetanurae and may offer more information to clarify the systematic allocation of *Megaraptor* itself.

Carcharodontosauridae

Carcharodontosaurids may be among the most surprising hunters of the dinosaurian world (Fig. 6.31). They attained fabulous sizes for a predator

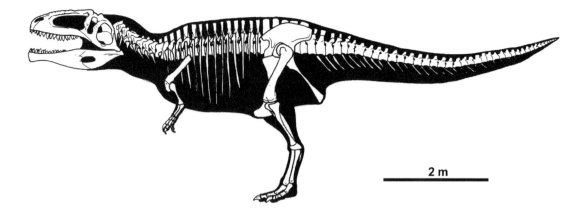

Fig. 6.31. Reconstructed skeleton of *Giganotosaurus*.

(approximately 13 meters long), surpassing in length and possibly in weight *Tyrannosaurus rex*. For example, the femur of *Giganotosaurus* is 143 centimeters long; the largest *Tyrannosaurus* femur is 135 centimeters long (P. Currie, pers. comm.).

The name of this theropod family was coined by German paleontologist Ernst Stromer in 1931 and derives from the type genus and species *Carcharodontosaurus saharicus*, which was found in Cenomanian localities in northern Africa. However, the most complete and informative carcharodontosaurid that is currently known is the Patagonian *Giganotosaurus carolinii*, represented by an almost-complete skeleton that includes cranial material. More recently, two specimens of a new carcharodontosaurid, *Tyrannotitan chubutensis*, were discovered in central Patagonia, adding valuable information about the early history of the group (Novas, de Valais, Vickers-Rich, and Rich 2005).

The Cretaceous Carcharodontosauridae constitute the latest clade of Carnosauria, one of the three tetanuran subclades cited above. Carnosauria also includes the Late Jurassic *Allosaurus* and *Sinraptor* and the Aptian *Acrocanthosaurus*. Derived carcharodontosaurids evolved large skulls, reaching lengths in *Giganotosaurus* of 190 centimeters from premaxilla to quadrate, thus surpassing the length of the largest skull (at 140 centimeters long) of *Tyrannosaurus rex* (Larson 1991). The skull of carcharodontosaurids is characterized by its highly fenestrated condition due to the enlarged antorbital fenestra (probably the widest among theropod dinosaurs) and an expanded infratemporal fenestra. Strongly developed rugosities decorate the nasals and the maxilla. An unusual feature among theropods is the contact of the lacrimal and postorbital above the orbit, a feature that was convergently acquired by abelisaurid ceratosaurians. In addition, the postorbital bone of carcharodontosauids is similar to that of abelisaurids in an orbital flange that divides the orbit into two main spaces. Important modifications occurred in the carchardodontosaurid braincase, as illustrated by *Giganotosaurus* and *Carcharodontosaurus*, in which the parietal crest is transversely wide but rostrocaudally short and the entire occipital surface of the skull is deeply inclined dorsorostrally (Fig. 6.32).

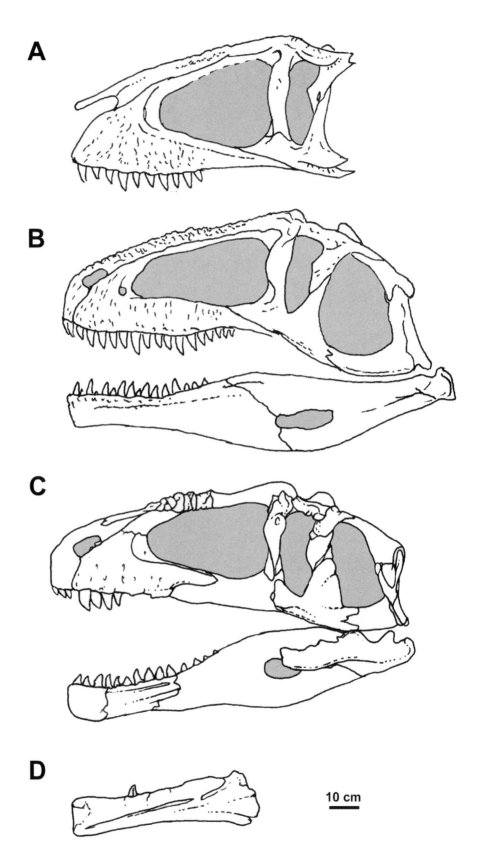

The dentary of carcharodontosaurids is distinguished by a deep square-shaped rostral end with a ventral process, or chin. Carcharodontosaurid teeth are blade-like and have marginal arcuate enamel wrinkles on the labial side of the caudal carina.

The postaxial cervical vertebrae are strongly opisthocoelous. A sharp ridge separates the dorsal from the lateral surface of the cervical neural arch, as is the case for abelisaurids. The presacral vertebrae have well-developed pneumatic foramina and fossae, in particular a pair of pleurocoels on the cervical and dorsal vertebrae. The neural spines of dorsal vertebrae are craniocaudally extended, dorsoventrally deep, and transversely thick, thus indicating a strongly developed epaxial musculature. Carcharodontosaurid dorsals, sacrals, and proximal caudals are characterized by their tall neural spines, which are nearly twice as tall as their respective centra.

The pelvic girdle of carcharodontosaurids has distinctive tetanurine features, such as a sigmoid external margin of the pubis in cranial view, a pubic symphysis that is reduced to the central part of the bone, a ventrally open pubic obturator foramen, and a slender ischia with a proximodistally reduced obturator process. The hindlimb bones are massive and have two remarkable features: the femoral head is dorsally projected, and the tibia and fibula are proportionally short compared to the length of the femur. The proportionally short tibia and fibula of carcharodontosaurids (which are less than 70 percent of the length of the femur) presumably indicate a greater reduction of running abilities for these giant animals than for another group of bulky theropods, the tyrannosaurids.

Although the skeletal anatomy of carcharodontosaurids are spectacular for their size, most are poorly known, and their diversity is currently restricted to four Late Cretaceous genera: *Carcharodontosaurus*, *Giganotosaurus*, *Tyrannotitan*, and *Mapusaurus*. Indisputable members of Carcharodontosauridae are of Gondwanan distribution, where they constituted the dominant large predatory dinosaurs of Gondwana during Aptian through Turonian times, sharing the role of top predators with the similarly big spinosaurids.

Available information reveals that carcharodontosaurids were widely distributed geographically. With the possible exception of a tooth with carchardontosaurian traits from the Cenomanian of Japan (Chure, Manabe, Tanimoto, and Tomida 1999), no carcharodontosaurid teeth have yet been recovered from Aptian–Cenomanian beds of the northern hemisphere, thus suggesting that such theropod groups were unique to, or at least dominant in, Gondwana.

The carcharodontosaurid fossil record from South America is currently made up of three named taxa (*Giganotosaurus carolinii*, *Tyrannotitan chubutensis*, and *Mapusaurus roseae*) as well as several fragmentary specimens (consisting mainly of isolated teeth) from Upper Cretaceous beds of Patagonia and Brazil. Very large teeth with arcuate enamel wrinkles curved toward the marginal serrations (a set of features thought characteristic of Carcharodontosauridae; Sereno et al. 1996) have been recorded in

Fig. 6.32. Skulls and jaws of carcharodontosaurids:
(A) *Carcharodontosaurus*.
(B) *Giganotosaurus*.
(C) *Mapusaurus*.
(D) *Tyrannotitan*.

Fig. 6.33. *Tyrannotitan chubutensis:* (A) Right dentary in lateral view. (B) Detail of dental serrations. (C, D) Cervical 9? in (C) left lateral and (D) cranial views. (E) Articulated sequence of dorsal vertebrae 3 through 8 in right lateral view. *From Novas et. al. (2005). Reprinted from Naturwissenschaften 92: 226–230. A large Cretaceous theropod from Patagonia, Argentina, and the evolution of carcharodontosaurids. Authors: Novas, F. E., de Valais, S., Vickers-Rich, P. and Rich, T. Page 227 (fig. 1) and page 228 (fig. 2). © Copyright 2005. With kind permission of Springer Science and Business Media.*

Albian, Cenomanian, and Turonian beds of different localities in South America (e.g., Veralli and Calvo 2004; Rich, Vickers-Rich, Novas, Cúneo, Puerta, and Vacca 2000; Kellner and Campos 2000) and Africa (Sereno et al. 1996). However, large teeth with these kind of wrinkles have been also observed in the North American tyrannosaurid *Daspletosaurus* (J. Farlow, pers. comm.), thus indicating that isolated teeth may not constitute enough evidence for a particular group of theropods.

TYRANNOTITAN CHUBUTENSIS

Tyrannotitan comes from La Juanita farm, located 28 kilometers northeastern of Paso de Indios town in Argentina's Chubut Province (Fig. 4.13). Two skeletons of this animal were found in the red sandstones of the Cerro Castaño Member (Aptian–Albian) of the Cerro Barcino Formation. *Tyrannotitan* is the oldest and most basal member of Carcharodontosauridae, constituting the sister group of the Patagonian *Giganotosaurus* and the Saharan *Carcharodontosaurus*, both of which are of Cenomanian age.

Comparison with the almost-complete and larger skull of the carcharodontosaurid *Giganotosaurus* gives an estimated skull length of *Tyrannotitan* of nearly 150 centimeters (Fig. 6.32). The dentary is ornamented by oblique grooves and rugosities along the ventral half of its lateral surface that is different from a band of smooth bone surface along the dental margin (Fig. 6.33A), a pattern of ornamentation resembling that of *Giganotosaurus* and abelisaurids (e.g., *Carnotaurus*). Up to 16 alveoli are present on the dentaries. As in other carcharodontosaurids, the teeth have marginal arcuate enamel wrinkles on the labial side of the caudal carina (Sereno et al. 1996). However, the tooth denticles from the cranial carina are bilobate in side view, a characteristic that seems unique among theropods (Fig. 6.33B). The cervical vertebrae are strongly opisthocoelous. The dorsal vertebrae have strong ligament scars protruding both cranially and caudally (Fig. 6.33C–E). The slender shoulder girdle of *Tyrannotitan* corresponds to the configuration present in more-basal carnosaurs (e.g., *Allosaurus, Acrocanthosaurus*); it has a narrow scapular blade and acromial process rising abruptly from the scapula at an angle approaching 90° (Fig. 6.34B). Preserved portions of the humerus and ulna indicate that *Tyrannotitan* had forelimbs that were proportionally short and robust. The femur length of one of the specimens of *Tyrannotitan* is estimated to be 140 centimeters.

GIGANOTOSAURUS CAROLINII

This incredible beast was briefly described by Rodolfo Coria and Leonardo Salgado in 1997. The holotype specimen consists of most of a skull and many important pieces of the postcranial skeleton, which was preserved in articulation (Fig. 6.31). A second specimen, consisting of a single partially preserved dentary, was more recently described (Calvo and Coria 2000). Both specimens come from the Candeleros Formation.

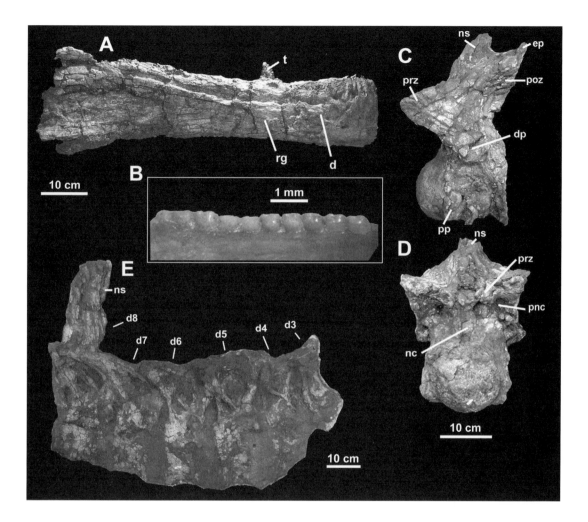

The holotype was excavated in El Chocón, while the referred specimen was found in the locality of Los Candeleros.

The skull of *Giganotosaurus* is elongate, measuring approximately 180 centimeters in the holotype but reaching an estimated length of 190 centimeters in the referred specimen (Calvo and Coria 2000; Fig.6.32B). More recently, Coria and Currie (2002a) provided valuable information regarding the anatomy of the braincase. The supratemporal opening is notably small in rostrocaudal dimension. The sagittal crest is absent, constituting a flat area between both supratemporal openings (Fig. 6.35). Moreover, *Giganotosaurus* lacks the depressions for temporal muscles over the caudal surface of the frontals, but the margins of these apertures overhang the openings. Another cranial peculiarity of *Giganotosaurus* is in the lacrimal bone, which has a dorsoventrally short ventral process, thus suggesting that the jugal had a dorsoventrally deep process for contact with the lacrimal. Thus, the articulation of lacrimal and jugal occurred at the orbital mid-height, constituting a bizarre feature that probably was autapomorphic of this Patagonian theropod. Above the orbits the skull roof is closed by an accessory ossification located between the

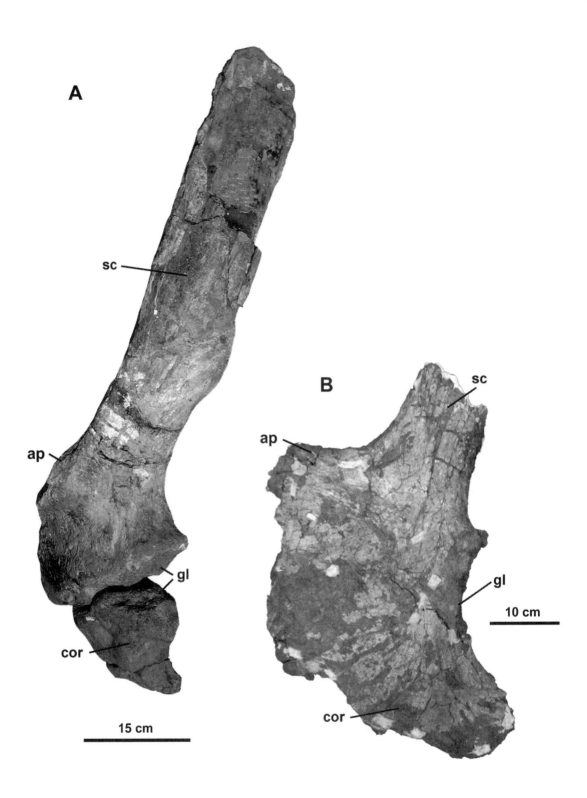

lacrimal and the postorbital, defining a thick supraorbital brow similar to that in *Sinraptor* and *Carcharodontosaurus*.

The dorsal vertebrae of *Giganotosaurus* are so big that they might be confused with sauropod vertebrae: one of the mid-dorsals is 63 centimeters tall and the centrum has a transverse diameter of 15 centimeters. Dorsal vertebrae are distinguished by their high neural spines. The lateral surfaces of the neural spines are deeply excavated, as are the cranial and caudal margins, thus creating an H-shaped contour in cross-section. This feature is also present in *Tyrannotitan* and was acquired by other giant animals like sauropods. The dorsal neural spines of *Giganotosaurus* are devoid of the ligament scars of the kind described above for *Tyrannotitan*. The prezygapophyses of some of the dorsals have elliptical excavations on their cranial surface resembling these of *Tyrannotitan*. The caudal vertebrae have two small foramina (approximately 4 millimeters in diameter). The transverse processes are very elongate but they are not distally expanded, as they are in abelisaurids.

The basal tetanuran pattern of the pectoral girdle of *Giganotosaurus* (Calvo 1999) is highly modified (Fig. 6.34A): the longitudinal axis of the scapular blade is caudally inclined with respect to its proximal margin. The scapula is remarkably robust; it is 6 centimeters thick immediately above the glenoid cavity, compared to 11 centimeters in craniocaudal width. The coracoid is heart-shaped in side view and is strongly reduced compared to the plate-like condition present in most basal tetanurans, including the basal carcharodontosaurid *Tyrannotitan* (Fig. 6.34B). The biceps tubercle is reduced and the glenoid cavity is mainly developed on the scapula. The unusually robust and morphologically bizarre pectoral girdle of *Giganotosaurus* supports the idea that important transformations occurred in derived carcharodontosaurids that probably also involved forelimb anatomy.

Some bones of the pelvic girdles are known (Calvo 1999), including a pubes with a completely opened obturator notch and very big distal feet and a slender ischia. The femur of the holotype specimen of *Giganotosaurus carolinii* is 143 centimeters long (Coria and Salgado 1995). A developed trochanteric shelf is present on the lateral surface of proximal femur.

Fig. 6.34. Carcharodontosaurid left scapulae and coracoids in lateral view: (A) *Giganotosaurus*. (B) *Tyrannotitan*.

MAPUSAURUS ROSEAE

This new carcharodontosaurid theropod was described by Coria and Currie in 2006. It comes from the Huincul Formation (Cenomanian) of Neuquén Province, the same unit that has yielded giant sauropods, including *Argentinosaurus huinculensis*. *Mapusaurus* (Fig. 6.36) is known from a minimum of seven individuals of different sizes that range from about 5 to 11 meters long. These specimens were recovered from a single bone bed at the Cañadón del Gato fossil site, 20 kilometers southwest of Plaza Huincul. *Mapusaurus* has carcharodontosaurid features, such as narrow blade-like teeth with wrinkled enamel, heavily sculptured facial

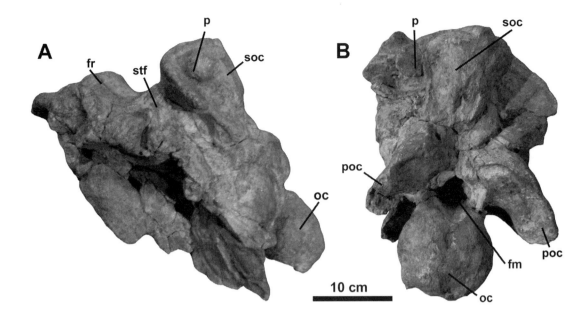

Fig. 6.35. Basicranium of *Giganotosaurus carolinii* in (A) left lateral and (B) caudal views. *Photographs by Fernando Novas.*

Fig. 6.36. *Mapusaurus roseae.* (A–E) Selected cranial bones of the left side: (A) Postorbital in lateral view. (B) Lacrimal in lateral view. (C, D) Nasal in (C) dorsal and (D) lateral views. (E) Maxilla in lateral view. (F–H) Anterior dorsal in (F) cranial, (G) caudal, and (H) right lateral views. (I, J) Axis in (I) caudal and (J) lateral views. (K–M) Posterior dorsal vertebra in (K) cranial, (L) left lateral,

bones, a supraorbital shelf formed by a postorbital/palpebral complex, and a dorsomedially directed femoral head.

Mapusaurus is characterized by a deep, short, and narrow skull with relatively large triangular antorbital fossae and narrow, unfused rugose nasals. The orbit is subdivided into upper and lower regions by the lacrimal processes and possibly by the postorbital as well (Fig. 6.36A and B). Overall, the skull of *Mapusaurus* appears to be deeper and narrower than that of *Giganotosaurus*. *Mapusaurus* also differs from *Giganotosaurus* in having thick, rugose, and unfused nasals that are narrower at the rostral end (Fig. 6.36C). The nasals show the remarkable condition of having well-developed dorsolateral rugosities above the antorbital fossa. Although nasal rugosities are common in many large theropods (e.g., *Allosaurus, Acrocanthosaurus, Sinraptor*), they are never as prominent as they are in *Mapusaurus* (Coria and Currie 2006). As in *Giganotosaurus*, the orbits of *Mapusaurus* are roofed by a supraorbital shelf formed by the postorbital and a palpebral bone that rostrally contact the prefrontal and the lacrimal (Coria and Currie 2002). The neural spines of the dorsal vertebrae of *Mapusaurus* are taller and wider than in *Giganotosaurus* (Fig. 6.36 K–M). As is documented in other carcharodontosaurids (e.g., *Tyrannotitan*; Novas, de Valais, Vickers-Rich, and Rich 2005), the forelimb bones of *Mapusaurus* are considerably reduced in relation to the rest of the body. The humerus of *Mapusaurus* has a broad distal end with little separation between the articular condyles (Fig. 6.36N–P). Notably, the iliac blade (in particular the preacetabular wing) looks reduced in comparison with the acetabular diameter (Fig. 6.36Q).

The presence of a single carnivorous taxon with individuals of different ontogenic stages has been interpreted by Coria and Currie (2006) as evidence of variation within a single population and may also indicate some behavioral traits for *Mapusaurus*.

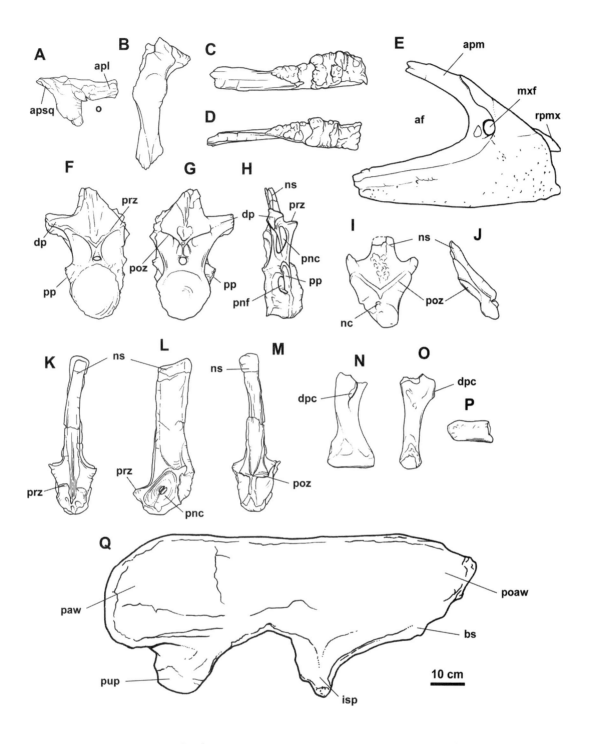

Coelurosauria

Coelurosauria is a diverse clade of tetanuran theropods that are phylogenetically defined as the clade composed of extant birds and all theropods that share a more recent common ancestor with extant birds than with *Allosaurus* (Holtz, Molnar, and Currie 2004). Among conspicuous extinct and (M) caudal views. (N–P) Right humerus in (N) cranial, (O) lateral, and (P) distal views. (Q) Left ilium in lateral aspect. *Redrawn from Coria and Currie (2006).*

coelurosaurs are the ostrich-like ornithomimids, the giant tyrannosaurids, the peculiar therizinosaurs and oviraptorosaurs, the bizarre alvarezsaurids, and the famous velociraptorines and kin (Fig. 6.37). Coelurosaurian theropods are recorded in Jurassic and Cretaceous rocks around the world; the outstanding discoveries have mainly come from beds of Cretaceous age in North America and Asia. Extensive collections of these animals were made in western North America as well as in the Gobi Desert in Mongolia. Particularly relevant are discoveries from the Early Cretaceous Yixian Formation at Liaoning, China, where hundreds of well-preserved specimens of tiny coelurosaurs still preserve traces of feathers and feather-like features. No doubt the fossil record of coelurosaurs from the northern hemisphere provide the most complete and informative source for understanding coelurosaurian phylogeny.

However, the record of coelurosaurian theropods from the southern continents has increased considerably since 1990, yielding valuable evidence about diversification of the group. Currently the remains of at least ten species of non-avian coelurosaurs have been recovered from different Cretaceous stages in South America, Africa, Madagascar, and Australia. They are members of a variety of coelurosaurian lineages that include the small-bodied basal coelurosaurs *Nqwebasaurus thwasi*, from the Early Cretaceous of Africa (De Klerk, Forster, Sampson, Chinsamy, and Ross 2000); *Santanaraptor placidus* (Kellner 1999) and *Mirischia asymmetrica* (Naish, Martill, and Frey 2004), both from the Aptian of Brazil; and *Aniksosaurus darwini*, from the Cenomanian of Patagonia (Martínez and Novas 1997, 2006). Possible oviraptorosaurian remains have been reported in Australia (Currie, Vickers-Rich, and Rich 1996), but more-derived maniraptorans are better represented in Gondwana through the highly specialized Patagonian alvarezsaurids *Alvarezsaurus calvoi* and *Patagonykus puertai* (Bonaparte 1991a; Novas 1996b, 1997d). Deinonychosaurians have been documented from the Cenomanian of northern Africa (Rauhut and Werner 1995), and from Turonian through Maastrichtian beds of Argentina, including *Unquillosaurus ceibalii, Unenlagia comahuensis, Unenlagia paynemili, Neuquenraptor argentinus, Buitreraptor gonzalezorum,* and *Austroraptor cabazai* (Novas 1999, 2004; Novas and Puerta 1997; Calvo, Porfiri, and Kellner 2004; Novas and Agnolín 2004; Novas and Pol 2005; Novas, Pol, Canale, Porfiri, and Calvo 2009) and from the Late Cretaceous of Madagascar (e.g., *Rahonavis ostromi;* Forster, Sampson, Chiappe, and Krause 1998). Finally, the list of southern coelurosaurians also includes avians from Argentina (e.g., several enantiornithine species, the flightless *Patagopteryx deferrariisi*, and neornithines of different affiliation; Chiappe 2007b) and Brazil (Alvarenga and Nava 2005).

Although evidence recovered from Gondwana is still patchy compared with the breadth of discoveries made in the northern continents, the specimens listed above increase the repertoire of adaptations and the geographical distribution of Coelurosauria as a whole and modify previous ideas about their phylogenetic relationships and paleobiogeographical evolution.

The phylogenetic relationships among the most important clades in Coelurosauria are depicted in Figure 6.37. Compsognathidae is interpreted as having a basal position within Coelurosauria (Holtz, Molnar, and Currie 2004) and represents the sister group of the remaining coelurosaurs (e.g., tyrannosaurids, ornithomimids, oviraptorosaurs, therizinosaurids, alvarezsaurids, deinonychosaurs, and birds). The Maniraptora, which group together Alvarezsauridae, Therizinosauroidea, Oviraptorosauria, Deinonychosauria, and Aves, are an outstanding group of derived coelurosaurians. Within Maniraptora (Fig. 6.37B), alvarezsaurids seem to occupy a basal position, constituting the sister group of two main maniraptoran branches: an unnamed group made up of oviraptorosaurs and theririzinosauroids; and another clade, Paraves, that includes deinonychosaurs (e.g., troodontids and dromaeosaurids) and birds. All of the groups of Maniraptora, except Therizinosauroidea, have been documented in the Cretaceous of South America.

I will here review the fossil record of basal coelurosaurians documented in South America (e.g., *Orkoraptor burkei*, *Aniksosaurus darwini*, *Santanaraptor placidus*, and *Mirischia asymmetrica*) and explore the available evidence about more-derived members of this clade.

ORKORAPTOR BURKEI

This new theropod was collected from the Upper Cretaceous (Maastrichtian) Pari Aike Formation of southern Patagonia (Novas, Ezcurra, and Lecuona 2008). It is based on a postorbital, quadratojugal, coronoid(?), several teeth, an atlantal intercentrum and neurapophysis, two caudal vertebrae, and the proximal half of tibia. The proximal half of the right tibia is preserved. It is 46 centimeters long as preserved, but it may have been approximately 70 centimeters when complete. *Orkoraptor burkei* measured approximately 6–7 meters long based on extrapolations from other large theropods, such as *Albertosaurus* and *Sinraptor*.

Orkoraptor exhibits characteristics of maniraptorans that include an upturned rostral process of the postorbital, the caudoventral corner of which is widely concave. Furthermore, several features resemble those of the maniraptoran clades Compsognathidae and Deinonychosauria: its teeth have no denticles and have carina in the mesial margin.

The new Patagonian theropod differs from other coelurosaurians (Ornithomimosauria, Compsognathidae, Alvarezsauridae, Dromaeosauridae, Aves) in that it has caudal vertebrae with a single pair of small pleurocoels on each side, and a median depression flanked by two longitudinal and narrow furrows on each tooth.

The crowns of *Orkoraptor* teeth are tall and have strongly curved caudal margins. Median longitudinal depressions are present on both the labial and lingual surfaces of the teeth, conferring cross-sections shaped like figure eights for most levels of the crown. This feature is reminiscent of some coelurosaurians, such as *Saurornitholestes* and *Deinonychus*, in which such longitudinal depressions are usually present. However, in

Orkoraptor the topography of the lingual surface of the crown is notably complex: the main median depression is rostrally and caudally flanked by two longitudinal and narrow furrows, the caudal one being deeper than the rostral furrow. The presence of furrows flanking that depression is probably an autapomorphy of *Orkoraptor*. Denticles are present only along the caudal carinae of the teeth.

The available tail vertebrae (probably caudals 14 to 18) have a small oval pleurocoel on each side of the centrum. The presence of pleurocoels in caudal vertebrae is not a common feature of theropods and has been reported only in some tetanurans, such as *Megaraptor, Carcharodontosaurus*, and Oviraptorosauria.

Interestingly, in its absence of mesial denticles, in the figure eight–shaped cross-sections of the teeth, and in the overall morphology of the postorbital *Orkoraptor* resembles deinonychosaurians. However, the Patagonian taxon differs from deinonychosaurians in that the proximal caudal centra have elliptical (instead of square) articular surfaces and are shaped like and hourglass in ventral view (instead of being box-like; Gauthier, 1986). Furthermore, dromaeosaurids are characterized by an inverted T-shaped quadratojugal (Currie, 1995), a condition that is absent in *Orkoraptor*. The fact that *Orkoraptor* retains these plesiomorphic features suggests that it is not a member of Deinonychosauria.

Orkoraptor is not a member of Abelisauridae, Carcharodontosauridae, Alvarezsauridae, and Deinonychosauria, theropod clades that were frequently recorded in Cretaceous beds of South America. The available evidence supports that *Orkoraptor* is a representative of a yet-unknown branch of large tetanurans that inhabited South America at the close of the Cretaceous.

ANIKSOSAURUS DARWINI

This animal comes from the Cenomanian Lower Member of the Bajo Barreal Formation of central Patagonia. It was discovered by Rubén D. Martínez and his crew from levels of the Bajo Barreal Formation that are exposed in southeastern Chubut Province (Martínez and Novas 1997, 2006). *Aniksosaurus* is represented by pieces of at least five individuals of approximately the same size (around 2 meters long), a fossil association suggesting that *Aniksosaurus* probably lived in groups. It is thought that *Aniksosaurus* constitutes a basal coelurosaur mainly on the basis of the widely expanded cuppedicus fossa and brevis shelf of the ilium (Fig. 6.38A and B). Despite its moderate size, *Aniksosaurus* is stoutly constructed. Its bones are proportionally short and robust, and the forelimbs end in ungual phalanges that are thick and strongly curved. *Aniksosaurus* resembles the Late Jurassic *Allosaurus fragilis* in the construction of its forelimbs and hindlimbs, especially the femur (Fig. 6.38C–F). Metatarsal IV of the foot and its corresponding digit are transversely narrow (Fig. 6.38G). Some traits of the ilium of *Aniksosaurus* are unique, such as an extremely expanded brevis shelf and fossa. This fossa expands transversely

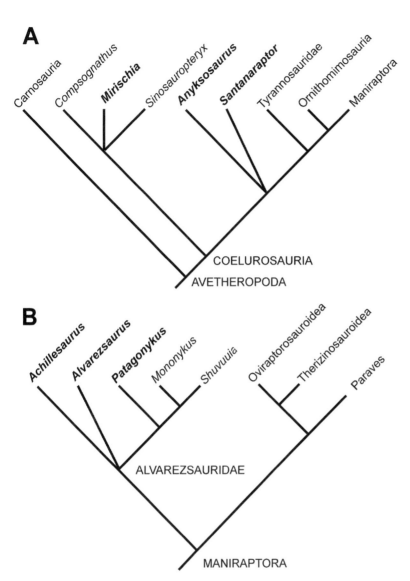

Fig. 6.37. Phylogenetic relationships of coelurosaurian theropods: (A) Cladogram of Avetheropoda and Coelurosauria. (B) Cladogram of Maniraptora. Genera recorded in South America are indicated in bold.

toward the rear; in ventral view its outline is paddle-shaped. Interestingly, the fossa resembles that of the Jurassic ornithopod *Dryosaurus* (Galton 1981) and is reminiscent of that in the bird *Patagopteryx* (Alvarenga and Bonaparte 1992).

SANTANARAPTOR PLACIDUS

This coelurosaur from the Romualdo Member of the Santana Formation was named by Alexander Kellner in 1999. It was a slender animal a little larger than a turkey and is known by a single specimen that consists of a partial pelvis, hindlimb elements (Fig. 6.39), and some caudal vertebrae. The soft tissues of the specimen have been preserved extensively, including muscle fibers (Kellner 1996a). The ischium has a triangular

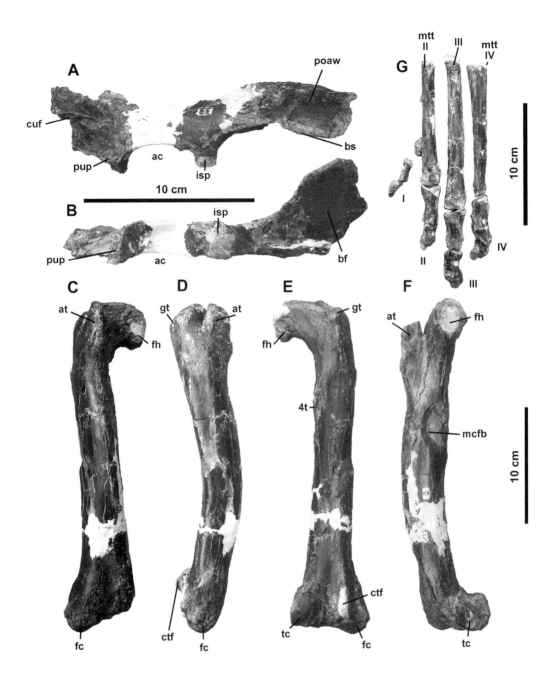

Fig. 6.38. *Aniksosaurus darwini:* (A, B) Left ilium in (A) lateral and (B) ventral views. (C–F) Right femur in (C) cranial, (D) lateral, (E) caudal, and (F) medial views. (G) Left foot in dorsal view. *Photographs courtesy of Rubén Martínez.*

obturator process, thus supporting the membership of *Santanaraptor* in Coelurosauria. However, the proximal position of this process indicates that this South American coelurosaur is not a member of Maniraptora, in which the obturator process is more distally located. The metatarsals of *Santanaraptor* are elongate (70 percent of the length of femur) and lack derived arctometatarsalian features (Holtz, Molnar, and Currie 2004). Interestingly, the femur has a triangular-shaped fibular condyle (as observed in distal view) that is separated from the inner condyle by a deep notch.

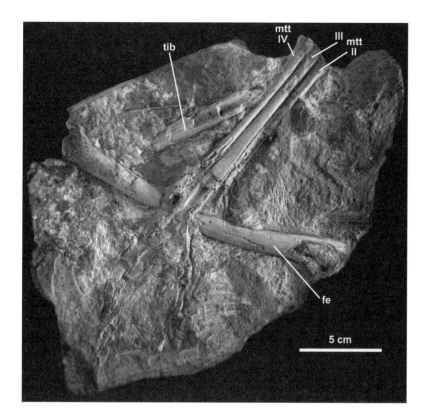

Fig. 6.39. *Santanaraptor placidus:* Block rock containing the femur, tibia, and metatarsals of the right hindlimb. *Photograph by Fernando Novas.*

This condition resembles that of tyrannosaurids and ornithomimids, thus suggesting phylogenetic ties with these two basal coelurosaurian clades. Some features that Kellner (1999, 2001) originally cited as autapomorphies of *Santanaraptor* are, in fact, more widely distributed among basal coelurosaurs, such as the ischial obturator notch and a foramen on the base of the anterior trochanter of the femur (Naish 2000). Similarly, the distally unfooted ischium (Kellner 2001) is a widely distributed feature among coelurosaurians.

MIRISCHIA ASYMMETRICA

This animal was found in Chapada do Araripina in Pernambuco in northeastern Brazil from beds of the Santana Formation (Martill, Frey, Sues, and Cruickshank 2000). It is known from a partial skeleton that was contained in a single calcium carbonate concretion that produced several bones in an excellent state of preservation (Fig. 6.40; Plate 14). In addition, Martill and colleagues (2000) indicated the presence of a large vacuity behind the pubic apron that was originally preserved in the rock matrix. Martill and colleagues (2000) and later Hutchinson (2001) felt that the structure occupying this vacuity was a postpubic air sac. *Mirischia* closely resembles the compsognathids *Compsognathus longipes* from the Late Jurassic of Germany, *Aristosuchus pusillus* from the Early Cretaceous of England, and *Sinosauropteryx prima* from the Early Cretaceous of China. Its small size (roughly 1 meter long) and general anatomy are congruent with this group of basal coelurosaurians. A possible synapomorphic feature of compsognathids that is documented in *Mirischia* is dorsal vertebrae with cranially and caudally concave neural spines such that the apex of the spines is between 63 percent and 67 percent longer than the base (Fig. 6.40A). The width of the pelvis shows that *Mirischia* had a very narrow body that was only approximately 30 millimeters wide across the sacrum. The pubis forms a long (145 millimeters) and slender shaft, the distal end of which has a large pubic boot (Fig. 6.40B and C). *Mirischia* is remarkable in that it has an obturator foramen on the caudally projecting rectangular flange of the right pubis, but a ventrally open notch in the corresponding position on the left pubis (Fig. 6.40B and C). An elongate pubic foramen is present distally and divides the cranial third of the foot. As Hutchinson (2001) has noted, this foramen may have accommodated a ventral pneumatic duct leading to a postpubic air sac. The ischia of *Mirischia* are asymmetrical: the left ischia is perforated by an oval foramen, while the right ischia has an open notch in the corresponding position.

Mirischia resembles *Santanaraptor* in the slender and apparently unfooted condition of the ischium, which has a triangular obturator process. Also, in both taxa, the femur is fairly elongate and gracile with a similarly shaped anterior trochanter on the proximal femur. Naish and colleagues (2004) considered *Mirischia* to be a different taxon from *Santanaraptor*, mainly based on the condition of the ischial obturator notch that in the later taxon fully separates the pubic pedicle from the obturator process. However, Naish and colleagues (2004) also note that "the femur of *Santanaraptor* has a large sulcus" that is not present in *Mirischia*. A sulcus of this type, which separates the fibular condyle from the outer condyle of the distal femur (Kellner 1999), corresponds to a portion of the femur that is not preserved in the holotype specimen of *Mirischia* (Naish, Martill, and Frey 2004). Thus, distinctions between *Mirischia* and *Santanaraptor* require review.

Alvarezsauridae

This clade of bizarre coelurosaurians includes small forms the size of a turkey (nearly 1 meter long) such as *Alvarezsaurus* and *Mononykus* and larger animals up to 2 meters long such as *Patagonykus* (Fig. 6.41). They seem to have been predators of small animals, presumably insects, a suspicion based on the small size of the head and teeth and the construction of the forelimbs (Perle, Chiappe, Barsbold, Clark, and Norell 1994; Senter 2005). At present, alvarezsaurids are represented by the Turonian–Campanian *Alvarezsaurus calvoi*, *Patagonykus puertai*, and *Achillesaurus manazzonei* from Patagonia (Bonaparte 1991a; Novas 1996b, 1997d; Martinelli and Vera 2007) and the Campanian–Maastrichtian *Mononykus olecranus*, *Shuvuuia deserti*, and *Parvicursor remotus* from Mongolia (Perle, Norell, Chiappe, and Clark 1993, 1994; Chiappe, Norell, and

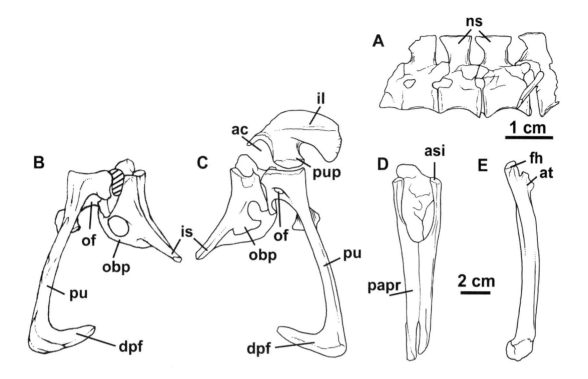

Fig. 6.40. *Mirischia asymmetrica*: (A) Sequence of dorsal vertebrae in right lateral view. (B) Articulated left pubis and ischium. (C) Composite reconstruction of preserved bones of pelvic girdle in right lateral view. (D) Paired pubes in cranial aspect. (D) Right femur in lateral view. Redrawn from Martill, Frey, Sues, and Cruickshank (2000).

Clark 2002), although some isolated remains from Maastrichtian beds of North America indicate that alvarezsaurids were also distributed there (Chiappe, Norell, and Clark 2002). The Mongolian forms are represented by complete cranial and postcranial skeletons. Although South American taxa are less well known, they are stratigraphically older and morphologically more plesiomorphic than their Asiatic cousins and offer valuable evidence about the phylogenetic relationships of these peculiar animals in the context of coelurosaurian evolution (Fig. 6.37B).

Alvarezsaurids are distinguished by markedly shortened forelimbs that have a robust digit I ending in a stout ungual (Fig. 6.42). Although they are short, the humerus, ulna, and radius are robust and have well-developed processes for muscle insertion, indicating that they were used in strenuous work. In alvarezsaurids, the length of the forelimbs are less than 20 percent of the length of the hindlimb, in contrast with most theropods, in which the length of the forelimbs are 40 to 53 percent of the length of the hindlimb. In Tyrannosauridae the forelimbs are 22–26 percent of the hindlimb length, but in *Mononykus* the forelimb is even shorter than in tyrannosaurids, since it is only 18 percent of the length of the hindlimb (Perle, Norell, Chiappe, and Clark 1993; Novas 1996b).

Alvarezsaurids are unique among Archosauria in the morphology of the first phalanx of manual digit I: this bone is wide and craniocaudally compressed, resulting in a curious proximal articulation that describes a horizontal "B" in proximal aspect. This morphology sharply contrasts with that seen in other theropods, in which the proximal contour of the first phalanx of digit I is triangular (e.g., *Deinonychus*; Ostrom 1969) or

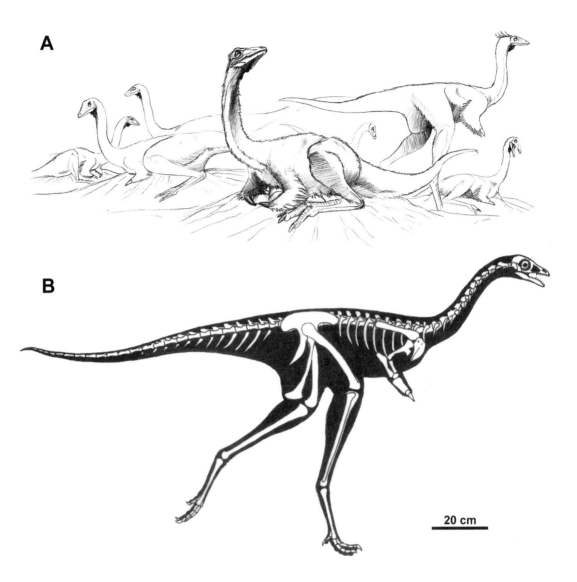

Fig. 6.41. *Patagonykus puertai:* (A) Life reconstruction. (B) Reconstructed skeleton. *Illustrations by J. González.*

looks like a vertical rectangle (as in *Allosaurus*; Madsen 1976). Another peculiarity is a pair of strongly developed proximal ridges on the palmar surface of the phalanx.

Chiappe (1995) speculated that alvarezsaurids used their claws to strip bark or perhaps the stems of low-growing vegetation. The unusual morphology of the forelimbs is not readily interpreted in reference to the animal's behavior, but some authors think that *Mononykus* (at least) was capable both of scratching and digging to make depressions in the ground and of hook-and-pull movements. The latter behavior is consistent with a diet of insects that make tough earthen nests or live inside plants. Phil Senter (2005) interpreted *Mononykus* as a dinosaurian anteater. Perle and colleagues (1993, 1994) commented that the sternum and forelimbs of *Mononykus* resemble those of moles; both have a keeled sternum, a humerus that is short and expanded, an ulna with an elongate olecranon process, and stout and slightly curved unguals.

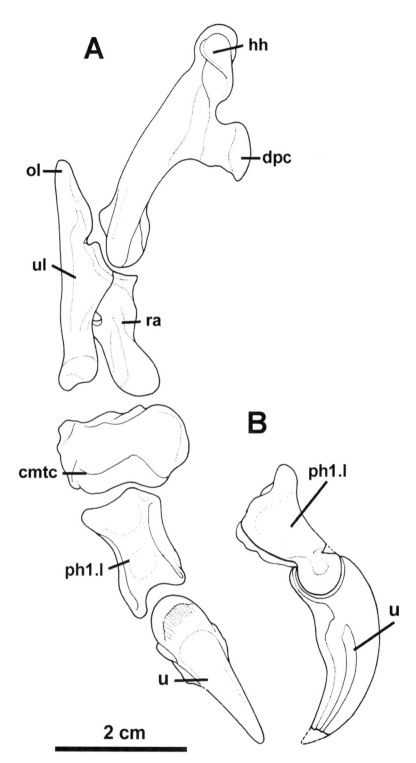

Fig. 6.42. Right forelimb of *Mononykus olecranus:* (A) Lateral view of the forelimb. (B) Phalanx 1 and ungual of digit I in caudal aspect. *From Novas (1996a). Reprinted with permission of the Queensland Museum.*

Aside from the highly modified and short but powerfully constructed forelimbs, the common alvarezsaurid ancestor evolved other synapomorphies pertaining to the vertebral column (e.g., a procoelous caudal centra), suggesting that the tail was highly mobile in contrast with that of most non-avian theropods, in which the distal end of the tail was more stiffened (Gauthier 1986; Gatesy 2001).

The peculiar adaptations of the forelimbs and vertebral column are combined with plesiomorphic features that reveal their basal position of *Mononykus* within Maniraptora. But what makes these animals so interesting phylogenetically is their surprisingly modern avian features. These include the absence of hyposphene-hypantrum articulation, a caudoventrally directed pubis, a distally closed femoral popliteal fossa, an accessory cnemial crest, a fibula that does not contact the tarsus, and a transversely narrow and laterally displaced astragalar ascending process.

This mixture of features has led to controversy about the phylogenetic relationships of Alvarezsauridae within Coelurosauria. Originally they were interpreted as non-volant avians that were more closely related to modern birds than *Archaeopteryx* is (Perle, Norell, Chiappe, and Clark 1993, 1994). Some researchers thought that alvarezsaurids were closely related to ornithomimosaurians (e.g., Bonaparte 1991a; Martin 1997; Sereno 2001), but most now agree that alvarezsaurids are basal maniraptorans that convergently evolved avian features (e.g., Novas and Pol 2002; Chiappe, Norell, and Clark 2002).

The current documentation of the Alvarezsauridae in both South America and Asia suggests that this clade successfully occupied a wide range of environmental conditions. For example, the Neuquén Group, which includes the Río Neuquén and Río Colorado formations, constitutes a succession of sandstones and mudstones deposited under fluvial and lacustrine conditions (Digregorio and Uliana 1980). At the other extreme of sedimentological conditions in which alvarezsaurids are recorded are the Asiatic Djadokhta (Tugrugeen Shireh) and Barun Goyot (Khermeen Tsav) formations, which were deposited under subaerial conditions of sand dunes, small lakes, and streams in hot and semi-arid climates in areas that lacked permanent fluvial systems (Gradzinsky, Kielan-Jawarowska, and Maryánska 1977; Osmólska 1980). It seems clear that alvarezsaurids inhabited a wide range of paleoenvironments, from desert environments like those indicated by the sedimentology of the Djadokhta Formation to more humid conditions like those suggested by the fluvial deposits of the Nemegt, Portezuelo, and Bajo de la Carpa formations.

ALVAREZSAURUS CALVOI

This was the first described member of the Alvarezsauridae group, and it provided the name for the entire family. José Bonaparte (1991a) described *Alvarezsaurus* on the basis of a single partially preserved specimen found in the Bajo de la Carpa Formation (Fig. 6.43). *Alvarezsaurus* is the size of a chicken and is characterized by paddle-shaped craniocaudally elongate postzygapophyses on the cervical vertebrae (Bonaparte 1991a; Fig. 6.43A and B). This condition contrasts with that of other theropods (e.g., *Piatnitzkysaurus, Archaeornithomimus, Ornitholestes, Deinonychus*), in which the postzygapophyses are rectangular-shaped, are not constricted at their bases, and have a convex dorsal surface with a prominent epipophysis. In the caudal cervicals of *Alvarezsaurus* the postzygapophyses have a strong craniocaudally oriented buttress running along the medial margin. Another vertebral feature present in *Alvarezsaurus* is related to the elongation of caudal centra: in the distalmost preserved caudal (presumably corresponding to the 15th through 18th sector of caudals), the centrum is 213 percent of the length of the most proximally preserved caudal. This condition resembles that of *Archaeopteryx*, in which the longest tail vertebrae (12th and 13th caudals) are 185 to 287 percent of the length of the proximal caudal vertebrae. The elongation of the distal caudal segments in *Alvarezsaurus* sharply contrasts with the proportions seen in non-avian maniraptorans (e.g., *Ornitholestes, Sinornithoides, Deinonychus*), in which the length of the distal caudals is no more than 175 percent of the length of the proximal caudals. *Alvarezsaurus* is also distinguished from other alvarezsaurids in which the length of caudal vertebrae is more or less similar along the tail (e.g., *Mononykus*) or the distal caudals are considerably smaller than the proximal ones, as occurs in *Patagonykus*.

The pectoral girdle of *Alvarezsaurus* is peculiar, even among alvarezsaurids. As Bonaparte (1991a) noted, the scapular blade of this animal is slender and reduced and is probably distally unexpanded (Fig. 6.43H). The shape and proportions of the scapular blade differ from the unreduced scapular blade of *Mononykus*, suggesting that a variety of morphologies of this part of the skeleton evolved within Alvarezsauridae. The stout morphology of the first digit ungual of the manus of *Alvarezsaurus* (Fig. 6.43E–G) suggests that the first manual digit of this taxon was powerfully constructed. On the basis of this evidence, it is also expected that *Alvarezsaurus* had extremely short forelimbs.

PATAGONYKUS PUERTAI

This taxon comes from the Portezuelo Formation, which is exposed at Portezuelo Hill in Neuquén Province in northwestern Patagonia. At present, *Patagonykus puertai* is the best-documented alvarezsaurid from South America (Figs. 6.41 and 6.44). It is also the oldest known member of this bizarre group of theropod dinosaurs (Novas 1996b, 1997d). Despite their close geographic and stratigraphic provenance, *Patagonykus puertai* and *Alvarezsaurus calvoi* do not share derived characteristics within Alvarezsauridae. On the contrary, some derived features (e.g., strongly spherical caudal articular surfaces of the centra of the last sacral and first caudal vertebrae, ventrally keeled sacral vertebrae, a supraacetabular crest, and a fourth femoral trochanter) unite *Patagonykus* with the Mongolian alvarezsaurids (Fig. 6.37B).

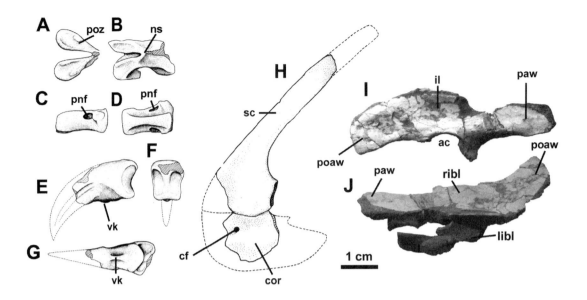

Fig. 6.43. *Alvarezsaurus calvoi:* (A, B) Neural arches of the cervical vertebrae in dorsal view (cranial is to the right). (C, D) Centra of the cervical vertebrae in right lateral view. (E–G) Manual ungual in (E) side, (F) proximal, and (G) ventral views. (H) Left scapula and coracoid in lateral view. (I) Right ilium in lateral view. (J) Right ilium, partial sacrum, and proximal end of the left femur in dorsal view. *(A–F) From Novas (1996a). Reprinted with permission of the Queensland Museum. (I, J) Photographs by Fernando Novas.*

Patagonykus is diagnosed by the following features: ventrally curved, tongue-shaped lateral margins of the postzygapophyses in the dorsal vertebrae; a bulge on the caudal bases of the neural arch of dorsal, sacral, and caudal vertebrae; a transversely narrow humeral articular facet of the coracoid; subcylindrical internal tuberosity of humerus, which is wider at its extremity than in its base; a conical and strongly medially projected humeral entepicondyle; proximomedial hook-like processes on the first phalanx of manual digit I; and a rectangular-shaped ectocondylar tuber of the femur in distal view.

The dorsal vertebral postzygapophyses of *Patagonykus* are concave ventrally, similar to those of *Deinonychus*, for example. However, this concavity is exaggerated by the ventral projection of the lateral margin of the postzygapophyses (Fig. 6.44C). This lateroventral flange of the postzygapophyses has not been reported in other coelurosaurs, including *Mononykus* (Perle, Chiappe, Barsbold, Clark, and Norell 1994). The resemblances between the proximal caudals of *Patagonykus* and eusuchian crocodiles (e.g., *Caiman*) are noteworthy (Fig. 6.44A): a ball-shaped caudal articular surface; robust, craniocaudally short neural spines; transversely thin ligamentary scars at the base of the neural spines; and a deep excavation between the prezygapophyses and postzygapophyses bases of the neural spines. It is not possible to say whether these osteological resemblances between *Patagonykus* and extant crocodiles correspond with similarities in the distribution and development of the epaxial musculature, but the existence of a strong procoelous condition suggests a high degree of movement all along the tail, in contrast with the more reduced degree of movement of the distal two-thirds of the tail in most tetanurines, including birds (Gauthier 1986).

The forelimbs of *Patagonykus* were as short as those of *Mononykus*. The first phalanx of manual digit I has a proximopalmar hook-like process that is greatly developed, more so than in *Mononykus* (Figs. 6.42 and 6.44D, E, H–M). The ungual of manual digit I is a stout bone with a quadrangular proximal end that matches the shape of the distal ginglymoid of the first phalanx. Deep lateral grooves incise the sides of the ungual phalanx and open ventrally into distinct notches, which may be the forerunners of the pair of ventral foramina that are present in the more-derived alvarezsaurids *Mononykus* and *Shuvuuia* (Chiappe and Coria 2003). The ungual is almost flat proximoventrally and across the surface between the latter ventral notches. Distal to these notches, however, the ventral surface of the ungual phalanx has a short central ridge that defines oval depressions laterally. This keeled condition of the flexor tuberosity is unique to *Patagonykus* and *Alvarezsaurus*.

In the pelvic girdle, *Patagonykus* is remarkable in the condition of the pubis: this long and slender bone is caudoventrally directed and ends distally in a massive pubic foot (Fig. 6.44F and G) that differs from the rod-like, caudally oriented, and distally unexpanded pubis of *Mononykus*.

ACHILLESAURUS MANAZZONEI

This new alvarezsaurid taxon comes from the Bajo de la Carpa Formation, which crops out in Paso Córdova, Río Negro Province (Martinelli and Vera 2007). *Achillesaurus* is known by a fragmentary postcranial skeleton that includes the last sacral vertebra, two caudals, part of an ilium, the proximal end of a femur, a distal tibia that is articulated with the astragalus, and the proximal portion of metatarsals II, III, and IV. Martinelli and Vera (2007) interpreted as autapomorphic of *Achillesaurus* the fact that one of the proximal caudals (presumably caudal 4) has a biconcave centrum, with its cranial surface 30 percent larger in diameter than the caudal surface of the centrum. However, the caudal surface of this vertebra is not well preserved and its condition may be artifactual. Moreover, a caudal of the same specimen of *Achillesaurus* that corresponds to a more distal position is procoelous (Martinelli and Vera 2007), a condition that is usual among alvarezsaurids.

Martinelli and Vera (2007) indicated that *Achillesaurus* is distinguished from *Alvarezsaurus* (which was documented in the same bed unit but from a different geographic location) mainly by its larger size and by some details of the ilium and caudal vertebrae. However, the holotype specimen of *Alvarezsaurus calvoi* is not a full-grown individual (Novas 1996b), raising doubts about whether such distinctions are due to ontogenetic variations. In my view, the preserved remains of *Achillesaurus manazzonei* do not offer enough anatomical evidence to support its taxonomic validity.

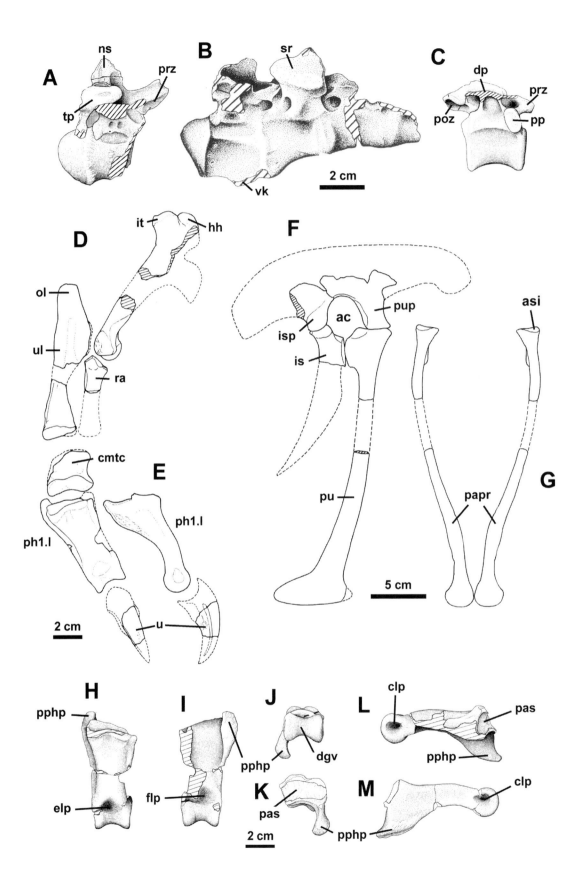

POSSIBLE OVIRAPTOROSAURIANS

Oviraptorosauria is a clade of coelurosaurian theropods that is defined as all the maniraptorans closer to *Oviraptor* than to birds (Osmólska, Currie, and Barsbold 2004). Oviraptorosaurs are considered to be the sister taxon of Therizinosauroidea (Holtz, Molnar, and Currie 2004). They are characterized by a short and deep skull and the lack of teeth. It is thought that the premaxilla and the rostral extremity of the dentary were covered by a horny beak. The forelimbs of oviraptorosaurians are elongate with three functional digits that end in curved and trenchant claws. Oviraptorosaurs are well documented from Late Cretaceous beds of North America and Asia. Scant and very fragmentary remains that putatively belong to this group have been reported from Early Cretaceous beds of Australia (Currie, Vickers-Rich, and Rich 1996) and Brazil (Frey and Martill 1995). An isolated manual ungual from the Bauru beds has been described as resembling these of elmisaurid oviraptorosaurs (Novas, Borges Ribeiro, and Carvalho 2005). A cervical vertebra originally interpreted as oviraptorosaurian by Frankfurt and Chiappe (1999) is now considered to belong to Abelisauroidea (Agnolín and Martinelli 2007).

A PURPORTED OVIRAPTOROSAUR FROM THE SANTANA FORMATION

Frey and Martill (1995) reported a fragmentary sacrum and partial ilium from the Lower Cretaceous Santana Formation of Brazil that they referred as to "oviraptorosaurid" (in current theropod nomenclature, the terms Oviraptorosauria and Oviraptoridae are used, but not "Oviraptorosauridae").

Based on the size of the preserved bones (for example, the caudal end of the iliac blade is estimated at 15 centimeters high and the sacral vertebrae are 6–7 centimeters in craniocaudal length), it is probable that the animal attained a total length of 4–5 meters, making it larger than known oviraptorosaurs (which were 2–3 meters long; Osmólska, Currie, and Barsbold 2004). The reasons Frey and Martill (1995) offered to support their conclusion that the specimen belongs to Oviraptorosauria are weak, and some authors (Makovicky and Sues 1998) have raised doubts about this systematic assignment. The postacetabular wing of the ilium of the Brazilian theropod looks deep and is caudally notched, presenting a square-shaped contour in side view. Moreover, a sharply pointed caudodorsal process is present. This kind of ilium is different from that of oviraptorosaurs (e.g., *Caudipteryx*, *Ingenia*, *Nomingia*, *Rinchenia*; Osmólska, Currie, and Barsbold 2004), in which the caudal margin of the postacetabular wing is shallower and more rounded (e.g., caudally convex). Instead, the shape of the caudal end of the ilium is more like that of abelisauroids, especially in a caudodorsal process.

Previous authors (e.g., Kellner 1999; Makovicky and Sues 1998) have argued that this specimen from the Santana beds does not have unequivocal features supporting its referral to Oviraptorosauria.

Fig. 6.44. *Patagonykus puertai*: (A–C) Vertebrae in right lateral view: (A) Proximal caudal. (B) Posterior sacrals. (C) Mid-dorsal vertebra. (D) Reconstructed right forelimb in lateral view. (E) Phalanx 1 and ungual of digit I in caudal aspect. (F) Reconstruction of the pelvic girdle in right lateral view. (G) Paired pubes in cranial aspect. (H–M) Right manual phalanx of digit 1 in (H) dorsal, (I) ventral, (J) distal, (K) proximal, (L) lateral, and (M) medial views. (A–C, F–M) From Novas (1997d). © Copyright 1997 the Society of Vertebrate Paleontology. Reprinted and distributed with permission of the Society of Vertebrate Paleontology. (D, E) From Novas (1996a). Reprinted with permission of the Queensland Museum.

The discovery of a peculiar manual ungual was reported from outcrops of the Marília Formation at Serra do Veadinho, Peirópolis, Minas Gerais State, Brazil (Novas, Borges Ribeiro, and Carvalho 2005; Fig. 4.5). The ungual, tentatively interpreted as corresponding to manual digits II or III, would have been approximately 60 millimeters long (along the outer curvature) when complete (Fig. 6.45). The size of the bone indicates that the animal that produced the ungual was comparatively small and slender, probably reaching 2 meters in total length. The ungual is elongate, dorsoventrally low, and sharply pointed in side view. A transversely wide lip is present on the proximodorsal margin of the ungual, and the flexor tubercle is block-like and ventrally flat. Deeply excavated grooves are present on both the medial and lateral surfaces, extending from the base of the proximal articular facet to the distal extremity of the phalanx. Except for its distal third, the ventral surface of the ungual is flat and is clearly demarcated from the sides of the bone by sharp edges. Notably, the distal third of the ventral surface forms a sharp cutting edge, or keel.

The set of characteristics described above is documented only in a restrictive group of derived maniraptoran theropods that includes oviraptorosaurs, *Microvenator celer*, troodontids, dromaeosaurids, and birds (Rauhut 2003a). For example, ungual lips are pronounced among Elmisauridae (e.g., *Chirostenotes*, *Elmisaurus*) and in the basal oviraptorosaur *Microvenator* (Currie 1990). Proximodorsal lips are also present in Troodontidae (Osmólska and Barsbold 1990) but not uniformly in all manual unguals. Lips are also present in early birds; *Archaeopteryx* is an outstanding example (Wellnhofer 1988, 1993). The manual unguals of digits II and III of *Deinonychus* (Ostrom 1969) have proximodorsal lips, although they are less developed than in the above-mentioned taxa. In the therizinosaurid *Alxasaurus* a lip is present on the ungual of digit II (Russell and Dong 1993). Proximal lips are absent in the manual unguals of *Compsognathus* (Ostrom 1978) and *Ornitholestes* (Rauhut 2003a). However, the flexor tuber in the Brazilian specimen is proportionally lower than the proximal articular surface, in contrast with troodontids, dromaeosaurids, and birds, in which the height of the tuber exceeds half the height of the articular facet (Rauhut 2003a). Consequently, the specimen under study may be a non-paravian maniraptoran that is more derived than *Ornitholestes* and Alvarezsauridae. The ungual from Brazil superficially resembles that of some oviraptorosaurs (e.g., elmisaurids) in being elongate and dorsoventrally depressed, but the absence of proximally bifurcated grooves, which characteristically occur in elmisaurids (Currie 1990), prevent referral of the isolated ungual to this theropod clade.

The morphology of the manual ungual reveals that the taxonomic diversity of South American theropod faunas was greater than expected. It may be a new example of derived maniraptorans that has not been recorded before on this continent.

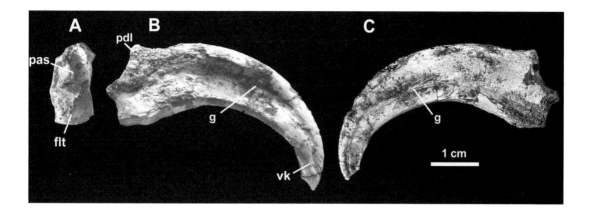

Fig. 6.45. Elmisaurid-like manual ungual from the Marília Formation (Brazil) in (A) proximal, (B) lateral?, and (C) medial? views. From Novas, Borges Ribeiro, and Carvalho (2005). Reprinted with permission of Revista del Museo Argentino de Ciencias Naturales.

Deinonychosauria

This group of coelurosaurians is of considerable interest due to its phylogenetic and functional implications for the origin of birds. Deinoncyhosauria gathers all the theropods that are closely related to birds, grouping them together within the clade Paraves (Fig. 6.46). As their name implies, deinonychosaurian theropods evolved specialized talons on pedal digit II that presumably served as both a defensive and an offensive device. Dromaeosauridae is one of the main clades into which Deinoncyhosauria is split (the other being Troodontidae). Famous dromaeosaurid deinonychosaurians are *Velociraptor mongoliensis* and *Deinonychus antirrhopus*, but the group is composed at present of a dozen genera mainly from Cretaceous beds of North America and Asia. Particularly relevant are the dromaeosaurids from the Early Cretaceous Jehol Group of Liaoning Province, northeast China, including *Sinornithosaurus* and *Microraptor*, which shed light on feather evolution and the early acquisition of wings (Xu, Wang, and Wu 1999; Xu, Zhou, and Wang 2000; Xu, Zhou, and Prum 2001; Ji, Norell, Gao, Ji, and Ren 2001; Hwang, Norell, Ji, and Gao 2002; Norell, Ji, Gao, Yuan, Zhao, and Wang 2002).

Phylogenetically, Dromaeosauridae is defined as all descendants of the most recent common ancestor of *Microraptor zhaoianus*, *Sinornithosaurus millenii*, and *Velociraptor mongoliensis* (Norell and Makovicky 2004). The main anatomical features of dromaeosaurids include short T-shaped frontals with a sinusoidal edge that demarcates the rostral boundary of the supratemporal fenestra; a squamosal shelf that overhangs caudolaterally; a lateral process of the quadrate that contacts the quadratojugal dorsally above the enlarged quadrate foramen; raised stalk-like parapophyses on the dorsal vertebrae; a modified raptorial pedal digit II (which is also present in troodontids); elongate chevrons and prezygapophyses of the caudal vertebrae that span several vertebrae; and a subglenoid fossa on the coracoid.

Since the first dromaeosaurids were described at the beginning of the 1920s (e.g., *Dromaeosaurus* and *Velociraptor*), their discoveries have been predominantly restricted geographically to the northern hemisphere.

However, this situation changed in recent years with novel finds in Argentina, including the first dromaeosaurids named from the southern continents: *Unenlagia comahuensis, Unenlagia paynemili, Neuquenraptor argentinus, Austroraptor cabazai, Buitreraptor gonzalezorum,* and possibly also *Unquillosaurus ceibalii.* Although incomplete, the Cretaceous fossil record of Gondwanan dromaeosaurids suggests that an important adaptive radiation of derived maniraptoran theropods took place in Gondwana at the end of the Mesozoic. The presence of these taxa in the Cretaceous of South America and Africa (the Cenomanian Wadi Milk Formation in Sudan; Rauhut and Werner 1995) eliminates previous hypotheses that deinonychosaurians were endemic to Laurasia. Moreover, the Gondwanan record of non-avian maniraptorans also includes elmisaurid-like forms, alvarezsaurids, and bizarre representatives of large size. This strongly suggests that an important adaptive radiation of maniraptoran theropods took place in the southern continents during the Cretaceous Period.

UNQUILLOSAURUS CEIBALII

This taxon is represented by a large (i.e., 51 centimeters long) and isolated but almost complete pubis from the Maastrichtian Los Blanquitos Formation, which outcrops at Sierra de la Candelaria in northwestern Argentina's Salta Province. *Unquillosaurus ceibalii* was described by Jaime Powell (1979), who interpreted this large theropod as a member of "Carnosauria." However, the pubic anatomy of *Unquillosaurus* (Fig. 6.47) is more congruent with that of maniraptoran coelurosaurians (Novas and Agnolín 2004). The proximal end of the pubis has a thick edge defining a deep groove, which Powell (1979) originally interpreted as diagnostic of *Unquillosaurus ceibalii*. However, this abnormal morphology (which is not recorded in other saurischian dinosaurs) is better explained as the broken pubic pedicle of the ilium adhering to the external surface of the pubis. In fact, this piece of bone resembles the pubic pedicle of the ilium of *Deinonychus* and *Unenlagia* (Ostrom 1969; Novas and Puerta 1997); it is craniocaudally extended and ventrally concave and its cranial and ventral margins form an angle exceeding 90 degrees in side view. Derived features present in *Unquillosaurus* include an opisthopubic pelvis, a craniocaudally wide pubic pedicle of the ilium, a strongly concave ventral margin of the pubic pedicle, a short cranial process of the pubic foot, and pubic foot whose length is less than 30 percent of total length of the pubis. Moreover, some traits shared with early avians (e.g., a proximodistally tall and craniocaudally short pubic foot) suggest that this dinosaur may be more closely related to birds than suspected. Its large size (which is comparable to the large Mongolian dromaeosaurid *Achillobator*; Perle, Norell, and Clark 1999) makes *Unquillosaurus* one of the largest known maniraptorans. *Unquillosaurus* adds to the list of bird-like theropods discovered in Gondwana, such as *Unenlagia* and *Rahonavis*, but constitutes a giant member of the group. A radiation of large-sized maniraptorans apparently occurred in South America during the Late Cretaceous and probably includes other

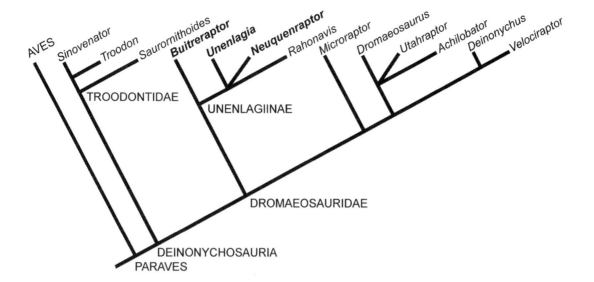

Fig. 6.46. Phylogenetic relationships of the main paravian groups. Genera recorded in South America are indicated in bold.

recent discoveries (e.g., Novas, Canale, and Isasi 2004; Novas, Pol, Canale, Porfiri, and Calvo 2009). The unique pubic features of *Unquillosaurus* also point to the conclusion that this taxon was part of a lineage of predatory dinosaurs that was endemic to South America.

AN UNNAMED DROMAEOSAURID
FROM THE PORTEZUELO FORMATION

Jorge Calvo and Juan Porfiri discovered an almost articulated (albeit incomplete) skeleton and partial skull of a new and small (approximately 2 meters long) but notably robust coelurosaurian that has remarkable dromaeosaurid features (Porfiri, Calvo, and Novas 2005). The specimen was recovered in the Portezuelo Formation in outcrops along the coast of Lake Barreales (which is man-made) from the same spot that yielded the new specimens of *Megaraptor namunhuaiquii* (Calvo, Porfiri, Veralli, Novas, and Poblete 2004) and *Unenlagia paynemili* (Calvo, Porfiri, and Kellner 2004). The present specimen includes premaxillae; maxillae with teeth; nasals; cervical, dorsal, and sacral vertebrae; ribs; the scapulocoracoid; and both humeri. The preserved portions of the premaxilla and maxilla suggest that the narial opening was large and elongate. The maxillae are low and long in front of the antorbital opening. The maxillary dental margin is almost straight and is different in this regard from the dental margin of other dromaeosaurids, which is convex (e.g., *Velociraptor*). Two fenestrae are laterally exposed on the rostral corner of the preorbital depression. The maxillary teeth have caudally curved crowns, which are caudally (but not rostrally) serrated. The roots of the teeth are longer than their respective crowns.

The complete sequence of cervicals requires more preparation, but the centra appear to be dorsoventrally depressed and transversely wide

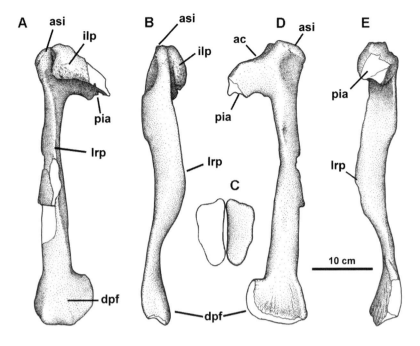

Fig. 6.47. *Unquillosaurus ceibali:* Left pubis in (A) lateral, (B) cranial, (C) distal, (D) medial, and (E) caudal views. *From Novas and Agnolín (2004). Reprinted with permission of* Revista del Museo Argentino de Ciencias Naturales.

and have cranial articular surfaces that form obtuse angles with their ventral surfaces. The dorsal vertebrae have interesting characteristics, including centra that are as deep as they are craniocaudally long and posterior dorsals with large and double pleurocoels and tall neural spines that resemble these of *Unenlagia*. The thoracic ribs are distinguished by their robustness, especially in the transverse expansion of the lamina that connects the capitulum and the tuberculum.

The scapula of this new creature differs from that of other dromaeosaurids; it is notably wider proximally. The acromial process is low and rounded, and the glenoid cavity is caudally oriented (instead of laterally, as in *Unenlagia*, *Buitreraptor* and other paravians). The coracoid is quadrangular-shaped in lateral view and has a prominent biceps tuber.

Unenlagiinae

This term was coined by José Bonaparte (1999d) as a way to emphasize the morphological distinctions of *Unenlagia* with the remaining theropods and to unite this Patagonian taxon with the purported ornithomimosaurian *Timimus*, from the Early Cretaceous of Australia (Rich and Vickers-Rich 1994). Unenlagiinae is the name currently applied to the clade of Gondwanan dromaeosaurids formed by *Unenlagia*, *Neuquenraptor*, *Buitreraptor*, *Rahonavis*, and *Austroraptor* (Makovicky, Apesteguía, and Agnolín 2005; Novas, Pol, Canale, Porfiri, and Calvo 2009; Fig.6.46).

As all deinonychosaurs, unenlagiines have a modified pedal digit II, allowing hyperextension. However, they clearly differ from their Laurasian relatives in a longirostrine skull that exceeded the length of the femur by at least 25 percent. In *Buitreraptor* and *Austroraptor* (the only two unenlagiines in which the skull is partially known), the external surface of the dentary has a deep longitudinal groove occupied by mental foramina, a condition resembling that of troodontids. The teeth of unenlagiines are minute, widely spaced, and devoid of serrations, in contrast to the larger and serrated teeth of Laurasian dromaeosaurids. Unenlagiines are also characterized by a proximally pinched metatarsal III, a condition they also share with basal dromaeosaurids, troodontids, ornithomimosaurians, and tyrannosaurids, among other maniraptorans.

The available evidence indicates that the Unenlagiinae diversified during the Late Cretaceous in the southern continents. They evolved into long-limbed forms that varied in size from that of a pigeon (*Rahonavis*) and a turkey (*Buitreraptor*) to rhea-sized forms with shorter forelimbs (*Unenlagia*) as well as large animals with strongly reduced forelimbs (e.g., *Austroraptor*).

UNENLAGIA COMAHUENSIS

The type specimen of this theropod was discovered by the author and assistants in the Turonian–Coniacian Portezuelo Formation exposed at Sierra del Portezuelo, Neuquén Province, Argentina. The partial but almost articulated skeleton corresponds to that of an ostrich-size carnivorous dinosaur roughly 2 meters long. Once envisaged as phylogenetically closer to birds than the remaining dromaeosaurids (Novas and Puerta 1997; Forster, Sampson, Chiappe, and Krause 1998), *Unenlagia* is now interpreted as a member of Dromaeosauridae, a clade deeply nested within Deinonychosauria, the sister taxon of Aves (Novas and Pol 2005; Makovicky, Apesteguía, and Agnolín 2005). However, *Unenlagia* and their close relatives have outstanding avian-like features that help us understand better the origin of birds and their flight (Novas 1999, 2004). Moreover, study of these theropods generate intriguing hypothesis about how non-avian maniraptorans acquired flying capabilities (Makovicky, Apesteguía, and Agnolín 2005).

Unenlagia is characterized by tall neural spines on the posterior dorsal and anterior sacral vertebrae; the spine is nearly twice the height of the centrum (Fig. 6.48A). Deep lateral pits occur on the base of the neural spines of these vertebrae, constituting an autapomorphic trait of *Unenlagia* (Fig. 6.48A; lp). Six fused sacrals are present, although the ilia extend the length of at least 9 vertebrae (i.e., 6 sacrals, 2 dorsals, and 1 caudal) instead of 7 vertebrae (i.e., 5 sacrals, 1 dorsal, and 1 caudal), as in *Deinonychus* and *Archaeopteryx*.

The scapula of *Unenlagia* is strap-like (Fig. 6.48B), closely resembling that of *Buitreraptor* and *Archaeopteryx*. As in these taxa, the acromion is triangular in lateral aspect and is sharply cranioventrally projected. As in other dromaeosaurids and avians, the humeral articulation of the scapula of *Unenlagia* is laterally oriented. However, the scapula of *Unenlagia* is "twisted" such that the acromial process and glenoid cavity are exposed

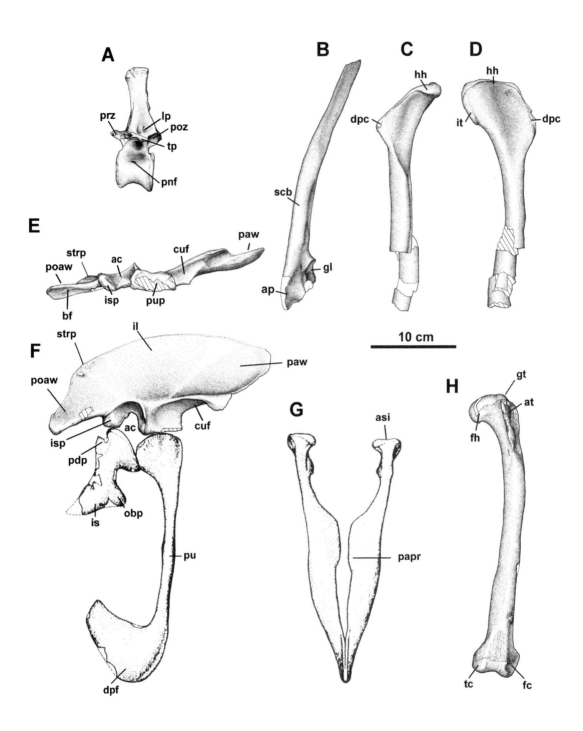

laterally while the scapular blade is exposed dorsally. Notably, the same condition is documented in the new Patagonian dromaeosaurid *Buitreraptor* (Makovicky, Apesteguía, and Agnolín 2005), dismissing recent interpretations about the orientation of the glenoid cavity in *Unenlagia* (Carpenter 2002). The humerus of *Unenlagia* (Fig. 6.48C and D) is a slender but not particularly elongate bone that is 71 percent as long as the femur. As in other paravians, the internal tuberosity (a homologue of the bicipital crest of modern birds) is proximodistally extended. A remarkably avian feature of the humerus is the laterally oriented deltopectoral crest, a condition that *Unenlagia* shares with *Buitreraptor* and that contrasts with the cranially projected crest present in *Deinonychus*, for example.

The pelvic girdle has paravian features (Novas 2004; Fig. 6.48E and F): the ilium is cranially extensive but caudally short, the pubis is ventrally oriented, and the ischium is short, plate-like, and triangular in side view. Such osteological changes (and their muscular correlates) may indicate that these animals had increased jumping and landing capabilities. The pelvic girdle of *Unenlagia* seems to be intermediate in many anatomical respects between that of dromaeosaurid maniraptorans and that of early birds (e.g., *Archaeopteryx*, *Confuciusornis*). The fossa for the M. cuppedicus is well developed, a condition that is widely present among Coelurosauria, including basal birds (Fig. 6.48E; cuf). In contrast to non-avian coelurosaurs (i.e., *Deinonychus*), the acetabulum of *Unenlagia* tends to close off medially, resembling that of *Archaeopteryx*, *Patagopteryx*, and hesperornithiforms. The postacetabular blade is low and sharp, as in *Archaeopteryx* and enantiornithines, and the brevis fossa is considerably more reduced than in other maniraptorans. In the ilia of *Saurornitholestes*, *Deinonychus*, and *Velociraptor*, a precursor of the avian supratrochanteric process is expressed as a faint transverse enlargement of the posterodorsal iliac margin (Fig. 6.48E and F; strp). In *Unenlagia*, this process is more prominent and triangular, indicating that this taxon is more derived toward the avian condition. Also, it is associated with a ridge that runs posterodorsally along the lateral surface of the ilium above the acetabulum. The rounded preacetabular wing of *Unenlagia* resembles that of early birds in extending anteriorly, in contrast with less-derived coelurosaurs in which the cranial margin of the ilium is straight (e.g., oviraptorosaurs, ornithomimids) or notched (e.g., *Tyrannosaurus*, most dromaeosaurids).

The pubis is slightly shorter than the femur (Fig. 6.48F and G); it is oriented ventrally, as in other maniraptorans, and distally it has a caudally projected foot. The pubic shaft expands transversely into a pubic apron that is widely extended proximodistally, as in *Deinonychus* but in contrast with the more-reduced symphysis of *Archaeopteryx* and more-derived birds. As Calvo, Porfiri, and Kellner (2004) noted, the pubis of *Unenlagia* is not straight in lateral view (as originally reconstructed; Novas and Puerta 1997) but is cranially convex, a condition that Calvo and colleagues interpreted as autapomorphic for this taxon. The ischium is a short plate-like bone that is triangular in side view. The obturator notch is enclosed distally by a triangular obturator process, a primitive condition that is lost in *Archaeopteryx* and more-derived birds (Fig. 6.48F; obp). However, the dorsal edge of the ischium has a prominent proximodorsal process that is separated from the rest of the bone by a deep notch, a condition that is also present in *Buitreraptor*, *Rahonavis*, *Archaeopteryx*, and basal birds (Novas and Puerta 1997; Makovicky, Apesteguía, and Agnolín 2005).

Fig. 6.48. *Unenlagia comahuensis*: (A) Posterior dorsal vertebra in left lateral view. (B) Left scapula in lateral view (cranial points downward). (C, D) Left humerus in (C) lateral and (D) cranial views. (E) Right ilium in ventral aspect. (F) Pelvic girdle in right lateral view. (G) Paired pubes in cranial view. (H) Left femur in cranial view.

The hindlimb bones (e.g., femur, tibia) are long and slender, suggesting that *Unenlagia* was a fleet-footed animal. The femur (Fig. 6.48H) resembles *Buitreraptor*, *Rahonavis*, and *Archaeopteryx* (among other paravians) in that it is long and slender and has a proportionally small head and a proximally projected anterior trochanter. Both the fourth trochanter and the posterior trochanter are absent. The tibia is long and slender and is 13 percent longer than the femur.

Unenlagia offers insight into anatomical transformations that led to the acquisition of flapping flight among avians (Novas 1999). The glenoid cavity faces laterally, which has important consequences for forelimb postures and movements. In most dinosaurs the glenoid cavity is orientated posteroventrally, restricting the range of forelimb movements to motion that occurred laterally and below the body. In *Unenlagia*, in contrast, the orientation of the glenoid socket closely resembles that of birds. This anatomical modification, which is recognized in other dromaeosaurids (Norell and Makovicky 1999; Xu 2002; Makovicky, Apesteguía, and Agnolín 2005), has two significant functional implications. One is the ability to elevate the humerus almost vertically during maximum upstroke, thus increasing the dorsoventral excursion of the forelimbs; the other is the ability of the forelimbs to fold against the body, a unique postural activity that in living birds is usually related to both the possession of and protection of feathers (Novas and Puerta 1997). Although no epidermal structures were preserved in the sandstone in which *Unenlagia* was found, the almost avian folding of its forelimbs deduced from the morphology of the shoulder girdle opens the possibility that the Patagonian dinosaur was also feathered.

Unenlagia probably used its arms not only to hold and tear prey, as other theropods did, but also to control the body position while running, turning, and leaping. *Unenlagia* had the ability to extend the elevatory range of humeral movements that could have prepared the arms for a powerful downstroke: in other words, incipient flapping activity. These anatomical and functional transformations of the shoulder girdle and forelimbs are related to a paramount aspect of bird evolution: the acquisition of flapping, powered flight. The advanced morphology of the shoulder girdle of dromaeosaurids, as illustrated by *Unenlagia*, clearly indicates that critical functional attributes present in modern birds (namely, the wide arc of dorsoventral forelimb movements) evolved prior to the acquisition of flight (Novas and Puerta 1997; Novas 1999). Although *Unenlagia* was unable to fly, its extensive avian-like forelimb elevation (the upstroke) illustrates a crucial adaptation that would enable a smaller-sized winged theropod to generate enough thrust to lift off from the ground. Birds inherited this condition, suggesting that the extensive forelimb elevation constituted a prerequisite for powered flight. Although debate about the emergence of birds and flight is far from closed, any discussion concerning this topic must explain the functional significance of the increased upstroke and forelimb folding described for non-flying theropods that are ancestral to birds (Novas and Puerta 1997; Novas 1999).

So what are the differences between flying (e.g., *Archaeopteryx*) and flightless (e.g., *Unenlagia*) dinosaurs? No major osteological distinctions involved with flight are recognized, except for a reduction in body size and mass (facilitating body maneuverability) and changes in relative proportions of the forelimbs and their integuments. Basal birds inherited almost the same kind of shoulder girdle as non-flying theropods and apparently retained the wide arc of forelimb movements and flapping abilities that were present in their dinosaurian ancestors. Birds experienced both the elongation of their forelimbs and the enlargement of their feathers, thus increasing the power of their wing thrusts (Novas 1999).

In sum, the contribution of *Unenlagia* to our understanding of avian origins is that flapping abilities preceded the acquisition of flight, an idea embryonically developed by Baron Franz Nopsca in 1907. Proto-wings of theropods like *Unenlagia* that have an increased arc of the downstroke can be envisaged as aerodynamically formed to increase speed on land. Later in this group's phylogeny, this evolved into enlarged airfoils that allowed the animals to take off from the ground. Consequently, the evolution of flying capabilities can be envisaged as an aerial extension of displacement on the ground using forelimbs as propellers. In this regard, the evidence afforded by *Unenlagia* and the remaining maniraptorans allows us to explain the origin of powered flight without invoking deep modifications of the behavior and function of the forelimbs.

UNENLAGIA PAYNEMILI

This species of *Unenlagia* was described by Jorge Calvo, Juan Porfiri, and Alexander Kellner in 2004. Its holotype specimen comes from the same levels as *U. comahuensis* (i.e., the Portezuelo Formation) but from a different fossil locality (Barreales Lake, Neuquén Province). *U. paynemili* differs from *U. comahuensis* in a number of characteristics. Its bones are more gracile, the angle between the anterior rim of the deltopectoral crest and the humerus shaft is about 116 degrees (as opposed to 140 degrees in *U. comahuensis*), the anterior process on the distal end of pubis is small, the distal end of the postacetabular blade of ilium is broader and rounded, and the brevis fossa is more shallow. The type specimen of *U. paynemili* is smaller than *U. comahuensis*. The humeral shaft is elongate and almost straight except for the proximal third, which is caudally inflected. As Calvo, Porfiri, and Kellner (2004) noted, this feature distinguishes *Unenlagia* from other dromaeosaurids (e.g., *Deinonychus*), in which the humerus is more sigmoid and lacks this posterior inflection.

NEUQUENRAPTOR ARGENTINUS

This animal was described by Diego Pol and the present author in 2005. The holotype specimen of *Neuquenraptor* consists mainly of most of a left foot and some other bones of the hindlimb (Fig. 6.49) that were discovered while digging up the rib cage of a titanosaurid sauropod in beds of the Turonian–Coniacian Portezuelo Formation that are exposed at Portezuelo Hill, Neuquén Province, Argentina (Novas and Pol 2005; Fig. 4.7). *Neuquenraptor* is approximately 2 meters long.

Neuquenraptor shares with *Buitreraptor* the presence on metatarsal II of a lateral expansion over the caudal surface of metatarsal III. This feature combines with a proximally pinched metatarsal III, a distal end of metatarsal III that is incipiently ginglymoid, a pedal digit II in which phalanges 1 and 2 are sub-equal in length, and a trenchant digit II ungual phalanx. Metatarsal II of *Neuquenraptor* is transversely wider than metatarsal IV (contrasting with troodontids, in which metatarsal IV is robust) and ends distally in a well-developed ginglymoid articulation, a condition that is present in other Dromaeosauridae. The distal end of metatarsal III expands over the cranial surfaces of metatarsals II and IV but is caudally hidden by the lateral and medial projections of metatarsals II and IV, respectively. This peculiar metatarsal articulation (which is also present in *Buitreraptor*) corresponds with the arctometatarsalian condition, which characteristically occurs in troodontids, ornithomimids, caenagnathids, tyrannosaurids, and basal dromaeosaurids (e.g., *Sinornithosaurus* and *Microraptor*). Metatarsal IV of *Neuquenraptor* has a prominent and sharp longitudinal ridge that is posteriorly directed along its posterolateral margin (similar to the crista lateralis plantaris of living birds; Fig. 6.49B and C; plr). Pedal digit II has distinctive deinonychosaurian features: phalanges 1 and 2 have expanded distal ginglymoidal joints and phalanx 2 has a strong proximoventral process for extensive dorsoventral excursions. The ungual phalanx of digit II is enlarged and strongly curved and has a sharp cutting edge. This ungual phalanx is grooved on both sides, with the lateral groove occupying a more dorsal position than the medial one, as is usual among dromaeosaurids.

Makovicky and colleagues (2005) indicated the possibility that *Neuquenraptor* is a junior synonym of *Unenlagia*. In support of their suspicion, these authors noted the near-identical dimensions and proportions of the phalanges in the raptorial second toe of *Neuquenraptor* and the material assigned to *Unenlagia paynemili* (Calvo, Porfiri, and Kellner 2004). However, the isolated phalanx that Calvo, Porfiri, and Kellner (2004) referred to *Unenlagia paynemili* was not found in association with the holotype specimen of this species. Thus it should be considered Deinonychosauria indet. Moreover, an ungual phalanx also referred to *Unenlagia paynemili* was misinterpreted by Calvo, Porfiri, and Kellner (2004) and later also by Makovicky and colleagues (2005) as belonging to the foot, but this ungual corresponds to the manus, a portion of the skeleton that remains unknown for *Neuquenraptor*. In short, the available evidence neither supports nor dismisses the notion that *Neuquenraptor* is a synonym of *Unenlagia*.

The remains of a small deinonychosaurian have been documented in James Ross Island, located in the northeastern Antarctic Peninsula, from beds of the Snow Hill Island Formation (Maastrichtian). The fragmentary specimen, which Judd Case and colleagues (2007) described, has a sharp longitudinal ridge on the caudal surface of metatarsal III,

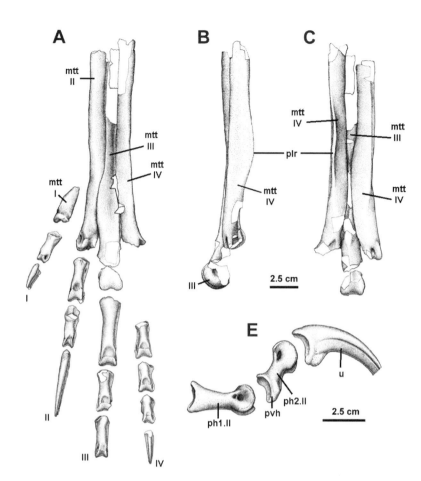

Fig. 6.49. *Neuquenraptor argentinus:* (A) Left hindfoot in dorsal view. (B) Metatarsals IV and III in lateral aspect. (C) Metatarsals II–IV in plantar view. (E) Pedal digit 2 in medial view. *From Novas and Pol (2005).*

resembling that of *Neuquenraptor* and *Buitreraptor*, suggesting that the Antarctic form could be a polar representative of Unenlagiinae.

BUITRERAPTOR GONZALEZORUM

This spectacular theropod was reported by Peter Makovicky and colleagues in 2005 and constitutes the most complete and earliest undoubted maniraptoran collected from Gondwanan continents. Two specimens of *Buitreraptor* are available from the Cenomanian Candeleros Formation, one consisting of a nearly complete articulated adult skeleton and a second consisting of an articulated pelvic girdle, a right hindlimb, and a sacrum. The complete fusion of neurocentral sutures suggests that these specimens had reached their adulthood. *Buitreraptor* was less than 1 meter long.

The skull of *Buitreraptor* (Fig. 6.50A) is long and low and is considerably longer than the femur. The nasals are flat and dorsally narrow, suggesting a muzzle that is more elongate and thin than in *Velociraptor*. The teeth are small and have no serrations (Fig. 6.50B). Low ridges that form the ventrolateral corners of the last cervical centrum end caudally in small tubers, a feature unique to *Buitreraptor*. The dorsal neural spines are tall and rectangular but are more laminar (e.g., axially expanded) than in *Unenlagia*, for example, in which the neural spines are stouter. The caudals of *Buitreraptor* are curious in that the prezygapophyses overlap only half of the preceding vertebra, unlike Laurasian dromaeosaurids, in which prezygapophyses are considerably more elongate. *Buitreraptor* has a broad and robust furcula that has a low ridge instead of a hypocleideum (Makovicky, Apesteguía, and Agnolín 2005). The scapular blade closely resembles that of *Unenlagia* in being long and strap-like with a pronounced curvature near the glenoid, in consequence being laterally oriented, as in other paravians (Fig. 6.50C). The humerus of *Buitreraptor* is 30 percent longer than the scapula (Fig. 6.50D), an unusual ratio matched or surpassed only in avians and some dromaeosaurids from Liaoning (Makovicky, Apesteguía, and Agnolín 2005).

The pelvis (Fig. 6.50E) has highly derived features, in particular in the shape of the brevis shelf and fossa. The fossa is lobate as seen from below and is dorsoventrally flattened, in contrast with the excavated condition present in other dinosaurs in which this fossa is developed. The pubis of *Buitreraptor* is vertically oriented, but the pubic shaft has a strong anteriorly convex curvature, as in dromaeosaurids, including *Unenlagia paynemili* (Calvo, Porfiri, and Kellner 2004). The ischium has a proximally positioned dorsal process, as in many paravians; the obturator process is elongate and strongly pointed; and the lateral surface has a remarkable crescentic ridge that is unreported in other paravians (including *Unenlagia*).

The foot is arctometatarsal, although the third metatarsal is visible along the plantar and extensor faces and metatarsals II and IV are symmetrical around metatarsal III, as in *Neuquenraptor*. Notably, metatarsals II and III terminate in ginglymoid articulations (Makovicky, Apesteguía, and Agnolín 2005).

According to Peter Makovicky and colleagues (2005), Unenlagiinae also includes the sickle-toed *Rahonavis ostromi* from the Late Cretaceous of Madagascar. *Rahonavis* has been usually interpreted as a basal bird that is more derived than *Archaeopteryx* (Forster et al. 1996; Chiappe 2007a), especially because of its greatly elongated forearm bones and quill knobs along the ulna that indicate well-developed flight feathers in the wing (Forster et al. 1996). However, if the hypothesis that depicts *Rahonavis* as deeply nested within Unenlagiinae is accepted, it must be concluded that the elongation of forelimbs in unenlagiines and the development of wing-like forelimbs in *Rahonavis* may have occurred independently from birds. In other words, flight capability may have originated independently in derived unenlagiines and Aves.

AUSTRORAPTOR CABAZAI

The holotype specimen of this taxon was discovered by the author and his crew at the Bajo de Santa Rosa fossil site in Río Negro Province, where extensive outcrops of the Campanian–Maastrichtian Allen Formation

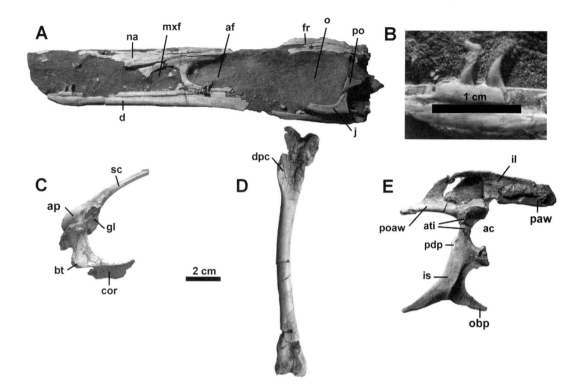

Fig. 6.50. *Buitreraptor gonzalezorum:* (A) Skull in left lateral view. (B) Detail of left dentary teeth. (C) Left scapulocoracoid in lateral view. (D) Right humerus in cranial view. (E) Articulated right ilium and ischium in lateral view. *Photographs by Fernando Novas.*

occur (e.g., Martinelli and Forasiepi 2004). *Austroraptor* is known from several cranial and postcranial elements that belong to a large (~5 meters long) but gracile deinonychosaurian (Fig. 6.51). Preserved skull bones allow reconstruction of a low and elongate head for this animal (Novas, Pol, Canale, Porfiri, and Calvo 2009). The tibia of *Austroraptor* is 55 centimeters long, which is longer than the tibia of *Achillobator giganticus* (49 centimeters; Perle, Norell, and Clark 1999) and *Utahraptor ostrommaysi* (50.5 centimeters; Kirkland, Burge, and Gaston 1993).

The frontals resemble these of *Troodon* and *Sinornithosaurus*, for example, in being triangular-shaped in dorsal view, and together with the postorbital these bones define a wide and rounded orbital cavity. The jaws are elongate and low and have 25 teeth alveoli on the dentary. Bizarre traits of *Austroraptor* occur in the morphology of the teeth, which are proportionally small, cone-shaped, circular in cross-section, and have no serrations or carinae. As in other deinonychosaurians, the centrum of the cervical vertebrae has a kidney-shaped cranial articular surface that is almost in the same plane as the ventral surface (Fig. 6.51A–C). In addition, the neural spines of the dorsal vertebrae are transversely enlarged at their distal extremities. The humerus has a well-developed (i.e., proximodistally elongate) internal tuberosity, a common feature among paravians. However, *Austroraptor* shares with dromaeosaurids a caudally projected internal tuberosity that creates a strong sigmoid curvature for the entire proximal end of the bone in proximal view. Notably, the humerus of *Austroraptor* is approximately 50 percent of the femur length; in most paravians, this ratio is greater (e.g., Xu 2002). The manual ungual phalanges are strongly curved and have prominent extensor tubercles. Metatarsal III is proximally pinched, as in *Neuquenraptor* and other basal deinonychosaurs. Pedal phalanx 2 of digit II shows a constricted neck between both the proximal and distal articular surfaces as well as a caudoventrally projected heel (Fig. 6.51D–F; pvh); these are characteristics usually found among dromaeosaurid theropods (albeit not exclusively). A notable disparity in the transverse width of the pedal phalanges is evident between digits II and IV. Digit IV is broader than digit II (Fig. 6.51G–I), indicating a marked asymmetry in foot construction.

The presence of large-sized unenlagiines at the end of the Late Cretaceous further expands the recently noted trend in increase in body size for Gondwanan dromaeosaurids (Turner, Pol, Clarke, Erickson, and Norell 2007): *Austroraptor* is estimated to be approximately 1.5 times as large as *Unenlagia*, the largest previously known southern dromaeosaur known from the earlier beds of Patagonia. In this phylogenetic context, the large size that southern dromaeosaurids acquired matches that convergently developed by the largest Laurasian dromaeosaurids (i.e., *Utahraptor* and *Achillobator*; Novas, Pol, Canale, Porfiri, and Calvo 2009).

The forelimbs of *Austroraptor* were considerably reduced compared to the length of the hindlimbs, constituting an exceptional case among the characteristically long-armed dromaeosaurids, including other known unenlagiines. The morphology of *Austroraptor* contrasts with that of the large dromaeosaurids from Laurasia (for example *Utahraptor*), in which the manual unguals are considerably much larger than in *Austroraptor*, despite their similar body size.

The available evidence supports the theory that by the end of the Cretaceous, South America was inhabited by large-sized coelurosaurians of different phylogenetic lineages including *Austroraptor* and the bizarre maniraptoran *Unquillosaurus* (Novas and Agnolín 2004). These large coelurosaurians shared the role of big predators with the well-diversified abelisaurid ceratosaurians.

Aves

The South American fossil record of Mesozoic birds is comprised of widely dispersed but informative evidence from Argentina, Chile, and Brazil. Aside from several feather impressions from the Early Cretaceous (Aptian) Santana Formation in Brazil (Kellner and Campos 2000), most bird discoveries are currently restricted to the Coniacian through Maastrichtian interval of the Late Cretaceous and include skeletal remains, eggs, and footprints. The record of Cretaceous birds has been summarized and thoroughly considered by Luis Chiappe in a number of authoritative publications (e.g., Chiappe 1992, 1993, 1995, 2001, 2002, 2007a, 2007b). This evidence, albeit partial, reveals a very diverse avifauna in this continent at the end of the Mesozoic that is

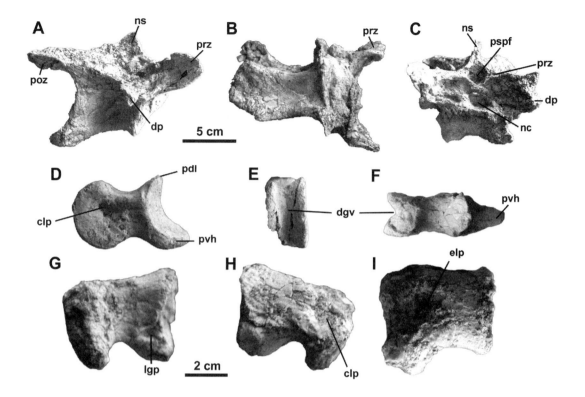

Fig. 6.51. *Austroraptor cabazai:* (A–C) Anterior cervical vertebra in (A) lateral, (B) ventral, and (C) cranial views. (D–F) Pedal phalanx II.2 in (D) lateral, (E) distal, and (F) dorsal views. (G–I) Pedal phalanx IV.2 in (G) lateral, (H) medial, and (I) dorsal views. *Photographs by Fernando Novas.*

currently represented by enantiornithines, *Patagopteryx*, *Limenavis*, and neornithines.

Perhaps the most significant aspect of the South American Cretaceous avifauna is the diversity and numerical dominance of the Enantiornithes (Walker 1981; Chiappe 1996, 2007a), an extinct group distributed worldwide that is considered to be the sister taxon of Ornithuromorpha (i.e., *Patagopteryx* plus Ornithurae; Chiappe 2001). The remains of neornithine birds are scarce in the Cretaceous of South America but include representatives of Gaviidae and Galliformes, which supports the view that South America was a cradle of the early evolution of modern avians.

Disarticulated enantiornithine skeletons were first discovered by José Bonaparte and his assistants at northwestern Argentina's El Brete fossil site in beds of the latest Cretaceous Lecho Formation. These remains (Fig. 6.52A) were first described by Cyril Walker in 1981. The enantiornithine birds are represented by *Neuquenornis volans* (Chiappe and Calvo 1994), which was discovered in beds of the Bajo de la Carpa Formation exposed at Neuquén City. *Neuquenornis* is characterized by a gracile tarsometatarsus, a wing-like posterior trochanter on the proximal end of the femur, subequal widths of the major and minor metacarpals, and a pronounced sternal keel that projects cranially beyond the margin of the sternum. The morphology of the wing (which has a stouter ulna for the stronger attachment of flying feathers), shoulder (which has a delimitation of the triosseal canal at the junction of coracoid, scapula, and furcula for funneling the supracoracoid tendon, a muscle largely responsible for the upstroke of the

Fig. 6.52. Avian bones. (A) Articulated right ilium and ischium of an enantiornithine bird from El Brete, Salta Province. (B–D) Proximal end of a right coracoid of a neornithine bird from the Portezuelo Formation, Neuquén Province, in (A) ventral, (B) medial, and (C) dorsal views. *(A) Photograph by Fernando Novas. (B–D) From Agnolín, Novas, and Lío (2006). Reprinted with permission of* Ameghiniana.

wing), and sternum (which has a deeper ventral keel for the origin of the larger pectoral muscles) of *Neuquenornis* also highlights the aerodynamic capabilities of enantiornithines, which included flight at low speeds and a high degree of maneuverability (Chiappe 2007a).

Abundant enantiornithine remains, including incomplete skulls, were recently discovered in beds of the Adamantina Formation in the region of Presidente Prudente in the State of São Paulo (Alvarenga and Nava 2005).

Patagopteryx deferrariisi is a peculiar adaptive kind of South American bird. This very interesting flightless bird was described by Herculano Alvarenga and José Bonaparte in 1992 on the basis of several specimens recovered from the Campanian Bajo de la Carpa Formation in Neuquén City. *Patagopteryx* is the best-represented Mesozoic bird from Patagonia and was the subject of a detailed monograph about its anatomy and phylogenetic relationships (Chiappe 2002). *Patagopteryx*, which is about the size of a hen, is characterized by reduced forelimbs but robust hindlimbs. Among the most outstanding autapomorphies of this taxon are a quadrate fused to the pterygoid, biconvex articular facets of the fifth thoracic vertebra, a procoelous condition and very broad centra of the six vertebrae following this biconvex element (thoracics 6–11), a minor metacarpal that is more robust than the major metacarpal, a strap-like morphology and distal cranioventral curvature of the pubis, an ischium that is shaped like a paddle, and pamprodactyl feet (four toes facing forward; Chiappe 2002). *Patagopteryx* is the basalmost member of ornithuromorph birds to lose the capability to fly.

The more-derived birds include *Limenavis patagonica* and *Neogaeornis wetzeli*. *Limenavis* was described by Julia Clarke and Luis Chiappe in 2001 and was discovered in the Early Maastrichtian Allen Formation, which is exposed in Salitral Moreno, 20 kilometers south of the city of General Roca in Río Negro Province, Argentina. It is represented by portions of several wing bones. *Limenavis* is a member of Carinatae, as demonstrated by the anatomically modern characteristics of its wing bones, and is phylogenetically closer to neornithines than the North American bird *Ichthyornis* is. Sedimentological information suggests that *Limenavis* inhabited the Atlantic shores of Patagonia at the end of the Cretaceous.

Neogaeornis wetzeli is a gaviid represented by a tarsometatarsus collected from marine beds of the Maastrichtian Quiriquina Formation in Bahía San Vicente in the Province of Concepción in Chile (Lambrecht 1929).

Finally, Federico Agnolín, Fernando Novas, and Gabriel Lío (2003, 2006) announced the discovery of a tiny isolated coracoid collected in beds of the Early Turonian Cerro Lisandro Formation in the Province of Neuquén. The coracoid is small; its total estimated length is 3 centimeters (Fig. 6.52B–D). The bone has derived traits considered to be diagnostic of Neornithes as well as some of the apomorphic traits of Galliformes. The importance of this find is that it may constitute one of the oldest known Neornithes yet recorded. Few examples that could be

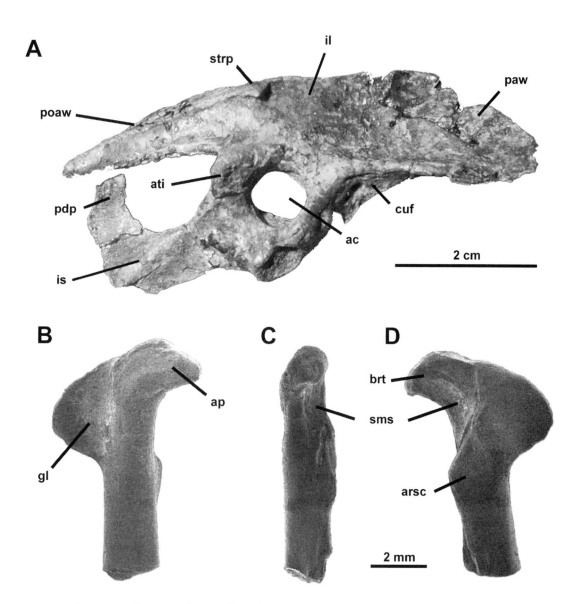

interpreted as neornithines are known from the Mesozoic (Chiappe and Dyke 2002; Dyke and van Tuinen 2004), and those are mainly restricted to the Upper Cretaceous (i.e., Campanian, Maastrichtian) from America, Europe, Asia, and Antarctica (Hope 2002). The discovery of neornithine remains in the Cerro Lisandro Formation (estimated to be 93 million years old; Leanza, Apesteguía, Novas, and de la Fuente 2004) supports the interpretation that the diversification of modern groups of birds was well in progress during the Turonian (e.g., Hedges, Parker, Sibley, and Kumar 1996; Dyke and van Tuinen 2004) or even earlier (Cooper and David 1997). Another remarkable aspect of the coracoid from Neuquén is that it constitutes one of the few terrestrial neornithines recorded in the Late Mesozoic, in contrast with the majority of the Cretaceous neornithines, which correspond to aquatic groups (e.g., Charadriiformes, Anseriformes; Olson and Parris 1987; Elzanowski 1995).

The presence of such derived birds in Patagonia as well as the diversified true neornithine birds described for the Late Cretaceous of Antarctica (e.g., *Polarornis, Vegavis*; Noriega and Tambussi 1995; Case and Tambussi 1999; Chatterjee 2002; Clarke, Tambussi, Noriega, Erikson, and Ketcham 2004) support the hypothesis that the earliest divergences of neornithines could have taken place in Gondwana (Cracraft 2001; Chatterjee 2002). The southern origin of neornithines is also supported by the fact that many primitive lineages—paleognaths, galloanserines, gruiforms, and others—are mostly distributed across the former Gondwanan landmasses (Cracraft 2001).

Cretaceous Theropod Footprints and Eggs

The footprint record of theropod dinosaurs is both geographically and stratigraphically extensive, with finds in Colombia, Ecuador, Perú, Bolivia, Brazil, and Argentina (Leonardi 1994; Jaillard et al. 1993; Calvo 1991, 1999, 2007; Figs. 4.5 and 6.1). However, in contrast with the numerous studies devoted to analyzing the morphology and paleobiology of South American sauropod and ornithischian footprints, few attempts have yet been made (e.g., Calvo 1991, 1999, 2007) to correlate the diversity of theropod trackways with the lineages documented from skeletal remains in this continent.

Abundant dinosaur footprints have been found in the bedding planes of the extensive outcrops of the Candeleros Formation around Lake Ezequiel Ramos Mexía (Fig. 6.53). Jorge Calvo dedicated several articles to analyzing this evidence (Calvo 1991, 1999, 2007), describing the following theropod ichnotaxa: *Abelichnus astigarrae* (a tridactyl whose tracks are around 50 centimeters long and have prominent claw impressions, which Calvo inferred were produced by a theropod the size of *Giganotosaurus*), *Bressanichnus patagonicus* (a tridactyl whose track is between 20 and 25 centimeters long), *Deferrariischnium mapuchensis* (a tridactyl whose track is approximately 20 centimeters long and has an elongate digit III that is more than 50 percent of the track length and has a sharp claw impression), and *Picunichnus benedettoi* (a tetradactyl whose track is around 22 centimeters long and has marks of digit I).

The ichnological evidence for Cretaceous birds seems more encouraging. A large number of footprints of the Maastrichtian bird *Patagonichnornis venetiorum* have been reported by Rodolfo Casamiquela (1987) from Ingeniero Jacobacci in the Province of Río Negro. Although referral of these footprints to birds is based on their small size, slender digital impressions, and wide angles of divarication (Lockley, Yang, Matsukawa, Fleming, and Lim 1992), their taxonomic position within Aves is far from settled. Coria, Currie, Eberth, and Garrido (2002) reported the discovery of numerous footprints that represent at least three avian ichnotaxa from the Campanian–Santonian Anacleto Formation, which is exposed at Barrosa Hill near Plaza Huincul in the Province of Neuquén (Coria, Chiappe, and Dingus 2002). Among these, the most common footprints

Fig. 6.53. Cretaceous theropod footprints from the Candeleros Formation exposed around the shorelines of Lake

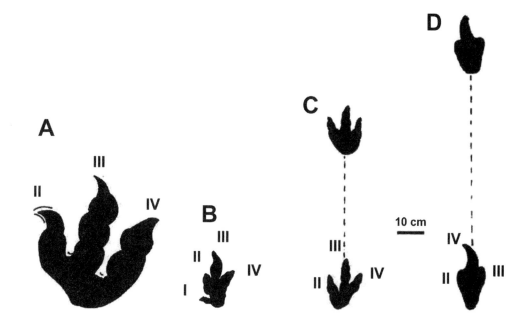

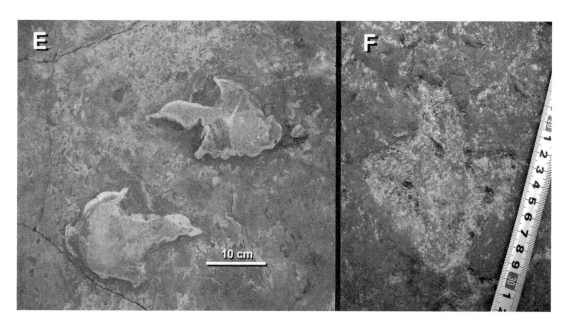

are morphologically comparable to the ichnotaxa *Aquatilavipes* from the Lower Cretaceous of Canada (Currie 1981). Other footprints from Barrosa Hill are similar in size to those referred to *Aquatilavipes* but have a distinct hallux impression. Coria and colleagues (2002) identified these footprints as the ichnotaxon *Ignotornis*. They named a different and much smaller type of footprint *Barrosopus slobodai*. Although these footprints document a variety of birds that inhabited the riverine environments of northwestern Patagonia, it is difficult to relate their pedal morphology to any known lineage of birds.

Ezequiel Ramos Mexía: (A) *Abelichnus astigarrae*. (B) *Picunichnus benedettoi*. (C) *Bressanichnus patagonicus*. (D, E) *Deferrariischnium mapuchensis*. (F) Theropoda indet. (A–D) Redrawn from Calvo (2007). (E, F) Photographs by Fernando Novas.

In contrast with the abundant record of titanosaurid eggs, the oological evidence for theropod dinosaurs in South America is much more fragmentary and is restricted to some isolated eggs and eggshells of the oofamily Elongaloolithidae from the Allen Formation in Río Negro Province (Manera de Bianco 2000b; Manera de Bianco and Tomasini 2003; Magalhães Ribeiro, Simón, Fernández, Salgado, and Coria 2005).

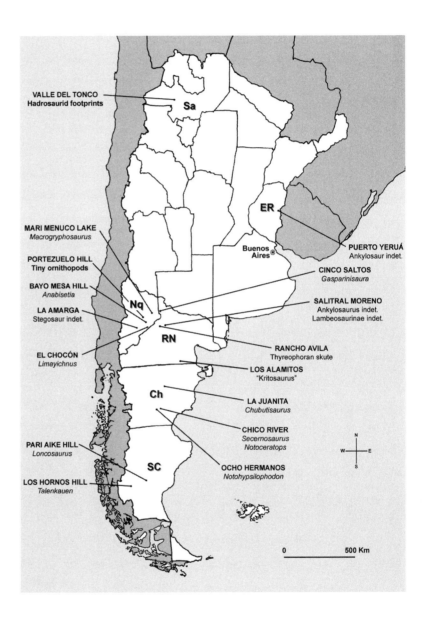

Fig. 7.1. Map of Argentina indicating Cretaceous fossil localities and a list of ornithischian genera discovered in each of them. The provinces of Argentina that yielded ornithischian remains are Salta (Sa), Entre Ríos (ER), Neuquén (Nq), Río Negro (RN), Chubut (Ch), and Santa Cruz (SC).

CRETACEOUS ORNITHISCHIANS 7

When Friedrich von Huene visited Argentina in 1923, only one fragmentary Cretaceous ornithischian was known: *Notoceratops bonarelli*, found by Augusto Tapia in 1918 and interpreted by him as a member of Ceratopsia. Von Huene had an opportunity to examine this and other isolated bones collected at different places in Patagonia that he recognized as ornithischian dinosaurs. Aside from corroborating *Notoceratops* as a ceratopsian, von Huene identified a small coracoid from Cinco Saltos (Río Negro Province) as corresponding to this subgroup of horned herbivores. He also coined the name *Loricosaurus scutatus* for the first South American ankylosaur, based on several dermal plates discovered in the highly fossiliferous deposits of Cinco Saltos. In addition, von Huene assigned other isolated pieces to this group of armored ornithischians, including a flattened spike, two metapodials, and a fragmentary sacrum. The German paleontologist also had the chance to describe (von Huene 1929b, 1934a) some specimens collected in Cretaceous beds from Uruguay consisting of isolated teeth that he identified as "psittacosaurids" or "camptosaurines" as well as some ornithischian footprints in Cretaceous beds from Brazil (von Huene 1931b).

This small set of poorly known ornithischians remained unchanged until 1964, when Rodolfo Casamiquela described the first hadrosaurids from the southern hemisphere. The discovery was made as early as 1949 in rocks of the Coli Toro Formation, exposed at the center of Río Negro Province in northern Patagonia, and although no generic or specific names were given to this new animal, the remains were enough to demonstrate that duck-billed dinosaurs colonized South America at the end of the Cretaceous period (Casamiquela 1964b). Further documentation of these animals was later provided by the discovery at higher latitudes in Patagonia (Chubut Province) of the first named southern hadrosaurid, *Secernosaurus koerneri*, described by Michael Brett-Surman in 1979 (the holotype specimen of this taxon was collected in 1923 by Elmer Riggs's expedition from Chicago's Field Museum of Natural History).

This was the state of knowledge about South American ornithischians at the beginning of the 1980s. During this decade new discoveries and studies conducted by José Bonaparte and Jaime Powell amplified the diversity of Cretaceous dinosaurs in general and modified previous ideas about ornithischian diversification in the southern hemisphere. In particular, the discovery of the armored titanosaurian *Saltasaurus loricatus* (Bonaparte and Powell 1980; Powell 1980) made possible a reinterpretation of the dermal scutes of the purported ankylosaurian *Loricosaurus scutatus* as corresponding to titanosaurid sauropods, so the name for the only armored ornithischian from South America became a nomen nudum. In addition, a better knowledge of Patagonian hadrosaurids was made possible by the discovery of nearly 20 specimens of the new taxon *Kritosaurus australis*, constituting the most significant discovery from anatomical, biostratigraphical, and paleobiogeographical points of view. This find was one of the most important pieces of evidence that led José Bonaparte to hypothesize that a great biotic interchange occurred between the Americas at the end of the Mesozoic (Bonaparte 1984b,c; Bonaparte Franchi, J. Powell, and E. Sepúlveda 1984). Supporting this conclusion was the discovery by Jaime Powell of an incomplete albeit wonderful specimen of a possible lambeosaurine hadrosaurid in Salitral Moreno, Río Negro, an animal that may have also colonized South America as part of such a faunal interchange.

The skeletal evidence of Late Cretaceous ornithopods amassed during the 1980s was complemented by a rich footprint record from Lower Cretaceous beds in Brazil and Upper Cretaceous strata in Bolivia. This ichnological evidence was extensively studied by the renowned paleoichnologist Giuseppe Leonardi, who demonstrated that ornithischians were present and diverse in South America during most of the Cretaceous, prior to the Campanian–Maastrichtian arrival of hadrosaurids (see Leonardi 1989 for a thorough account of ichnological evidence from this continent).

Since the beginning of the 1990s, a wealth of discoveries of new and more-complete ornithischian dinosaurs has come to light, mainly from Patagonia, that constitutes the bony counterpart of the diverse footprint record. The list of Cretaceous ornithischians from South America has increased rapidly thanks to discoveries of Early Cretaceous stegosaurs (Bonaparte 1996b), Maastrichtian ankylosaurs (Coria and Salgado 2001), and basal ornithopods (Calvo, Porfiri, and Novas 2005; Coria and Salgado 1996a; Coria and Calvo 2002; R. D. Martínez 1998b; Novas 1997d; Novas, Cambiaso, and Ambrosio 2004), most of them older than the Maastrichtian and consequently earlier than the late arrival of hadrosaurids at the end of the Mesozoic.

At present, the fossil record of South American ornithischians includes teeth and skeletal remains of stegosaurs, ankylosaurs, putative ceratopsians, and ornithopods of different affiliation. This record, mainly coming from the Upper Cretaceous of Patagonia, is enriched by an abundant and geographically widespread ichnological record that includes ornithopodan and ankylosaurian footprints documented in several places in Brazil, Bolivia, and Argentina (Fig. 7.1). So far, no osteological remains of ornithischians have been reported from South American countries other than Argentina. In this regard, an isolated and fragmentary skeletal piece found in the Aptian Santana Formation that was originally interpreted as an ornithischian ischium (Leonardi and Borgomanero 1981) has been recently re-identified as a thoracic rib of a spinosaurid theropod (Machado and Kellner 2007).

The new discoveries have changed our perception about the evolutionary history of this group of plant-eating dinosaurs. This evidence, albeit patchy, hints that a hidden diversity of ornithischian clades is still waiting to be discovered. Current knowledge suggests that these reptiles played an important role as small- to medium-sized dinosaurian herbivores in South American Cretaceous communities. However, it is still true that in Gondwana, ornithischians were less abundant and diverse than sauropod dinosaurs (in particular titanosaurs), in sharp contrast with dinosaurs of the Laurasian Cretaceous, whose record shows that ornithischian dinosaurs were abundant and highly diversified (especially hadrosaurids and ceratopsians; Bonaparte 1986d, 1996b; Bonaparte and Kielan-Jawarowska 1987).

Ornithopods are the best represented ornithischians in the Cretaceous of South America. They include small basal iguanodontians (e.g., the Cenomanian *Notohypsilophodon comodorensis* and the Santonian *Gasparinisaura cincosaltensis*; R. D. Martínez 1998b; Coria and Salgado 1996a) as well as larger forms such as the Cenomanian *Anabisetia saldiviai* (Coria and Calvo 2002), the Turonian–Coniacian *Macrogryphosaurus gondwanicus* (Calvo, Porfiri, and Novas 2005, 2008), and the presumably Maastrichtian *Talenkauen santacrucensis* (Novas, Cambiaso, and Ambrosio 2004). Hadrosaurids of large size are known from Campanian–Maastrichtian beds, where they are represented by hadrosaurines and lambeosaurines, the bones of which have been found at different localities in Argentina (e.g., Mendoza, La Pampa, Río Negro, and Chubut Provinces; Apesteguía and Cambiaso 1999; González Riga and Casadío 2000; Martinelli and Forasiepi 2004).

Following current phylogenetic hypotheses (e.g., Weishampel 2004), two main lineages of genasaurian ornithischians are recognized (Fig. 7.2): Thyreophora (mainly including ankylosaurs and stegosaurs) and Cerapoda (including ornithopods, ceratopsians, and pachycephalosaurs). To date, no remains of pachycephalosaurians have been recorded in the southern hemisphere, and it must be remembered that the purported pachycephalosaur *Majungasaurus atopus* (Sues and Taquet 1979) from the Late Cretaceous of Madagascar has been reassigned to abelisaurid theropods (Sampson et al. 1998).

In contrast with the amount of information available for South American theropods and sauropods, ornithischians currently known from this continent are numerically rarer and their phylogenetic relationships are far from settled. For this reason, this chapter does not include as many details about the anatomy and functions of these dinosaurs as it does for Cretaceous sauropods and theropods.

Stegosauria

Stegosaurs are medium-sized to large herbivorous dinosaurs with proportionally small heads, short forelimbs, and columnar hindlimbs. They have parasagittal rows of plates and spikes along their strongly curved backs and tails (Galton and Upchurch 2004b). They diversified during

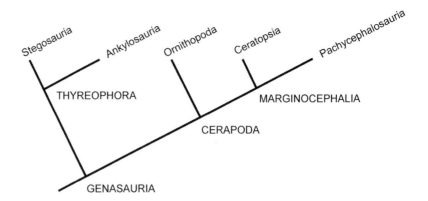

Fig. 7.2. Phylogenetic relationships of the main clades of ornithischians considered in the text.

the Jurassic, especially at the end of the period, but diminished in abundance and diversity during the Cretaceous. Stegosaurs were widely distributed in Gondwana, as documented by hundreds of bones of the African *Kentrosaurus aethiopicus* from the Upper Jurassic Tendaguru Formation (Tanzania); by *Paranthodon africanus* from the Early Cretaceous Kirkwood Formation (South Africa); and by *Dravidosaurus blanfordi* from Coniacian beds of India (Galton and Upchurch 2004b).

In South America, the record of stegosaurians consists of fragmentary but conclusive evidence from the Barremian–Early Aptian La Amarga Formation in Neuquén Province of Argentina (Fig. 7.3). This small stegosaur, similar in size to *Kentrosaurus* (i.e., less than 5 meters long), is represented by three cervicals and some isolated scutes described by Bonaparte in 1996. The available neck vertebrae of this Patagonian stegosaur resemble the cranial cervicals of *Stegosaurus* (Ostrom and McIntosh 1966) in that they are amphiplatyan with polygonal-shaped articular surfaces (Fig. 7.3A–D). Laterally the centra have a pair of elliptical depressions, and the ventral one of these depressions is deeper than the dorsal one. The ventral surface of the centrum is flat, a feature that also applies to one of the most posterior cervicals. The neural canal is quite wide, resembling the widening of the neural canals of other stegosaurs (e.g., *Stegosaurus*). The diapophyses, located high on the neural arch, are laterally projected. The parapophyses, in contrast, occupy the craniodorsal corner of centrum. As Bonaparte (1996b) noted, the cervicals of the La Amarga stegosaur are craniocaudally shorter than those of the African *Kentrosaurus*. Some of the scutes are conical and small (4.5 centimeters in diameter). Two other incomplete ossifications are flattened and bigger; the largest diameter is nine centimeters (Fig. 7.3E and F). Both kinds of scutes have an excavated ventral surface, but the flattened scutes have a net of well-marked grooves (canaliculi).

Ankylosauria

Ankylosaurs constitute a thyreophoran subgroup characterized by flat and wide skulls that are densely decorated with dermal ossifications; an armored, flattened body; and a quadrupedal stance (e.g., Carpenter 2001;

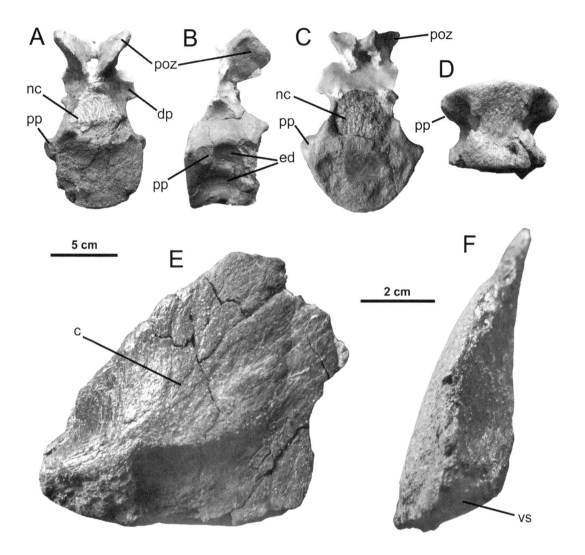

Fig. 7.3. Stegosauria indet. from the La Amarga Formation: (A–D) Cervical vertebra in (A) posterior, (B) lateral, (C) anterior, and (D) ventral views. (E, F) Plate-like dermal ossification in (E) side and (F) cranial? views. *Photographs by Fernando Novas.*

Vickaryous, Maryańka, and Weishampel 2004). Two main ankylosaurian clades are usually distinguished—Nodosauridae and Ankylosauridae—but recent reviews (Carpenter 2001) also recognize a third family, the Polacanthidae, construed as more closely related to Ankylosauridae than to Nodosauridae. Nodosaurids are recorded from the Aptian through the Maastrichtian and attained a wide geographic distribution, including most of the northern hemisphere, South America, Antarctica, and Australia. In contrast, ankylosaurids (mainly recorded in the Upper Cretaceous) and polacanthids (documented from the Mid-Jurassic to the Late Early Cretaceous) are known only in North America, Europe, and Asia.

The fossil record of southern ankylosaurs is still scarce and is restricted to *Minmi paravertebra* from the Lower Cretaceous (Aptian) of Australia (Molnar 1980a, 1996b) and *Antarctopelta oliveroi* from the Maastrichtian of the northeastern Antarctic peninsula (Gasparini and Olivero 1989; Gasparini, Olivero, Scasso, and Rinaldi 1987; Gasparini, Pereda-Suberbiola, and Molnar 1996; Salgado and Gasparini 2006) and some isolated bones of an unnamed ankylosaur recovered from Maastrichtian beds of northern Patagonia (Coria and Salgado 2001). Trackways from Bolivia and Brazil may belong to this clade of thyreophoran dinosaurs (von Huene 1931b; Leonardi 1984, 1994; Thulborn 1990; Meyer, Hippler, and Lockley 2001; McCrea, Lockley, and Meyer 2001).

The phylogenetic relationships of the available Gondwanan ankylosaurs are far from clear. For example, *Minmi* (by far the best-known ankylosaur from the southern continents) has been considered by some authors (e.g., Vickaryous, Maryańka, and Weishampel 2004) to be the oldest and most basal member of Ankylosauridae, while others (e.g., Carpenter 2001) put *Minmi* as the sister taxon of the remaining ankylosaurs (i.e., Nodosauridae, Polacanthidae, and Ankylosauridae). The forms from Antarctica and Patagonia do not have derived features of Polacanthidae or Ankylosauridae in their dermal armor and postcranial skeletons, suggesting possible membership in Nodosauridae. In sum, the available fossil record from the southern hemisphere suggests that nodosaurid ankylosaurs were already distributed across Gondwana from at least Albian times.

The first clear evidence supporting the presence of ankylosaurs in Patagonia was reported by Rodolfo Coria and Leonardo Salgado, who discovered isolated bones of these animals in Salitral Moreno, 25 kilometers south of General Roca, Río Negro Province, from beds of the Campanian–Maastrichtian Allen Formation (Coria 1994a; Salgado and Coria 1996; Coria and Salgado 2001). The materials include a tooth, some dorsal and caudal vertebrae, a femur, and several scutes collected from an area no larger than 50 square meters that corresponds to a fluvial paleochannel that also yielded hadrosaur and titanosaurid remains (Fig. 7.4A–I). In spite of this taphonomic association, Coria and Salgado (2001) exercised caution and did not refer these ankylosaurian bones to a single individual. However, all the available materials congruently correspond to an ankylosaur of small size, so their association into a single individual is not improbable. The tooth has a crown four millimeters tall and a root ten millimeters long. The crown has cingula on both the lingual and labial surfaces that are asymmetrically placed: on one side of the crown the cingulum is straight, but on the opposite side the cingulum arches toward the crown apex. The dorsal vertebrae have high neural arches with high and narrow neural canals (Fig. 7.4C and D). The prezygapophyses are ventrally fused, forming a U-shaped structure enclosing a broad prespinal basin, as characteristically occurs among ankylosaurs. The transverse processes of dorsal vertebrae are directed upward, as is usual among these dinosaurs. The caudal centra (Fig. 7.4E) are craniocaudally short, and some of them are deeply grooved ventrally. The transverse processes of the caudal vertebrae are elongate and laterally projected. The femur is slightly more than 25 centimeters long. It is proportionally short, stout, and transversely expanded (Fig. 7.4A and B). The femoral head is large, rounded, and dorsomedially projected, suggesting a closed acetabulum, which apomorphically characterizes the pelvic girdle of ankylosaurs.

Both the greater and anterior trochanters form a blunt prominence, as in most nodosaurids. These trochanters are positioned low in relation to the femoral head, as also occurs among nodosaurids, and the fourth trochanter is represented by a low ridge located proximally on the caudolateral side of the femur. Distally the femur ends in two well-developed rounded condyles. Some dermal plates are conical and roughly elliptical and have a blunt spike, while others are more massive and have acute and prominent spikes (Fig. 7.4F–I).

As Coria and Salgado (2001) note, the ankylosaur material from Salitral Moreno does not have derived features diagnostic of Nodosauridae, with the probable exception of the asymmetrically arranged cingula on the tooth crown. However, available postcranial elements show features that are congruently found among these ankylosaurs, including the proximally placed fourth trochanter (a plesiomorphic condition retained by nodosaurids) and small body size (a feature that is more congruent with nodosaurids than with ankylosaurids, which are bigger). Following Coria and Salgado (2001), this Patagonian ankylosaur is tentatively referred as to Nodosauridae.

Friedrich von Huene (1929a) described a portion of what appears to be a flattened spike (Fig. 7.4J–L), which he took as ankylosaurian based on similarities with the spikes of the Barremian *Hoplitosaurus* from North America. The plate is 8 centimeters deep (as preserved) and 2.5 centimeters wide; it is triangular in side view and has sharp margins. The spike in question was found at the Rancho Avila fossil locality (not far from Cerro Policía, Río Negro Province) from unknown beds of the Upper Cretaceous Neuquén Group (Powell 2003). The material is currently lost in the collections of the La Plata Museum, but based on the only available source of information (i.e., the description and illustrations von Huene offered in 1929a), the material may correspond to a dermal ossification of a thyreophoran dinosaur, pending more information that would enable us to refer it either to Ankylosauria or to Stegosauria.

An isolated conical scute referred to an indeterminate ankylosaur (Fig. 7.4M and N) was recovered in Entre Ríos Province, northeastern Argentina, from beds of the Puerto Yeruá Formation (Upper Cretaceous), which is lithostratigraphically equivalent to the Guichón Formation of Uruguay (De Valais, Apesteguía, and Udrizar Sauthier 2003). It consists of a small ossification (its major diameter is 5.5 centimeters) with a prominent and sharp dorsal keel and a ventral concave elliptical surface. The describers of the fossil indicate that the scute is almost identical in shape and size to one of the osteoderms collected in Salitral Moreno by Coria and Salgado (2001).

Von Huene (1929a) also cited a partial sacrum that he interpreted as corresponding to an ankylosaurid dinosaur. The sacrum was found at the same fossil locality that yielded the titanosaurian *Antarctosaurus wichmannianus* (15 kilometers southwest of the city of General Roca, Río Negro), and for this reason its correspondence with this sauropod is not improbable (it should be noted that no complete titanosaurid sacra

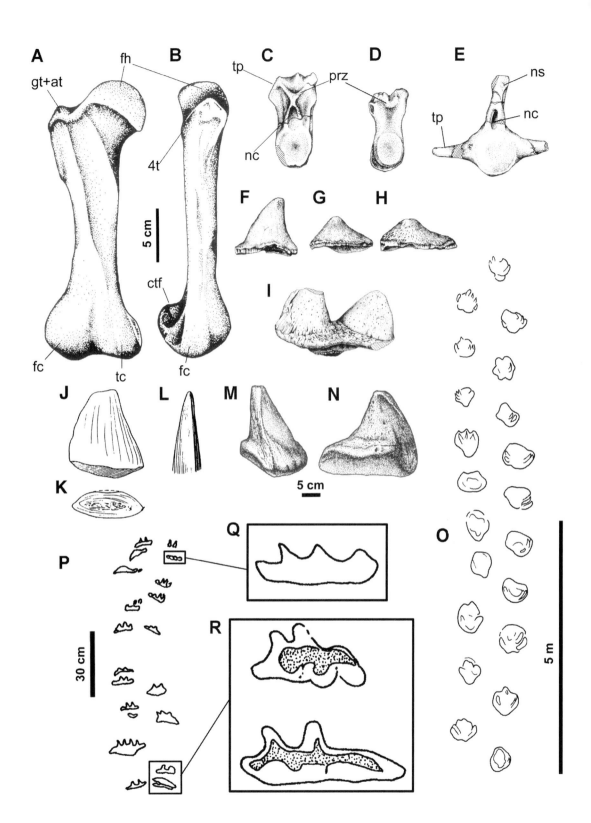

articulated with ilia were available to von Huene for study). The sacrum is represented by some transverse processes that are dorsoventrally deep and distally fused, defining an extensive lateral surface for articulation with the ilium. This fragmentary sacrum, which corresponds to a large animal (matching fairly well with the size of *Antarctosaurus*), requires a comparative survey not just with the sacra of ankylosaurs but also with the sacra of titanosaurids (e.g., *Saltasaurus*, *Epachthosaurus*), which also have similar transverse processes.

The bone record of ankylosaurs is complemented by trackways found in Upper Cretaceous beds of Bolivia that may belong to Ankylosauria (Leonardi 1984; McCrea, Lockley, and Meyer 2001). The footprints were named *Ligabueichnium bolivianum* by Giuseppe Leonardi in 1984, who saw the possible ankylosaurian nature of the material (Fig. 7.4O). The footprints come from the Campanian Toro Toro Formation, which is exposed in Huayllas Hill, close to Toro Toro village, north of the Department of Potosí (Leonardi 1994; Meyer, Hippler, and Lockley 2001). Each footprint of *Ligabueichnium* is tetradactyl, measures roughly 50–60 centimeters long, and has a pace length of 1.20–1.70 meters. Another set of probable ankylosaurian footprints (Fig. 7.4P–R) was discovered at the Cal Orcko fossil site, near the town of Sucre but from the overlying El Molino Formation (Maastrichtian; Meyer, Hippler, and Lockley 2001). Notably, the trackways from both Toro Toro and El Molino correspond to large animals, presumably 6–7 meters long (the size of *Euoplocephalus*, for example), thus exceeding the size of the remaining ankylosaurs recorded in Gondwana (e.g., *Minmi* is approximately 3 meters long, while the estimated length for the ankylosaurs discovered in Antarctica and Salitral Moreno is about 2 meters). Although these trackways from Bolivia are congruent in morphology with ankylosaurian footprints found elsewhere (e.g., *Tetrapodosaurus*; McCrea, Lockley, and Meyer 2001), it is important to note that the authors who studied these trace fossils (Leonardi 1984; McCrea, Lockley, and Meyer 2001) were not completely confident about their ankylosaurian nature, suggesting the possibility (albeit more remote) that such footprints could belong to ceratopsians. If *Ligabueichnium* and the trackways from El Molino do turn out to be ankylosaurian, this means that very large members of the group inhabited South America along with smaller members of the clade.

Unnamed trackways from the Lower Cretaceous of Brazil, probably from the Piranhas Formation (Rio do Peixe Group), were described by von Huene in 1931b. Each of these footprints is about 33 centimeters long and was tentatively identified by von Huene as ankylosaurian, though Haubold (1971) suggested that the trackmaker might have been a ceratopsian. More recently, Thulborn (1990) indicated that these trackways possibly constitute pseudo-bipedal tracks (i.e., manus prints that are overtrodden and obliterated by pes prints) produced by an ankylosaur.

Fig. 7.4. Ankylosaurs from South America. (A–I) Ankylosauria indet. from the Maastrichtian Allen Formation in northwest Patagonia: (A, B) Right femur in (A) cranial and (B) lateral views. (C, D) Dorsal vertebrae in cranial view. (E) Caudal vertebrae in cranial view. (F–I) Dermal ossifications. (J–L) Flattened spike found in Rancho Avila, northwest Patagonia, in (J) side, (L) front?, and (K) basal views. (M, N) Dermal scute from Puerto Yeruá, northeastern Argentina, in (M) front? and (N) side views. (O) Footprints of *Ligabueichnium bolivianum*, from the Campanian Toro Toro Formation, Bolivia. (P–R) Ankylosaur trackway from the El Molino Formation, Cal Orko, Bolivia: (P) Complete trackway. (Q, R) Detail of manus and pes imprints. *(A–I) From Coria and Salgado (2001). Reprinted from* The Armored Dinosaurs. *Ed. Kenneth Carpenter. Page 162 (fig. 8.2) and page 163 (fig. 8.3). With permission of Indiana University Press: Bloomington. (J–L) From von Huene (1929a). (M, N) From De Valais, Apestiguía, and Sauthier (2003). Reprinted with permission of Ameghiniana. (O–R) From McCrea, Lockley, and Meyer (2001). Reprinted from* The Armored Dinosaurs. *Editor: Kenneth Carpenter. Page 442 (fig. 20.26) and page 443 (fig. 20.27). With permission of Indiana University Press: Bloomington.*

Ornithopoda

Ornithopoda is defined as cerapodans that are closer to *Edmontosaurus* than to *Triceratops* (Norman, Sues, Witmer, and Coria 2004). Ornithopods are characterized by a ventral offset of the premaxillary occlusal margin relative to the maxillary tooth row, a jaw articulation that is offset ventral to the maxillary tooth row, and a premaxilla that contacts the lacrimal on the external surface of the snout (Norman, Sues, Witmer, and Coria 2004). The most distinctive ornithopod clade is Iguanodontia, which includes Iguanodontidae, Hadrosauridae, and a wide arrange of Jurassic and Cretaceous taxa. Current phylogenetic analyses (e.g., Norman, Sues, Witmer, and Coria 2004) do not support a monophyletic Hypsilophodontidae, into which most small ornithopods more primitive than Iguanodontia previously were usually lumped. The fossil record of South American ornithopods is comprised of a diversity of basal iguanodontians and hadrosaurids.

Documentation of Cretaceous basal iguanodontians mainly comes from beds of Cenomanian through Campanian ages. Nevertheless, recent discoveries in the Maastrichtian Allen and Pari Aike formations (Coria, Salgado, Currie, Paulina Carabajal, and Arcucci 2004; Novas, Cambiaso, and Ambrosio 2004) demonstrate that basal iguanodontians survived until the end of the Mesozoic on this continent. Several trackways have been recovered from the Lower Cretaceous and Cenomanian of Brazil and the Cenomanian and Maastrichtian of Argentina that are referred to as "Ornithopoda"; some are referred to as "Hadrosauridae." South American basal ornithopods range from the size of a chicken up to the size of a Indian elephant, and ichnological evidence indicates that big iguanodontians (e.g., *Limayichnus*; Calvo 1991) were approximately 9 meters long.

Most of the currently recorded basal ornithopods from Gondwana (e.g., *Anabisetia*, *Valdosaurus*, *Lurdusaurus*, *Muttaburrasaurus*, *Kangnasaurus*, *Macrogryphosaurus*, and *Talenkauen*) can be placed within Iguanodontia (e.g., Calvo, Porfiri, and Novas 2005, 2006; Coria and Calvo 2002; Norman 2004; Novas, Cambiaso, and Ambrosio 2004; Cambiaso 2007). However, their phylogenetic relationships are far from settled (Fig. 7.5). For example, the Australian *Muttaburrasaurus* and the Patagonian *Gasparinisaura* are variously interpreted as basal iguanodontians that are more derived than *Tenontosaurus* (e.g., Coria and Salgado 1996a; Coria and Calvo 2002; Novas, Cambiaso, and Ambrosio 2004) or as basal non-iguanodontian ornithopods that, in the case of *Gasparinisaura*, may have diverged prior to the origin of *Thescelosaurus* (e.g., Norman, Sues, Witmer, and Coria 2004; Weishampel, Jianu, Csiki, and Norman 2003; Cambiaso 2007).

Several features support the notion that at least some of the ornithopods recorded in the southern continents formed part of an iguanodontid clade endemic to Gondwana (Cambiaso 2007). For example, the derived characteristics that are widely present among southern basal ornithopods include a greatly reduced deltopectoral crest on the humerus, a

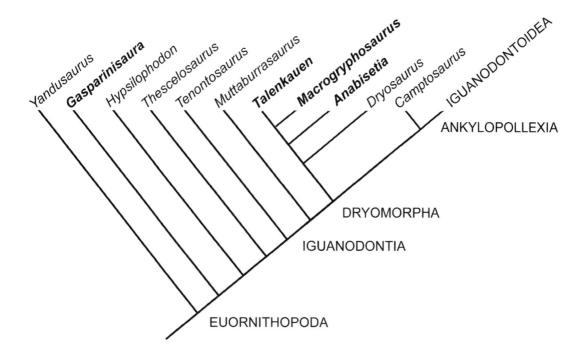

Fig. 7.5. Cladogram depicting the phylogenetic relationships of selected basal ornithopods. Genera recorded in South America are indicated in bold.

feature shared by *Talenkauen*, *Anabisetia* (Coria and Calvo 2002), and *Notohypsilophodon* (R. D. Martínez 1998b), as well as a transversely compressed second metatarsal, which is present in the Patagonian *Talenkauen*, *Anabisetia*, and *Gasparinisaura* (Salgado, Coria, and Heredia 1997); the South African *Kangnasaurus* (Cooper 1985); and an unnamed Late Cretaceous iguanodontian from the Antarctic Peninsula (Novas, Cambiaso, and Ambrosio 2004; Cambiaso 2007). Also remarkable are the apomorphic similarities between the boat-shaped chevrons of *Gasparinisaura* and *Macrogryphosaurus* (Calvo, Porfiri, and Novas 2005, 2008). The tooth morphology of *Gasparinisaura*, *Talenkauen*, and *Anabisetia* is similar; all have a prominent primary ridge on both maxillary and dentary teeth and several strong secondary ridges. This kind of tooth is also documented in the Australian "hypsilophodontids" *Atlascopcosaurus* and *Quantasaurus* (Rich and Vickers-Rich 1989). Nevertheless, not all of the basal ornithopods documented in the southern landmasses have this feature; it is absent, for example, in *Muttaburrasaurus* (Bartholomai and Molnar 1981; Molnar 1996a).

At the least, available information supports the notion that the Patagonian ornithopods *Anabisetia*, *Talenkauen*, and *Macrogryphosaurus* are joined in a clade of southern iguanodontians (Cambiaso 2007). In this regard, Calvo and collaborators (Calvo, Porfiri, and Novas 2005, 2008) recognized derived features that are uniquely shared by *Talenkauen* and *Macrogryphosaurus*, leading these authors to coin the name Elasmaria (Greek: thin plate) for the group containing these two Patagonian taxa (Fig. 7.5).

Fig. 7.6. *Gasparinisaura cincosaltensis:* (A) Reconstructed skeleton. (B) Skull and jaws in lateral view. (C) Detail of maxillary teeth. (D) Distal caudal vertebrae and respective haemal arches in right lateral view. *(A) Illustration by J. González. (B, D) Courtesy of Andrea Cambiaso. (C) Redrawn from Coria and Salgado (1996a).*

GASPARINISAURA CINCOSALTENSIS

This interesting animal was described in detail by Rodolfo Coria and Leonardo Salgado in 1996 and constituted the first find of a basal ornithopod in South America. The discovery of this dinosaur stimulated the debate about phylogenetic relationships among basal ornithopods (e.g., Coria and Salgado 1996a; Salgado, Coria, and Heredia 1997; Coria and Calvo 2002; Norman, Sues, Witmer, and Coria 2004). The holotype specimen of *Gasparinisaura cincosaltensis* was recovered from beds of the Anacleto Formation, which is exposed at the town of Cinco Saltos in Río Negro Province (Figs. 4.7 and 7.1). It includes an almost-complete skull with atlas and axis in articulation, the sacrum, the pectoral and pelvic girdles, both humeri, and almost-complete hindlimbs (Fig. 7.6A). Moreover, a dozen other specimens of *Gasparinisaura* offer valuable additional information about the anatomy of this small and fleet-footed ornithischian. The length of one of the specimens of *Gasparinisaura* is approximately 1.50 meters (Coria and Salgado 1996a; Salgado, Coria, and Heredia 1997).

Gasparinisaura is diagnosed by modifications of the jugal bone, which is rostrally wedged between the maxilla and lacrimal, and an ascending process that is rostrocaudally wide and in contact with the ventral postorbital process. *Gasparinisaura* is also unique among ornithopods in the shape of the infratemporal fenestra, the dorsal portion of which is positioned caudal to the mandibular articulation and the ventral margin of which forms a slot bordered by the quadratojugal (Fig. 7.6B). As in other basal ornithopods, *Gasparinisaura* has a short and deep skull with a proportionally large orbit and antorbital fossa (Norman, Sues, Witmer, and Coria 2004). The teeth of *Gasparinisaura* resemble other basal ornithopods in that they have an asymmetrical crown, but they look highly derived in the development of a prominent vertical ridge; they resemble iguanodontians in this feature (Coria and Salgado 1996a; Norman, Sues, Witmer, and Coria 2004; Fig. 7.6C).

The shape of the haemal arches of the distal caudal vertebrae of *Gasparinisaura* are curious: they are subtriangular in side view and the ventral margin is proximodistally expanded (Fig. 7.6D), a highly derived condition that is absent in other ornithopods, with the exception of the Patagonian *Macrogryphosaurus* (Calvo, Porfiri, and Novas 2005, 2008). This modification of distal caudal chevrons is accompanied in *Gasparinisaura* by a reduction of transverse processes as well as by a certain degree of elongation of the prezygapophyses. This set of modifications calls to mind the transition point described for derived coelurosaurian theropods (e.g., ornithomimids, tyrannosaurids), and suggests that *Gasparinisaura* convergently attained advanced modifications in caudal musculature and tail function with these theropods.

The scapular girdle looks conservative in the distally expanded scapular blade, the caudal orientation of glenoid cavity, and the elliptical coracoid. The humerus is slender, with a prominent triangular

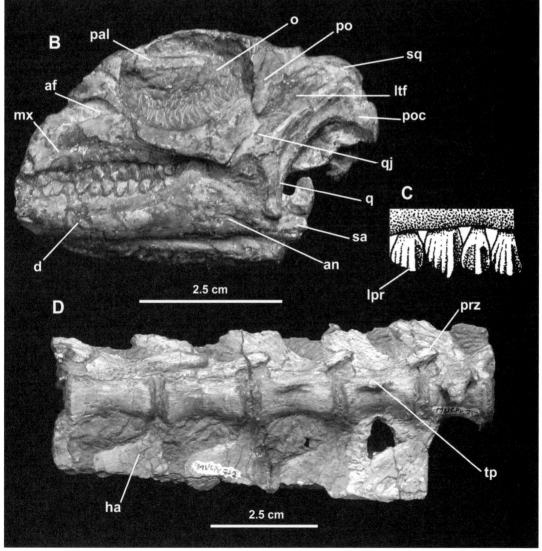

deltopectoral crest. The ilia of *Gasparinisaura* have a laterally expanded brevis shelf. The ilia contact a sacrum composed of six fused vertebrae. The pubis is a slender bone, with an elongate, rod-like prepubic process (Fig. 7.7A), contrasting with the transversely flattened prepubic process of more derived iguanodontians (e.g., *Anabisetia*; Fig. 7.7B). Both the greater and anterior trochanters of the proximal femur are fused, a condition that Coria and Salgado documented among all juvenile and adult individuals of *Gasparinisaura*. The distal tarsals are apparently fused into a single piece, unlike the paired condition seen in most basal ornithopods (Norman, Sues, Witmer, and Coria 2004). The foot of *Gasparinisaura* is long and slender and contrasts with feet of other basal ornithopods in the atrophy of metatarsal I, which has lost its phalanges. Proximally, metatarsal II is transversely compressed, representing less than 15 percent of the transverse width of the three metatarsals (Coria and Salgado 1996a). This reduction of metatarsal II of *Gasparinisaura* and other South American ornithopods (e.g., *Talenkauen*) parallels a similar trend in the "eurymetatarsal" feet of abelisauroid theropods (see also R. Molnar, quoted in Salgado, Coria, and Heredia 1997).

In their original description, Coria and Salgado (1996a) interpreted *Gasparinisaura* as a basal iguanodontian that shared several synapomorphies with the remaining members of this ornithopod clade (e.g., Dryosauridae and Ankylopollexia), including a lateral primary ridge on the maxillary teeth, a wide brevis shelf on the ilium, and a metatarsal I that is reduced or absent. A similar opinion was offered by Coria and Calvo (2002) in their study of *Anabisetia*. However, recent reviews of the phylogenetic relationships of basal ornithopods (e.g., Weishampel, Jianu, Csiki, and Norman 2003; Norman, Sues, Witmer, and Coria 2004; Cambiaso 2007) put *Gasparinisaura* outside Iguanodontia. Although the anatomy of *Gasparinisaura* conforms well with that of *Hypsilophodon* and other ornithopods that are less derived than iguanodontians (Norman, Sues, Witmer, and Coria 2004), it is important to emphasize some of its derived characteristics (e.g., maxillary teeth with lateral primary ridges, boat-shaped mid-caudal chevrons, a proximally narrow metatarsal II, an ilium with a sigmoidal dorsal margin), which may have been convergently acquired with other southern ornithopods.

ANABISETIA SALDIVIAI

This basal iguanodontian is currently known from four partially articulated skeletons recovered at Bayo Mesa Hill, 30 kilometers south of Plaza Huincul, Neuquén Province, from rocks of the Turonian Lisandro Formation (Coria and Calvo 2002). *Anabisetia* is larger (2 meters long) and older than *Gasparinisaura*, albeit anatomically more derived than this taxon. *Anabisetia* is diagnosed by a caudoventrally oriented occipital condyle, a scapula with a strong acromial process, a flattened fifth metacarpal, an ilium with a preacetabular wing that is longer than 50 percent of the total ilium length, and a preacetabular wing that extends cranially

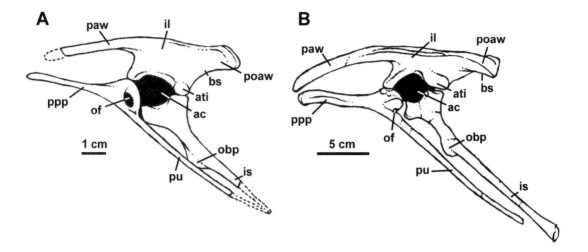

Fig. 7.7. Comparison of the pelvic girdles of two South American ornithopods in left lateral view: (A) *Gasparinisaura cincosaltensis*. (B) *Anabisetia saldiviai*. (A) Redrawn from Coria and Salgado (1996). (B) Redrawn from Coria and Calvo (2002).

at the same level as the prepubic process (Figs. 7.7B and 7.8A–C). The maxillary teeth are leaf-shaped and have a conspicuous primary ridge on the labial side and nine smaller denticles on the tooth apex, a morphology that is present in other ornithopods such as *Gasparinisaura*, *Anabisetia*, and *Talenkauen*.

The cervical centra are long and low with sharp ventral keels, a condition that may unite *Anabisetia* with the similarly long-necked *Talenkauen* and *Macrogryphosaurus* (Cambiaso 2007; see below). The proximal caudals have slender neural arches and transverse processes that are strongly upwardly directed. Unfortunately, no haemal arches have been documented yet for *Anabisetia* (Cambiaso 2007), so future discoveries will be needed to determine if this taxon also had boat-shaped chevrons like those described for *Gasparinisaura* and *Macrogryphosaurus*.

The scapula expands distally, as is usual among ornithopods, although the acromial process is better developed than in any other member of this clade. The humerus is slender and almost straight and has a low deltopectoral crest, as in *Notohypsilophodon* and *Talenkauen*. This is different from the stronger crest documented in *Gasparinasaura*. The ulna has a well-developed olecranon process, as in *Dryosaurus* and most ornithopods. Some elements of the manus of *Anabisetia* were disarticulated when they were found. The metacarpals are more slender than in *Hypsilophodon* and *Dryosaurus*. They are rounded in cross-section except for metacarpal V, which is flat and has sharp lateral borders. The manual phalanges are proportionally long and slender, suggesting a hand with grasping ability (Coria and Calvo 2002).

The ilium of *Anabisetia* is low and long and its dorsal margin is slightly sigmoidal, resembling that of *Gasparinisaura*, *Talenkauen*, *Macrogryphosaurus*, *Dryosaurus*, and more-derived iguanodontians (Coria and Salgado 1996a). The preacetabular wing is elongate, more than half of the whole length of the ilium. The preacetabular wing reaches the same cranial level as the prepubic process (Figs. 7.7B and 7.8B). *Anabisetia* is therefore the ornithopod with the longest preacetabular wing

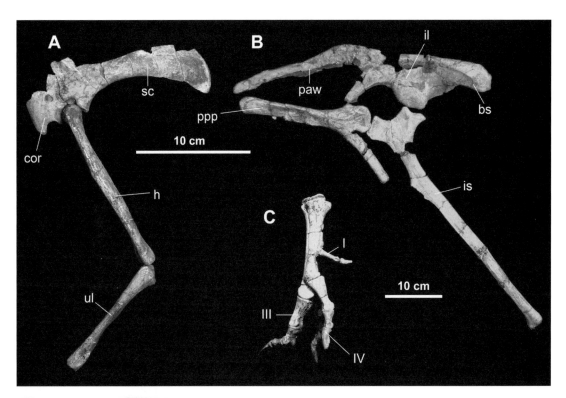
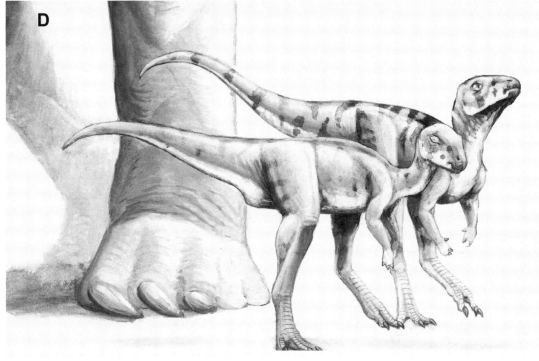

currently known. Posteriorly, the dorsal border of the ilium is transversely thick and its dorsal surface flares laterally, as in *Dryosaurus* (Galton 1981). In contrast, *Gasparinisaura* has an iliac blade with a thinner dorsal border, a condition that may constitute a plesiomorphic feature of this taxon. The postacetabular process of *Anabisetia* is low and long and has a wide brevis shelf. The pubis has a transversely compressed prepubic process that expands distally, as in most Dryomorpha. The ischium has two features that are present in the remaining Dryomorpha: a proximally placed obturator process and a small distal foot that is craniocaudally oriented.

The femur of *Anabisetia* has a well-developed anterior trochanter that is proximally at the same level as the greater trochanter. The distal end of the femur is cranially and caudally grooved, a condition that is known for most iguanodontians. The tibia is slender and longer than the femur. Almost-complete hindlimbs in perfect articulation were discovered, demonstrating that the foot had a reduced metatarsal I and its corresponding digit, which did not reach the ground during normal progression (Figs. 7.8C and 7.9). Metatarsals II to III are robust and articulated with stout toes. Metatarsal V, in contrast, is very reduced, as in most basal ornithopods. One of the available specimens of *Anabisetia* has partial fusion between the distal tibia-fibula and the astragalus-calcaneum as well as between the distal tarsal and the proximal metatarsals, a feature documented in a few other ornithopods (e.g., *Heterodontosaurus*).

Preliminary cladistic analysis performed by Coria and Calvo (2002) resulted in two equally parsimonious trees that placed *Anabisetia* either as a monophyletic group with *Gasparinisaura* or as the plesiomorphic sister taxon of Dryomorpha. Coria and Calvo explained that this ambiguity in the analysis relates to the limited cranial material available for *Anabisetia*. The strict consensus tree of their analysis produced an unsolved trichotomy composed of *Gasparinisaura*, *Anabisetia*, and Dryomorpha. *Anabisetia* is nested within Iguanodontia because it has a dentary with parallel dorsal and ventral borders, an ilium with a sigmoid dorsal edge, and a femur with a deep cranial intercondylar groove (Coria and Calvo 2002). *Anabisetia* shares with iguanodontians that are more derived than *Tenontosaurus* one primary ridge on each maxillary tooth, a broad brevis shelf, a transversely flattened prepubic process, an ischium with a craniocaudally oriented distal foot, and a reduced metatarsal I (Coria and Salgado 1996a). More recently, Andrea Cambiaso's (2007) studies of ornithopods from Argentina resulted in her recognition of an iguanodontian group composed by *Talenkauen* and *Anabisetia* based on derived features shared by these two Patagonian taxa (Fig. 7.5), mainly due to the morphology of the cervical vertebrae (e.g., epipophyses and an elongate and low centrum).

Fig. 7.8. *Anabisetia saldiviai:* (A) Left scapular girdle, humerus and ulna in lateral view. (B) Pelvic girdle in left lateral view. (C) Right foot in medial aspect. (D) Life restoration. *(A–C) Courtesy of Andrea Cambiaso. (D) Illustration by J. González.*

Fig. 7.9. Metatarsals of *Anabisetia* in plantar view. *Redrawn from Coria and Calvo 2002.*

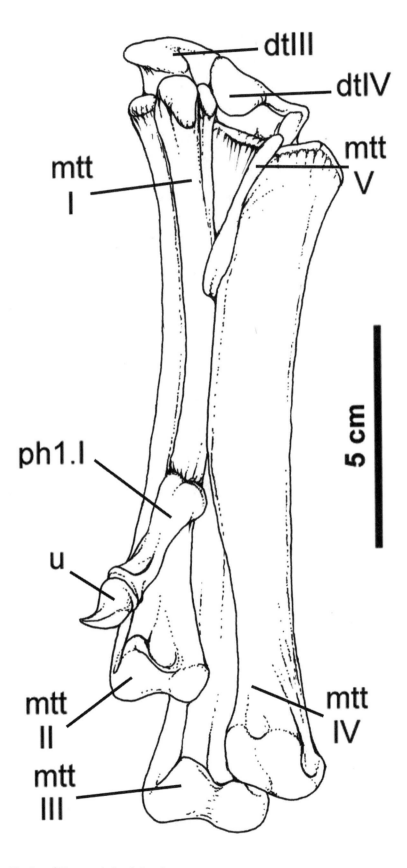

TALENKAUEN SANTACRUCENSIS

The holotype specimen of this creature was recovered from Late Cretaceous beds exposed in the southwestern part of Patagonia, a region of South America that has been poorly explored for dinosaur remains (Novas, Bellosi, and Ambrosio 2002; Novas, Cambiaso, and Ambrosio 2004; Cambiaso 2007; Fig. 4.14). The specimen is represented by a partially articulated skeleton that preserved the rostrum, the jaws and teeth, the precaudal vertebral column and ribs, the pectoral and pelvic girdles, and forelimb and hindlimb bones (Fig. 7.10A), which were found on the southern coast of Lake Viedma, Santa Cruz Province, Argentina. The skeleton comes from beds of the Maastrichtian Pari Aike Formation. *Talenkauen santacrucensis* is a basal iguanodontian distinguished by the odd presence of plate-like uncinate processes on both sides of the thorax, the functional meaning of which is not clear. In the original description of this animal (Novas, Cambiaso, and Ambrosio 2004) two autapomorphic characteristics of *Talenkauen* were recognized: well-developed epipophyses on cervical 3 and the plate-like uncinate processes on the rib cage. However, the new ornithopod *Macrogryphosaurus gondwanicus* from the Portezuelo Formation, which was discovered by Jorge Calvo and his assistants in Neuquén Province, also shows the same features, thus leaving open the question about which derived features are diagnostic of *Talenkauen*.

The holotypic specimen of *Talenkauen santacrucensis* measures no more than 4 meters long and represents one of the larger non-hadrosaurian ornithopods reported from South America. The head looks comparatively small compared to the size of the body; it is slightly larger than the head of the Jurassic *Dryosaurus*. Two empty alveoli indicate premaxillary teeth, a plesiomorphic characteristic for an iguanodontian. The worn maxillary teeth are rhomboid in side view and have a prominent primary ridge on the surface of the labial crown (Fig. 7.10B). The V-shaped predentary has, as in most iguanodontians, a pair of elongate tapering processes and a bilobate ventral process for articulation with the dentaries (Fig. 7.10C). The dentary is primitive and has anteriorly tapering dorsal and ventral margins. The presacral vertebral column of *Talenkauen* is composed of 9 cervicals and 16 dorsals, a normal count for a basal iguanodontian. A net of ossified tendons extends along both sides of the neural spines from the first dorsal though at least the first four preserved sacral vertebrae. The cervical vertebrae have moderately developed neural spines and slightly down-curved postzygapophyses, characteristics that are less derived than in dryomorph ornithopods. However, cervical 3 has prominent epipophyses, a feature that is infrequent among ornithischian dinosaurs. As a whole, the neck of *Talenkauen* looks elongate compared to those of *Hypsilophodon* and *Dryosaurus* (Sues and Norman 1990).

The humerus of *Talenkauen* (Fig. 7.10D) is slender and weakly expanded at its extremities and has a slightly marked deltopectoral crest, differing from the ornithopods in which this crest, albeit reduced, is

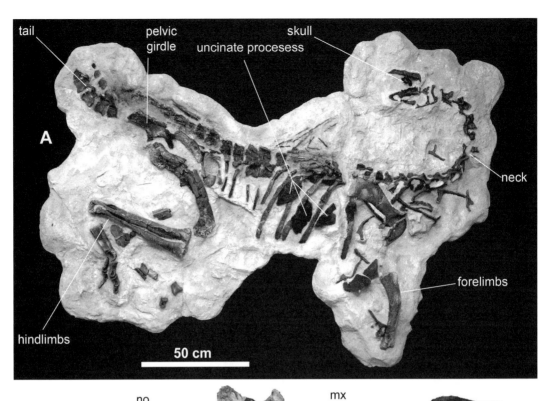
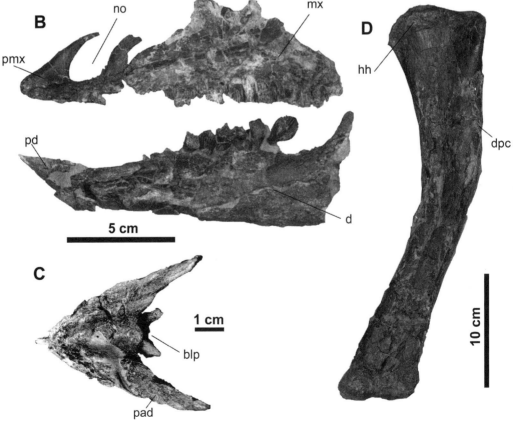

present (e.g., *Thescelosaurus, Dryosaurus, Camptosaurus, Iguanodon*). This peculiar morphology is shared with the purported hypsilophodontid *Notohypsilophodon* and the iguanodontian *Anabisetia*, both from the Cenomanian of Patagonia (Cambiaso 2007).

The ilium of *Talenkauen* is slender and dorsoventrally low and shorter than the femur. The dorsal margin of the ilium is sigmoid, as in other iguanodontians (Coria and Salgado 1996a; Coria and Calvo 2002). The prepubic process of the pubis is cranially elongate, ending at the level of the 13th dorsal vertebra. It is lateromedially flat and dorsoventrally deep, a condition shared with derived iguanodontians (Norman, Sues, Witmer, and Coria 2004; Coria 1999). The femur and tibia are massive and the tibia is shorter than the femur, as is usual for graviportal ornithopods. Metatarsal III of *Talenkauen* is a robust bone, whereas metatarsal II is notably narrow transversely, departing from the more robust proportions of metatarsal II seen in most ornithopods (e.g., *Tenontosaurus, Hypsilophodon, Dryosaurus, Camptosaurus, Iguanodon*). It is interesting to note that an eurymetatarsal foot developed among some Cretaceous iguanodontians from Gondwana, including *Talenkauen, Anabisetia, Kangnasaurus*, and *Gasparinisaura* (Fig. 7.9). Whether this condition is a derived feature that evolved in the common ancestor of all these southern ornithopods is a matter of discussion (Cambiaso 2007).

An outstanding feature of *Talenkauen* is the plate-like polygonal structures on both sides of the thorax (Fig. 7.10A). The plates are thin (no more than 3 millimeters thick) and dorsoventrally elongate (the major diameter is 180 millimeters). The external surfaces of the plates are smooth, lacking foramina or grooves, and no muscle scars are apparent. The plates are located along the mid-length of dorsal ribs 1 through 8, but they are not fused to the ribs. Like the Maastrichtian basal ornithopod *Thescelosaurus neglectus* from North America (Fischer, Russell, Stoskopf, Barrick, Hammer, and Kuzmitz 2000) and the newly described *Macrogryphosaurus gondwanicus* from Patagonia (Calvo, Porfiri, and Novas 2008), *Talenkauen* has polygonal plates on both sides of the thorax.

The plate-like structures of *Talenkauen* contrast with the thick and profusely ornamented dermal ossifications of armored dinosaurs (e.g., ankylosaurs, stegosaurs, titanosaurids). The absence of vascular grooves on their surfaces, their proximity to the caudal margins of ribs, and their serial arrangement on the thorax support their homology with the uncinate processes present in living and extinct diapsids (Novas, Cambiaso, and Ambrosio 2004).

Fig. 7.10. *Talenkauen santacrucensis:* (A) Reconstructed skeleton. (B) Premaxillary, maxillary predentary, and dentary in lateral aspect. (C) Predentary bone in dorsal view. (D) Left humerus in cranial view. *Photographs by Hernán Canutti.*

MACROGRYPHOSAURUS GONDWANICUS

The almost-complete and articulated skeleton of this new iguanodontian was recovered by Jorge Calvo and collaborators from beds of the Turonian–Coniacian Portezuelo Formation on the coast of Lake Mari Menuco, approximately 50 kilometers northwest of the city of Neuquén. The skeleton (which lacks the skull and limb bones) includes 8 cervicals,

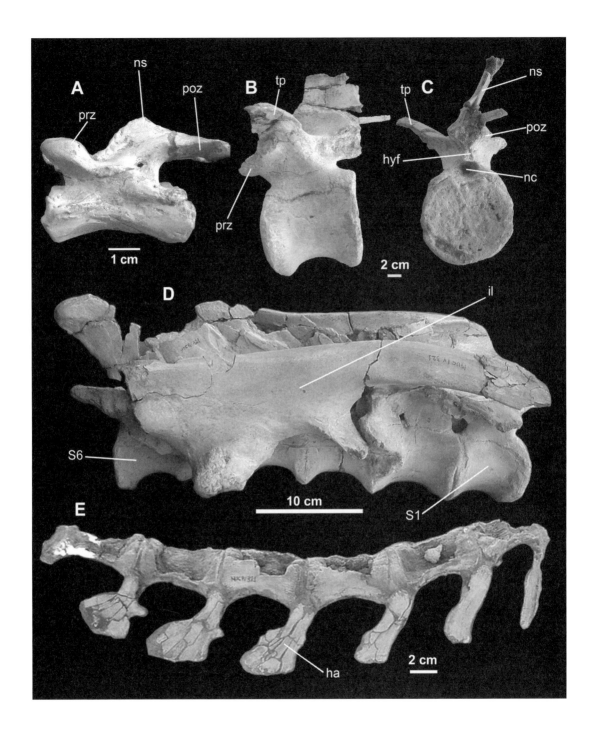

14 dorsals, 6 sacrals, 13 caudals, both ilia, the pubes and ischia, the sternal plates, some plate-like uncinate processes, ribs, and ossified tendons (Fig. 7.11). The specimen measures 6 meters long and is thus the largest basal iguanodontian known from bone remains. However, the holotype specimen of *Macrogryphosaurus* is not a full-grown individual, suggesting that adult sizes probably grew to more than 6 meters long.

As in *Talenkauen*, the neck of *Macrogryphosaurus* is elongate (Fig. 7.11A) and cervical 3 has epipophyses, features that are unusual for ornithopod dinosaurs (Calvo, Porfiri, and Novas 2005, 2008). There are 14 dorsal vertebrae, which is different from the 16 dorsals we see in *Talenkauen*. Thus, *Macrogryphosaurus* is a rare example of a form with a lower number of dorsal vertebrae than is typical of ornithopod dinosaurs. Another curious feature is that the last dorsal (dorsal 14; Fig. 7.11B and C) has a well-developed hyposphene that is ventral to the postzygapophyses. The tail vertebrae are distinguished by boat-shaped haemal arches on the distal caudals (Fig. 7.10E), a derived condition it shares with *Gasparinisaura* (it is important to remember that distal caudal vertebrae were not preserved in *Talenkauen* and *Anabisetia*). This set of derived features in different portions of the vertebral column of *Macrogryphosaurus* resembles features of saurischian dinosaurs (e.g., Gauthier 1986) and suggests a more flexible neck and stiffened dorsal and distal caudal regions.

Macrogryphosaurus has thoracic plates similar to those described for *Talenkauen* (Novas, Cambiaso, and Ambrosio 2004) and the hypsilophodontid *Thescelosaurus* from the Maastrichtian of North America (Fischer, Russell, Stoskopf, Barrick, Hammer, and Kuzmitz 2000). However, in the new Neuquenian dinosaur, the plates are subcircular and are present not only on the lateral sides of thorax but also on the chest, cranial to the sternal plates (Calvo, Porfiri, and Novas 2005, 2008).

Fig. 7.11. *Macrogryphosaurus gondwanicus:* (A) Cervical vertebra in left lateral view. (B, C) Last dorsal (dorsal 14) in (B) left lateral and (C) caudal views. (D) Right ilium and sacrum in lateral view. (E) Mid-caudal vertebrae and respective haemal arches in right lateral view. *Photographs courtesy of Jorge Calvo.*

NOTOHYPSILOPHODON COMODOROENSIS

The single available specimen of this small ornithopod (around 1.6 meters long) was found in levels of the Lower Member of the Cenomanian Bajo Barreal Formation, 28 kilometers northeast of the town of Buen Pasto in south-central Chubut Province (R. D. Martínez 1998b). The holotype of *Notohypsilophodon comodoroensis* includes isolated vertebrae and some forelimb and hindlimb bones (Fig. 7.12). The lack of cranial materials plus the fragmentary nature of the specimen conspire against recognizing diagnostic characteristics for *Notohypsilophodon*. Features originally cited by Rubén D. Martínez (1998b) as diagnostic of this taxon (e.g., an anteromedial knob on the proximal extreme of the tibia, a pronounced narrowing of the fibular shaft, an astragalus whose proximal surface occurs at two levels, a calcaneum with a pronounced posterodistal projection, ungual pedal phalanges with a flat ventral surface) are dubious at the least.

Notohypsilophodon was originally referred as to Hypsilophodontidae, but this referral was made on the basis of plesiomorphic features that are widely present among ornithopod dinosaurs (e.g., an anterior trochanter below the greater trochanter, the absence of an extensor groove on the distal femur, caudal flexure of humerus at the level of the deltopectoral crest). On the contrary, some derived characteristics (e.g., a transversely expanded medial condyle of the distal femur, a humerus with a slightly marked deltopectoral crest) suggest that *Notohypsilophodon* may be more

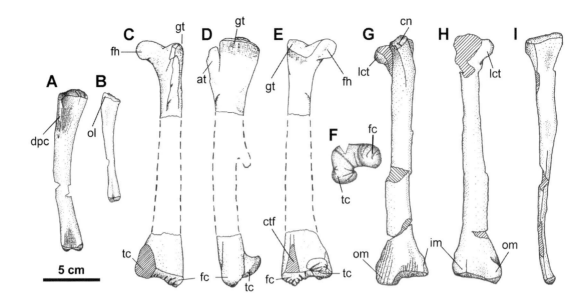

Fig. 7.12. *Notohypsilophodon comodoroensis*: (A) Right humerus in cranial view. (B) Left ulna in medial view. (C–F) Left femur (lacking most of its shaft, reconstructed in dotted lines) in (C) cranial, (D) lateral, (E) caudal, and (F) distal views. (G, H) Right tibia in (G) cranial and (H) caudal views. (I) Left fibula in lateral view. *From Martínez (1998b). Reprinted from* Acta Geológica Leopoldensia *21:119–135.*

closely related to other Patagonian ornithopods, such as *Anabisetia* and *Talenkauen* (Cambiaso 2007).

Although the anatomy and phylogeny of *Notohypsilophodon* require review, its discovery is relevant because it may constitute the oldest known skeletal remains of an ornithopod dinosaur recorded in Cretaceous beds in South America, supporting the view that these ornithischians were highly diverse prior to end of the Cretaceous (Novas 1997c; Coria 1999).

"LONCOSAURUS ARGENTINUS"

This taxon was recognized by Florentino Ameghino in 1899 on the basis of a fragmentary proximal femur (Fig. 7.13) and an associated isolated tooth crown (Ameghino 1921b). The materials were recovered from the Santonian–Campanian Mata Amarilla Formation, which is exposed near Pari Aike Hill, Santa Cruz Province, southern Patagonia. On the basis of the hollow bone and the serrated and curved tooth, Ameghino considered *Loncosaurus* to be a megalosaurid theropod, and later von Huene (1929a) interpreted this taxon as a probable ornithomimid. However, in 1980 Ralph Molnar recognized the ornithopod nature of *Loncosaurus* on the basis of the prominent fourth trochanter and the caudal projection of the greater trochanter. More recently, Coria and Salgado (1996b) published a review of this taxon. The femur of *Loncosaurus* has a pendant fourth trochanter and a well-marked basitrochanteric fossa (i.e., the site of insertion of M. caudifemoralis brevis), the shape of which closely matches that of other basal ornithopods (although, as Coria and Salgado correctly indicate, this type of fossa is present in most ornithischians, with the exception of large ornithopods such as *Iguanodon* and quadrupedal stegosaurs, ankylosaurs, and ceratopsians). However, no other features of *Loncosaurus* support a less inclusive assignment beyond Ornithopoda

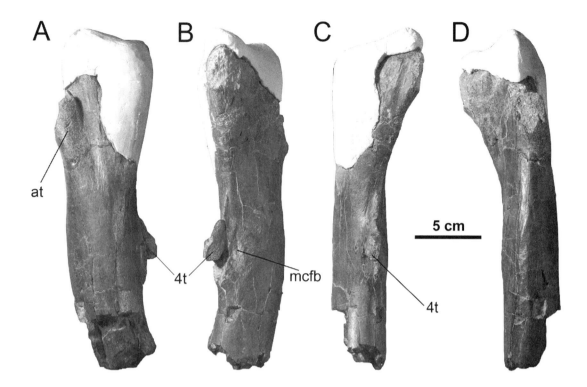

Fig. 7.13. *Loncosaurus argentinus*. Proximal portion of left femur in (A) lateral, (B) medial, (C) caudal, and (D) cranial views. *Photographs by Fernando Novas.*

(Coria 1999). The estimated size of the entire animal (roughly 3 meters long, based on a reconstructed femur length of 40 centimeters) would suggest iguanondontian affinities for *Loncosaurus*. Finally, because of the impossibility of recognizing autapomorphic features in this fragmentary specimen, Coria and Salgado (1996b) considered *Loncosaurus argentinus* to be a nomen vanum.

AN *OURANOSAURUS*-LIKE CAUDAL VERTEBRAE FROM NORTHEASTERN BRAZIL

Ouranosaurus nigeriensis is a large derived iguanodontian that has been documented in Aptian beds of Gadoufaua, Niger (Taquet 1976). The caudal vertebrae of this animal are notable for their transversely wide and cranially excavated neural spines and elongate, proximodorsally directed prezygapophyses. Caudal vertebrae of this peculiar shape were described by Ernst Stromer (1931) as belonging to the theropod "Spinosaurus B" from the Cenomanian of Egypt and were more recently referred by Dale Russell to the enigmatic theropod *Sigilmassasaurus brevicollis* (Russell 1996). However, a review of this topic by Novas, Dalla Vecchia, and Pais (2005) led to the conclusion that such mid-to-distal caudals are in fact ornithopod in nature.

Medeiros and Schultz (2001b, 2002; see also Medeiros, Freire, and Pereira 2007) described an isolated caudal vertebra from South America that is closely similar to those of the African iguanodontian *Ouranosaurus* as belonging to the theropod *Sigilmassasaurus brevicollis* (Fig. 7.14).

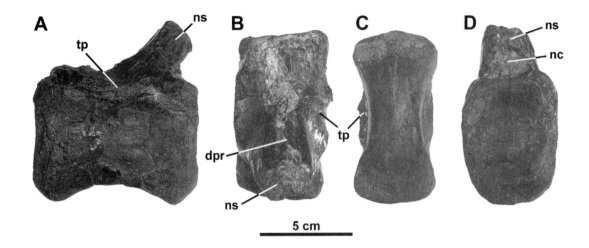

Fig. 7.14. *Ouranosaurus*-like caudal vertebrae from northeastern Brazil in (A) left lateral, (B) dorsal, (C) ventral, and (D) distal views. *Photographs courtesy of Manuel Medeiros.*

The material was collected in the Aptian–Albian Alcântara Formation in Ilha do Cajual in the state of Maranhão in northeastern Brazil. More recently, Avilla, Candeiro, and Abrantes (2003) described another distal caudal vertebra from the same area and formation that they recognized as closely similar to *Ouranosaurus*. As in the African ornithopod, the caudal centrum is amphiplatyan with flat lateral surfaces, a proximal articular surface that is hexagonal in outline, and a semicircular pit cranial to the neural spine.

TINY ORNITHOPODS FROM PATAGONIA

Remains of hen-sized ornithopods consisting of isolated vertebrae, femora, and tibiae were recovered by the present author and assistants at the Sierra del Portezuelo fossil site from different levels of the Portezuelo Formation in Neuquén Province. One of these bones (found in association with the maniraptoran theropod *Patagonykus puertai*) is a femur, the intact length of which is estimated to be 8 centimeters. It has a well-developed, craniocaudally convex greater trochanter with a markedly sharp external edge. The proximocaudal margin of the head is strongly sinuous, as occurs in most derived ornithopods. The shape of the proximal femur resembles that of *Dryosaurus*, for example, in that the caudal surface of the femoral head has an obturator ridge formed by a posterodistal projection of the greater trochanter and the groove for ligament attachment. The latter is externally bounded by a strong proximodistally short ridge (Novas 1997d). The remaining bone fragments come from levels of the Cerro Lisandro Formation and were briefly described by Coria (1999). Although the phylogenetic relationships of these small ornithischians remain almost unexplored, they are important in showing that tiny ornithopods were the primary consumers of low-stature vegetation.

From the Upper Cretaceous Gichón Formation, which is exposed at the Department of Paysandú, Uruguay, von Huene (1934a) described small teeth that yielded the skeleton of the crocodyliform *Uruguaysuchus* (Rusconi 1933). Von Huene originally considered these teeth to be similar to those of "psittacosaurine" and "camptosaurine" ornithopods, but studies in progress by Matías Soto provide support for the theory that this dental material has a basal iguanodontid nature (M. Soto, pers. comm.).

Ornithopod Footprints

Several Lower Cretaceous and basal Upper Cretaceous localities in Brazil and Argentina have yielded ornithopod trackways. They mainly correspond to large bipedal animals and a few quadrupedal forms. Among the most frequently cited ornithischian footprints are those of *Iguanodonichnus frenki* and *Camptosaurichnus fasolae* from the Baños del Flaco Formation, which is exposed in Colchagua, Chile (Casamiquela and Fasola 1968). However, recent studies of these prints (e.g., Sarjeant, Delair, and Lockley 1998; Lockley and Wright 2001) demonstrated that *Iguanodonichnus* is a sauropod trace fossil.

Most of the ichnological sites in Brazil are Lower Cretaceous in age, including the productive outcrops of the Rio do Peixe Group (Sousa Basin), and the Cariri Formation (Araripe Basin; Figs. 4.3 and 4.5). The Rio do Peixe Group (Berriasian–Hauterivian) is exposed in the State of Paraiba in northeastern Brazil and has yielded abundant dinosaur footprints mainly from the lower (Antenor Navarro Formation) and middle (Sousa Formation) levels (Leonardi 1984, 1989, 1994; Carvalho 2000). From the Antenor Navarro Formation, Giuseppe Leonardi described in 1984 the footprints of a large quadrupedal ornithopod (foot length about 50 centimeters) that he named *Caririchnium magnificum* (Fig. 7.15A). Originally, this author interpreted the *Caririchnium*-maker as a stegosaurian, but later it was reinterpreted as a large quadrupedal ornithopod (Leonardi 1989, 1994; Lockley and Wright 2001). *Caririchnum* constitutes the only report of a quadrupedal ornithopod trackway from South America (Lockley and Wright 2001). In the overlying Sousa Formation, Leonardi (1979a, 1979b, 1989, 1994) reported ornithopod trackways named *Sousaichnium pricei* (footprint about 56 centimeters long). An isolated ornithopod footprint around 20 centimeters wide and 20 centimeters long (Carvalho, Viana, and Lima Filho 1995) was found in the Cariri Formation (Late Jurassic–Early Cretaceous), which is exposed at Milagres in the Araripe Basin and reveals the tracks of smaller bipedal ornithopods.

Few fossil sites with ornithischian tracks of Cenomanian age are reported from Brazil (one is the Alcântara Formation in Ilha do Medo in the state of Maranhão; Carvalho 2001) and Argentina. Jorge Calvo (1991,

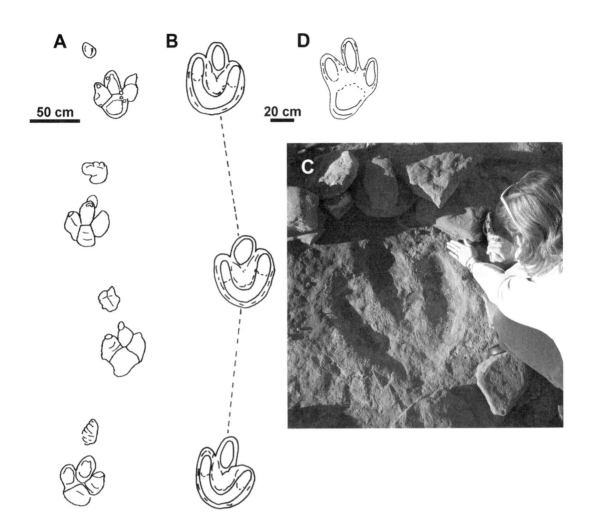

Fig. 7.15. Cretaceous ornithopod footprints. (A) *Caririchnum magnificum*. (B, C) *Sousaichnium monettae*. In the photograph Liliana R. Lo Coco (D) *Limayichnus major*. (A) Redrawn from Lockley and Wright (2001). (B, D) Redrawn from Calvo (1991). (C) Photograph by Fernando Novas.

1999, 2007) described ornithopod footprints there from the Cenomanian Candeleros Formation (which is widely exposed around Lake Ezequiel Ramos Mexía, near El Chocón in Neuquén Province), including *Limayichnus major* (Fig. 7.15B and C) and *Sousaichnium monettae* (Fig. 7.15D). The first ichnite is the most abundant in the area; it is 60–65 centimeters long and corresponds to an ornithopod roughly 9 meters long. *Sousaichnium monettae* is similar in size but differs from *Limayichnus* in being more asymmetrical with a marked heel and long digits.

Hadrosauridae

Hadrosaurid iguanodontians, the so-called duck-billed dinosaurs, are characterized by their transversely wide and flattened edentulous beaks and complex dentitions organized into dental batteries (Horner, Weishampel, and Forster 2004). The clade is phylogenetically defined as the most recent common ancestor of *Telmatosaurus*, *Parasaurolophus*, and all the

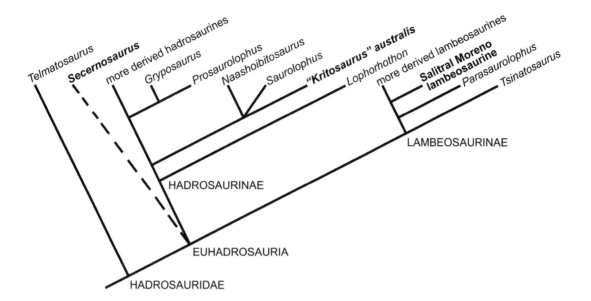

Fig. 7.16. Cladogram depicting possible phylogenetic relationships among genera of hadrosaurids. South American representatives indicated in bold.

descendants of this common ancestor (Horner, Weishampel, and Forster 2004; Fig. 7.16). Hadrosaurids were large animals, up to 12 meters long. The conspicuous diagnostic features of Hadrosauridae are related to the masticatory apparatus, including a dental battery composed of three or more replacement teeth per tooth family, distal extension of the dentary tooth row to terminate caudal to the apex of the coronoid process, and caudalmost termination of the dentary well behind the coronoid process. Hadrosaurids are also distinguished from other ornithopods in modifications of the vertebral column, such as cervical prezygapophyses that are elevated well above the level of the neural canal, long and dorsally arched postzygapophyses, and an increase in the number of sacral vertebrae to eight or more. The pectoral girdle and hindlimbs of hadrosaurids also have peculiar features, such as a proximally narrow scapula, a horizontally projected acromial process, and a deep extensor groove on the distal femur (Horner, Weishampel, and Forster 2004).

Hadrosaurids are mainly known from the northern hemisphere, especially from North America, where they constitute approximately 60 percent of the dinosaur specimens collected from Late Cretaceous beds (Russell 1967). More than 30 hadrosaur genera are currently recognized, ranging from Santonian through Maastrichtian in age. The earliest and oldest known members of Hadrosauridae are recorded from Asia, where they probably originated. Southern records of hadrosaurids are currently available from South America (Casamiquela 1964b; Brett-Surman 1979; Bonaparte Franchi, J. Powell, and E. Sepúlveda 1984; Bonaparte and Rougier 1987; Powell 1987c) and presumably Antarctica (Case et al. 2000; Rich, Vickers-Rich, Fernández, and Santillana 1999), but the group is unrecorded in the rest of Gondwana.

Hadrosaurs figure prominently in interpretations of the paleobiogeographic evolution of South American Cretaceous dinosaurs (e.g.,

Brett-Surman 1979; Bonaparte 1986d; Bonaparte and Kielan-Jawarowska 1987). The presence of hadrosaurs in South America has been interpreted by Casamiquela (1964b), Brett-Surman (1979), and Bonaparte (1984b, 1984c) as evidence of a colonization event of these dinosaurs from North America.

Late Cretaceous beds of South America in which hadrosaurids are documented constitute different sedimentary units (e.g., the Los Alamitos, La Colonia, Allen, Loncoche, and Paso del Sapo formations) deposited in littoral paleoenvironments associated with a transgressive pulse of marine waters from the Atlantic Ocean that covered most of Patagonia at the end of the Mesozoic. South American hadrosaurs were numerically conspicuous in these lowlands close to the seacoast (e.g., Casamiquela 1980a; Bonaparte Franchi, J. Powell, and E. Sepúlveda 1984; González Riga and Casadío 2000; Martinelli and Forasiepi 2004).

Although abundant and relatively well preserved, the Patagonian hadrosaurs have not been described in detail, and the only available studies are those of the holotype specimen of *"Kritosaurus" australis* by Bonaparte and co-authors (Bonaparte Franchi, J. Powell, and E. Sepúlveda 1984; Bonaparte and Rougier 1987). Probably the main consequence of this lack of detailed analyses is the lack of agreement about the phylogenetic relationships of Patagonian hadrosaurs with their North American counterparts (Horner, Weishampel, and Forster 2004; Fig. 7.16).

"KRITOSAURUS" AUSTRALIS

"Kritosaurus" australis constitutes the best represented hadrosaurid from South America. This southern form, which was more than 5 meters long, is known from several specimens collected from a single fossil site at Los Alamitos Farm, near Arroyo Verde in southeastern Río Negro Province. It is represented by some fragmentary bones from the skull and jaw; isolated teeth; cervical, dorsal, sacral, and caudal vertebrae; ribs; a scapula; a sternum; forelimb bones; and pelvic and hindlimb bones (Fig. 7.17).

"Kritosaurus" australis was originally interpreted (e.g., Bonaparte Franchi, J. Powell, and E. Sepúlveda 1984; Bonaparte 1986d; Bonaparte and Rougier 1987) as closely related to the then-named *Kritosaurus incurvimanus* from the Campanian of North America. However, the name of this northern taxon has been recently replaced by the name *Gryposaurus incurvimanus*. Moreover, recent studies (e.g., Horner, Weishampel, and Forster 2004) suggest that *"Kritosaurus" australis* is not closely related to *Gryposaurus incurvimanus* but instead is related to other hadrosaurine hadrosaurids from North America (i.e., *Saurolophus* and *Naashoibitosaurus*). This is why the name *"Kritosaurus" australis* is cited in quotation marks.

Bonaparte and Rougier (1987) noted strong resemblances between *"Kritosaurus" australis* and *Gryposaurus* from North America in the construction of the frontal and occipital regions as well as in the postcranial skeleton. The differences these authors acknowledged occur in the

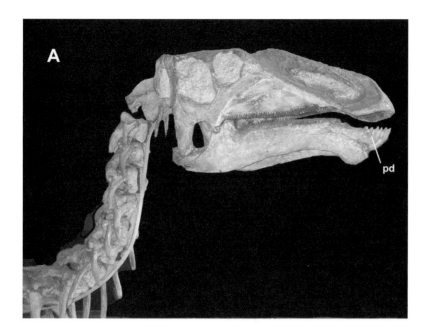

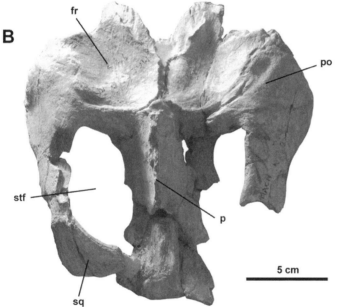

Fig. 7.17. *Kritosaurus australis:* (A) Reconstructed skull and neck of a mounted skeleton, Museo Argentino de Ciencias Naturales Bernardino Rivadavia. (B) Posterior portion of skull in dorsal aspect (occipital side to the bottom of the page). *Photographs by Fernando Novas.*

morphology of the basicranium (e.g., the exoccipitals rise higher and are more robust in the Patagonian form, thus producing larger condylar protuberances), the coracoid (in the Patagonian form it is craniocaudally shorter and more hooked than in the North American *Gryposaurus*), and the distal end of femur (the Patagonian form has a roofed supracondylar canal instead of one that is cranially open, as in *Gryposaurus*).

The predentary of "*Kritosaurus*" *australis* has well-defined large and quadrangular denticles along the oral border of the bone (Fig. 7.17A). Predentary denticles are present in other hadrosaurids, but they are triangular (Horner, Weishampel, and Forster 2004). Furthermore, *Gryposaurus* lacks denticles on the oral margin of the predentary (Bonaparte and Rougier 1987). The frontals form a modest dome, a feature that Horner, Weishampel, and Forster (2004) interpreted as a reversion to a basal hadrosaurid condition. The pelvic morphology of *"Kritosaurus" australis* closely resembles that of *Gryposaurus incurvimanus* in the shape of the prepubic process of the ischium. Differences occur in the iliac supraacetabular process (it is more lateroventrally projected in *"Kritosaurus"* than in *Gryposaurus*) and in the elongation of both pre- and postacetabular processes: in the South American hadrosaurid they look shorter in proportion with the iliac depth.

Horner, Weishampel, and Forster (2004) found no compelling characteristics that would place *"Kritosaurus" australis* within *Kritosaurus*. Moreover, Horner and colleagues (2004) depicted *"Kritosaurus" australis* as sharing a common ancestor with the Campanian *Naashoibitosaurus ostromi* from the southwestern United States and the Campanian–Maastrichtian *Saurolophus* from Asia and North America. This phylogenetic arrangement is based on the shared presence of a pointed and symmetrically triangular rostral jugal process. Horner, Weishampel, and Forster (2004) argued that *"Kritosaurus" australis* might be most closely related to *Saurolophus* because these taxa share low zygapophyseal peduncles on the cervical vertebrae.

SECERNOSAURUS KOERNERI

This taxon was the first hadrosaur named from South America. It is known from a partial braincase, caudal vertebrae, a scapula, an ilia, and a fibula, which were collected about 3 kilometers east of the head of the Chico River in southeastern Chubut Province. The lithostratigraphic provenance of the specimen is matter of discussion: in the original description of *Secernosaurus*, Brett-Surman (1979) cited the Maastrichtian–Paleocene San Jorge Formation, a unit that is currently named the Salamanca Formation (Uliana and Legarreta 1999). But Bonaparte (1996b) and Novas (1997c) cited the Maastrichtian Laguna Palacios Formation as the unit that yielded *Secernosaurus*. However, neither the Salamanca nor Laguna Palacios formations are present in the area where the specimen was found (R. D. Martínez, pers. comm.). Rubén D. Martínez and his assistants explored the area around the Chico River and found a partially preserved ilium that may belong to a hadrosaurid (Luna, Casal, Martínez, Lamanna, Ibiricu, and Ivany 2003), thus strengthening the possibility that this is the correct area (and lithostratigraphic unit) from which *Secernosaurus* was excavated. Following Martínez (1998b, and pers. comm.), the holotype specimen of *Secernosaurus koerneri* almost probably comes from the upper levels of the Bajo Barreal Formation, the age of which is Campanian–Maastrichtian in this region of Chubut.

On the basis of the ilium length (50 centimeters), the total length of *Secernosaurus* is roughly estimated at 4.50 meters. The ilium of this hadrosaur has some interesting features (Fig. 7.18A). For example, the postacetabular process is deflected dorsomedially, and the supraacetabular process is relatively smaller than in any hadrosaur of the same size. Brett-Surman (1979) noted a primitive condition in the ilium of *Secernosaurus* consisting of a dorsomedially twisted postacetabular wing; this contrasts with the transversely flattened condition of the postacetabular blade of most hadrosaurids. However, Bonaparte (1996b) suggested that this may be an artifact of preservation (contra Brett-Surman 1972, 1979).

As originally noted by Brett-Surman (1972), the ilium of *Secernosaurus* looks primitive because it has a rudimentary supraacetabular process and a postacetabular wing that is dorsoventrally depressed instead of being vertical, as in most hadrosaurids. Emphasizing the primitive aspect of *Secernosaurus*, Brett-Surman (1972) noted that this taxon resembles other basal hadrosaurs such as *Gilmoreosaurus* and *Bactrosaurus*. However, *Secernosaurus* is more derived than these Asiatic forms in that the Patagonian taxon has a sigmoid dorsal margin of the ilium (instead of being continuously convex, as in *Gilmoreosaurus* and *Bactrosaurus*); a pubic pedicle that is short and less defined than in the Asiatic forms; a cranioventrally curved preacetabular spine of the ilium that passes ventrally beyond an imaginary line uniting both the pubic and ischiadic pedicles (in contrast with the less ventrally projected preacetabular spine of *Gilmoreosaurus* and *Bactrosaurus*); and a postacetabular wing that is proportionally longer than in the Asiatic forms and has a straight ventral margin that is almost on the same line as the articular surface for the ischium (Fig. 7.18A; isp).

Horner, Weishampel, and Forster (2004) depicted *Secernosaurus* as having unresolved relationships within Hadrosauridae, noting that, like *Bactrosaurus*, *Tanius* and *Telmatosaurus*, it lacks the features that would commit it to a higher level within the clade. Nevertheless, on the basis of the iliac features enumerated above, one may conclude that *Secernosaurus* was phylogenetically closer to the euhadrosaurids (i.e., the clade that combines Hadrosaurinae and Lambeosaurinae) than to the more-primitive hadrosaurids.

José Bonaparte (pers. comm.) thinks that *Secernosaurus* is different from "*Kritosaurus*" *australis*, an interpretation I accept. The preacetabular wing of the ilium in *Secernosaurus* is considerably longer than in "*Kritosaurus*" *australis* (Fig. 7.18B) and laterally overlaps the prepubic process of pubis. The latter process is fairly similar in these two taxa, thus suggesting that *Secernosaurus* and "*Kritosaurus*" *australis* may constitute members of the subfamily Hadrosaurine.

SALITRAL MORENO LAMBEOSAURINE

The lambeosaurines include famous hadrosaurids such as *Tsintaosaurus*, *Parasaurolophus*, *Corythosaurus*, *Hypacrosaurus*, and *Lambeosaurus* (Fig. 7.16), which are united by cranial synapomorphies (mainly related with the nasal passage) and postcranial features such as elongate neural

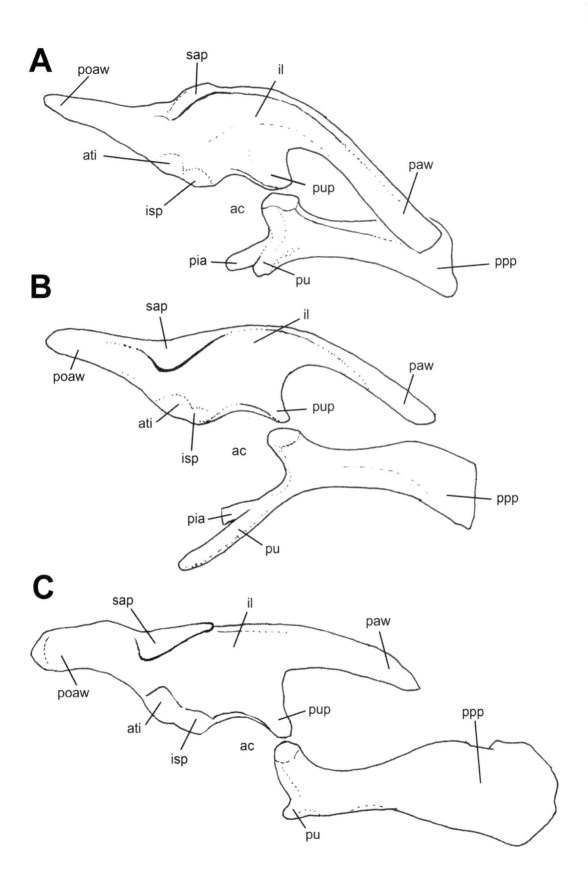

spines on the caudal dorsal and sacral vertebrae and spines that are more than three times the height of the centrum (Horner, Weishampel, and Forster 2004).

The first lambeosaurine recorded outside Laurasia was found in northern Patagonia by Jaime Powell (Figs. 7.18B and 7.19). It consists of a partially articulated skeleton recovered from the Lower Member of the Allen Formation (Maastrichtian) exposed at Salitral Moreno, about 20 kilometers south of the town of General Roca in Río Negro Province. The material was found almost completely articulated and included posterior dorsals, 8 sacrals, 25 caudal vertebrae, fragments of ossified ligaments in their original position, several hemapophyses and fragments of ribs, an incomplete scapula, an ilium and prepubis, femora, tibiae, and fibulae, and tarsal and foot bones. It is a hadrosaur of great size—its femur is 90 centimeters long—implying an animal around 6–7 meters long, thus surpassing the length of "Kritosaurus" australis (5.70 m) and Secernosaurus koerneri (4.50 meters).

The yet-unnamed lambeosaurine from Patagonia (Powell 1987c) is characterized by its sacral and caudal neural spines, which are transversely narrow at their bases but progressively thicker toward their distal extremities (Fig. 7.19A and B). The proximal caudal vertebrae have short centra and very elongate neural spines, as much as 4.8 times the height of the centrum (this ratio applies for caudal 1). Regarding this last condition, Horner, Weishampel, and Forster (2004) indicate that lambeosaurines that are more derived than *Tsintaosaurus* (i.e., *Parasaurolophus*, *Corythosaurus*, *Hypacrosaurus*, and *Lambeosaurus*) are united by unambiguous characteristics, including dorsal and sacral neural spines that elongate to more than three times the height of the centrum. Following this, the Patagonian specimen may be a member of this subgroup of derived lambeosaurines (Fig. 7.16).

The ilium is almost complete; it lacks only the extremity of the preacetabular wing (Figs. 7.18C and 7.19C). Its general aspect is intermediate between that of *Lambeosaurus* and *Parasaurolophus* (Powell 1987c). The ilium of the Patagonian form is proportionally deep, thus resembling the condition in lambeosaurines. Its axis is slightly recurved ventrally, as in *Lambeosaurus* (Powell 1987c). The supraacetabular process is massive, as characteristically occurs among lambeosaurines (Horner, Weishampel, and Forster 2004), and its lateral margin is straight. The postacetabular wing is ventrally curved, so its ventral margin is more concave than in "Kritosaurus" australis, and its dorsal border is markedly convex. The prepubic process is paddle-shaped and more expanded than in "Kritosaurus" australis (Fig. 7.18B).

The femur has an anterior intercondylar canal that is almost closed by the foremost portion of the condyles, whose borders are separated by only 2 millimeters (Powell 1987c). This feature, however, has been also reported for "Kritosaurus" australis and is variably present in adult hadrosaurids in general (Horner, Weishampel, and Forster 2004). The left foot is almost completely preserved and articulated. It shows a bony callus in metatarsal II that surely developed as the result of a fracture.

Fig. 7.18. Right pelvic girdles of South American hadrosaurids in lateral view. The ilia are aligned at the level of the acetabular antitrochanter and are oriented taking as horizontal an imaginary line that unites the pubic and ischiadic pedicles. (A) *Secernosaurus*. (B) "Kritosaurus" australis. (C) Lambeosaurinae indet. Note the cranioventral inflexion of the preacetabular process of the ilium in *Secernosaurus* and the cranially projected and paddle-like prepubic process in Lambeosaurine indet.

Fig. 7.19. Lambeosaurine indet. from the Allen Formation. (A) Sacral vertebrae in right lateral view. (B) Proximal caudal vertebra in caudal view. (C) Right ilium articulated with cranial portion of pubis in lateral view. *Photographs by Fernando Novas.*

Jaime Powell considered this dinosaur to be a possible lambeosaurine because of the curious morphology of its axial skeleton. The sacrum is composed of eight vertebrae, the most anterior of which has a ventral crest, as occurs in lambeosaurines (Brett-Surman 1979). In addition, the neural spines of sacral vertebrae are very tall, longer than in the lambeosaurines *Hypacrosaurus* and *Barsboldia*. The specimen differs from hadrosaurines (including "*Kritosaurus*" *australis* and *Secernosaurus koerneri*), which have relatively short neural spines and ventral depressions in the sacral centra. The ilium of the Patagonian lambeosaurine is also dorsoventrally deep, as in other members of Lambeosaurinae, in contrast with the shallower ilium of "*Kritosaurus*" *australis*. However, not all of the features present in the Patagonian lambeosaurine match well with North American and Asian members of the subfamily. For example, the ischium does not have the distal expansion that is plesiomorphically retained among lambeosaurines and thus it is more similar to the hadrosaurines (e.g., *Edmontosaurus*, "*Kritosaurus*" *australis*). In addition, apart from sacral 1, the remaining sacrals are ventrally grooved (as in hadrosaurines) instead of being keeled, as in lambeosaurines.

Hadrosaurid Footprints and Eggshells

Alonso (1980, 1989) and Alonso and Marquillas (1986) described dinosaur tracks from the Maastrichtian Yacoraite Formation in Salta Province in northwestern Argentina that they interpreted as belonging to Hadrosauridae. The ichnites were named *Hadrosaurichnus australis* (Alonso 1980), *Taponichnus donottoi*, and *Telosichnus saltensis* (Alonso and Marquillas 1986). However, later authors (e.g., Leonardi 1989; Lockley and Wright 2001) reinterpreted some of these footprints (e.g., *Hadrosaurichnus*, *Taponichnus*) as theropodan, as suggested by their elongate shape and long step. Nevertheless, footprints named *Telosichnus saltensis* may well correspond to a large ornithopod (the footprint length is about 52 centimeters), as suggested by the rounded digit impressions and transversely wide footprint (Alonso and Marquillas 1986; Thulborn 1990). With reservations, I follow these authors in referring *Telosichnus* to the Hadrosauridae.

Described footprints from beds of the Upper Vilquechico Formation (Upper Campanian–Early Maastrichtian), which are exposed near the village of Vilquechico in southern Perú, were named *Hadrosaurichnus titicacaensis* (P. Ellemberger, quoted in Jaillard et al. 1993) and were referred to Hadrosauridae because of their arrangement in a trackway of a bipedal dinosaur, with footprints showing a short impression of digit III, among other features.

Probable hadrosaurid eggshells have been reported by Manera de Bianco and Khöler (2003) from Upper Cretaceous beds that are probably equivalent with the Maastrichtian Allen Formation (T. Manera de Bianco, pers. comm.) and that crop out at Blanco Hill, close to Yaminué, Río Negro Province. The eggshells are delicate (1 millimeter

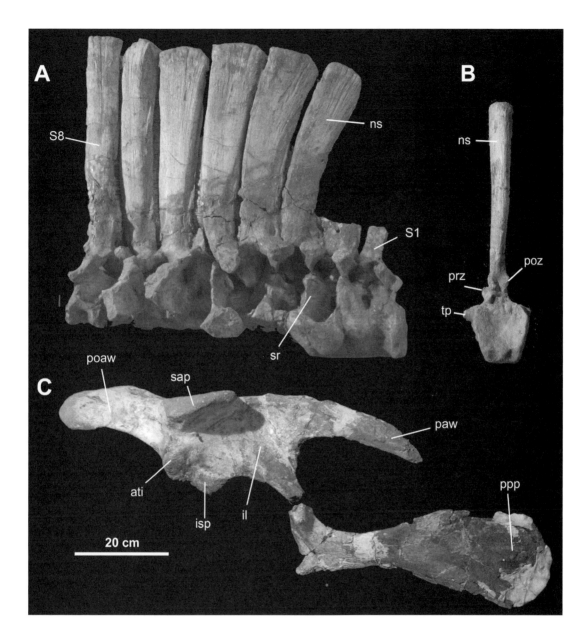

thick) and have features that are present in the oofamily Sphaeroolithidae (Mikhailov 1997; Carpenter 1999), including an external surface that is sculptured with a netlike pattern of nodes and ridges with pits and grooves between them (i.e., sagenotuberculate ornamentation), a prolatocaniculate pore system (i.e., the canals vary in width along their length), and a prolatospherulitic morphotype (i.e., the shell units are fan-shaped).

The presence of ceratopsids in Gondwana is weakly based on very fragmentary and partial material. One of these specimens consist of an isolated ulna recovered from Aptian beds in Australia for which the name

Ceraptopsia

Fig. 7.20. Purported South American ceratopsids: (A–D) Left dentary of *Notoceratops bonarelli* in (A) ventral, (B) lateral, (C) dorsal, and (D) ventrolateral views. (E) Isolated right coracoid collected in Cinco Saltos that was originally referred by von Huene (1929a) to Ceratopsia but probably belongs to Sauropoda. (F, G) Quadrupedal footprints of an ornithischian dinosaur from the Cenomanian Itapecurú Formation, northeastern Brazil: (F) entire trail; (G) detail of a footprint. *(A–E) From von Huene (1929a). (F) From Carvalho (2001). (G) Photograph courtesy of Ismar de Sousa Carvalho.*

Serendipaceratops arthurclarkei was coined (Rich and Vickers-Rich 1994, 2003). The other element is a jaw that geologist Augusto Tapia described and named *Notoceratops bonarelli* in 1918. This discovery was made north of the Chico River, east of Lake Colhué Huapi in Chubut Province, roughly in the same geographic region where the hadrosaurid *Secernosaurus* was discovered. As in the case of the latter animal, the stratigraphic provenance of *Notoceratops* may be from the upper levels of the Bajo Barreal Formation (which is Campanian–Maastrichtian in this region of Chubut). Unfortunately, the holotype is currently lost, precluding close scrutiny and comparisons of this important specimen. Interpretations about the systematic of *Notoceratops* are diverse: Tapia (1918), von Huene (1929a), and Bonaparte (1978, 1996b) agreed on the ceratopsid nature of *Notoceratops*. Molnar (1980b) and Powell (2003) pointed out that the jaw could belong to a hadrosaurid, a group that is positively known in South America.

The jaw of *Notoceratops bonarelli* is 24.5 centimeters long as preserved; it is medially broken and its teeth are unknown (Fig. 7.20A–D). An outstanding feature of this dentary is a prominent bump, with associated scars, located below and rostral to the coronoid process, a feature that I was unable to detect in any other ornithischian dinosaur currently known. Whether this bump is natural or pathologic is impossible to resolve. A large blind excavation is located on the "ventral surface" of the dentary, rostral to the level of the bump. Von Huene (1929a) emphasized that the jaw of *Notoceratops* absolutely agrees with that of ceratopsids and is clearly distinguished from that of hadrosaurids, as mainly evidenced by the rostral point of the bone, where it articulates with the predentary. However, von Huene did not explicitly cite which are the features *Notoceratops* has in common with ceratopsids. Hadrosaurid dentaries show notable differences from *Notoceratops*: in the latter form, the dentary looks transversely thicker and curved (i.e., laterally concave, as ventrally seen), unlike in hadrosaurids, in which the dentary is narrower and straight (as seen from below). Also, the rostral end is transversely thicker and rounded instead of being dorsoventrally flattened, as in hadrosaurids. Having noted the distinctions of *Notoceratops* from these ornithopods, it must also be said that the morphology of the dentary of *Notoceratops* is not easily matched with that of ceratopsians. For example, the dentary lacks the strong longitudinal ridge along the external surface extending from the coronoid process toward the rostral tip of the bone, which is seen in many ceratopsians. In conclusion, I will accept with reservations the referral of *Notoceratops* as a Ceratopsia incertae sedis.

A small coracoid from Cinco Saltos measuring 17 centimeters in diameter was originally referred by von Huene (1929a) to Ceratopsia (Fig. 7.20E). It is quadrangular in side view and has a prominent process on the craniodistal corner of the bone. Although I agree with von Huene that this coracoid is not referable to titanosaurids (which are abundantly represented at the Cinco Saltos fossil site), it does not show features of Ceratopsia. For example, the coracoid from Cinco Saltos lacks the hooked

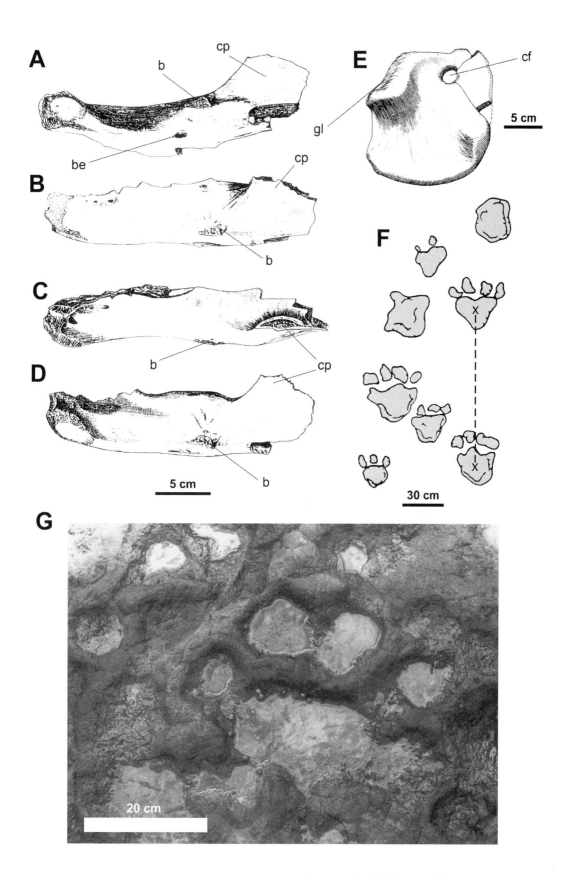

process on the caudodistal corner that is present among ceratopsians (e.g., *Leptoceratops*, *Microceratops*, *Triceratops*). Similarities with this coracoid are seen in some Jurassic sauropods (e.g., *Apatosaurus*; Ostrom and McIntosh 1966), suggesting that it might belong to Diplodocimorpha. Whether the coracoid is sauropod or not cannot be asserted, but at least it is not referable to Ceratopsia.

Bony or teeth remains of ceratopsians remain elusive in South America, and the ichnological evidence is also inconclusive. Footprints described from Bolivia are more confidently referred to Ankylosauria than to Ceratopsia (McCrea, Lockley, and Meyer 2001). However, Ismar de Souza Carvalho (2001) described a very interesting trackway of large quadrupedal ornithischians from the Cenomanian Itapecurú Formation in Ilha do Medo in the state of Maranhão in northeastern Brazil (Fig. 7.20F and G). The tracks indicate a tridactyl manus and a tetradactyl pes in a large quadrupedal animal (the largest footprints are 30 centimeters wide and 45 centimeters long). The tracksite surface is irregular, which may be the result of bioturbation caused by the path of more than one individual. Except for the tridactyl condition of the manus prints, these footprints resemble those of the trackways of the ankylosaur named *Tetrapodosaurus* (McCrea, Lockley, and Meyer 2001). Nevertheless, the Brazilian footprints more closely match unnamed tracks questionably attributed to ceratopsians from the Upper Cretaceous of North America (Thulborn 1990). In fact, the digit count of the Brazilian tracks fits better with the digit number and proportions of the pes and manus of ceratopsians than with those of ankylosaurs (Thulborn 1990). If these footprints are shown to be ceratopsian, that would indicate that these animals were in South America at the beginning of the Late Cretaceous.

Fig. 8.1. Cretaceous dinosaurian faunal succession based on evidence obtained in the Neuquina Basin. *Redrawn from Leanza, Apesteguía, Novas, and de la Fuente (2004).*

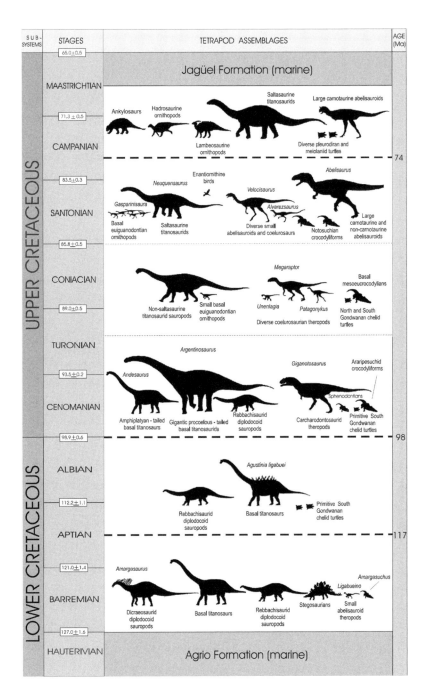

A SUMMARY OF THE CRETACEOUS DINOSAURS

8

Cretaceous faunas of South America show important taxonomical differences from those of the northern hemisphere. As noted by Bonaparte (1986d) and Bonaparte and Zofía Kielan-Jaworowska (1987), South American dinosaur faunas were dominated by herbivorous sauropods of different clades (e.g., dicraeosaurids, rebbachisaurids, titanosaurs) that were very different from the Cretaceous herbivorous dinosaur faunas of North America, which mainly constituted a diversity of ornithischian dinosaurs (e.g., hadrosaurids, ceratopsians, ankylosaurs, pachycephalosaurs). Among theropod dinosaurs that diversified on the southern landmasses were abelisauroids, carcharodontosaurids, and spinosaurids, which remain undocumented in North America and Asia (although spinosaurids and abelisauroids have been recorded in Europe). As a counterpoint, the large Laurasian predators, Tyrannosauridae, are unknown in South America and the remaining Gondwanan continents. This has led to the hypothesis that a prolonged paleobiogeographical isolation between Gondwana and Laurasia occurred for most of the Cretaceous (Bonaparte 1986d). This hypothesis finds strong support from Cretaceous reptiles, especially crocodyliforms, which underwent a dramatic evolutionary radiation in Gondwana. Notosuchians, araripesuchids, baurusuchids, and sarcosuchids predominate in the Cretaceous of the southern hemisphere, in contrast with the much more modest diversification of crocodiles that occurred in the northern continents for the same period. Outstanding differences are also detected in patterns of turtle distribution (M. de la Fuente, pers. comm.): cryptodirans are poorly documented in the southern continents (where they are represented only by bizarre meiolanids from Patagonia and Australia), in contrast with the northern hemisphere, where cryptodirans are highly diversified. On the other hand, pleurodiran turtles were considerably more abundant and diverse in Gondwana than in North America. Finally, sphenodontid lepidosaurs become extinct in the northern hemisphere at the beginning of the Cretaceous, contrasting with the diversity, abundance, and large size attained by eilenodontine sphenodontids during the Late Cretaceous in Patagonia (Apesteguía and Novas 2003).

In general, southern ornithischian dinosaurs were less abundant and less diverse than in Laurasia. For example, the recorded ankylosaurs from Gondwana belong to Nodosauridae and no members of the more-derived Ankylosauridae have yet been discovered in the southern landmasses. South America was populated for most of the Cretaceous by small to large ornithischians (e.g., stegosaurs, ankylosaurs, ornithopods), which shared the role of plant-eaters with the larger sauropods. Some dinosaur clades from Gondwana (e.g., abelisaurids, dicraeosaurids, rebbachisaurids) show remarkable autapomorphies that may be interpreted as the result of a prolonged biogeographical endemism (Bonaparte 1986d; Novas 1997c). In contrast, the effects of physical isolation are weakly manifested in the morphology of the southern ornithopods. With the exception of some bizarre features (e.g., the developed cranial bulla of the Australian *Muttaburrasaurus*, Bartholomai and Molnar 1981; and the elongate neural spines of the African *Ouranosaurus*, Taquet 1976), the anatomy of basal iguanodontians from Gondwana resembles that of their Jurassic and Cretaceous relatives from Laurasia. Beyond that, no suprageneric groups of exclusively Gondwanan ornithopods have yet been recognized, except for the elasmarian iguanodontians *Talenkauen* and *Macrogryphosaurus*, which share derived features in the vertebral column and chest (Calvo, Porfiri, and Novas 2005, 2008; Novas, Cambiaso, and Ambrosio 2004).

Dinosaur Diversification in the Cretaceous of South America

The Cretaceous Period constituted the last 80 million years of the Mesozoic, during which there were profound modifications in the environment, including the dismembering of Pangea, the spread of new epeiric seas, the remodeling of oceanic and atmospheric circulation patterns, changes in climatic conditions, and the origination and diversification of new kinds of plants (angiosperms). Warm temperatures and relatively dry conditions prevailed during much of the Cretaceous in South America (and Africa), to which intense volcanic activity (as manifested by recurrent ash falls and lava flows) was added as a source of ecological disturbance.

It is expected that such physical, climatic, and ecological transformations affected or controlled the geographic distribution of the dinosaurs evolving at that time and selected for their anatomical features. Although this as a general statement may be true, discerning particular evolutionary patterns of diversification (let alone identifying the specific environmental changes responsible for them) is a hard (and many times discouraging) task that in most cases result in highly speculative hypotheses.

A few attempts have been made to recognize "evolutionary stages" (Bonaparte 1979c, 1986d, 1996b), "intervals" (Novas 1997c), or "faunal assemblages" (Leanza, Apesteguía, Novas, and de la Fuente 2004) corresponding to different portions of the Cretaceous (Fig. 8.1). Building on these studies, we might somewhat arbitrarily lump dinosaurs recorded in Cretaceous rocks of South America into successive "evolutionary stages," each one characterized by a particular taxonomic composition, with extinction events at the end of each of these stages and diversification processes at the beginning of the following stage. However, boundaries are not always clear cut but are instead rather diffuse, and future discoveries may alter the timing of the proposed evolutionary events. In sum, the scenario described in the following pages must be taken as a preliminary synthesis.

Three main Cretaceous faunal assemblages are recognized: a) an Early Cretaceous stage composed of diplodocimorphs, stegosaurs, and abelisauroids; b) a mid-Cretaceous stage in which abundant and diverse diplodocimorphs were associated with spinosaurids and carcharodontosaurids; and c) a Late Cretaceous stage in which abelisauroids and derived titanosaurids predominated, with an important presence of hadrosaurids. During stages b and c, basal iguanodontians and dromaeosaurids also were present (Fig. 8.1).

South American Dinosaur Faunas during the Early Cretaceous (Berriasian through Barremian)

Our understanding of the Early Cretaceous evolutionary stage of dinosaur faunas is based on discoveries mainly from the Botucatu, La Amarga, and La Paloma formations. The Late Jurassic–Early Cretaceous Botucatu ichnofauna from Brazil provides the only evidence about dinosaur faunal composition at the beginning of the Cretaceous in South America, and the main peculiarity of the Brazilian ichnofauna is the absence of sauropod dinosaurs in an arid environment. The paleoenvironmental conditions associated with the evolutionary history of Early Cretaceous dinosaurs in South America should be noted. It is particularly interesting that a hothouse climatic regime characterized most of the Mesozoic, especially the Early Cretaceous. Current paleoclimatic models (Chumakov et al. 1995) reconstruct South America–Africa as a large, mainly arid landmass for the Berriasian through the Aptian (Skelton 2003). The large paleodesert of Botucatu, which came into existence in Late Jurassic times, continued into the beginning of the Cretaceous, when significant fissure volcanism (the Serra Geral basalts) took place. Aeolian deposits of the Caiuá Formation, which overlie the Serra Geral basalts, indicate that desert conditions continued to predominate for most of the Early Cretaceous in the Paraná Basin (Soares 1981).

Later dinosaur faunas of the Early Cretaceous (based on evidence from the La Amarga and La Paloma formations) are characterized by the occurrence of dicraeosaurids, basal titanosauriforms, basal abelisauroid theropods, and stegosaurs (Fig. 8.1). The presence of Early Cretaceous stegosaurians in South America is consistent with the record of this group in other parts of the world, in particular southern Africa, where *Paranthodon africanus* is found in association with diplodocoid sauropods (i.e., the rebbachisaurid *Algoasaurus*) in the Valanginian Kirkwood Formation. After the beginning of the Cretaceous (Barremian) and at least into the Cenomanian, two major sauropod clades are recorded in South America: the Titanosauriformes and the Diplodocimorpha. Barremian sauropod faunas include bizarre dicraeosaurids (*Amargasaurus*) and titanosauriforms (as evidenced by discoveries made in the La Paloma Formation; Rauhut, Cladera, Vickers-Rich, and Rich 2003), suggesting that sauropods were much more diverse than is currently thought.

The faunal structure described for the Early Cretaceous of South America contrasts with that described by Robert Bakker (1978) for North

America, in which high-browser sauropods were replaced by low-browser ornithischians over the course of the Early Cretaceous. Two observations emerge from this comparison. First, in South America the faunal composition and structure remained similar to that of the Late Jurassic (made up of stegosaurs, ornithopods, diplodocoids, and basal titanosauriforms); second, the origin and diversification of angiosperms in this continent were not triggered by faunal changes like those that occurred in North America (see Barrett and Willis 2001). Although the South American faunal structure may look conservative, the survival of gargantuan herbivores in the terrestrial ecosystems of Gondwana may have been necessary to support a diversity of large theropods. Notably, the herbivore faunal turnover described for the Early Cretaceous of North America (Bakker 1977) was accompanied by a decrease in diversity among big-game predators, including the Early Cretaceous allosauroids (e.g., *Acrocanthosaurus*) and the Late Cretaceous tyrannosaurids, along with a small contingent of large dromaeosaurids (e.g., *Utahraptor*). This situation sharply contrasts with the greater taxonomic diversity reached in Gondwana (as evidenced by Cenomanian and later Cretaceous dinosaur faunas from South America and Africa; Farlow and Holtz 2002), where a greater diversity of large theropod clades is known to have evolved that included spinosaurids, *Megaraptor*, carcharodontosaurids, abelisaurids, large basal coelurosaurians (*Orkoraptor*), and large dromaeosaurids (e.g., *Unquillosaurus, Austroraptor*).

Although most of the evidence supporting the isolation of Gondwanan from Laurasian tetrapod faunas comes from Late Cretaceous rocks, the presence in the Early Cretaceous in South America of several taxa not recorded in Berriasian–Hauterivian formations of the northern hemisphere (e.g., the Cloverly Formation; Russell 1993) suggests that the isolation between Laurasia and Gondwana was already complete during the Hauterivian. Such taxa include the above-mentioned dinosaurs from the Barremian–Early Aptian La Amarga Formation (e.g., dicraeosaurids, basal abelisaurs, and stegosaurs; Bonaparte 1994; Salgado and Bonaparte 1991). However, differences between southern South America (e.g., Patagonia) and Laurasia are probably due not only to the physical separation mentioned above but also to more or less effective barriers that were more local to West Gondwana (e.g., extensive deserts in the Paraná and Chaco Basins in northeastern South America; Soares 1981; Zambrano 1981; Uliana and Biddle 1988).

South American Dinosaur Faunas of the Mid to Early Late Cretaceous Interval (Aptian through Cenomanian)

Available geological data indicate that during the Albian and Cenomanian, a humid belt developed close to the equator (Skelton 2003). However, an arid climate predominated across most of the South American continent, as during the Early Cretaceous. Subsequent to Aptian times, salt deposits that accumulated in the Neuquina Basin as well as in the proto-Atlantic indicate high rates of evaporation rather than precipitation. The environment surrounding the Araripe Basin in northeastern Brazil is generally considered to have been arid or semi-arid: thick sequences of evaporites occurred in the Ipubi Formation immediately below the Albian Santana Formation. The dinosaur assemblage of the Santana Formation is unusual in that only theropods (small and large) have been recorded (Naish, Martill, and Frey 2004). Although ornithischians and sauropods have not yet been documented in this formation, members of these two dinosaur lineages have been found in the Cenomanian Itapecurú Formation along with footprints in the Lower Cretaceous pre-Aptian Antenor Navarro Formation that indicate that these groups were present in mid-Cretaceous environments of Brazil. According to Naish, Martill, and Frey (2004), it is likely that only small numbers of sauropods and large ornithopods inhabited the Araripe region during the Aptian–Albian. Thus, in the absence of these primary consumers, large theropods (e.g., the spinosaurid *Irritator*) may have preyed on other vertebrates that are abundantly recorded in those beds (e.g., fishes, pterosaurs).

In South America (and Africa), dicraeosaurid sauropods and stegosaurs are not recorded in beds younger than Barremian, and it is probable that these two dinosaur lineages became extinct before Aptian times. Titanosaurians and rebbachisaurids are recorded in the Cenomanian Candeleros and Huincul Formations of northern Patagonia and in the Cenomanian–Coniacian Lower Bajo Barreal Formation of central Patagonia (Fig. 8.1). Titanosauriforms currently recorded for the Aptian–Cenomanian interval include *Andesaurus*, *Ligabuesaurus*, and *Chubutisaurus*, while South American representatives of Diplodocimorpha for the same time span are *Limaysaurus*, *Rayososaurus*, *Cathartesaura*, and *Amazonsaurus* and presumably also *Agustinia*. They are the most frequently found dinosaurs in Lower Cretaceous (Aptian–Albian) strata in Patagonia, surpassing the basal titanosauriforms of the same age in the number of collected specimens and species diversity (e.g., *Ligabuesaurus*). Moreover, morphological disparity observed among South American diplodocimorphs suggests that the group underwent an important evolutionary radiation during the Early Cretaceous and the beginning of the Late Cretaceous. This supports the idea that basal diplodocoids, particularly rebbachisaurids, played a significant role in Patagonian dinosaur communities during the pre-Turonian Cretaceous (Salgado, Garrido, Cocca, and Cocca 2004; Lamanna, Martínez, Luna, Casal, Dodson, and Smith 2001).

The beds in which the earliest records of *Limaysaurus* were reported (the Aptian–Albian Lohan Cura Formation) also yielded remains of the huge titanosaurian *Ligabuesaurus*. Salgado and colleagues (2004) pointed out that while rebbachisaurids (*Limaysaurus*, for example) have humeri that are much shorter than their femora (as in other diplodocoids), in the titanosaurian *Ligabuesaurus* the forelimbs are extremely elongate. Due to their strongly differing limb proportions, Salgado and colleagues argued that *Ligabuesaurus* and *Limaysaurus* had different feeding habits. The fact that titanosauriforms recorded in the Lower and Mid-Cretaceous of Patagonia (including *Ligabuesaurus*; Bonaparte, González Riga, and Apesteguía 2006) have relatively broad teeth (unlike diplodocoids, which are characterized by their cylindrical dentition; Wilson 2002) supports this suspicion.

Documentation of ankylosaurs in the Aptian of Australia (i.e., *Minmi paravertebra*; Molnar 1980a) demonstrates that the group was present in Gondwana from at least the Early Cretaceous. Footprints from the Lower Cretaceous of Brazil support the idea that ankylosaurs may have been present in South America during and after the Early Cretaceous (von Huene 1931b).

During Aptian–Cenomanian times, spinosaurid theropods diversified and dominated numerically, as has been recorded in Saharan Africa and northeastern South America. Abundant gigantic carcharodontosaurids also formed part of this assemblage, along with medium-sized and large abelisauroids. Small coelurosaurian theropods were also present and diverse, as documented in the Aptian–Albian Santana Formation.

Notably, the available fossil record from the Aptian–Cenomanian includes the largest known representatives for each dinosaurian clade: *Argentinosaurus* for Titanosauria, *Ekrixinatosaurus* for Abelisauroidea, and *Giganotosaurus* for Carcharodontosauridae. Ornithopods also reached considerable size in the Aptian of northern Africa (e.g., *Ouranosaurus*, approximately 6 meters long) and the Cenomanian of South America (approximately 9 meters long, based on *Limayichnus* footprints). Of course, gigantism among dinosaurs is not limited to this interval, because big animals are also recorded in younger intervals of the Cretaceous.

The Early Cretaceous dinosaur faunas from South America correspond to the interval (Barremian through Albian) when this continent was geographically connected with Africa. From Aptian through Santonian times, South America became progressively more isolated from other continents (Scotese and Golonka 1992). Complete isolation between South America and Africa occurred at some point between 106 million years (Albian) and 84 million years ago (Campanian) with the full development of a marine barrier between the two continents (Pitman, Cande, LaBrecque, and Pindell 1993).

As a result of such close paleogeographic relationships, some dinosaur taxa are shared by both continents. Abundant literature exists about the Mid-Cretaceous vertebrate taxa that Africa and South America share (see for example, Bonaparte 1986d; Calvo and Salgado 1995; De Broin 1980; Buffetaut and Taquet 1977; Medeiros and Schultz 2001b; Medeiros, Freire, and Pereira 2007). The Aptian through Cenomanian reptiles recorded on both sides of the Atlantic include theropods of different lineages (e.g., Abelisauroidea, Spinosauridae, Carcharodontosauridae, Dromaeosauridae) as well as a diversity of sauropods (e.g., Rebbachisauridae, Titanosauria), crocodiles (e.g., *Sarcosuchus*, Araripesuchidae, Notosuchia, and presumably Trematochampsidae), and turtles (e.g., Araripemyidae,

Pelomedusidae). The presence of these animals on both continents may suggest that a diverse faunal assemblage was uniformly distributed across the southern continents. Nevertheless, little study has been made about the faunal diversity within Gondwanan continents. Some distinctions in taxonomic composition become apparent when northern Gondwanan (e.g., the African Sahara plus northeastern Brazil) reptile faunas are compared with those recorded at higher latitudes (e.g., Patagonia). As defended in several papers by the Brazilian paleontologist Manuel Medeiros and collaborators (e.g., Medeiros 2001; Medeiros and Schultz 2001a, 2001b, 2002, 2004; Medeiros, Freire, and Pereira 2007), the dinosaurian fauna from northern Brazil resembles more that of northern Africa than that of Patagonia, an idea later adopted by Novas, Dalla Vecchia, and Pais (2005). Cenomanian beds of northeastern Brazil yielded remains of spinosaurid teeth and caudal vertebrae that closely resemble those of the iguanodontian *Ouranosaurus*. Also, remains of large pholidosaurid crocodiles (Buffetaut and Taquet 1977) have been found in the Lower Cretaceous of Brazil, suggesting that a fairly uniform faunal assemblage was distributed across Gondwanan lands around the paleoequator. In contrast, relevant reptile taxa that have frequently been recorded in mid-Cretaceous beds of northern Gondwana (e.g., spinosaurids, sarcosuchids, *Ouranosaurus*-like iguanodontians) have not been documented so far in the productive Cretaceous outcrops of Patagonia. In contrast with the trans-Saharan-Brazilian reptile assemblages, the Patagonian reptiles were mainly composed of sauropods, followed by carcharodontosaurid and abelisaurid theropods (Fig. 8.2) and some non-pholidosaurid crocodiles (Novas, Dalla Vecchia, and Pais 2005).

Carcharodontosaurids inhabited both low (e.g., Sahara) and high paleolatitudes (e.g., Patagonia; Novas, Martínez, de Valais, and Ambrosio 1999; Novas, de Valais, Vickers-Rich, and Rich 2005; Novas, Dalla Vecchia, and Pais 2005). Spinosaurids, instead, seem to have preferred regions closer to the equator. In northern Gondwanan faunas, spinosaurids and carcharodontosaurids are abundant while abelisauroids are numerically rare. It is difficult to explain this uneven distribution of these theropod clades across Gondwana. Presumably some kind of latitudinal factor (e.g., temperature, humidity) controlled the geographic distribution of reptile taxa (Novas, Dalla Vecchia, and Pais 2005; Medeiros, Freire, and Pereira 2007).

Different types of dinosaurs recorded in Patagonia had spiny projections on their backs (Leanza, Apesteguía, Novas, and de la Fuente 2004). Such devices evolved independently and in different ways in a variety of Early Cretaceous dinosaurs: in *Amargasaurus* they consist of elongations of bifid neural spines, in *Agustinia* they consist of long stick-shaped osteoderms, and in *Rayososaurus*, as in all rebbachisaurids, they have remarkably tall neural spines, about seven times the height of the centrum. Stegosaurians also have spine-like and plate-like dermal projections (Fig. 8.1). These non-homologous features, which are especially noteworthy in sauropods but also in Early to Mid-Cretaceous African theropods and

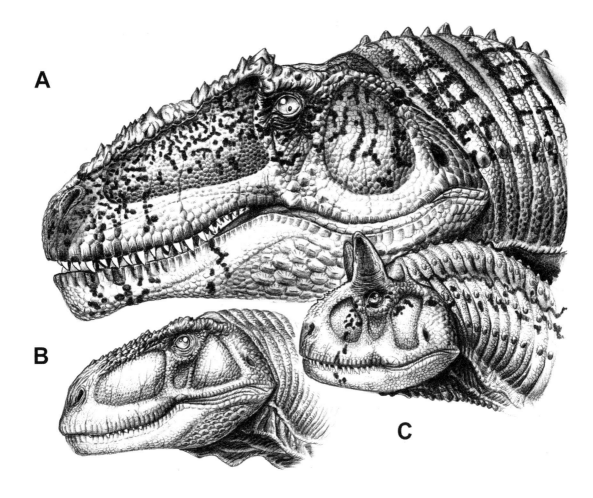

Fig. 8.2. Theropod heads. (A) *Giganotosaurus*. (B) *Abelisaurus*. (C) *Carnotaurus*.

ornithopods (e.g., *Spinosaurus, Ouranosaurus*), suggest non-phylogenetic causes such as special devices to cope with environmental conditions (Bailey 1997; Leanza, Apesteguía, Novas, and de la Fuente 2004). Such bony outgrowths may have supported skin membranes (in the neck of *Amargasaurus*) as well as a humped back above the dorsal vertebrae, so they may have played crucial adaptive roles (e.g., regulating body temperature or storing energy) to cope with the arid environments that prevailed for most of the Cretaceous Period.

From Albian through Cenomanian times, Gondwana was inhabited by large theropods that included carcharodontosaurids, spinosaurids, and the bizarre tetanuran *Bahariasaurus* (Stromer 1934; Sereno et al. 1996). This Mid-Cretaceous fauna was also composed of huge titanosaurs (e.g., *Argentinosaurus, Argyrosaurus, Paralititan*), basal diplodocimorphs (e.g., dicraeosaurids and rebbachisaurids), and, in northern Africa, crocodiles that were up to 12 meters long (e.g., *Stomatosuchus* and *Sarcosuchus*; Stromer 1936; Smith et al. 2001; Salgado 2003a; Sereno, Larsson, Sidor, and Gado 2001). In the post-Turonian, carcharodontosaurids and spinosaurids become rare or were absent in South America and were replaced by smaller abelisauroids. The fossil record of carcharodontosaurid teeth in Patagonia seems to extend to the Turonian Huincul Formation (Canale, Novas, and Simón 2005), although the teeth of abelisaurids are much more common than those belonging to Dromaeosauridae and Carcharodontosauridae. After the sharp decline of carcharodontosaurids and the virtual extinction of spinosaurids at the end of the Cenomanian, Gondwanan theropod assemblages consisted mainly of abelisauroids and secondarily of a wide array of tetanurans (e.g., *Megaraptor*, coelurosaurians; Fig. 8.1). Although abelisauroid diversification had been under way since the Early Cretaceous, they became abundant during Late Cretaceous times in South America, India, and Madagascar.

Carcharodontosaurid theropods went extinct almost at the same time as diplodocimorphs. Concurrently, the abelisaurian theropods increased in number. The available evidence indicates that dicraeosaurids had become extinct by the Aptian, while rebbachisaurids survived in South America until the Cenomanian. Salgado (2003a) has speculated that the diplodocoid extinction may have facilitated the expansion and diversification of derived titanosaurians during the Late Cretaceous.

Ornithopod dinosaurs also appear to have participated in this Cenomanian-Turonian faunal turnover: at least during Aptian through Cenomanian times, northern Africa was populated by a diversity of bulky ornithopods (e.g., *Ouranosaurus, Lurdusaurus*). Lamanna and colleagues (2004) have noted the apparent absence of African ornithischians in post-Cenomanian times. Although this is probably the result of limited sampling, as these authors admit, nevertheless a comparable absence of ornithischians from coeval sediments in Madagascar and possibly Indo-Pakistan constitutes a notable faunal similarity among these landmasses, so it is not improbable that regional extinction of ornithischian dinosaurs occurred in this western portion of Gondwana. If this distinction proves to be true, then Cretaceous faunas from South America differentiated from those of Africa, Madagascar, and Indo-Pakistan after the complete geographical isolation of these Gondwanan landmasses in post-Cenomanian times.

Other huge reptiles such as pholidosaurid crocodiles were absent in the southern landmasses after the Turonian. Interestingly, a change in dominance from araripesuchid to notosuchid crocodyliforms and from chelid to podocnemidoid turtles also occurred at that time (Apesteguía 2002; M. De la Fuente, pers. comm.). Furthermore, the notable abundance of araripesuchid crocodyliforms and large sphenodontians in beds of the Candeleros Formation at the La Buitrera fossil site is absent in younger sedimentary units, where notosuchians are more abundant and diverse.

Notably, mid-Cretaceous faunal transformations in South America may parallel the faunal replacement that occurred in North America, where Aptian–Albian acrocanthosaurids, brachiosaurids, titanosaurids, and large basal iguanodontians were replaced in the Cenomanian by hadrosaurs, ceratopsians, and tyrannosaurids (Bakker 1977; Kirkland, Lucas, and Estep 1998). This suggests that an extinction event took place probably on a global scale at the same general time (e.g., Cenomanian–Turonian). It is unknown whether the extinction of Early Cretaceous groups and the radiation of new terrestrial vertebrate groups occurred in a stepwise manner or was a response to a single catastrophic event (Novas, de Valais, Vickers-Rich, and Rich 2005).

South American Dinosaur Faunas of the Late Senonian Interval (Turonian through Campanian)

In beds of Upper Cretaceous age, the titanosaurians clearly surpass other sauropod groups in number of specimens and species diversity. The absence of diplodocoids in beds of Senonian age is at least as significant as the abundance of titanosaurs (Leanza, Apesteguía, Novas, and de la Fuente 2004). During this time, sauropods reached their greatest diversity and morphological disparity, including the first appearance in Patagonia of saltasaurines and large and gracile basal titanosaurs. The latter include medium-sized and gigantic forms (Leanza, Apesteguía, Novas, and de la Fuente 2004). In South America, sauropods still attained gigantic sizes, as documented in beds of the Río Neuquén Subgroup and the Chubut Group by *Argentinosaurus*, *Argyrosaurus*, and *Antarctosaurus? giganteus*. Titanosaurs with narrow tooth crowns are more abundant than taxa with broad tooth crowns (Fig. 8.3). As Leonardo Salgado (2003a) has noted, titanosaurs with cylindrical teeth radiated after the extinction of the diplodocoids. This modification in faunal composition, although not entirely correlated in time, may have been ecologically related to the diversification and numerical abundance of angiosperms (Salgado 2003a). Angiosperms originated around Valanginian times in dry and warm equatorial regions and dispersed later toward the poles, reaching Patagonia by Aptian times. However, angiosperms did not become important in plant biomasses before the Turonian, and this may have had an impact on herbivore faunas. The large size and softness of the leaves and fruits may have made them more palatable than the coriaceous leaves and hard cones of gymnosperms. Probably the change in tooth shape documented among titanosaurs is related to a greater abundance of angiosperm leaves and fruits during the Turonian; titanosaur teeth changed from a coarse, broad-crowned kind of tooth to a more slender, pencil-like kind of tooth.

Although available information indicates that ornithopods had been present in South America since the Berriasian and coelurosaurian theropods since at least the Aptian, the most extensive fossil samples of these two dinosaur clades come from Patagonian beds from the Cenomanian through the Maastrichtian. A diversity of iguanodontian ornithopods proliferated from Cenomanian through Maastrichtian times (e.g., *Notohypsilophodon*, *Gasparinisaura*, *Anabisetia*, *Talenkauen*, and large forms represented by the producer of *Limayichnus* footprints). Prior to Maastrichtian times, basal iguanodontians were widely distributed and

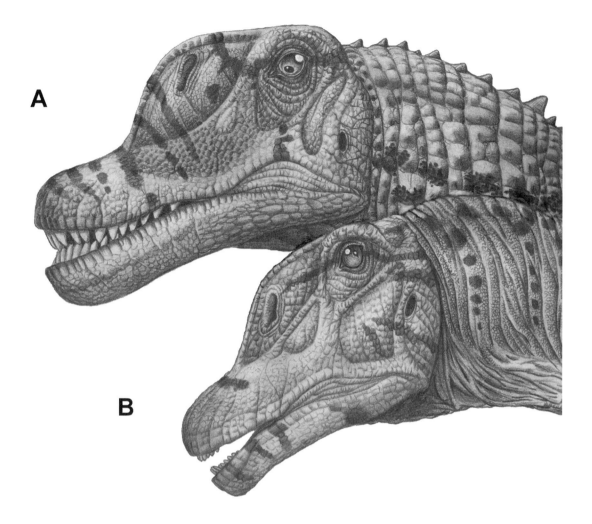

Fig. 8.3. Titanosaurian heads. (A) Yet-unnamed basal titanosaurian from the Lower Bajo Barreal Formation. (B) *Antarctosaurus*.

diversified in South America. This is congruent with the Cretaceous fossil record from Africa, Australia, and Antarctica, in which basal members of Iguanodontia have been found (Taquet 1976; Taquet and Russell 1999; Bartholomai and Molnar 1981; Cooper 1985; Rich and Vickers-Rich 1989; Calvo, Porfiri, and Novas 2005, 2008; Coria and Salgado 1996a; Coria and Calvo 2002; Hooker, Milner, and Sequeira 1991; Coria 1999; Novas, Cambiaso, and Ambrosio 2004; Salgado, Coria, and Heredia 1997). Southern ornithopods included minute forms that together with some abelisauroids (e.g., *Velocisaurus*), and maniraptoran theropods (e.g., *Alvarezsaurus*, *Buitreraptor*) constituted the smallest non-avian dinosaurs recorded in South America. Mid-sized and slender ornithopods (e.g., *Anabisetia*, *Macrogryphosaurus*) as well as bulkier iguanodontians evolved (Calvo 1991; Coria and Cambiaso 2007) alongside these minute ornithopods.

Turonian–Santonian theropods include derived maniraptorans (Novas and Puerta 1997), such as dromaeosaurids and alvarezsaurids (Novas 1996b; Chiappe and Coria 2003) and the large basal tetanuran *Megaraptor* (presumably related to spinosaurids) (Coria, Currie, Eberth, Garrido, and Koppelhus 2001; Coria and Currie 2002b; Calvo, Porfiri Veralli, Novas, and Poblete 2004). Carcharodontosaurid theropods were still present in the Turonian (Veralli and Calvo 2004) but not abundant.

The presence of alvarezsaurids in Late Cretaceous rocks of Mongolia and Patagonia is puzzling, mainly because they are among the few taxa (the others are the Titanosauridae, represented in Mongolia by *Opisthocoelicaudia*, and the Dromaeosauridae) that also found among the otherwise sharply different Late Cretaceous faunas of South America and Asia. The shared presence of alvarezsaurids in South America and Asia may indicate that *Alvarezsaurus* and *Patagonykus* were endemic taxa from Gondwana (e.g., South America; Bonaparte 1991a), which evolved in isolation during Cenomanian to Santonian times. If so, alvarezsaurids may be interpreted as later emigrants to Asia (via North America?) when continental connections were established during the Campanian (Bonaparte 1986d). This is consistent with paleogeographic reconstructions (e.g., Scotese and Golonka 1992) and with paleobiogeographic interpretations of the evolution of the vertebrate faunas of Gondwana as a whole (e.g., Bonaparte 1986d; Bonaparte and Kielan-Jaworowska 1987).

But both alvarezsaurids and dromaeosaurids in Gondwana may be explained in the context of a vicariance model (Novas and Pol 2005): these lineages were descendants of Middle to Late Jurassic maniraptorans that attained a worldwide distribution and that later produced vicariant taxa when Gondwana isolated from Laurasia. Although dispersal of maniraptoran lineages cannot be dismissed, vicariance is the more conservative explanation for maniraptoran distribution. *Archaeopteryx* in the Tithonian demonstrates that maniraptoran diversification was well under way at the end of the Jurassic period. This leads us to reconsider the paleobiogeographical history of Cretaceous theropod faunas recorded in the southern landmasses, which have traditionally been interpreted as being markedly different from those of the northern continents. This difference still seems to apply to the large theropods (for example, abelisaurids, carcharodontosaurids, spinosaurids, tyrannosaurids), but the distinctions are less marked when theropods of small body size are compared (for example, deinonychosaurs, alvarezsaurids, basal avialans). The gracile abelisauroid noasaurids are an exception; so far they have not been recorded in Laurasia.

Several groups of inland birds are recorded for the Coniacian–Santonian interval, including enantiornithines (e.g., *Neuquenornis*), flightless birds (e.g., *Patagopteryx*), and neornithines (as represented by a coracoid of a possible galliform; Agnolín, Novas, and Lío 2006).

Bonaparte (1991a) noted important distinctions between the fossil record of vertebrates of the Santonian Bajo de la Carpa Formation with that of the unconformably overlying Campanian–Maastrichtian Allen Formation, which he interpreted as showing a faunal replacement. However, the Bajo de la Carpa Formation is mainly composed of aeolian deposits that contain partially articulated skeletons of land-living tetrapods such as crocodyliforms, birds, and non-avian theropods. These deposits differ sedimentologically and taphonomically from the overlying Allen Formation, which consists of fluvial and lacustrine facies that preserved isolated teeth and bones of fishes and mammals buried far from their original sources. Moreover, titanosaurid diversity remained almost unaffected in the Santonian-Maastrichtian transition. At Cinco Saltos and Pellegrini Lake, three titanosaur taxa are recorded in the Río Colorado Subgroup: *Laplatasaurus araukanicus*, *Titanosaurus australis*, and *Pellegrinisaurus powelli*. In the overlying Allen Formation (in the locality of Salitral Moreno), three sauropod species are known: *Aeolosaurus sp.*, *Rocasaurus muniozi*, and *Bonatitan reigii*. Abelisaurids are likewise well represented in both.

The expansion of lacustrine and marine-influenced environments during the Campanian-Maastrichtian (Fig. 8.4) obviously reduced the extent of terrestrial habitats and at least some reptiles were probably affected by these changes. For example, no information is currently available about Maastrichtian notosuchians and araripesuchians in Patagonia and a reduction in non-hadrosaurid ornithopods is apparent.

South American Dinosaur Faunas of the Maastrichtian Interval

In contrast with Senonian pre-Maastrichtian herbivore faunas, for which titanosaurids occur in the largest numbers, followed by basal iguanodontians, Maastrichtian dinosaur assemblages (Fig. 8.1) are characterized by a diversity of titanosaurs—especially the small-sized saltasaurids—alongside abundant hadrosaurs, which at some localities become numerically dominant, such as Los Alamitos and Ingeniero Jacobacci (Río Negro Province). Ankylosaurs, although not numerically important, have been reported from widely dispersed fossil sites in South America (i.e., Bolivia, Patagonia, Entre Ríos Province in Argentina).

Abelisauroids were numerically dominant and taxonomically diverse in the Late Cretaceous of Patagonia and Baurú (Brazil). The dominance of abelisauroids in Maastrichtian faunas of South America is congruent with the information available from other Gondwanan localities of Late Cretaceous age (i.e., Madagascar, India). An unexpected radiation of big dromaeosaurids is documented in Maastrichtian beds of South America, represented by *Unquillosaurus ceibalii* (Los Blanquitos Formation, northwestern Argentina; Novas and Agnolín 2004) and *Austroraptor cabazai* (Allen Formation, Patagonia; Novas, Pol, Canale, Porfiri, and Calvo 2009), as well as derived maniraptorans with manual ungual phalanges resembling those of elmisaurids (Novas, Borges Ribeiro, and Carvalho 2005).

At the end of the Mesozoic, basal iguanodontians were still present (Coria, Salgado, Currie, Paulina Carabajal, and Arcucci 2004; Novas, Cambiaso, and Ambrosio 2004), but hadrosaurids were dominant. At least three different genera inhabited Patagonia: *Secernosaurus*, "*Kritosaurus*" *australis*, and an as-yet unnamed lambeosaurine (Powell 1987c), which probably arrived from North America (Bonaparte 1986d; see Salinas,

Juárez Valieri, and Fiorelli 2005 for a different view on this topic). The fact that hadrosaurids suddenly appear in the fossil record of Patagonia supports this paleobiogeographic hypothesis, in contrast with the underlying Río Colorado Subgroup—for example at the Cinco Saltos fossil site—which has provided abundant collections of titanosaurids but no hadrosaurid bones.

The record of hadrosaurids is geographically widespread in Argentina, including central and north Patagonia and La Pampa, Mendoza, and (possibly) Salta provinces. In Patagonia at least, hadrosaurids shared the landscape with titanosaurids and ankylosaurs. The association of titanosaurids and hadrosaurids has been reported from different Late Campanian and Early Maastrichtian units in northwestern Patagonia, such as the Angostura Colorada Formation at Ingeniero Jacobacci (Casamiquela 1980a); the Los Alamitos Formation at Los Alamitos Farm (Bonaparte, Franchi, Powell, and Sepúlveda 1984), where hadrosaurids are more abundant than titanosaurids; and the Allen Formation at Salitral Moreno, where titanosaurids and hadrosaurids are about equally common (Powell 1987b).

The Late Cretaceous Bauru Group of Brazil has not yet yielded ornithopod remains. Kellner and Campos (2000) suggested that hadrosaurs are most likely to be found in Brazil, particularly in Upper Cretaceous units of the Bauru Group. However, the notable absence of hadrosaurs, so common in northern Patagonia, opens the possibilities that the Bauru Group was deposited prior to the Late Campanian–Maastrichtian or that the hadrosaurids that colonized South America avoided the inland environmental conditions under which the Bauru beds were deposited. Notably, footprints of ornithopod dinosaurs are absent in the ichnologically productive Maastrichtian beds from Bolivia, where tracks of sauropods, theropods, and ankylosaurs are preserved.

In contrast to earlier times, Campanian–Maastrichtian titanosaurids include not only medium-sized members but also the smallest known adult sauropods (e.g., *Neuquensaurus*, *Saltasaurus*). However, the discovery of the huge *Puertasaurus* in the Campanian-Maastrichtian Pari Aike beds of southwestern Patagonia demonstrates that gigantic South American sauropods continued up to the end of the Mesozoic Era and that they constituted an abundant element in these southern ecosystems (Novas, Salgado, Calvo, and Agnolín 2005; Lacovara, Harris, Lamanna, Novas, Martínez, and Ambrosio 2004). Beyond southernmost Patagonia, other Maastrichtian fossil sites with sauropods in South America (Powell 2003), Madagascar (Curry Rogers and Forster 2001), and Europe (Jianu and Weishampel 1999) preserve titanosaurs that are considerably smaller than *Puertasaurus*. Thus far, the biggest Cretaceous dinosaurs (e.g., *Argentinosaurus huinculensis*, *Antarctosaurus? giganteus*, and *Puertasaurus reuili*) have been documented in South America. Presence of such huge animals in the Late Cretaceous occurred in periods when continental habitats of Patagonia were severed by extensive sea-water flooding, a pattern that counters that depicted for large-size extinct mammals, whose body mass increased with increasing land area (Burness, Diamond, and Flannery 2001). The situation described before for South America resembles that of Laurasia, where large-bodied dinosaurs diversified during the Campanian, when the western part of North America was separated from the eastern part of the continent by a shallow seaway, thus restricting dinosaur distributions. In other words, dinosaurs attained larger sizes than did terrestrial mammalian species, but they did so on a small landmass (Farlow, Dodson, and Chinsamy 1995).

Available knowledge suggests that two different morphotypes of titanosaurs coexisted in Patagonia during the Late Cretaceous: the robust and proportionally small saltasaurines and the larger but slender titanosaurs represented by *Laplatasaurus*. The development of dermal ossifications in titanosaurids was common during the Maastrichtian: there is greater morphological diversity of such osteoderms in collections from the Allen Formation at Salitral Moreno than in the older (i.e., Santonian) Río Colorado beds cropping out in the Cinco Saltos–Pellegrini Lake region.

The reported presence of ankylosaurs in association with hadrosaurs from levels of the Allen Formation has been explained (Salgado and Coria 1996; Coria and Calvo 2002) as the result of faunal interchange between South and North America during Campanian–Maastrichtian times (Bonaparte 1986d). Although hadrosaurs probably arrived from North America at the end of the Cretaceous, no evidence exists to support the idea that ankylosaurs were also Late Cretaceous immigrants, because ankylosaurs had already been present in Gondwana (e.g., Australia) since Aptian times. However, the present evidence demonstrates that by the latest Cretaceous, in addition to small-sized ankylosaurs (Coria and Salgado 2001; De Valais, Apesteguía, and Udrizar Sauthier 2003) big representatives also existed, as indicated by footprints from Bolivia (Meyer, Hippler, and Lockley 2001; McCrea, Lockley, and Meyer 2001).

No pachycephalosaurians have been found in the southern landmasses, and evidence to support the presence or absence of ceratopsians in the south is ambiguous at the moment.

On the basis of the available vertebrate fossil record and according to current paleogeographic reconstructions (Scotese and Golonka 1992), South America renewed contact with North America from Campanian to Maastrichtian times, when intense orogeny caused continentalization of the Caribbean region, forming a Central American land bridge or a string of islands between North and South America (Bonaparte 1986d; Zambrano 1981; Pitman, Cande, LaBrecque, and Pindell 1993). The record of Laurasian taxa in the Late Campanian–Maastrichtian of South America (e.g., Hadrosaurinae, Lambeosaurinae; Bonaparte 1987; Powell 1987b) can be explained as the result of such a faunal interchange (Bonaparte 1986d).

During Campanian-Maastrichtian times, Patagonia remained close to the Antarctic Peninsula, and because Antarctica was also attached to Australia and New Zealand, a common evolutionary history of a wide variety of terrestrial and marine organisms is shared by South America,

Fig. 8.4. Map of southern Argentina indicating the extent of Maastrichtian marine transgressions.

Antarctica, and Australia. However, the similarities are not total: in southern Patagonia, remains of titanosaurs are quite abundant (Novas, Cambiaso, Lirio and Núñez 2002; Novas, Fernández, Gasparini, Lirio, Núñez, and Puerta 2002; Novas, Cambiaso, and Ambrosio 2004), but contrary to expectations, the Cretaceous record of Antarctic dinosaurs does not include at the moment any sauropod remains, while ornithischians are common. A remarkable disparity in the relative abundance and taxonomic diversity of sauropods and ornithischians is apparent in the Gondwanan realm. While in South America, Madagascar, and India the Cretaceous record of sauropods overwhelms that of ornithischians, in southern Gondwana (e.g., Antarctica, Australia, and New Zealand; Bartholomai and Molnar 1981; Gasparini, Olivero, Scasso, and Rinaldi 1987; Rich and Vickers-Rich 1989; Hooker, Milner, and Sequeira 1991; Wiffen 1996; Rich, Vickers-Rich, Fernández, and Santillana 1999; Case et al. 2000; Novas, Cambiaso, and Ambrosio 2004) the sauropod record is sparse. Instead, ornithischians of different groups (e.g., iguanodontians, hypsilophodontids, nodosaurids) are frequently found at the higher paleolatitudes of Gondwana. It is probable that herbivorous dinosaurs were not evenly distributed across the Gondwanan landmasses; instead, a certain degree of provincialism is apparent.

Explanations for such dissimilarities in faunas may include differences in daylight and temperatures that are implied by the respective paleolatitudinal positions. Cretaceous plant communities of Antarctica experienced a unique combination of climatic conditions not seen on the earth today: polar warmth coupled with a strongly seasonal light regime (Cantrill 2001). Nevertheless, other possibilities (or combinations thereof) cannot be dismissed (e.g., the geographic isolation of Antarctic dinosaurs from those of Patagonia; sedimentological and taphonomic processes).

Convincing evidence (e.g., Zinsmeister 1982; Macellari 1985; Huber and Watkins 1992) suggests that by Late Cretaceous times a number of shallow-water, cool-temperature marine organisms (e.g., foraminifers, bivalves, gastropods, ammonites) were endemic to Patagonia, the Antarctic Peninsula, southeast Australia, and New Zealand. This region and its fauna are known as the Weddellian Province (e.g., Zinsmeister 1982). Marine vertebrate members of this single broad paleobiogeographic unit include bizarre elasmosaurids such as the Patagonian *Aristonectes* and its close Antarctic relative, *Morturneria* (Chatterjee and Small 1989). The peculiar mosasaurs *Lakumasaurus*, *Taniwhasaurus*, and *Moanasaurus* may also be considered part of this unique faunal assemblage (Novas, Fernández, Gasparini, Lirio, Núñez, and Puerta 2002). The available evidence suggests that the paleobiogeography of marine reptiles was more complex than previously thought. Because of their relative geographic isolation, southern marine reptile faunas may have followed evolutionary pathways different from those of their northern (e.g., North American) counterparts.

The concept of the Weddellian realm has been enlarged to include terrestrial plants (the "Nothofagus flora" characteristic of high southern

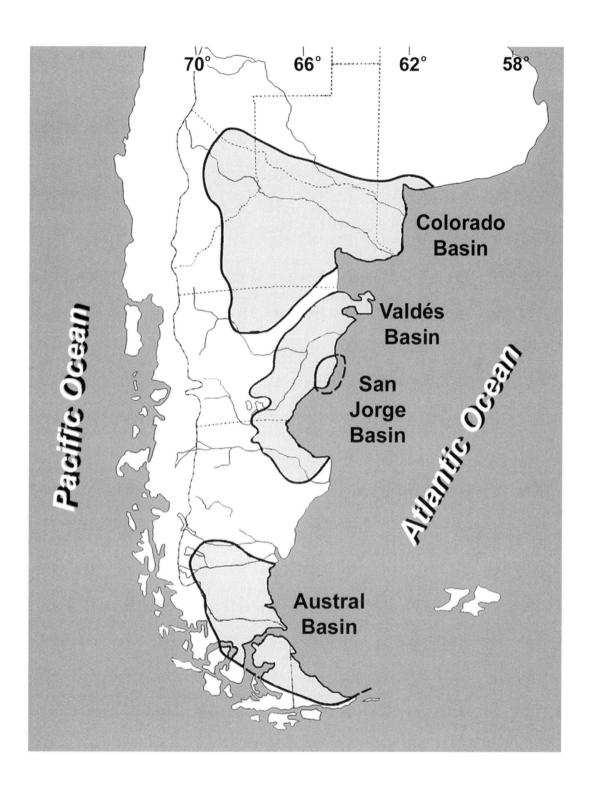

latitudes) and mammals (e.g., marsupials and monotremes). The list of Weddellian organisms may be enlarged by including ornithischian dinosaurs (e.g., basal iguanodontids, hadrosaurs, nodosaurs), which in Antarctica, Australia, and New Zealand appear to have been more successful than titanosaurids, which were the group of herbivorous dinosaurs that were characteristically more abundant in Cretaceous faunal assemblages of other parts of Gondwana (e.g., Patagonia, Madagascar, India).

With regard to the mass extinction at the Cretaceous–Paleocene boundary, South America does not at the moment offer clear evidence about the demise of dinosaurs on this continent. The problem remains almost unstudied from the point of view of terrestrial ecosystems, mainly because the youngest dinosaur-bearing beds are regarded as Early Maastrichtian and sedimentary units that correspond to the latest Maastrichtian are marine in origin. Thus, interest in the extinction has focused on marine vertebrates and invertebrates.

Titanosaur-hadrosaur communities lived close to lacustrine environments that were related to a major Maastrichtian transgressive phase over most of South America (Uliana and Biddle 1988; Fig. 8.4). This Late Campanian–Maastrichtian transgression of the Atlantic extended around the Somuncura Massif, also flooding the San Jorge, Austral, and Neuquén basins in Patagonia. This transgressive phase promoted diversification of tetrapods adapted for life in lowlands and littoral regions, such as hadrosaurs and neornithine birds. Probably related to the expansion of areas periodically affected by flooding is the catastrophic burial of dinosaur nesting sites (principally visited by titanosaurids), which took place more than once in northern Patagonia. How this mortality of dinosaur hatchlings may have affected the population growth structure of these animals is unknown, but at the very least it seems obvious that the places that once served for titanosaurid nesting had to be abandoned as shallow marine waters flooded extensive areas during the Late Maastrichtian through Early Paleocene. Habitat changes consequent to this end-Cretaceous marine transgression cannot be excluded as contributing to the K-T extinction of some dinosaurian groups.

WORKS CITED

Accarie, H., B. Beaudoin, J. Dejax, G. Friès, J. G. Michard, and P. Taquet. 1995. Découverte d'un dinosaure théropode nouveau (*Genusaurus sisteronis* n. g., n. sp.) dans l'Albien marin de Sisteron (Alpes de Haute-Provence, France) et extension au Crétacé Inférieur de la lignée cératosaurienne. *Comptes Rendus de l'Académie des Sciences de Paris* (Série IIa) 320: 327–334.

Agnolín, F. L., S. Apesteguía, and P. Chiarelli. 2004. The end of a myth: The mysterious ungual claw of *Noasaurus leali*. *Journal of Vertebrate Paleontology* 24(Suppl. 3): 33A.

Agnolín, F. L., M. D. Ezcurra, and D. Pais. 2005. Systematic reinterpretation of the pigmy *Allosaurus* from the Lower Cretaceous of Victoria (Australia). *Ameghiniana* 42(Suppl. 4), 13R.

Agnolín, F. L., and A. Martinelli. 2007. Did oviraptorosaurs (Dinosauria; Theropoda) inhabit Argentina? *Cretaceous Research* 28(5): 785–790.

Agnolín, F. L., F. E. Novas, and G. Lío. 2003. Restos de un posible galliforme (Aves: Neornithes) del Cretácico Tardío de Patagonia. *Ameghiniana* 40(Suppl. 4): 49R.

———. 2006. Neornithine bird coracoid from the Upper Cretaceous of Patagonia. *Ameghiniana* 43(1): 245–248.

Aguilera, E., H. Salas, and E. Peña. 1989. La Formación Cajones: Cretácico terminal del subandino central de Bolivia. *Revista Técnica de Yacimientos Petrolíferos Fiscales Bolivia* 10(3–4): 131–148.

Allain, R., R. Tykoski, N. Aquesbi, N. Jalil, M. Monbaron, D. Russell, and P. Taquet. 2007. An abelisauroid (Dinosauria: Theropoda) from the Early Jurassic of the High Atlas Mountains, Morocco, and the radiation of ceratosaurs. *Journal of Vertebrate Paleontology* 27(3): 610–624.

Alonso, R. N. 1980. Icnitas de dinosaurios (Ornithopoda, Hadrosauridae) en el Cretácico Superior del Norte de Argentina. *Acta Geológica Lilloana* 15(2): 55–63.

———. 1989. Late Cretaceous dinosaur trackways in Northern Argentina. In D. D. Gillette and M. G. Lockley (Eds.), *Dinosaur tracks and traces*, 223–228. Cambridge: Cambridge University Press.

Alonso, R. N., and R. A. Marquillas. 1986. Nueva localidad con huellas de dinosaurios y primer hallazgo de huellas de aves en la Formación Yacoraite (Maastrichtiano) del norte argentino. In *Actas del IV Congreso Argentino de Paleontología y Bioestratigrafía*, 2:33–42. Argentina: Mendoza.

Alvarenga, H. M. F., and J. F. Bonaparte. 1992. A new flightless land bird from the Cretaceous of Patagonia. In K. E. Campbell (ed.), *Papers in avian paleontology honoring Pierce Brodkorb*, 51–64. Los Angeles: Natural History Museum of Los Angeles County Science.

Alvarenga, H. M. F., and W. R. Nava. 2005. Aves enanthiornithes do Cretáco Superior da Formacão Adamantina do Estado de São Paulo, Brasil. In *Boletim de Resumos: II Congresso Latino-Americano de Paleontologia de Vertebrados*, 20. Rio de Janeiro: Museo Nacional.

Ameghino, F. 1921a. Sinopsis geológico paleontólogica de la Argentina. In A. J. Torcelli (comp.), *Obras Completos y Correspondencia Científica*, 12: 485–734. La Plata: Taller de Impresiones Oficiales.

———. 1921b. Nota preliminar sobre el "*Loncosaurus argentinus*." In A. J. Torcelli (comp.), *Obras Completos y Correspondencia Científica*, 12: 751–752. La Plata: Taller de Impresiones Oficiales.

Apesteguía, S. 2002. Successional structure in continental tetrapods faunas from Argentina along the Cretaceous. In *Boletim do VI° Simpósio sobre o Cretáceo do Brasil—II° Simposio sobre el Cretácico de América del Sur*, 135–141A. São Pedro, Brasil: UNESP.

———. 2004. *Bonitasaura salgadoi* gen. et sp. nov.: A beaked sauropod from the Late Cretaceous of Patagonia. *Naturwissenschaften* 91(10): 493–497.

Apesteguía, S., and A. V. Cambiaso. 1999. Hallazgo de hadrosaurios en la Formación Paso del Sapo (Campaniano-Maastrichtiano, Chubut): otra localidad del "Senoniano Lacustre." *Ameghiniana* 36(Suppl. 4): 5R.

Apesteguía, S., S. de Valais, J. A. Gonzáles, P. A. Gallina, and F. L. Agnolín. 2001. The tetrapods fauna of "La Buitrera," new locality from the basal Late Cretaceous of North Patagonia, Argentina. *Journal of Vertebrate Paleontology* 21(Suppl. 3): 29A.

Apesteguía, S., and F. E. Novas. 2003. A Late Cretaceous sphenodontian from Patagonia provides insight into lepidosaur evolution in Gondwana. *Nature* 425: 609–612.

Araújo, D. C., and T. D. Gonzaga. 1980. Uma nova especie de *Jachaleria*, Therapsida, Dicynodontia do Triassico do Brasil. In *Actas II Congreso Argentino de Paleontología y Bioestratigrafía y I Congreso Latinoamericano de Paleontología*, 2:159–174. Buenos Aires.

Arcucci, A. 1986. Nuevos materiales y reinterpretación de *Lagerpeton chanarensis* Romer (Thecodontia, Lagerpetonidae nov.) del Triasico Medio de La Rioja, Argentina. *Ameghiniana* 23(3–4): 233–242.

———. 1987. Un nuevo Lagosuchidae (Thecodontia–Pseudosuchia) de la Fauna de Los Chañares (Edad Reptil Chanarense, Triásico Medio), La Rioja, Argentina. *Ameghiniana* 24(1–2): 89–94.

———. 1997. Dinosauromorpha. In P. J. Currie and K. Padian (Eds.), *Encyclopedia of dinosaurs*, 179–183. San Diego: Academic Press.

———. 2005. Una reevaluación de los dinosauromorfos basales y el origen de Dinosauria. In *Boletim de Resumos: II Congresso Latino-Americano de Paleontología de Vertebrados*, 33. Rio de Janeiro: Museo Nacional.

Arcucci, A., and R. A. Coria. 1997. Primer registro de Theropoda (Dinosauria-Saurischia) de la Formación Los Colorados (Triásico Superior, La Rioja, Argentina). *Ameghiniana* 34(4): 531.

———. 1998. Skull features of a new primitive theropod from Argentina. *Journal of Vertebrate Paleontology* 18(Suppl. 3): 24–25A.

———. 2003. A new Triassic dinosaur. *Ameghiniana* 40(2): 217–228.

Arcucci, A., C. Forster, F. Abdala, C. May, and C. Marsicano. 1995. "Theropod" tracks from the Los Rastros Formation (Middle Triassic), La Rioja Province, Argentina. *Journal of Vertebrate Paleontology* 15(Suppl. 3): 16A.

Arcucci, A., C. Marsicano, and A. T. Caselli. 2004. Tetrapod association and paleoenvironment of Los Colorados Formation (Argentina): A significant sample from western Gondwana at the end of the Triassic. *Geobios* 37(5): 557–568.

Avilla, L. S., C. R. A. Candeiro, and E. A. L. Abrantes. 2003. Ornithischian remains from the Lower Cretaceous of Brazil and its paleobiogeographic implications. In *Boletim do 3° Simpósio Brasileiro de Paleontologia de Vertebrados*, 14. Rio de Janeiro.

Azevedo, S., and A. Kellner. 1998. A titanosaurid (Dinosauria, Sauropoda) osteoderm from the Upper Cretaceous of Minas Gerais, Brazil. *Boletim do Museu Nacional-Geologia* 44: 1–6.

Báez, A. M. 1981. Redescription and relationships of *Saltenia ibanezi*, a Late Cretaceous pipid frog from northwestern Argentina. *Ameghiniana* 18(3–4): 127–154.

Báez, A. M., and C. Marsicano. 2001. A heterodonotosaurid ornithischian dinosaur from the Upper Triassic of Patagonia. *Ameghiniana* 38(3): 271–279.

Báez, A. M., and L. Nicoli. 2004. A new look at an old frog: The Jurassic *Notobatrachus* Reig from Patagonia. *Ameghiniana* 41(3): 257–270.

Bailey, J. B. 1997. Neural spine elongation in dinosaurs: Sailbacks or buffalo backs? *Journal of Paleontology* 71(6): 1124–1146.

Bakker, R. T. 1977. Tetrapod mass extinctions—A model of the regulation of speciation rates and immigration by cycles of topographic diversity. In A. Hallam (ed.), *Patterns of evolution*, 439–468. Amsterdam: Elsevier Scientific Publishing Co.

———. 1978. Dinosaur feeding behavior and the origin of flowering plants. *Nature* 274: 661–663.

Bakker, R. T., and P. M. Galton. 1974. Dinosaur monophyly and a new class of vertebrates. *Nature* 248: 169–172.

Barberena, M., D. Araujo, and E. Lavina. 1985. Late Permian and Triassic tetrapods of Southern Brazil. *National Geographic Research* 1(1): 15–20.

Barrett, P. M., R. J. Butler, F. E. Novas, S. C. Moore-Fay, J. M. Moody, J. M. Clark, and M. R. Sánchez-Villagra. 2008. Dinosaur remains from the La Quinta Formation (Lower or Middle Jurassic) of the Venezuelan Andes. *Palaeontologische Zeitschrift* 82(2): 163–177.

Barrett, P. M., and K. J. Willis. 2001. Did dinosaurs invent flowers? Dinosaur-angiosperm coevolution revisited. *Biological Reviews* 76: 411–447.

Barrio, R., J. Panza, and F. Nullo. 1999. Jurásico y Cretácico del Macizo del Deseado, provincia de Santa Cruz. In R. Caminos (ed.), *Geología Argentina*, 511–527. Buenos Aires: Instituto de Geología y Recursos Minerales.

Bartholomai, A., and R. E. Molnar. 1981. *Muttaburrasaurus*, a new iguanodontid (Ornithischia: Ornithopoda) dinosaur from the Lower Cretaceous of Queensland. *Memoirs of the Queensland Museum* 20: 319–349.

Baur, G. 1891. Remarks on the reptiles generally called dinosaurs. *American Naturalist* 25(293): 434–454.

Bell, C. M., and M. Suárez. 1989. Vertebrate fossils and trace fossils in Upper Jurassic–Lower cretaceous red beds in the Atacama region, Chile. *Journal of South American Earth Sciences* 2(4): 351–357.

Benedetto, J. 1973. Herrerasauride, nueva familia de saurisquios triasicos. *Ameghiniana* 10(1): 89–102.

Benton, M. J. 1991. What really happened in the Late Triassic? *Historical Biology* 5: 263–278.

———. 1994. Late Triassic to Middle Jurassic extinctions among continental tetrapods: Testing the pattern. In N. C. Fraser and H. D. Sues (Eds.), *In the shadow of the dinosaurs: Early Mesozoic tetrapods*, 366–397. Cambridge: Cambridge University Press.

Bertini, R. J. 1996. Evidéncias de Abelisauridae (Carnosauria: Saurischia) do Neocretáceo da Bacia do Paraná. In *Boletim do IV° Simposio sobre o Cretáceo do Brasil*, 267–271. São Pedro: UNESP.

Bertini, R. J., L. G. Marshall, M. Gayet, and P. M. Brito. 1993. Vertebrate faunas from the Adamantina and Marília formations (Upper Bauru Group, Late Cretaceous, Brazil) in their stratigraphic and paleobiogeographic contexts. *Neues Jahrbuch für Geologie und Paläontologie Abhandlungen* 188(1): 71–101.

Bertini, R., R. Santucci, L. Ribeiro, and A. Arruda-Campos. 2000. *Aeolosaurus* (Sauropoda: Titanosauria) from the Upper Cretaceous of Brazil. *Ameghiniana* 37(Suppl. 4): 19R.

Bigarella, J. J., and R. Salamuni. 1961. Early Mesozoic wind patterns as suggested by dune bedding in the Botucatu Sandstone of Brazil and Uruguay. *Bulletin of the Geological Society of America* 72(7): 1089–1106.

Bittencourt, J. S., and A. W. A. Kellner. 2002. Abelisauria (Theropoda, Dinosauria) teeth from Brazil. *Boletim do Museu Nacional* 63: 1–8.

———. 2004. On a sequence of sacrocaudal theropod dinosaur vertebrae from the Lower Cretaceous Santana Formation, Northeastern Brazil. *Aquivos do Museu Nacional* 62(13): 309–320.

Bonaparte, J. F. 1967. Dos nuevas "faunas" de reptiles triásicos de Argentina. In *Gondwana Symposium*, 283–306. Mar del Plata: UNESCO.

———. 1972. Los tetrápodos del sector superior de la Formación Los Colorados, La Rioja, Argentina (Triásico Superior). Parte I. *Opera Lilloana* 22: 1–183.

———. 1973. Edades/Reptil para el Triásico de Argentina y Brasil. In *Actas V Congreso Geológico Argentino* 3: 93–129. Buenos Aires.

———. 1975. Nuevos materiales de *Lagosuchus talampayensis* Romer (Thecodontia, Pseudosuchia) y su significado en el origen de los Saurischia. Chañarense inferior, Triásico Medio de Argentina. *Acta Geológica Lilloana* 13: 5–90.

———. 1976. *Pisanosaurus mertii* Casamiquela and the origin of the Ornithischia. *Journal of Paleontology* 50(5): 808–820.

———. 1978. El Mesozoico de America del Sur y sus Tetrápodos. *Opera Lilloana* 26: 1–569.

———. 1979a. *Coloradia brevis* n. g. et n. sp. (Saurischia-Prosauropoda), dinosaurio Plateosauridae de la Formación Los Colorados, Triásico Superior de La Rioja, Argentina. *Ameghiniana* 15(3): 327–332.

———. 1979b. Dinosaurs: A Jurassic assemblage from Patagonia. *Science* 205(4413): 1377–1379.

———. 1979c. Faunas y paleobiogeografía de los tetrápodos mesozoicos de América del Sur. *Ameghiniana* 16(3–4): 217–246.

———. 1981. Dinosaurios del Jurásico de América del Sur. *Investigación y Ciencia* 63: 110–121.

———. 1982. Faunal replacement in the Triassic of South America. *Journal of Vertebrate Paleontology* 2(3): 362–371.

———. 1984a. Locomotion in rauisuchid thecodonts. *Journal of Vertebrate Paleontology* 3(4): 210–218.

———. 1984b. El intercambio faunístico de vertebrados continentales entre América del Sur y del Norte a fines del Cretácico. In *III Congreso Latinoamericano de Paleontología*, 38–450. México: D. F.

———. 1984c. Nuevas pruebas de la conexión fisica entre Sudamerica y Norteamerica en el Cretacico Tardío (Campaniano). In *III Congreso Argentino de Paleontología y Bioestratigrafía*, 3:141–149. Corrientes.

———. 1985. A horned Cretaceous carnosaur from Patagonia. *National Geographic Research* 1(1): 149–151.

———. 1986a. The early radiation and phylogenetic relationships of the Jurassic sauropod dinosaurs, based on vertebral anatomy. In K. Padian (ed.), *The beginning of the age of dinosaurs*, 245–258. Cambridge: Cambridge University Press.

———. 1986b. Les Dinosaures (Carnosaures, Allosauridés, Sauropodes, Cétiosauridés) du Jurassique Moyen de Cerro Cóndor (Chubut, Argentine). *Annales de Paléontologie* 72(3): 247–289.

———. 1986c. Les Dinosaures (Carnosaures, Allosauridés, Sauropodes, Cétiosauridés) du Jurassique Moyen de Cerro Cóndor (Chubut, Argentine). 2nd Part. *Annales de Paléontologie* 72(4): 326–386.

———. 1986d. History of the terrestrial Cretaceous vertebrates of Gondwana. In *IV Congreso Argentino de Paleontología y Bioestratigrafía*, 2:63–95. Mendoza: Editorial Inca.

———. 1987. The Late Cretaceous fauna of Los Alamitos, Patagonia, Argentina. *Revista del Museo Argentino de Ciencias Naturales Bernardino Rivadavia* 3(3): 103–179.

———. 1991a. Los vertebrados fósiles de la Formación Rio Colorado, de la ciudad de Neuquén y cercanías, Cretácico Superior, Argentina. *Revista del Museo Argentino de Ciencias Naturales 'Bernardino Rivadavia* 4(3): 17–123.

———. 1991b. The Gondwanan theropod families Abelisauridae and Noasauridae. *Historical Biology* 5: 1–25.

———. 1994. Patagonia: The world's smallest dinosaur. A new discovery. *Ligabue Magazine* 24: 136–137.

———. 1996a. *Dinosaurios de America del Sur.* Buenos Aires: Museo Argentino de Ciencias Naturales Bernardino Rivadavia.

———. 1996b. Cretaceous tetrapods of Argentina. *Münchner Geowissenschaftliche Abhandlungen* 30: 73–130A.

———. 1997a. *El Triasico de San Juan—La Rioja Argentina y sus Dinosaurios.* Buenos Aires: Museo Argentino de Ciencias Naturales Bernardino Rivadavia.

———. 1997b. *Rayososaurus agrioensis* Bonaparte 1995. *Ameghiniana* 34(1): 116.

———. 1999a. *Rebbachisaurus tessonei* Calvo y Salgado 1996 is not *Rebbachisaurus* Lavocat 1954. *Ameghiniana* 36(1): 96.

———. 1999b. An armored sauropod from the Aptian of Northern Patagonia, Argentina. In Y. Tomida, T. Rich, and P. Vickers-Rich (Eds.), *Proceedings of the Second Gondwanan Dinosaur Symposium*, 1–12. Tokyo: National Science Museum.

———. 1999c. Evolución de las vértebras presacras en Sauropodomorpha. *Ameghiniana* 36(2): 115–187.

———. 1999d. Tetrapod faunas from South America and India: A paleobiogeographic interpretation. *Proceedings of the Indian National Science Academy* 65: 427–437.

Bonaparte, J. F., G. Brea, C. L. Schultz, and A. G. Martinelli. 2007. A new specimen of *Guaibasaurus candelariensis* (basal Saurischia) from the Late Triassic Caturrita Formation of southern Brazil. *Historical Biology* 19(1): 73–82.

Bonaparte, J. F., and R. A. Coria. 1993. Un nuevo y gigantesco saurópodo titanosaurio de la Formación Limay (Albiano-Cenomaniano) de la Provincia del Neuquén, Argentina. *Ameghiniana* 30(3): 271–282.

Bonaparte, J. F., J. Ferigolo, and A. M. Ribeiro. 1999. A new early Late Triassic Dinosaur from Rio Grande do Sul State, Brazil. In Y. Tomida, T. H. Rich, and P. Vickers-Rich (Eds.), *Proceedings of the Second Gondwanan Dinosaur Symposium*, 89–109. Tokyo: National Science Museum.

Bonaparte, J. F., M. Franchi, J. Powell, and E. Sepúlveda. 1984. La Formación Los Alamitos (Campaniano-Maastrichtiano) del sudeste de Río Negro, con descripción de *Kritosaurus australis* n. sp. (Hadrosauridae). Significado paleogeográfico de los vertebrados. *Revista de la Asociación Geológica Argentina* 39(3–4): 284–299.

Bonaparte, J. F., and Z. Gasparini. 1979. Los saurópodos de los Grupos Neuquén y Chubut, y sus relaciones cronológicas. In *Actas del V Congreso Geológico Argentino*, 2:393–406. Neuquén.

Bonaparte, J. F., B. J. González Riga, and S. Apesteguía. 2006. *Ligabuesaurus leanzai* gen. et sp. nov. (Dinosauria, Sauropoda), a new titanosaur from the Lohan Cura Formation (Aptian, Lower Cretaceous) of Neuquén, Patagonia, Argentina. *Cretaceous Research* 27(3): 364–376.

Bonaparte, J. F., and Z. Kielan-Jaworowska. 1987. Late Cretaceous dinosaur and mammal faunas of Laurasia and Gondwana. In P. M. Currie and E. H. Koster (Eds.), *Fourth symposium on Mesozoic terrestrial ecosystems: Short papers*, 24–29. Drumheller: Tyrrell Museum of Paleontology Occasional Papers.

Bonaparte, J. F., A. G. Martinelli, C. L. Schultz, and R. Rubert. 2003. Sister group of mammals: Small cynodonts from the Late Triassic Southern Brazil. *Revista Brasileira de Paleontologia* 5(1): 5–27.

Bonaparte, J. F., and F. E. Novas. 1985. *Abelisaurus comahuensis*, n. g. et n. sp., Carnosauria del Cretácico tardío de Patagonia. *Ameghiniana* 21(2–4): 259–265.

Bonaparte, J. F., F. E. Novas, and R. A. Coria. 1990. *Carnotaurus sastrei* Bonaparte, the horned, lightly built carnosaur from the Middle Cretaceous of Patagonia. *Contributions in Science (Natural History Museum of Los Angeles County)* 416: 1–42.

Bonaparte, J. F., and J. Powell. 1980. A continental assemblage of tetrapods from the Upper Cretaceous beds of El Brete, northwestern Argentina (Sauropoda-Coelurosauria-Carnosauria-Aves). *Société Géologique de France* 59(139): 19–28.

Bonaparte, J. F., and J. Pumares. 1995. Notas sobre el primer cráneo de *Riojasaurus incertus* (Dinosauria, Prosauropoda, Melanorosauridae) del Triasico Superior de La Rioja, Argentina. *Ameghiniana* 32(4): 341–349.

Bonaparte, J. F., and G. W. Rougier. 1987. The Late Cretaceous fauna of Los Alamitos, Patagonia, Argentina. Part VII—The hadrosaurs. *Revista del Museo Argentino de Ciencias Naturales Bernardino Rivadavia* 3: 155–161.

Bonaparte, J. F., and H. D. Sues. 2006. A new species of *Clevosaurus* (Lepidosauria: Rhynchocephalia) from the Upper Triassic of Rio Grande do Sul, Brazil. *Palaeontology* 49(4): 917–923.

Bonaparte, J. F., and M. Vince. 1979. El hallazgo del primer nido de dinosaurios triásicos (Saurischia, Prosauropoda) Triásico Superior de Patagonia, Argentina. *Ameghiniana* 16(1–2): 173–182.

Borsuk-Bialynicka, M. 1977. A new camarasaurid sauropod *Opisthocoelicaudia skarzynskyi*, gen n. sp. nov. from the Upper Cretaceous of Mongolia. *Palaeontologia Polonica* 37: 5–64.

Bossi, G. 1990. Triásico. In J. F. Bonaparte, A. J. Toselli, and F. G. Aceñolaza (Eds.), *Geología de América del Sur*, 3:15–87. Tucumán: Série Correlación Geológica Nro 2, Universidad Nacional de Tucumán.

Brea, G., J. F., Bonaparte, C. L. Schultz, and A. Martinelli. 2005. A new specimen of *Guaibasaurus candelariensis* (basal Saurischia) from the Late Triassic Caturrita Formation of Southern Brazil. In *Boletim de Resumos: II Congreso Latino-Americano de Paleontología de Vertebrados*, 33. Rio de Janeiro: Museo Nacional.

Brett-Surman, M. K. 1972. The appendicular anatomy of hadrosaurian dinosaurs. Master's thesis. University of California at Berkeley.

———. 1979. Phylogeny and palaeobiogeography of hadrosaurian dinosaurs. *Nature* 277: 560–562.

Bridge, J. S., G. A. Jalfing, and S. M. Georgieff. 2000. Geometry, lithofacies, and spatial distribution of Cretaceous fluvial sandstone bodies, San Jorge basin, Argentina: Outcrop analog for the hydrocarbon-bearing Chubut Group. *Journal of Sedimentary Research* 70(2): 341–359.

Brinkman, D., and H. D. Sues. 1987. A staurikosaurid dinosaur from the Upper Triassic Ischigualasto Formation of Argentina and the relationships of the Staurikosauridae. *Palaeontology* 30(3): 493–503.

Buffetaut, E., and J. le Loeuff. 1995. Commentaire à la note *Découverte d'un dinosaure théropode nouveau* (Genusaurus sisteronis n. g., n. sp.) dans l'Albien marin de Sisteron (Alpes de Haute-Provence, France) et extension au Crétacé Inférieur de la lignée cératosaurienne. *Comptes Rendus de l'Académie des Sciences* (Série IIa) 321: 79–80.

Buffetaut, E., V. Suteethorn, G. Cuny, H. Tong, J. le Loeuff, S. Khansubha, and S. Jongautchariyakul. 2000. The earliest known sauropod dinosaur. *Nature* 407: 72–74.

Buffetaut, E., V. Suteethorn, J. le Loeuff, G. Cuny, H. Tong, and S. Khansubha. 2002. The first giant dinosaurs: A large sauropod from the Late Triassic of Thailand. *Comptes Rendus Palevol* 1(2): 103–109.

Buffetaut, E., and P. Taquet. 1977. The giant crocodilian *Sarcosuchus* in the Early Cretaceous of Brazil and Niger. *Palaeontology* 20(1): 203–208.

Burness, G. P., J. Diamond, and T. Flannery. 2001. Dinosaurs, dragons, and dwarfs: The evolution of maximal body size. *Proceedings of the National Academy of Science* 98(25): 14518–14523.

Cabaleri, N., and C. Armella. 1999. Facies lacustres de la Formación Cañadón Asfalto (Caloviano-Oxfordiano), en la quebrada Las Chacritas, Cerro Cóndor, provincia de Chubut. *Revista de la Asociación Geológica Argentina* 54: 375–388.

Cabrera, A. 1947. Un saurópodo nuevo del Jurásico de Patagonia. *Notas del Museo de La Plata, Paleontología* 12(95): 1–17.

Calvo, J. O. 1991. Huellas de dinosaurios en la Formación Río Limay (Albiano-Cenomaniano), Picún Leufú, Provincia de Neuquén, República Argentina (Ornithischia, Saurischia: Sauropoda-Theropoda). *Ameghiniana* 28(3–4): 241–258.

———. 1994. Jaw mechanics in sauropod dinosaurs. *Gaia* 10: 183–193.

———. 1999. Dinosaurs and other vertebrates of the Lake Ezequiel Ramos Mexía area, Neuquén-Patagonia, Argentina. In Y. Tomida, T. H. Rich, and P. Vickers-Rich (Eds.), *Proceedings of the Second Gondwanan Dinosaur symposium*, 13–45. Tokyo: National Science Museum.

———. 2002. Dinosaurios del Neuquén, principales hallazgos. El yacimiento del Lago Barreales: su importancia paleontológica y evaluación turística. In *1° Congreso Latinoamericano de Paleontología de Vertebrados*, 10–11R. Santiago de Chile.

———. 2007. Ichnology. In Z. Gasparini, L. Salgado, and R. Coria (Eds.), *Patagonian Mesozoic reptiles*, 314–334. Bloomington: Indiana University Press.

Calvo, J. O., and J. F. Bonaparte. 1991. *Andesaurus delgadoi* gen. et sp. nov. (Saurischia-Sauropoda), dinosaurio Titanosauridae de la Formación Río Limay (Albiano-Cenomaniano), Neuquén, Argentina. *Ameghiniana* 28(3–4): 303–310.

Calvo, J. O., and R. Coria. 2000. New specimen of *Giganotosaurus carolinii* (Coria and Salgado, 1995), supports it as the largest theropod ever found. *Gaia* 15: 117–122.

Calvo, J. O., R. A. Coria, and L. Salgado. 1997. Uno de los más completos Titanosauridae (Dinosauria, Sauropoda) registrados en el mundo. *Ameghiniana* 34(4): 534A.

Calvo, J. O., and B. J. González Riga. 2003. *Rinconsaurus caudamirus* gen. et. sp. nov., a new titanosaurid (Dinosauria, Sauropoda) from the Late Cretaceous of Patagonia, Argentina. *Revista Geológica de Chile* 30(2): 333–353.

Calvo, J. O., and G. V. Mazzetta. 2004. Nuevos hallazgos de huellas de dinosaurios en la Formación Candeleros (Albiano-Cenomaniano), Picún Leufú, Neuquén, Argentina. *Ameghiniana* 41(4): 545–554.

Calvo, J. O., J. D. Porfiri, B. González Riga, and A. W. A. Kellner. 2007. A new Cretaceous terrestrial ecosystem from Gondwana with the description of a new sauropod dinosaur. *Anais da Academia Brasileira de Ciências* 79(3): 529–541.

Calvo, J. O., J. D. Porfiri, and A. W. A. Kellner. 2004. On a new maniraptoran dinosaur (Theropoda) from the Upper Cretaceous of Neuquén, Patagonia, Argentina. *Arquivos do Museu Nacional do Río de Janeiro* 62(4): 549–566.

Calvo, J. O., J. D. Porfiri, and F. E. Novas. 2005. Un gigantesco Euiguanodontia (Dinosauria: Ornithischia) con delgadas placas en la región torácica, Cretácico Tardío, Patagonia. In *Boletim de Resumos: II Congreso Latino-Americano de*

Paleontologia de Vertebrados, 62. Rio de Janeiro: Museo Nacional.

———. 2008. Discovery of a new ornithopod dinosaur from the Portezuelo Formation (Upper Cretaceous), Neuquén, Patagonia, Argentina. *Arquivos do Museu Nacional do Rio de Janeiro* 65(4): 471–483.

Calvo, J. O., J. D. Porfiri, C. Veralli, F. E. Novas, and F. Poblete. 2004. Phylogenetic status of *Megaraptor namunhuaiquii* Novas based on a new specimen from Neuquén, Patagonia, Argentina. *Ameghiniana* 41(4): 565–575.

Calvo, J. O., J. D. Porfiri, C. Veralli, and F. Poblete. 2001a. A giant titanosaurid sauropod from the Upper Cretaceous of Neuquén, Patagonia, Argentina. *Ameghiniana* 38(Suppl. 4): 5R.

———. 2001b. One of the largest titanosaurid sauropods ever found, Upper cretaceous, Neuquén, Patagonia, Argentina. *Journal of Vertebrate Paleontology* 21(Suppl. 3): 37A.

Calvo, J. O., D. Rubilar-Rogers, and K. Moreno. 2004. A new Abelisauridae (Dinosauria: Theropod) from northwest Patagonia. *Ameghiniana* 41(4): 555–563.

Calvo, J. O., and L. Salgado. 1991. Posible registro de *Rebbachisaurus* Lavocat (Sauropoda) en el Cretácico Tardío de Patagonia. *Ameghiniana* 28(3–4): 404.

———. 1995. *Rebbachisaurus tessonei* sp. nov. a new Sauropoda from the Albian-Cenomanian of Argentina: New evidence on the origin of the Diplodocidae. *Gaia* 11: 13–33.

———. 1998. Nuevos restos de titanosauridae (Sauropoda) en el Cretácico Inferior de Neuquén, Argentina. In *VII Congreso Argentino de Paleontología y Bioestratigrafía*, 59R. Bahía Blanca, Argentina.

Cambiaso, A. 2007. Los ornitópodos e iguanodontes basales (Dinosauria, Ornithischia) del Cretácico de Argentina y Antártida. Tesis Doctoral, Facultad de Ciencias Exactas y Naturales, Universidad de Buenos Aires.

Campos, D. de A., and A. W. A. Kellner. 1991. Dinosaurs of the Santana Formation with comments on other Brazilian occurrences. In J. G. Maisey (Ed.), *Santana fossils: An illustrated atlas*, 372–375. Neptune City, N.J.: T. F. H. Publications.

———. 1999. On some sauropod (titanosauridae) pelves from the continental Cretaceous of Brazil. In Y. Tomida, T. Rich, and P. Vickers-Rich (Eds.), *Proceedings of the Second Gondwanan Dinosaur Symposium*, 143–166. Tokyo: National Science Museum.

Campos, D. de A., A. W. A. Kellner, R. J. Bertini, and R. M. Santucci. 2005. On a titanosaurid (Dinosauria, Sauropoda) vertebral column from the Bauru Group, Late Cretaceous of Brazil. *Aquivos do Museu Nacional do Rio de Janeiro* 63(3): 565–593.

Canale, J. D., F. E. Novas, and A. C. Scanferla. 2006. A complete skeleton of an abelisaurid theropod from the Late Cretaceous of Patagonia, Argentina. *Ameghiniana* 43(Suppl. 4): 27R.

Canale, J. D., F. E. Novas, and M. E. Simón. 2005. Dientes de terópodos de la Formación Huincul (Turoniano Inferior), provincia del Neuquén, Argentina. *Ameghiniana* 42(Suppl. 4): 63R.

Canale, J. I., C. A. Scanferla, F. L. Agnolín, and F. E. Novas. 2009. New carnivorous dinosaur from the Late Cretaceous of NW Patagonia and the evolution of abelisaurid theropods. *Naturwissenschaften* 96: 409–414.

Candeiro, R., C. T. Abranches, E. A. Abrantes, L. S. Avilla, V. C. Martins, A. L. Moreira, S. R. Torres, and L. P. Bergqvist. 2004. Dinosaur remains from western São Paulo state, Brazil (Bauru Basin, Adamantina Formation, Upper Cretaceous). *Journal of South American Earth Sciences* 18(1): 1–10.

Cantrill, D. J. 2001. Cretaceous high-latitude terrestrial ecosystems: an example from Alexander Island, Antarctica. *Asociación Paleontológica Argentina, Publicación Especial N°* 7: 39–44.

Canudo, J. I., and L. Salgado. 2003. Los dinosaurios del Neocomiense (Cretácico inferior) de la Península Ibérica y Gondwana occidental: implicaciones paleobiogeográficas. In F. Perez Lorente (ed.), *Dinosaurios y otros reptiles mesozoicos de España*, 26:251–268. La Rioja: Instituto de Estudios Riojanos.

Canudo, J. I., L. Salgado, J. Barco, R. Bolatti, and J. Ruiz-Omeñaca. 2004. Dientes de dinosaurios terópodos y saurópodos de la Formación Cerro Lisandro (Cenomaniense superior-Turoniense inferior, Cretácico Superior) en Río Negro (Argentina). *Geo-Temas* 6(5): 31–34.

Carpenter, K. 1999. *Eggs, nests, and baby dinosaurs: A look at dinosaur reproduction*. Bloomington: Indiana University Press.

———. 2001. Phylogenetic analysis of the Ankylosauria. In K. Carpenter (ed.), *The armored dinosaurs*, 455–483. Bloomington: Indiana University Press.

———. 2002. Forelimb biomechanics of nonavian theropod dinosaurs in predation. *Senckenbergiana Lethaea* 82(1): 59–76.

Carrano, M. T. 2007. The appendicular skeleton of *Majungasaurus crenatissimus* (Theropoda: Abelisauridae) from the Late Cretaceous of Madagascar. In S. D. Sampson and D. W. Krause (Eds.), *Majungasaurus crenatissimus from the Late Cretaceous of Madagascar*, 163–179. [Northbrook, Ill.]: Society of Vertebrate Paleontology.

Carrano, M. T., and S. D. Sampson. 1999. Evidence for a paraphyletic "Ceratosauria" and its implications for theropod dinosaur evolution. *Journal of Vertebrate Paleontology* 19(3): 36A.

Carrano, M. T., S. D. Sampson, and C. A. Forster. 2002. The osteology of *Masiakasaurus knopfleri*, a small abelisauroid (Dinosauria: Theropoda) from the Late Cretaceous of Madagascar. *Journal of Vertebrate Paleontology* 22(3): 510–534.

Carrano, M. T., and J. A. Wilson. 2001. Taxon distributions and the tetrapod track record. *Paleobiology* 27(3): 564–582.

Carvalho, I. S. 1994. As ocorrencias de icnofósseis de vertebrados na Bacia de São Luis, Cretáceo Superior, Estado do Maranhão. In *Boletim do III Simpósio sobre o Cretáceo do Brasil*, 119–122. Rio Claro: UNESP.

———. 2000. Geological environments of dinosaur footprints in the intracratonic basins of northeast Brazil during the Early Cretaceous opening of the South Atlantic. *Cretaceous Research* 21(2–3): 255–267.

———. 2001. Pegadas de dinossauros em depósitos estuarinos (Cenomaniano) da Bacia de São Luis (MA), Brasil. In D. F. Rossetti, A. M. Góes, and W. Truckenbrodt (Eds.), *O Cretaceo na Bacia de São Luis-Grajau*, 245–264. Belem: Museu Paraense Emílio Goeldi.

———. 2004. Dinosaur footprints from Northeastern Brazil: Taphonomy and environmental setting. *Ichnos* 11(3–4): 311–321.

Carvalho, I. S., L. S. Avilla, and L. Salgado. 2003. *Amazonsaurus maranhensis* gen. et sp. nov. (Sauropoda, Diplodocoidea) from the Lower Cretaceous (Aptian-Albian) of Brazil. *Cretaceous Research* 24(6): 697–713.

Carvalho, I. S., and R. A. Goncalves. 1994. Pegadas de dinossauros neocretáceas da Formação Itapecuru, Bacia de São Luis (Maranhao, Brasil). *Anais da Academia Brasileira de Ciências* 66(3): 279–292.

Carvalho, I. S., M. S. S. Viana, and M. F. Lima Filho. 1995. Os icnofósseis de dinosauros da Bacia do Araripe (Cretáceo Inferior, Ceará-Brazil). *Anais da Academia Brasileira de Ciências* 67(4): 433–442.

Casadío, S., T. Manera, A. Parras, and C. Montalvo. 2002. Huevos de dinosaurios (Faveoloolithidae) del Cretácico Superior de la cuenca del Colorado, provincia de La Pampa, Argentina. *Ameghiniana* 39(3): 285–293.

Casadío, S., T. Manera, A. Parras, C. Montalvo, and G. Cornachione. 2000. Primer registro en superficie de sedimentitas continentales del Cretácico Superior de la Cuenca del Colorado, sureste de La Pampa. *Revista de la Asociación Geológica Argentina* 55(1–2): 129–133.

Casal, G., R. D. Martínez, M. Luna, J. C. Sciutto, and M. C. Lamanna. 2007. *Aeolosaurus colhuehuapensis* sp. nov. (Sauropoda, Titanosauria) de la Formacion Bajo Barreal, Cretacico Superior de Argentina. *Revista Brasileira de Paleontologia* 10(1): 53–62.

Casamiquela, R. M. 1963. Consideraciones acerca de *Amygdalodon* Cabrera (Sauropoda, Cetiosauridae) del Jurásico Medio de Patagonia. *Ameghiniana* 3(3): 79–95.

———. 1964a. *Estudios icnológicos. Problemas y métodos de la icnología con aplicación al estudio de pisadas Mesozoicas (Reptilia, Mammalia) de la Patagonia*. Viedma: Ministerio de Asuntos Sociales de la Provincia de Río Negro.

———. 1964b. Sobre un dinosaurio hadrosáurido de la Argentina. *Ameghiniana* 3(9): 285–312.

———. 1967. Un nuevo dinosaurio ornitisquio triásico (*Pisanosaurus mertii*; Ornithopoda) de la Formación Ischigualasto, Argentina. *Ameghiniana* 4(2): 47–64.

———. 1977. The presence of the genus *Plateosaurus* (Sauropodomorpha) in the Upper Triassic of the El Tranquillo Formation, Patagonia. In *4th International Gondwana Symposium, Abstracts*, 30–31. Calcutta.

———. 1980a. Considérations écologiques et zoogéographiques sur les vertébrés de la zone littorale de la mer du Maastrichtien dans le Nord de la Patagonie. *Mémoires de la Société Géologique de France* 159(139): 53–55.

———. 1980b. La presencia del género *Plateosaurus* (Prosauropoda) en el Triásico Superior de la Formación El Tranquilo, Patagonia. In *Actas II Congreso Argentino de Paleontología y Bioestratigrafía y I Congreso Latinoamericano de Paleontología* 1: 143–158. Buenos Aries.

———. 1987. Novedades en ichnología de vertebrados en la Argentina. In *Anais do 10° Congresso Brasileiro de Paleontologia*, 445–456. Rio de Janeiro.

Casamiquela, R. M., J. Corvalán, and F. Franquesa. 1969. Hallazgo de dinosaurios en el Cretácico Superior de Chile. Su importancia cronológica-estratigráfica. *Instituto de Investigaciones Geológicas de Chile* 25: 3–31.

Casamiquela, R. M., and A. Fasola. 1968. Sobre pisadas de dinosaurios del Cretácico Inferior de Colchagua (Chile). *Publicaciones del Departamento Geológico de la Universidad de Chile* 30: 1–24.

Case, J. A., J. Martin, D. Chaney, M. Reguero, S. Marenssi, S. Santillana, and M. Woodburne. 2000. The first duck-billed dinosaur (family Hadrosauridae) from Antarctica. *Journal of Vertebrate Paleontology* 20: 612–614.

Case, J., J. Martin, and M. Reguero. 2007. A dromaeosaur from the Maastrichtian of James Ross Island and the Late Cretaceous Antarctic dinosaur fauna.

In A. Cooper, C. Raymond, and the 10th ISAES Editorial Team (Eds.), *Antarctica: A keystone in a changing world—Online proceedings for the 10th International Symposium on Antarctic Earth Sciences*. Reston, Va.: U.S. Geological Service and the National Academies Press. Available at http://pubs.usgs.gov/of/2007/1047/srp/srp083/index.html.

Case, J. A., and C. P. Tambussi. 1999. Maastrichtian record of neornithine birds in Antarctica: comments on a Late Cretaceous radiation of modern birds. *Journal of Vertebrate Paleontology* 19(3): 37A.

Caselli, A. T., C. A. Marsicano, and A. B. Arcucci. 2001. Sedimentología y paleontología de la Formación Los Colorados, Triásico Superior (Provincias de La Rioja y San Juan, Argentina). *Revista de la Asociación Geológica Argentina* 56: 173–188.

Charig, A. J., and A. C. Milner. 1986. *Baryonyx*, a remarkable new theropod dinosaur. *Nature* 324: 359–361.

Chatterjee, S. 1993. *Shuvosaurus*, a new theropod. *Research & Exploration: A Scholarly Publication of the National Geographic Society* 9(3): 274–285.

———. 2002. The morphology and systematics of *Polarornis*, a Cretaceous loon (Aves: Gaviidae) from Antarctica. In Z. Zhou and F. Zhang (Eds.), *Proceedings of the 5th Symposium of the Society of Avian Paleontology and Evolution*, 125–155. Beijing: Science Press.

Chatterjee, S., and D. Rudra. 1996. KT events in India: Impact, rifting, volcanism and dinosaur extinction. In F. E. Novas and R. E. Molnar (Eds.), *Proceedings of the Gondwanan dinosaur symposium*, 489–532. Brisbane: Queensland Museum.

Chatterjee, S., and B. J. Small. 1989. New plesiosaurs from the Upper Cretaceous of Antarctica. In A. Clarke and J. A. Crame (Eds.), *Origin and evolution of the Antarctic biota*, 197–215. London: Geological Society of London.

Chiappe, L. M. 1992. Osteología y sistemática de *Patagopteryx deferrariisi* Alvarenga y Bonaparte (Aves) del Cretácico de Patagonia. Filogenia e historia biogeográfica de las aves Cretácicas de América del Sur. Tesis Doctoral, Facultad de Ciencias Exactas y Naturales, Universidad de Buenos Aires.

———. 1993. *Enantiornithine (Aves) tarsometatarsi from the Cretaceous Lecho Formation of northwestern Argentina*. American Museum Novitates no. 3083. New York: American Museum of Natural History.

———. 1995. A diversity of early birds. *Natural History* 6: 52–55.

———. 1996. Early avian evolution in the Southern Hemisphere: The fossil record of birds in the Mesozoic of Gondwana. In F. E. Novas and R. E. Molnar (Eds.), *Proceedings of the Gondwanan dinosaur symposium*, 533–555. Brisbane: Queensland Museum.

———. 2001. Phylogenetic relationships among basal birds. In J. Gauthier and G. Gall (Eds.), *New perspectives on the origin and early evolution of birds: Proceedings of the International symposium in honor of John H. Ostrom*, 125–139. New Haven, Conn.: Peabody Museum of Natural History, Yale University.

———. 2002. Osteology of the flightless *Patagopteryx deferrariisi* from the Late Cretaceous of Patagonia. In L. M. Chiappe and L. M. Witmer (Eds.), *Mesozoic birds: Above the heads of dinosaurs*, 281–316. Berkeley: University of California Press.

———. 2007a. *Glorified dinosaurs: The origin and early evolution of birds*. Hoboken, N.J.: John Wiley and Sons.

———. 2007b. Aves. In Z. Gasparini, L. Salgado, and R. Coria (Eds.), *Patagonian Mesozoic Reptiles*, 257–270. Bloomington: Indiana University Press.

Chiappe, L. M., and J. O. Calvo. 1994. *Neuquenornis volans*, a new Enanthiornithes (Aves) from the Upper Cretaceous of Patagonia (Argentina). *Journal of Vertebrate Paleontology* 14(2): 230–246.

Chiappe, L. M., and R. Coria. 2000. Auca Mahuevo: un extraordinario sitio de nidación de dinosaurios saurópodos del Cretácico Superior de Patagonia. *Ameghiniana* 37(Suppl. 4): 13R.

———. 2003. A new specimen of *Patagonykus puertai* (Theropoda, Alvarezsauridae) from the Late Cretaceous of Patagonia. *Ameghiniana* 40(1): 119–122.

Chiappe, L. M., R. A. Coria, L. Dingus, F. Jackson, A. Chinsamy, and M. Fox. 1998. Sauropod embryos from the Late Cretaceous of Patagonia. *Nature* 396: 258–261.

Chiappe, L. M., R. A. Coria, L. Dingus, L. Salgado, and F. Jackson. 2001. Titanosaur eggs and embryos from Auca Mahuevo (Patagonia, Argentina): Implications for sauropod reproductive behavior. *Journal of Vertebrate Paleontology* 21(Suppl. 3): 40A.

Chiappe, L. M., and L. Dingus. 2001. *Walking on eggs: The astonishing discovery of thousands of dinosaur eggs in the badlands of Patagonia*. New York: Scribner.

Chiappe, L. M., L. Dingus, F. Jackson, G. Grellet-Tinner, R. Aspinall, J. Clarke, R. A. Coria, A. Garrido, and D. Loope. 2000. Sauropod eggs and embryos from the Upper Cretaceous of Patagonia. In *I Symposium of dinosaur eggs and embryos*, 23–29. Isona, Spain.

Chiappe, L. M., and G. Dyke. 2002. The Mesozoic radiation of birds. *Annual Review of Ecology and Systematics* 33: 91–124.

Chiappe, L. M., F. Jackson, L. Dingus, G. Grellet-Tinner, and R. A. Coria. 1999. Auca Mahuevo: An extraordinary dinosaur nesting ground from the Late Cretaceous of Patagonia. *Journal of Vertebrate Paleontology* 19(Suppl. 3): 37A.

Chiappe, L. M., M. A. Norell, and J. M. Clark. 2002. The Cretaceous, short-armed Alvarezsauridae: *Mononykus* and its kin. In L. M. Chiappe and L. M. Witmer (Eds.), *Mesozoic birds: Above the heads of dinosaurs*, 87–120. Berkeley: University of California Press.

Chiappe, L. M., L. Salgado, and R. A. Coria. 2001. Embryonic skulls of titanosaur sauropod dinosaurs. *Science* 293(5539): 2444–2446.

Chong Díaz, G. 1985. Hallazgo de restos óseos de dinosaurios en la Formación Hornitos, Tercera Región (Atacama, Chile). In *Actas: IV Congreso Geológico Chileno*, 152–159. [Antofagasta, Chile]: Universidad del Norte.

Chong Díaz, G., and Z. Gasparini. 1976. Los vertebrados mesozoicos de Chile y su aporte geo-paleontológico. *Actas: VI Congreso Geológico Argentino*, 1:45–67. Buenos Aires: La Asociación Geológico Argentino.

Chong Díaz, G., M. Vargas, D. Rubilar, M. Suárez, R. P. Cáceres, and A. W. A. Kellner. 2000. Cretaceous dinosaur locality at the Atacama Desert, northern Chile. *Ameghiniana* 37(Suppl. 4): 23R.

Christiansen, P. 2000. Dinosaur biomechanics. In G. Paul (ed.), *The Scientific American Book of Dinosaurs*, 64–75. New York: St. Martin's Press.

Chumakov, N. M., M, A. Zharkov, A. B. Herman, M. P. Doludenko, N. N. Kalandadze, E. L. Lebedev, A. G. Ponomarenko, and A. S. Rautian. 1995. Climatic zones in the middle of the Cretaceous Period. *Stratigraphy and Geological Correlation* 3: 3–14.

Chure, D. J., M. Manabe, M. Tanimoto, and Y. Tomida. 1999. An unusual theropod tooth from the Mifune Group (Late Cenomanian to Early Turonian), Kunamoto, Japan. In Y. Tomida, T. Rich, and P. Vickers-Rich (Eds.), *Proceedings of the Second Gondwanan Dinosaur Symposium*, 291–296. Tokyo: National Science Museum.

Cisneros, J. C., and C. L. Schultz. 2003. *Soturnia caliodon* n. g. n. sp., a procolophonid reptile from the Upper Triassic of southern Brazil. *Neues Jahrbuch für Geologie und Paläontologie, Abhandlungen* 227: 365–380.

Clarke, J. A., and L. M. Chiappe. 2001. *A new carinate bird from the Late Cretaceous of Patagonia (Argentina)*. American Museum Novitates no. 3323. New York: American Museum of Natural History.

Clarke, J. M., C. Tambussi, J. Noriega, P. Erikson, and R. Ketcham. 2004. Definitive fossil evidence for the extant avian radiation in the Cretaceous. *Nature* 433: 305–308.

Colbert, E. H. 1964. *Relationships of the saurischian dinosaurs*. American Museum Novitates no. 2181. New York: American Museum of Natural History.

———. 1970. *A saurischian dinosaur from the Triassic of Brazil*. American Museum Novitates no. 2405. New York: American Museum of Natural History.

———. 1981. A primitive ornithischian dinosaur from the Kayenta Formation of Arizona. *Museum of Northern Arizona Bulletin* 53: 1–69.

———. 1989. The Triassic dinosaur *Coelophysis*. *Museum of Northern Arizona Bulletin* 57: 1–160.

Cooper, A., and P. David. 1997. Mass survival of birds across the Cretaceous-Tertiary boundary: Molecular evidence. *Science* 275(5303): 1109–1113.

Cooper, M. R. 1985. A revision of the ornithischian dinosaur *Kangnasaurus coetzeei* Haughton, with a classification of the Ornithischia. *Annals of the South African Museum* 95: 280–317.

Coria, R. A. 1994a. Sobre la presencia de dinosaurios ornitisquios acorazados en Sudamérica. *Ameghiniana* 31(4): 398.

———. 1994b. On a monospecific assemblage of sauropod dinosaurs from Patagonia: Implications for gregarious behavior. *Gaia* 2: 209–213.

———. 1999. Ornithopod dinosaurs from the Neuquén Group, Patagonia, Argentina: Phylogeny and Biostratigraphy. In Y. Tomida, T. Rich, and P. Vickers-Rich (Eds.), *Proceedings of the Second Gondwanan Dinosaur Symposium*, 47–60. Tokyo: National Science Museum.

———. 2001. New theropod from the Late Cretaceous of Patagonia. In D. H. Tanke and K. Carpenter (Eds.), *Mesozoic vertebrate life*, 3–9. Bloomington: Indiana University Press.

———. 2007. Nonavian Theropods. In Z. Gasparini, L. Salgado, and R. Coria (Eds.), *Patagonian Mesozoic reptiles*, 229–256. Bloomington: Indiana University Press.

Coria, R. A., and J. Calvo. 2002. A new iguanodontian ornithopod from Neuquén Basin, Patagonia, Argentina. *Journal of Vertebrate Paleontology* 22(3): 503–509.

Coria, R. A., and A. Cambiaso. 2007. Ornithischia. In Z. Gasparini, L. Salgado, and R. Coria (Eds.), *Patagonian Mesozoic reptiles*, 167–187. Bloomington: Indiana University Press.

Coria, R., and L. M. Chiappe. 2000. Un nuevo terópodo abelisaurio de la Formación Río Colorado (Cretácico superior) de la Provincia del Neuquén. *Ameghiniana* 37(Suppl. 4): 22R.

———. 2001. Tooth replacement in a sauropod premaxilla from the Upper Cretaceous of Patagonia, Argentina. *Ameghiniana* 38(4): 463–466.

Coria, R. A., L. M. Chiappe, and L. Dingus. 2002. A new close relative of *Carnotaurus sastrei* Bonaparte,

1985 (Theropoda: Abelisauridae) from the Late Cretaceous of Patagonia. *Journal of Vertebrate Paleontology* 22(2): 460–465.

Coria, R. A., and P. J. Currie. 2002a. The braincase of *Giganotosaurus carolinii*. *Journal of Vertebrate Paleontology* 22(4): 802–811.

———. 2002b. Un gran terópodo celurosaurio en el Cretácico Superior de Neuquén. *Ameghiniana* 39(Suppl. 4): 9R.

———. 2006. A new carcharodontosaurid (Dinosauria, Theropoda) from the Upper Cretaceous of Argentina. *Geodiversitas* 28(1): 71–118.

Coria, R. A., P. J. Currie, D. Eberth, and A. Garrido. 2002. Bird footprints from the Anacleto Formation (lower Campanian, Upper Cretaceous). *Ameghiniana* 39(4): 453–463.

Coria, R. A., P. J. Currie, D. Eberth, A. Garrido, and E. Koppelhus. 2001. Nuevos vertebrados fósiles del Cretácico Superior de Neuquén. *Ameghiniana* 38(Suppl. 4): 6R.

Coria, R. A., P. J. Currie, and A. Paulina Carabajal. 2007. A new abelisaur theropod from northwestern Patagonia. *Canadian Journal of Earth Sciences* 43(9): 1283–1289.

Coria, R. A., and A. Paulina Carbajal. 2004. Nuevas huellas de Theropoda (Dinosauria, Saurischia) del Jurásico de Patagonia, Argentina. *Ameghiniana* 41(3): 393–398.

Coria, R. A., and J. Rodríguez. 1993. Sobre *Xenotarsosaurus bonapartei* Martínez, Giménez, Rodriguez y Bochatey, 1986; un problemático Neoceratosauria (Novas, 1989) del Cretácico de Chubut. *Ameghiniana* 30(3): 326–327.

Coria, R. A., and L. Salgado. 1995. A new giant carnivorous dinosaur from the Cretaceous of Patagonia. *Nature* 377: 224–226.

———. 1996a. A basal iguanodontian (Ornithischia: Ornithopoda) from the Late Cretaceous of South America. *Journal of Vertebrate Paleontology* 16(3): 445–457.

———. 1996b. "*Loncosaurus argentinus*" Ameghino, 1899 (Ornithischia, Ornithopoda): A revised description with comments on its phylogenetic relationships. *Ameghiniana* 33(4): 373–376.

———. 1999. Nuevos aportes a la anatomía craneana de los saurópodos titanosáuridos. *Ameghiniana* 36(1): 98R.

———. 2000. A basal Abelisauria Novas, 1992 (Theropoda-Ceratosauria) from the Cretaceous of Patagonia, Argentina. *Gaia* 15: 89–102.

———. 2001. South American ankylosaurs. In K. Carpenter (ed.), *The armored dinosaurs*, 159–168. Bloomington: Indiana University Press.

Coria, R. A., L. Salgado, P. J. Currie, A. Paulina Carabajal, and A. B. Arcucci. 2004. Nuevos registros de iguanodontes basales en el Cretácico de Patagonia. *Ameghiniana* 41(Suppl. 4): 42R.

Costa Franco-Rosas, A. C., L. Salgado, I. S. Carvalho, and C. F. Rosas. 2004. Nuevos materiales de titanosaurios (Sauropoda) en el Cretácico Superior de Mato Grosso, Brasil. *Revista Brasileira de Paleontologia* 7(3): 329–336.

Cowen, R. 1995. History of life. Boston: Blackwell Scientific Publications.

Cracraft, J. 2001. Avian evolution, Gondwana biogeography and the Cretaceous-Tertiary mass extinction event. *Proceedings of the Royal Society of London B* 268: 459–469.

Currie, P. J. 1981. Bird footprints from the Gething Formation (Aptian, Lower Cretaceous) of northeastern British Columbia, Canada. *Journal of Vertebrate Paleontology* 1(3–4): 257–264.

———. 1990. Elmisauridae. In D. B. Weishampel, P. Dodson, and H. Osmólska (Eds.), *The Dinosauria*, 245–248. Berkeley: University of California.

Currie, P. J. 1995. New information on the anatomy and relationship of *Dromaeosaurus albertensis* (Dinosauria: Theropoda). *Journal of Vertebrate Paleontology* 15: 576–591.

Currie, P. J., P. Vickers-Rich, and T. H. Rich. 1996. Possible oviraptorosaur (Theropoda, Dinosauria) specimens from the Early Cretaceous Otway Group of Dinosaur Cove, Australia. *Alcheringa* 20: 73–79.

Curry Rogers, K., and C. A. Forster. 2001. The last of the dinosaur titans: A new sauropod from Madagascar. *Nature* 412: 530–533.

———. 2004. The skull of *Rapetosaurus krausei* (Sauropoda: Titanosauria) from the Late Cretaceous of Madagascar. *Journal of Vertebrate Paleontology* 24(1): 121–144.

Czercas, S. 1997. Skin. In P. Currie and K. Padian (Eds.), *Encyclopedia of dinosaurs*, 669–675. San Diego: Academic Press.

Da Silva, R., I. Souza Carvalho, and C. Schwanke. 2007. Vertebrate dinoturbation from the Caturrita Formation (Late Triassic, Paraná Basin), Río Grande do Sul State, Brazil. *Gondwana Research* 11: 303–310.

Dal Sasso, C., S. Maganuco, E. Buffetaut, and M. A. Mendez. 2005. New information on the skull of the enigmatic theropod *Spinosaurus*, with remarks on its size and affinities. *Journal of Vertebrate Paleontology* 25(4): 888–896.

De Broin, F. 1980. Les tortues de Gadoufaoua (Aptian du Niger) apercu sue le paleobiogeographie des Pelomedusida (Pleurodira). *Memoires de la Société Géologique de France* 139: 39–46.

De Klerk, W. J., C. A. Forster, S. D. Sampson, A. Chinsamy, and C. F. Ross. 2000. A new coelurosaurian dinosaur from the Early Cretaceous of South Africa. *Journal of Vertebrate Paleontology* 20(2): 324–332.

De Valais, S., S. Apesteguía, and D. Udrizar Sauthier. 2003. Nuevas evidencias de dinosaurios de la Formación Puerto Yeruá (Cretácico), Provincia de Entre Ríos, Argentina. *Ameghiniana* 40(4): 631–635.

De Valais, S., and R. Melchor. 2003. Nuevos aportes sobre la icnofauna de la Formación La Matilde (Jurásico Medio), provincia de Santa Cruz, Argentina. *Ameghiniana* 40(Suppl. 4): 53–54R.

De Valais, S., R. Melchor, and J. Genise. 2003. *Hexapodichnus casamiquelai* isp. nov.: An insect trackway from the La Matilde Formation (Middle Jurassic), Santa Cruz, Argentina. *Asociación Paleontológica Argentina, Publicación Especial N° 9*: 35–41.

De Valais, S., F. E. Novas, and S. Apesteguía. 2002. Morfología del pie de los terópodos abelisaurios: nuevas evidencias del Cretácico de Patagonia. *Ameghiniana* 39(Suppl. 4): 9–10R.

Del Corro, G. 1975. Un nuevo saurópodo del Cretácico Superior, *Chubutisaurus insignis* gen. et sp. nov. (Saurischia, Chubutisauridae nov.) del Cretácico Superior (Chubutiano), Chubut, Argentina. *Actas del Primer Congreso Argentino de Paleontología y Bioestratigrafía*, 2:229–240. Tucumán: Asociación Paleontológica Argentina.

Depéret, C. 1896. Note sur les dinosauriens sauropodes et théropodes du Crétacé Supérieur de Madagascar. *Bulletin de la Société Géologique de France* 3(24): 176–196.

Dias-Brito, D., E. A. Musacchio, J. C. Castro, M. S. Maranhão, J. M. Suárez, and R. Rodrigues. 2001. Grupo Bauru: uma unidade continental do Cretácico no Brasil — concepsoes baseadas em dados micropaleontologicos, isotópicos e estratigráficos. *Revue de Paleobiologie* 20(1): 245–304.

Digregorio, J. H., and M. A. Uliana. 1980. Cuenca Neuquina. In J. C. M. Turner (coord.), *Segundo Simposio de Geología Regional Argentina*, 2:985–1032. Córdoba: Academia Nacional de Ciencias de Córdoba.

Dingus L., J. Clarke, G. Scott, C. Swisher, III, L. M. Chiappe, and R. Coria. 2000. *Stratigraphy and magnetostratigraphic/faunal constraints for the age of Sauropod embryo-bearing rocks in the Neuquén group (Late Cretaceous, Neuquén Province, Argentina)*. American Museum Novitates no. 3290. New York: American Museum of Natural History.

Dodson, P., D. W. Krause, C. A. Forster, S. D. Sampson, and F. Ravoavy. 1998. Titanosaurid (Sauropoda) osteoderms from the Late Cretaceous of Madagascar. *Journal of Vertebrate Paleontology* 18(3): 563–568.

Dyke, G., and D. van Tuinen. 2004. The evolutionary radiation of modern birds (Neornithes): Reconciling molecules, morphology and the fossil record. *Zoological Journal of the Linnean Society* 148(2): 153–177.

Dzik, J. 2003. A beaked herbivorous archosaur with dinosaur affinities from the early Late Triassic of Poland. *Journal of Vertebrate Paleontology* 23(3): 556–574.

Elzanowski, A. 1995. Cretaceous birds and avian phylogeny. *Courier Forschungsinstitut Senckenberg* 181: 37–53.

Engelmann, G., D. J. Chure, and A. R. Fiorillo. 2004. The implications of a dry climate for the paleoecology of the fauna of the Upper Jurassic Morrison Formation. *Sedimentary Geology* 167(3–4): 297–308.

Ezcurra, M. D. 2006. A review of the systematic position of the dinosauriform archosaur *Eucoelophysis baldwini* Sullivan and Lucas, 1999 from the Upper Triassic of New Mexico, USA. *Geodiversitas* 28(4): 649–684.

———. 2007. The cranial anatomy of the coelophysoid theropod *Zupaysaurus rougieri* from the Upper Triassic of Argentina. *Historical Biology* 19(2): 185–202.

Ezcurra, M. D., and F. E. Novas. 2005. Phylogenetic relationships of *Zupaysaurus rougieri* from NW Argentina. In *Boletim de Resumos: II Congresso Latino-Americano de Paleontología de Vertebrados*, 102–104. Rio de Janeiro: Museu Nacional.

———. 2007. Phylogenetic relationships of the Triassic theropod *Zupaysaurus rougieri* from NW Argentina. *Historical Biology* 19(1): 35–72.

Faccio, G. 1994. Dinosaurian eggs from the Upper Cretaceous of Uruguay. In K. Carpenter, K. F. Hirsch, and J. R. Horner (Eds.), *Dinosaur eggs and babies*, 47–55. Cambridge: Cambridge University Press.

Faccio, G., I. Ford, and F. Gancio. 1990. Primer registro fósil *in situ* de huevos de dinosaurios del Cretácico Superior del Uruguay (Fm. Mercedes). In *Boletin de investigacion, Universidad de la Republica, Facultad de Agronomia* 26: 1–12. Montevideo.

Farlow, J. O. 1992. Sauropod tracks and trackmakers: Integrating the ichnological and skeletal records. *Zubia* 10: 89–138.

———. 2007. A speculative look at the paleoecology of large dinosaurs of the Morrison Formation, or, Life with *Camarasaurus* and *Allosaurus*. In E. P. Kvale, M. K. Brett-Surman, and J. Farlow (Eds.), *Dinosaur paleoecology and geology: The life and times of Wyoming's Jurassic dinosaurs and marine reptiles*, 98–151. Shell, Wyo.: GeoScience Adventures Workshop.

Farlow, J. O., P. Dodson, and A. Chinsamy. 1995. Dinosaur biology. *Annual Review of Ecological Systematics* 26: 445–471.

Farlow, J. O., and T. R. Holtz. 2002. The fossil record of predation in dinosaurs. In M. Kowalewski and P. H. Kelley (Eds.), *The fossil record of predation*, 251–265. [Columbus, Ohio]: Paleontological Society.

Farlow, J. O., J. G. Pittman, and J. M. Hawthorne. 1989. *Brontopodus birdi*, Lower Cretaceous sauropod footprints from the U.S. Gulf Coastal Plain. In D. D. Gillette and M. G. Lockley (Eds.), *Dinosaur tracks and traces*, 372–394. New York: Cambridge University Press.

Ferigolo, J., and M. C. Langer. 2007. A Late Triassic dinosauriform from south Brazil and the origin of the ornithischian predentary bone. *Historical Biology* 19(1): 23–33.

Fernandes, L. A., and A. M. Coimbra. 1996. A Bacia Bauru (Cretáceo Superior, Brasil). *Anais da Academia Brasileira de Ciencias* 68(2): 195–205.

Fernandes, M., L. Fernandes Bueno dos Reis, and P. R. F. Souto. 2004. Occurrence of urolites related to dinosaurs in the Lower Cretaceous of the Botucatu Formation, Paraná Basin, São Paulo State, Brazil. *Revista Brasileira de Paleontologia* 7(2): 263–268.

Fernándes, M. A., and I. S. Carvalho. 2007. Pegadas fósseis da Formação Botucatu (Jurássico Superior-Cretáceo Inferior): o registro de un grande dinossauro Ornithopoda na Bacia do Paraná. In I. S. Carvalho et al. (Eds.), *Paleontologia: Cenarios de Vida*, 1:425–432. Río de Janeiro: Editora Interciencia.

Ferreira, C. S., S. A. Azevedo, I. S. Carvalho, R. A. Gonçalves, and M. A. Vicalvi. 1992. Os fósseis da Formação Itapecuru. In *Resumos Expandidos do 2do Simpósio sobre as Bacias Cretácicas Brasileiras*, 107–110. São Paulo.

Fischer, P. E., D. Russell, M. Stoskopf, R. Barrick, M. Hammer, and A. Kuzmitz. 2000. Cardiovascular evidence for an intermediate or higher metabolic rate in an ornithischian dinosaur. *Science* 288(5465): 503–505.

Flynn, J. J., J. M. Parrish, B. Rakotosamimanana, W. F. Simpson, R. L. Whatley, and A. R. Wyss. 1999. A Triassic fauna from Madagascar, including early dinosaurs. *Science* 286(5440): 763–765.

Forster, C. A. 1999. Gondwanan dinosaur evolution and biogeographic analysis. *Journal of African Earth Sciences* 28(1): 169–185.

Forster, C. A., S. D. Sampson, L. M. Chiappe, and D. Krause. 1998. The theropod ancestry of birds: New evidence from the Late Cretaceous of Madagascar. *Science* 279(5358): 1915–1919.

Frankfurt, N., and L. M. Chiappe. 1999. A possible oviraptorosaur from the Late Cretaceous of Northwestern Argentina. *Journal of Vertebrate Paleontology* 19(1): 101–105.

Frenguelli, J. 1951. Un huevo fósil del Rocanense. *Revista de la Asociación Geológica Argentina* 6(2): 108–112.

Frey, E., and D. Martill. 1995. A possible oviraptorosaurid theropod from the Santana Formation (Lower Cretaceous, Albian?) of Brazil. *Neues Jahrbuch fur Geologie und Palaeontologie* 7: 397.

Gaffney, E. S., and P. A. Meylan. 1991. Primitive pelomedusid turtle. In J. G. Maisey (ed.), *Santana fossils: An illustrated atlas*, 335–339. Neptune City, N.J.: T. F. H. Publications.

Gallina, P. A., and S. Apesteguía. 2005. *Cathartesaura anaerobica* gen. et sp. nov., a new rebbachisaurid (Dinosauria, Sauropoda) from the Huincul Formation (Upper Cretaceous), Río Negro, Argentina. *Revista del Museo Argentino de Ciencias Naturales Bernardino Rivadavia* 7(2): 153–166.

Gallina, P. A., S. Apesteguía, and F. E. Novas. 2002. ¿Un elefante bajo la alfombra? Los Rebbachisauridae (Sauropoda, Diplodocoidea) del Cretácico de Gondwana. Nuevas evidencias en "La Buitrera" (Formación Candeleros), provincia de Río Negro. XVIII Jornadas Argentinas de Paleontología de Vertebrados (Bahía Blanca). *Ameghiniana* 39(Suppl. 4): 10R.

Galton, P. M. 1976. Prosauropod dinosaurs (Reptilia: Saurischia) of North America. *Postilla* 169: 1–98.

———. 1977. On *Staurikosaurus pricei*, an early saurischian dinosaur from the Triassic of Brazil, with notes on the Herrerasauridae and Poposauridae. *Paläontologische Zeitschrift* 51(3–4): 234–245.

———. 1981. *Dryosaurus*, a hypsilophodontid dinosaur from the Upper Jurassic of North America and Africa: Postcranial skeleton. *Paläontologische Zeitschrift* 55(3–4): 271–312.

———. 1982. *Elaphrosaurus*, an ornithomimid dinosaur from the Upper Jurassic of North America and Africa. *Paläontologische Zeitschrift* 56(3–4): 265–275.

———. 2000. Are *Spondylosoma* and *Staurikosaurus* (Santa Maria Formation, Middle-Upper Triassic, Brazil) the oldest saurischian dinosaurs? *Paläontologische Zeitschrift* 74(3): 393–423.

Galton, P. M., and J. A. Jensen. 1979. A new large theropod dinosaur from the Upper Jurassic of Colorado. *Brigham Young University Geological Studies* 26(2): 1–12.

Galton, P. M., and P. Upchurch. 2004a. Prosauropoda. In D. B. Weishampel, P. Dodson, and H. Osmólska (Eds.), *The Dinosauria*, 232–258. Berkeley: University of California Press.

———. 2004b. Stegosauria. In D. B. Weishampel, P. Dodson, and H. Osmólska (Eds.), *The Dinosauria*, 343–362. Berkeley: University of California Press.

García, R., L. Salgado, and R. A. Coria. 2003. Primeros restos de dinosaurios saurópodos en el Jurásico de la cuenca Neuquina, Patagonia, Argentina. *Ameghiniana* 40(1): 123–126.

Gasparini, Z. 1992. Marine reptiles of the Circum-Pacific region. In G. E. G. Westermann (ed.), *The Jurassic of the circum-Pacific*, 361–364. Cambridge: Cambridge University Press.

Gasparini, Z., and E. Olivero. 1989. El dinosaurio antártico. *Ciencia Hoy* 1(4): 40–41.

Gasparini, Z., E. Olivero, R. Scasso, and C. Rinaldi. 1987. Un ankylosaurio (Reptilia, Ornithischia) Campaniano en el continente antártico. In *Anais do X Congreso Brasileiro de Paleontología* 1: 131–141. Rio de Janeiro.

Gasparini, Z., X. Pereda-Suberbiola, and R. E. Molnar. 1996. New data on the ankylosaurian dinosaur from the Late Cretaceous of the Antarctic Peninsula. In F. E. Novas and R. E. Molnar (Eds.), *Proceedings of the Gondwanan dinosaur symposium*, 39:583–594. Brisbane: Queensland Museum.

Gatesy, S. M. 2001. The evolutionary history of the theropod caudal locomotor module. In J. A. Gauthier and L. F. Gall, *New perspectives on the origin and early evolution of birds: Proceedings of the international symposium in honor of John H. Ostrom*, 333–346. New Haven, Conn.: Peabody Museum of Natural History.

Gauffre, F.-X. 1993. The prosauropod dinosaur *Azendohsaurus laaroussi* from the Upper Triassic of Marocco. *Palaeontology* 36(4): 897–908.

Gauthier, J. A. 1986. Saurischian monophyly and the origin of birds. In K. Padian (ed.), *The origin of birds and the evolution of flight*, 1–55. San Francisco, Calif.: California Academy of Sciences.

Gayet, M., L. G. Marshall, and T. Sempere. 1991. The Mesozoic and Paleocene vertebrates of Bolivia and their stratigraphic context: A review. In R. Suárez-Soruco (ed.), *Fósiles y Facies de Bolivia*, 12(3–4), *Vertebrados*, 393–433. Santa Cruz: Revista Técnica de Yacimientos Petrolíferos Fiscales Bolivia.

Gayet, M., L. G. Marshall, T. Sempére, F. J. Meunier, H. Cappetta, and J. C. Rage. 2001. Middle Maastrichtian vertebrates (fishes, amphibians, dinosaurs and other reptiles, mammals) from Pajcha Pata (Bolivia). Biostratigraphic, palaeoecologic and palaeobiogeographic implications. *Palaeogeography, Palaeoclimatology, Palaeoecology* 169(1): 39–68.

Gillette, D. 2003. The geographic and phylogenetic position of sauropod dinosaurs from the Kota Formation (Early Jurassic) of India. *Journal of Asian Earth Science* 21(6): 683–689.

Gilmore, C. W. 1920. Osteology of the carnivorous Dinosauria in the United States National Museum, with special reference to the genera *Antrodemus* (*Allosaurus*) and *Ceratosaurus*. *United States National Museum Bulletin* 110: 1–159.

Giménez, O. V. 1992. Estudio preliminar del miembro anterior de los saurópodos titanosáuridos. *Ameghiniana* 30(2): 154.

———. 1996. Hallazgo de impronta de piel de un dinosaurio saurópodo en la Provincia de Chubut. *Ameghiniana* 33(4): 465.

———. 2007. Skin impressions of *Tehuelchesaurus* (Sauropoda) from the Upper Jurassic of Patagonia. *Revista del Museo Argentino de Ciencias Naturales Bernardino Rivadavia* 9(2): 119–124.

Goldberg, K., and A. J. V. Garcia. 2000. Palaeobiogeography of the Bauru Group, a dinosaur-bearing Cretaceous unit, northeastern Paraná Basin, Brazil. *Cretaceous Research* 21(2–3): 241–254.

Gómez Omil, R. J., A. Boll, and R. M. Hernández. 1989. Cuenca cretácico-terciaria del Noroeste argentino (Grupo Salta). In G. A. Chebli and L. A. Spalletti (Eds.), *Cuencas sedimentarias argentinas*, 6:43–64. Tucumán: Universidad Nacional de Tucumán, Serie de Correlación Geológica.

González Riga, B. 2003. A new titanosaur (Dinosauria, Sauropoda) from the Upper Cretaceous of Mendoza Province, Argentina. *Ameghiniana* 40(2): 155–172.

———. 2005. Nuevos restos fósiles de *Mendozasaurus neguyelap* (Sauropoda, Titanosauria) del Cretácico Tardío de Mendoza, Argentina. *Ameghiniana* 42(3): 535–548.

González Riga, B., and J. O. Calvo. 2007. Huellas de dinosaurios saurópodos en el Cretácico de Argentina. 4th European Meeting on the Paleontology and Stratigraphy of Latin America. *Cuadernos del Museo Geominero* 8: 173–179.

González Riga, B., and S. Casadío. 2000. Primer registro de Dinosauria (Ornithischia, Hadrosauridae) en la provincia de La Pampa (Argentina) y sus implicancias paleobiogeográficas. *Ameghiniana* 37(3): 341–351.

Gradzinsky, R., Z. Kielan-Jawarowska, and T. Maryánska. 1977. Upper Cretaceous Djadokhta, Barun Goyot and Nemegt formations of Mongolia, including remarks on previous subdivisions. *Acta Geologica Polonica* 27: 281–318.

Gutierrez, F., and L. G. Marshall. 1994. Los primeros huesos de dinosaurios de Bolivia. Formación Cajones (Maastrichtiano) cerca de Santa Cruz de la Sierra. *Revista Técnica de Yacimientos Petrolíferos Fiscales Bolivia* 15(1–2): 131–139.

Haubold, H. 1971. Ichnia Amphibiorum et Reptiliorum fossilium. In O. Kühn (ed.), *Handbuch der Paläoherpetologie*, 18:1–124. Stuttgart: Gustav Fischer Verlag.

Hay, W. W., R. M. de Conto, C. N. Wold, K. M. Wilson, S. Volgt, M. Schulz, A. R. Wold, W. C. Dullo, A. B. Ronov, A. N. Balukhovsky, and E. Söding. 1999. Alternative global Cretaceous paleogeography. In E. Barrera and C. C. Johnson. (Eds.), *Evolution of the Cretaceous ocean-climate system*, 1–47. Boulder, Colo.: Geological Society of America.

Hedges, B., P. Parker, C. Sibley, and S. Kumar. 1996. Continental breakup and the ordinal diversification of birds and mammals. *Nature* 381: 226–229.

Heredia, S., and L. Salgado. 1999. Posición estratigráfica de los estratos supracretácicos portadores de dinosaurios en Lago Pellegrini, Patagonia septentrional, Argentina. *Ameghiniana* 36(2): 229–234.

Hirayama, R. 1998. Oldest known sea turtle. *Nature* 392: 705–708.

Holtz, T. R., Jr. 1994. The phylogenetic position of the Tyrannosauridae: Implications for theropod systematics. *Journal of Paleontology* 68(5): 1100–1117.

———. 1995. The arctometatarsalian pes, an unusual structure of the metatarsus of Cretaceous Theropoda (Dinosauria: Saurischia). *Journal of Vertebrate Paleontology* 14(4): 480–519.

———. 2000. A new phylogeny of the carnivorous dinosaurs. *Gaia* 15: 5–61.

Holtz, T. R., Jr., and M. K. Brett-Surman. 1997. The osteology of dinosaurs. In J. O. Farlow and M. K. Brett-Surman (Eds.), *The complete dinosaur*, 78–91. Bloomington: Indiana University Press.

Holtz, T. R., R. E. Molnar, and P. J. Currie. 2004. Basal tetanurae. In D. B. Weishampel, P. Dodson, and H. Osmólska (Eds.), *The Dinosauria*, 2nd ed., 71–110. Berkeley: University of California Press.

Hooker, J., A. Milner, and S. Sequeira. 1991. An ornithopod dinosaur from the late Cretaceous of West Antarctica. *Antarctic Science* 3(3): 331–332.

Hope, S. 2002. The Mesozoic radiation of Neornithes. In L. M. Chiappe and L. Witmer (Eds.), *Mesozoic birds: Above the heads of dinosaurs*, 339–388. California: University of California Press.

Horner, J. R., D. B. Weishampel, and C. A. Forster. 2004. Hadrosauridae. In D. B. Weishampel, P. Dodson, and H. Osmólska (Eds.), *The Dinosauria*, 438–463. Berkeley: University of California Press.

Huber, B. T., and D. K. Watkins. 1992. Biogeography of Campanian-Maastrichtian calcareous plankton in the region of the southern ocean: Paleogeographic and paleoclimatic implications. In J. P. Kennett and D. A. Warnke (Eds.), *The Antarctic paleoenvironment: A perspective on global change*, 31–60. Washington, D.C.: American Geophysical Union.

Hugo, C. A., and H. A. Leanza. 2001a. Hoja Geológica 3969–IV, General Roca, provincias del Neuquén y Río Negro. *Boletín del Instituto de Geología y Recursos Naturales* 308: 1–71.

———. 2001b. Hoja Geológica 3966–III, Villa Regina, provincia de Río Negro. *Boletín del Instituto de Geología y Recursos Naturales* 309: 1–53.

Hünicken, M., A. Tauber, and R. Leguizamón. 2001. Hallazgo de huevos y nidos de dinosaurios, asociados a vegetales silicificados: asignación al Cretácico de las secuencias portadoras aflorantes en Sanagasta, provincia de La Rioja. *Ameghiniana* 38(Suppl. 4): 10R.

Hunt, A. P. 1989. A new ornithischian dinosaur from the Bull Canyon Formation (Upper Triassic) of east-central New Mexico. In S. G. Lucas, and A. P. Hunt (Eds.), *Dawn of the age of dinosaurs in the American southwest*, 355–358. Albuquerque: New Mexico Museum of Natural History.

Hunt, A. P., S. G. Lucas, A. B. Heckert, R. M. Sullivan, and M. G. Lockley. 1998. Late Triassic dinosaurs from the western United States. *Geobios* 31(4): 511–531.

Hutchinson, J. 2001. The evolution of the pelvic osteology and soft tissues on the line to extant birds (Neornithes). *Zoological Journal of the Linnean Society* 131: 123–168.

Hwang, S. H., M. A. Norell, Q. Ji, and K. Gao. 2002. *New specimens of* Microraptor zhaoianus *(Theropoda: Dromaeosauridae) from northeastern China*. American Museum Novitates no. 3381. New York: American Museum of Natural History.

Iriarte, J., K. Moreno, D. Rubilar, and A. Vargas. 1999. A titanosaurid from the Quebrada La Higuera Formation (Upper Cretaceous), III Region, Chile. *Ameghiniana* 36(1): 102R.

Irmis, R., S. Nesbitt, and W. Parker. 2005. A critical review of the Triassic North American dinosaur record. In *Boletim de Resumos: II Congresso Latino-Americano de Paleontología de Vertebrados*, 139. Rio de Janeiro: Museu Nacional.

Jacobs, L., D. Winkler, and E. Gomani. 1993. New material of an Early Cretaceous titanosaurid dinosaur from Malawi. *Palaeontology* 36(3): 523–534.

Jaillard, E., H. Cappetta, P. Ellenberger, M. Feist, N. Grambast-Fessard, J. P. Lefranc, and B. Sigé. 1993. Sedimentology, palaeontology, biostratigraphy and correlation of the Late Cretaceous Vilquechico Group of southern Peru. *Cretaceous Research* 14(6): 623–661.

Jain, S. L., and S. Bandyopadhyay. 1997. New titanosaurid (Dinosauria: Sauropoda) from the Late

Cretaceous of central India. *Journal of Vertebrate Paleontology* 17(1): 114–136.

Jain, S. L., T. S. Kutty, T. Roy-Chowdury, and S. Chatterjee. 1979. Some characters of *Barapasaurus tagorei*, a sauropod dinosaur from the Lower Jurassic of Deccan, India. In B. Laskar and C. S. Raja Rao (Eds.), *Fourth International Gondwana Symposium: Papers*,1:204–216. Delhi: Hindustan Publishing Company.

Jalil, N. E., and F. Knoll. 2002. Is *Azendohsaurus laaroussii* (Carnian, Morocco) a dinosaur? *Journal of Vertebrate Paleontology* 22(Suppl. 3): 70A.

Janensch, W. 1914. Übersicht über die Wirbeltierfauna der Tendaguru-Schichten, nebst einer kurzen Charakterisierung der neu aufgeführten Arten von Sauropoden. *Archiv für Biontologie* 3: 81–110.

———. 1920. Über *Elaphrosaurus bambergi* und die Megalosaurier aus den Tendaguru-Schichten Deutsch-Ostafrikas. *Sitzungberichte Gessellschaft Natursforschende Freunde* 1920: 225–235.

———. 1925. Die Coelurosaurier und Theropoden der Tendaguru-Schichten Deutsch-Ostafrikas. *Palaeontographica* 1(Suppl. 7): 1–100.

———. 1929. Die Wirbelsäule der gattung *Dicraeosaurus*. *Palaeontographica* 3(Suppl. 7): 1–34.

———. 1935–1936. Die Schädel der Sauropoden *Brachisaurus*, *Barosaurus* und *Dicraeosaurus* aus den Tendaguru-Schichten Deutsch-Ostafrikas. *Palaeontographica* 2(Suppl. 7): 147–298.

Ji, Q., M. A. Norell, K. Q. Gao, S. A. Ji, and D. Ren. 2001. The distribution of integumentary structures in a feathered dinosaur. *Nature* 410: 1084–1088.

Jianu, C. M., and D. B. Weishampel. 1999. The smallest of the largest: A new look at possible dwarfing in sauropod dinosaurs. *Geologie en Mijnbouw* 78(3–4): 335–343.

Juárez Valieri, R., L. Fiorelli, and L. Cruz. 2007. *Quilmesaurus curriei* Coria, 2001 (Dinosauria, Theropoda). Su validez taxonómica y relaciones filogenéticas. *Revista del Museo Argentino de Ciencias Naturales Bernardino Rivadavia* 9(1): 59–66.

Kellner, A. W. A. 1995. Theropod teeth from the Late Cretaceous Bauru Group near Peirópolis, Minas Gerais, Brazil. In *Actas do XIV Congresso Brasileiro de Paleontologia*, 66–67. Minais Gerais.

———. 1996a. Fossilized theropod soft tissue. *Nature* 379(4): 2.

———. 1996b. Remarks on Brazilian dinosaurs. In F. E. Novas and R. E. Molnar (Eds.), *Proceedings of the Gondwanan dinosaur symposium*, 611–626. Brisbane: Queensland Museum.

———. 1999. Short note on a new dinosaur (Theropoda, Coelurosauria) from the Santana Formation (Romualdo Member, Albian), Northeastern Brasil. *Boletim do Museu Nacional, Geologia* 49: 1–8.

———. 2001. New information on the theropod dinosaurs from the Santana Formation (Aptian-Albian), Araripe Basin, northeastern Brazil. *Journal of Vertebrate Paleontology* 21(Suppl. 3): 67A.

Kellner, A. W. A., and S. A. K. de Azevedo. 1999. A new sauropod dinosaur (Titanosauria) from the Late Cretaceous of Brazil. In Y. Tomida, T. H. Rich, and P. Vickers-Rich (Eds.), *Proceedings of the Second Gondwanan Dinosaur Symposium*, 111–142. Tokyo: National Science Museum Monographs.

Kellner, A. W. A., S. A. Azevedo, L. Carvalho, D. Henriques, T. Costa, and D. Campos. 2004. Bones out of the jungle: on a dinosaur locality from Mato Grosso, Brazil. *Journal of Vertebrate Paleontology* 24(3) 78A.

Kellner, A. W. A., and D. A. Campos. 1996. First Early Cretaceous theropod dinosaur from Brazil. *Neues Jahrbuch für Geologie und Palaontologie, Abhandlungen* 199: 151–166.

———. 1997. The Titanosauridae (Sauropoda) of the Bauru Group, Late Cretaceous of Brazil. *Journal of Vertebrate Paleontology* 17(Suppl. 3): 56A.

———. 2000. Brief review of dinosaur studies and perspectives in Brazil. *Anais da Academia Brasileira de Ciências* 72(4): 509–538.

———. 2002. On a theropod dinosaur (Abelisauria) from the continental Cretaceous of Brazil. *Arquivos do Museu Nacional do Rio de Janeiro* 60(3): 163–170.

Kellner, A. W. A., D. A. Campos, S. A. K. de Azevedo, V. G. Silva, and L. B. Carvalho. 1995. Dinosaur localities in Cretaceous rocks of Mato Grosso, Brazil. *Journal of Vertebrate Paleontology* 15(3): 39A.

Kellner, A. W. A., D. A. Campos, S. A. Azevedo, and M. Trotta. 2005. Description of a titanosaurid caudal series from the Bauru Group, Late Cretaceous of Brazil. *Arquivos do Museu Nacional do Río de Janeiro* 63(3): 529–564.

Kellner, A. W. A., D. A. Campos, S. A. K. de Azevedo, M. Trotta, D. R. Henriques, M. T. Craik, and H. P. Silva. 2006. On a new titanosaur sauropod from the Bauru Group, Late Cretaceous of Brazil. *Boletim do Museu Nacional, Geologia* 74: 1–31.

Kellner, A. W. A., J. G. Maisey, and D. A. Campos. 1994. Fossil down feather from the Lower Cretaceous of Brazil. *Palaeontology* 37(3): 489–492.

Kellner, A. W. A., R. G. Martins Neto, and J. G. Maisey. 1991. Undeterminated feather. In J. G. Maisey (ed.), *Santana fossils: An illustrated atlas*, 376–377. Neptune City, N.J.: T. F. H. Publications.

Kerourio, P., and B. Sigé. 1984. L'apport des coquilles d'ouefs de dinosaures de Laguna Umayo à l'âge de la Formation Vilquechico (Pérou) et à la compréhension de *Perutherium altiplanense*. *Newsletters in Stratigraphy* 13(3): 133–142.

Kirkland, J. I., D. Burge, and R. Gaston. 1993. A large dromaeosaur (Theropoda) from the Lower Cretaceous of eastern Utah. *Hunteria* 2(10): 1–16.

Kirkland, J. I., S. G. Lucas, and J. W. Estep. 1998. Cretaceous dinosaurs of the Colorado Plateau. In S. G. Lucas, J. I. Kirkland, and J. W. Estep (Eds.), *Lower and Middle Cretaceous terrestrial ecosystems*, 14:79–89. Albuquerque: New Mexico Museum of Natural History and Science.

Kischlat, E. E. 1998. A new dinosaurian "rescued" from the Brazilian Triassic: *Teyuwasu barberenai*, new taxon. *Paleontologia em Destaque* 14(26): 58.

Kischlat, E. E., and M. C. Barberena. 1999. Triassic Brazilian dinosaurs: new data. *Paleontologia em Destaque* 14(26): 56.

Kischlat, E. E., and S. G. Lucas. 2003. A phytosaur from the Upper Triassic of Brazil. *Journal of Vertebrate Paleontology* 23(2): 464–467.

Kraemer, P., and A. Riccardi. 1997. Estratigrafía de la región comprendida entre los lagos Argentino y Viedma (49° 40'–50° 10' lst. S), Provincia de Santa Cruz. *Revista de la Asociación Geológica Argentina* 52: 333–360.

Lacovara, K., J. Harris, M. Lamanna, F. E. Novas, R. Martínez, and A. Ambrosio. 2004. An enormous sauropod from the Maastrichtian Pari Aike Formation of southernmost Patagonia. *Journal of Vertebrate Paleontology* 24(Suppl. 3): 81A.

Lamanna, M. C., R. D. Martínez, M. Luna, G. Casal, P. Dodson, and J. Smith. 2001. Sauropod faunal transition trough the Cretaceous Chubut Group of Central Patagonia. *Journal of Vertebrate Paleontology* 21(Suppl. 3): 70A.

Lamanna, M. C., R. D. Martínez, M. Luna, G. Casal, L. Ibiricu, and E. Ivany. 2004. New specimen of the problematic large theropod dinosaur Megaraptor from the Late Cretaceous of Central Patagonia. *Journal of Vertebrate Paleontology* 24(Suppl. 3): 81–82A.

Lamanna, M. C., R. D. Martínez, and J. B. Smith. 2002. A definitive abelisaurid theropod dinosaur from the early Late Cretaceous of Patagonia. *Journal of Vertebrate Paleontology* 22(1): 58–69.

Lamanna, M. C., J. B. Smith, P. Dodson, and Y. S. Attia. 2004. From dinosaurs to dyrosaurids (Crocodyliformes): Removal of the post-Cenomanian (Late Cretaceous) record of Ornithischia from Africa. *Journal of Vertebrate Paleontology* 24(3): 764–768.

Lambrecht, K. 1929. *Negaeornis wetzeli* n. g., n. sp., der erste Kredevogel der südlichen Hemisphäre. *Paläontologische Zeitschrift* 11: 121–129.

Langer, M. C. 2003. The pelvic and hind limb anatomy of the stem-sauropodomorph *Saturnalia tupiniquim* (Late Triassic, Brazil). *PaleoBios* 23(2):1–40.

———. 2004. Basal Saurischia. In D. B. Weishampel, P. Dodson, and H. Osmólska (Eds.), *The Dinosauria*, 25–46. Berkeley: University of California Press.

———. 2005a. Studies on continental Late Triassic tetrapod biochronology. I. The type locality of *Saturnalia tupiniquim* and the faunal succession in south Brazil. *Journal of South American Earth Sciences* 19(2): 205–218.

———. 2005b. Studies in continental Late Triassic tetrapod biochronology. II. The Ischigualastian and a Carnian global correlation. *Journal of South American Earth Sciences* 19(2): 219–239.

Langer, M. C., F. Abdala, M. Richter, and M. J. Benton. 1999. A sauropodomorph dinosaur from the Upper Triassic (Carnian) of southern Brazil. *Comptes Rendus de l'Académie des Sciences de Paris* (Série II) 329: 511–517.

Langer, M. C., and M. Benton. 2006. Early dinosaurs: a phylogenetic study. *Journal of Systematic Palaeontology* 4: 309–358.

Langer, M. C., and J. Ferigolo. 2005. The first ornithischian body-fossils in Brazil: Late Triassic (Caturrita Formation) of Rio Grande do Sul. In *Boletim de resumos: II Congreso Latino-Americano de Paleontología de Vertebrados*, 146. Rio de Janeiro: Museu Nacional.

Langer, M. C., M. A. Franca, and S. Gabriel. 2007. The pectoral girdle and forelimb anatomy of the stem-sauropodomorph *Saturnalia tupiniquim* (Upper Triassic, Brazil). *Special Papers in Palaeontology* No. 77: 113–137. London: The Palaeontological Association.

Langer, M. C., A. M. Ribeiro, C. L. Schultz, and J. Ferigolo. 2007. The continental tetrapod-bearing Triassic of South Brazil. In S. G. Lucas, and J. A Spielmann (Eds.), *The global Triassic*, 41:201–218,. Albuquerque: New Mexico Museum of Natural History and Science.

Langston, W., Jr., and J. W. Durham. 1955. A sauropod dinosaur from Colombia. *Journal of Paleontology* 29(6): 1047–1051.

Larson, P. L. 1991. The Black Hills Institute *Tyrannosaurus*—a preliminary report. *Journal of Vertebrate Paleontology* 11(3): 41–42A.

Lavocat, R. 1954. Sur les dinosauriens du Continental Intercalaire des Kem-Kem de la Daoura. In *Comptes Rendus de la Dix-Neuvième Session, Congrès Géologique International* 21, 65–68. Paris.

Lawver, L. A., L. M. Gahagan, and M. F. Coffin. 1992. The development of paleoseaways around Antarctica. *Antarctic Research Series* 56: 7–30.

Leal, L. A., S. A. K. Azevedo, A. W. A. Kellner, and A. A. S. da Rosa. 2004. A new early dinosaur

(Sauropodomorpha) from the Caturrita Formation (Late Triassic), Paraná Basin, Brazil. *Zootaxa* 690: 1–24.

Leanza, H. A. 1999. The Jurassic and Cretaceous terrestrial beds from southern Neuquén Basin, Argentina. Field Guide. *Instituto Superior de Correlación Geológica* (Série Miscelánea 4): 1–30.

———. 2003. Las sedimentitas huitrinianas rayosianas (Cretácico Inferior) en el ámbito central y meridional de la Cuenca Neuquina, Argentina. *Servicio Geológico Minero Argentino, Serie Contribuciones Técnicas—Geología* 2: 1–31.

Leanza, H. A., S. Apesteguía, F. E. Novas, and M. S. de la Fuente. 2004. Cretaceous terrestrial beds from the Neuquina Basin (Argentina) and their tetrapod assemblages. *Cretaceous Research* 25(1): 61–87.

Leanza, H. A., and C. A. Hugo. 1995. Revision estratigrafica del Cretacico Inferior continental en el ámbito sudoriental de la Cuenca Neuquina. *Revista de la Asociación Geológica Argentina* 50(1–4): 30–32.

Legarreta, L., and C. Gulisano. 1989. Análisis estratigráfico secuencial de la Cuenca Neuquina (Triásico Superior-Terciario Inferior). In G. Chebli and L. Spalletti (Eds.), *Cuencas Sedimentarias Argentinas. Serie de Correlación Geológica*, 6:221–243. Tucumán: Universidad Nacional de Tucumán.

Le Loeuff, J., E. Buffetaut, L. Calvin, M. Martin, V. Martin, and H. Tong. 1994. An armoured titanosaurid sauropod from the Late Cretaceous of Southern France and the occurrence of osteoderms in the Titanosauridae. *Gaia* 10: 155–159.

Leonardi, G. 1979a. Nota preliminar sobre seis pistas de dinossauros Ornithischia da Bacía do Rio do Peixe, em Sousa, Parnaíba, Brasil. *Anais da Academia Brasileira de Ciencias* 51(3): 501–516.

———. 1979b. New archosaurian trackways from the Rio do Peixe Basin, Paraiba, Brasil. *Annali dell'Universitá di Ferrara* (n. s.) 5(14): 239–249.

———. 1980a. On the discovery of an abundant ichnofauna (vertebrates and invertebrates) in the Botucatu Formation s.s. in Araraquara, São Paulo, Brasil. *Anais de la Academia Brasileira de Ciencias* 52(3): 559–567.

———. 1980b. Ornithischian trackways of the Corda Formation (Jurassic), Goias, Brazil. In *Actas II Congreso Latinoamericano de Paleontología y Biostratigrafía y I Congreso Latinoamericano de Paleontología*, 1:215–222. Buenos Aires.

———. 1981. As localidades com rastros fosseis de tetrapodes na America Latina. In *Anais II Congresso Latino-Americano Paleontologia*, 929–940. Porto Alegre, Brazil.

———. 1984. Le impronte fossili di dinosauri. In J. F. Bonaparte, E. H. Colbert, P. J. Currie, A. de Ricqlès, Z. Kielan-Jaworowska, G. Leonardi, N. Morello, and P. Taquet (Eds.), *Sulle orme dei dinosauri*, 163–186. Venezia, Erizzo: Centro Studi Richerche Ligabue.

———. 1989. Inventory and statistics of the South American dinosaurian ichnofauna and its paleobiological interpretation. In D. D. Gillette and M. G. Lockley (Eds.), *Dinosaur tracks and traces*, 165–178. Cambridge: Cambridge University Press.

———. 1994. Annotated Atlas of South American Tetrapod Footprints (Devonian to Holocene) with an appendix on Mexico and Central America. In *Companhia de Pesquisa de Recursos Minerais, Ministério de Minas e Energia*. Brasilia.

Leonardi, G., and G. Borgomanero. 1981. Sobre uma possível ocorrência de Ornitischia na Formação Santana, chapada do Araripe (Ceará). *Revista Brasileira de Geociências* 11: 1–4.

Leonardi, G., I. S. Carvalho, and M. A. Fernandes. 2007. The desert ichnofauna from Botucatu Formation (Upper Jurassic –Lower Cretaceous), Brazil. In I. S. Carvalho et al. (Eds.), *Paleontologia: Cenarios de vida*, 1:379–391. Río de Janeiro: Editora Interciencia.

Leonardi, G., and L. C. Godoy. 1980. Novas pistas de Tetrápodes de Formacao Botucatu no Estado de São Paulo. *Anais do XXXI Congresso Brasileiro de Geologia* 5: 3080–3089.

Leonardi, G., and F. H. L. Oliveira. 1990. A revision of the Triassic and Jurassic tetrapod footprints of Argentina and a new approach on the age and meaning of the Botucatu Formation footprints (Brazil). *Revista Brasileira de Geociências* 20(1–4): 216–229.

Lockley, M. G., A. S. Schulp, C. A. Meyer, G. Leonardi, and D. K. Mamani. 2002. Titanosaurid trackways from the Upper Cretaceous of Bolivia: evidence for large manus, wide-gauge locomotion and gregarious behaviour. *Cretaceous Research* 23(3): 383–400.

Lockley, M. G., S. Y. Yang, M. Matsukawa, F. Fleming, and S. K Lim. 1992. The track record of mesozoic birds: Evidence and implications. *Philosophical Transactions: Biological Sciences* 336(1277): 113–134.

Lockley, M. G., and J. Wright. 2001. Trackways of large quadrupedal ornithopods from the Cretaceous: A review. In D. R. Tanke and K. Carpenter (Eds.), *Mesozoic vertebrate life*, 428–442. Bloomington: Indiana University Press.

Long, J. A., and R. E. Molnar. 1998. A new Jurassic theropod dinosaur from western Australia. *Records of the Western Australian Museum* 19: 121–129.

Long, R. A., and P. A. Murry. 1995. Late Triassic (Carnian and Norian) tetrapods from the southwestern United States. *New Mexico Museum of Natural History and Sciences Bulletin* 4: 1–254.

Lucas, S. G., and M. J. Orchard. 2004. Triassic. In R. C. Selley, L. M. R. Cocks, and I. R. Plimer (Eds.), *Encyclopedia of Geology*, 344–351. Amsterdam: Elsevier.

Luna, M., G. Casal, R. D. Martínez, M. Lamanna, L. Ibiricu, and E. Ivany. 2003. La presencia de un Ornithopoda (Dinosauria: Ornithischia) en el Miembro Superior de la Formación Bajo Barreal (Campaniano-Maastrichtiano) del Sur del Chubut. *Ameghiniana* 40 (Suppl. 4): 61R.

Lydekker, R. 1893. Contributions to the study of the fossil vertebrates of Argentina. I. The dinosaurs of Patagonia. *Anales del Museo de La Plata, Paleontología* 2: 1–14.

Macellari, C. A., 1985. Paleobiogeografía y edad de la fauna de *Maorites-Gunnarites* (Ammonoidea) del Cretácico Superior de la Antártida y Patagonia. *Ameghiniana* 21(2–4): 223–242.

Machado, E. B., and A. W. A. Kellner. 2007. On a supposed ornithischian dinosaur from the Santana Formation, Araripe Basin, Brazil. In I. S. Carvalho et al. (Eds.), *Paleontologia: cenarios de vida*, 1:299–307. Río de Janeiro: Editora Interciencia.

Machado, E. B., A. W. A. Kellner, and D. A. Campos. 2005. Preliminary information on a dinosaur (Theropoda, Spinosauridae) pelvis from the Cretaceous Santana Formation (Romualdo Member) Brazil. In *Boletim de resumos: II Congresso Latino-Americano de Paleontología de Vertebrados*, 161–162. Rio de Janeiro: Museu Nacional.

Madsen, J. H. 1976. *Allosaurus fragilis*: A revised osteology. *Utah Geological and Mineral Survey Bulletin* 109: 1–163.

Madsen, J. H., and S. P. Welles. 2000. *Ceratosaurus* (Dinosauria, Theropoda). A revised osteology. *Miscellaneous Publication Utah Geological Survey* 2: 1–80.

Magalhães Ribeiro, C. M. 2002. Sauropod dinosaur eggs and eggshells from the Marilia Formation (Bauru Basin, Upper Cretaceous), Minas Gerais State, Brazil. *Journal of Vertebrate Paleontology* 22(3): 82A.

Magalhães Ribeiro, C. M., and L. C. Ribeiro. 2000. Nuevo hallazgo de huevo fósil y fragmentos de cáscaras de huevos en la Formación Marilia (Cretácico Tardío), Uberaba, Minas Gerais, Brasil. *Ameghiniana* 37(Suppl. 4): 27–28R.

Magalhães Ribeiro, C. M., M. E. Simón, M. Fernández, L. Salgado, and R. A. Coria. 2005. Dinosaur nesting sites from the Allen Formation (Upper Cretaceous), Rio Negro Province, Northern Patagonia, Argentina. In *Boletim de resumos: II Congresso Latino-Americano de Paleontología de Vertebrados*, 222. Rio de Janeiro: Museu Nacional.

Mahler, L. 2005. Record of Abelisauridae (Dinosauria: Theropoda) from the Cenomanian of Morocco. *Journal of Vertebrate Paleontology* 25(1): 236–239.

Makovicky, P. J., S. Apesteguía, and F. L. Agnolín. 2005. The earliest dromaeosaurid theropod from South America. *Nature* 437: 1007–1011.

Makovicky, P. J., and H. D. Sues. 1998. *Anatomy and phylogenetic relationships of* Microvenator celer *from the Lower Cretaceous of Montana*. American Museum Novitates no. 3240. New York: American Museum of Natural History.

Malumián, N., F. E. Nullo, and V. A. Ramos. 1983. Chapter 9. The Cretaceous of Argentina, Chile, Paraguay and Uruguay. In M. Moullade and A. E. M. Nairn (Eds.), *The Mesozoic, B*, 265–304. Amsterdam: Elsevier.

Manera de Bianco, T. 1996. Nueva localidad con nidos y huevos de dinosaurios (Titanosauridae) del Cretácico Superior, Cerro Blanco, Yaminué, Río Negro, Argentina. *Asociación Paleontológica Argentina, Publicación Especial* N° 4: 59–67.

———. 2000a. Nuevas observaciones acerca de las cáscaras de huevo de dinosaurio del Cretácico Superior del Cerro Blanco, Yaminué, Provincia de Río Negro, Argentina. *Ameghiniana* 37(Suppl. 4): 23R.

———. 2000b. Un nuevo tipo de cáscara del Cretácico Superior del Cerro Blanco, Yaminué, Provincia de Río Negro. *Ameghiniana* 37(Suppl. 4): 24R.

Manera de Bianco, T., and M. C. Bolognani. 2003. Tafonomía de huevos de dinosaurio del Cretácico Superior de Yaminué, provincia de Río Negro, Argentina. *Ameghiniana* 40(Suppl. 4): 101R.

Manera de Bianco, T., and G. Khöler. 2003. Primer registro de cáscaras de huevo de la oofamilia Spheroolithidae (Zhao, 1979) en Yaminué (Cretácico Superior), provincia de Río Negro, Argentina. *Ameghiniana* 40(Suppl. 4): 62R.

Manera de Bianco, T., C. Montalvo, S. Casadio, and A. Parras. 1999. Huevos de dinosaurios del Cretácico Superior en el departamento de Hucal, provincia de La Pampa, Argentina. In *Actes de VII Jornadas Pampeanas de Ciencias Naturales*, 46. Santa Rosa.

Manera de Bianco, T., A Parras, C. Montalvo, and S. Casadio. 2000. Paratoxonomía de huevos de dinosaurios del Cretácico Superior, cuenca del Colorado, La Pampa, Argentina. *Ameghiniana* 37 (Suppl. 4): 75R.

Manera de Bianco, T., and R. L. Tomassini. 2003. Alteración tafonómica de la superficie de cáscaras de huevos de estructura ornitoide del Cretácico Superior de Yaminué, provincia de Río Negro, Argentina. *Ameghiniana* 40(Suppl. 4):102R.

Marinho, T. S., and C. R. A. Candeiro. 2005. Titanosaur (Dinosauria: Sauropoda) osteoderms from the Maastrichtian of Uberaba, Minas Gerais State, Brazil. *Gondwana Research* 8(4): 473–477.

Marshall, L. 1989. El primer diente de dinosaurio en Bolivia. *Revista Técnica de Yacimientos Petrolíferos Fiscales Bolivia* 10(3–4): 129–130.

Marsicano, C. A., A. B. Arcucci, A. Mancuso, and A. T. Caselli. 2004. Middle Triassic tetrapod footprints of southern South America. *Ameghiniana* 41(2): 171–184.

Marsicano, C. A., and S. P. Barredo. 2004. Triassic tetrapod footprint assemblage from southern South America: palaeobiogeographical and evolutionary implications. *Palaeogeography, Palaeoclimatology, Palaeoecology* 203(3): 313–335.

Marsicano, C. A., N. S. Domnanovich, and A. C. Mancuso. 2005. Dinosaur origins: evidence from the footprint record. In *Boletim de Resumos: II Congreso Latino-Americano de Paleontología de Vertebrados*, 164–165. Rio de Janeiro: Museo Nacional.

Martill, D. M. 1993. *Fossils of the Santana and Crato Formations, Brazil.* Field Guides to Fossils, No. 5. Oxford: The Palaeontological Association.

Martill, D. M., A. R. I. Cruickshank, E. Frey, P. G. Small, and M. Clarke.1996. A new crested maniraptoran dinosaur from the Santana Formation (Lower Cretaceous) of Brazil. *Journal of the Geological Society* 153(1): 5–8.

Martill, D. M., and J. B. M. Figueira. 1994. A new feather from the Lower Cretaceous of Brazil. *Palaeontology* 37(3): 483–487.

Martill, D. M., and E. Frey. 1995. Colour patterning preserved in Lower Cretaceous birds and insects: The Crato Formation of NE Brazil. *Neues Jahrbuch für Geologie und Paläontologie Monatshefte* 2: 118–128.

Martill, D. M., E. Frey, H. D. Sues, and A. R. I. Cruickshank. 2000. Skeletal remains of the small theropod dinosaur with associated soft structures from the Lower Cretaceous Santana Formation of northeastern Brazil. *Canadian Journal of Earth Sciences* 37(6): 891–900.

Martin, L. D. 1997. The differences between dinosaurs and birds as applied to *Mononykus*. In D. Wolberg, E. Stump, and G. D. Rosenberg (Eds.), *Dinofest international: Proceedings of a symposium held at Arizona State University*, 337–343. [Philadelphia, Pa.]: Academy of Natural Sciences.

Martinelli, A. G., and A. M. Forasiepi. 2004. Late Cretaceous vertebrates from Bajo de Santa Rosa (Allen Formation), Río Negro province, Argentina, with the description of a new sauropod dinosaur (Titanosauridae). *Revista del Museo Argentino de Ciencias Naturales Bernardino Rivadavia* 6(2): 257–305.

Martinelli, A. G., and E. Vera. 2007. *Achillesaurus manazzonei*, a new alvarezsaurid theropod (Dinosauria) from the Late Cretaceous Bajo de la Carpa Formation, Río Negro Province, Argentina. *Zootaxa* 1582: 1–17.

Martínez, R. D. 1998a. An articulated skull and neck of Sauropoda (Dinosauria: Saurischia) from the Upper Cretaceous of central Patagonia, Argentina. *Journal of Vertebrate Paleontology* 18(Suppl. 3): 61A.

———. 1998b. *Notohypsilophodon comodorensis* gen. et sp. nov., un Hypsilophodontidae (Ornithischia: Ornithopoda) del Cretácico Superior de Chubut, Patagonia central, Argentina. *Acta Geologica Leopoldensia* 21: 119–135. São Leopoldo: Universidade do Vale do Rio dos Sinos.

———. 1999. Un cráneo y cuello articulado de Sauropoda (Dinosauria: Saurischia) del Cretácico Superior de Chubut. *Ameghiniana* 36(1): 105.

Martínez, R. D., O. Giménez, J. Rodríguez, and G. Bochatey. 1986. *Xenotarsosaurus bonapartei* nov. gen. et sp. (Carnosauria, Abelisauridae), un nuevo Theropoda de la Formación Bajo Barreal, Chubut, Argentina. In *Actas de IV Congreso Argentino de Paleontología y Bioestratigrafía*, 2:23–31. Mendoza.

Martínez, R. D., O. Giménez, J. Rodríguez, M. Luna, and M. Lamanna. 2004. An articulated specimen of the basal titanosaurian (Dinosauria: Sauropoda) *Epachthosaurus sciuttoi* from the early Late Cretaceous Bajo Barreal Formation of Chubut Province, Argentina. *Journal of Vertebrate Paleontology* 24(1): 107–120.

Martínez, R. D., M. Lamanna, J. Smith, G. Casal, and M. Luna. 1999. New Cretaceous theropod material from Patagonia. *Journal of Vertebrate Paleontology* 19(Suppl. 3): 62A.

Martínez, R. D., M. Lamanna, J. Smith, P. Dodson, and M. Luna. 2000. Dinosaurs from the Bajo Barreal Formation, Central Patagonia. *Journal of Vertebrate Paleontology* 20(Suppl. 3): 62A.

Martínez, R. D., and F. E. Novas. 1997. Un nuevo tetanuro (Dinosauria—Theropoda) de la Formación Bajo Barreal (Cretácico Superior), Patagonia. *Ameghiniana* 34(4): 538.

———. 2006. New coelurosaurian theropod from the early Late Cretaceous of central Patagonia, Argentina. *Revista del Museo Argentino de Ciencias Naturales Bernardino Rivadavia* 8(2): 243–259.

Martínez, R. D., F. E. Novas, and A. Ambrosio. 2004. Abelisaurid remains (Theropoda, Ceratosauria) from southern Patagonia. *Ameghiniana* 41(4): 577–585.

Martínez, R. N. 1999. The first South American record of *Massospondylus* (Dinosauria: Sauropodomorpha). *Journal of Vertebrate Paleontology* 19(Suppl. 3): 61A.

Martins Neto, R. G., and A. W. A. Kellner. 1988. Primeiro registro de pena na Formacão Santana (Cretáceo Inferior), bacia do Araripe, nordeste do Brasil. *Anais da Academia Brasileira de Ciências* 60(1): 61–68.

Mattley, C. A. 1923. Note on an armoured dinosaur from the Lameta beds of Jubbulpore. *Records of the Geological Survey of India* 55: 105–109.

Mazzoni, M. M., L. Spalletti, A. M. Iñiguez Rodriguez, and M. E. Teruggi. 1981. El Grupo Bahia Laura en el Gran Bajo de San Julián, Santa Cruz. In *Actas 8° Congreso Geológico Argentino*, 3: 485–507. San Luís.

McCrea, R., M. G. Lockley, and C. A. Meyer. 2001. Global distribution of purported ankylosaur track occurrences. In K. Carpenter (ed.), *The armored dinosaurs*, 413–454. Bloomington: Indiana University Press.

McIntosh, J. S. 1990. Sauropoda. In D. B. Weishampel, P. Dodson, and H. Osmólska (Eds.), *The Dinosauria*, 345–401. Berkeley: University of California Press.

Medeiros, M. A. 2001. A Laje do Coringa (Ilha do Cajual, Bacia de São Luís, Baía de São Marcos, MA): conteúdo fossilífero, bioestratinomia, diagênese e implicações na paleobiogeografia do Mesocretáceo do Nordeste brasileiro. Tese de Doutorado, Universidade Federal de Rio Grande do Sul, Programa de Pós-Graduação em Geociências.

Medeiros, M. A., P. C. Freire, and A. A. Pereira. 2007. Another African dinosaur recorded in the Eocenomanian of Brazil and a revision in the paleofauna of the Laje do Coringa site. In I. S. Carvalho et al. (Eds.), *Paleontologia: Cenarios de Vida*, 1:413–423. Río de Janeiro: Editora Interciencia.

Medeiros, M. A., and C. L. Schultz. 2001a. *Rebbachisaurus* (Sauropoda, Diplodocidomorpha) no Mesocretáceo do Nordeste do Brasil. In *Programa e Resumos do 13 Encontro de Zoologia do Nordeste*, 166. São Luis, Maranhão: Sociedade Nordestina de Zoologia/Universidade Federal do Maranhão.

———. 2001b. Uma paleocomunidade de vertebrados do Cretáceo Medio, Bacia de São Luis. In D. F. Rossetti, A. M. Goes, and W. Truckenbrodt (Eds.), *O Cretáceo na Bacia de São Luis-Grajaú*, 209–221. Belem: Museu Emilio Goeldi.

———. 2002. A fauna dinossauriana da "Laje do Coringa," Cretáceo médio do nordeste do Brasil. *Arquivos do Museu Nacional do Rio de Janeiro* 60(3): 155–162.

———. 2004. *Rayososaurus* (Sauropoda, Diplodocoidea) no Meso-Cretáceo do norte-nordeste brasileiro. *Revista Brasileira de Paleontologia* 7(2): 275–279.

Melchor, R. N., S. de Valais, and J. F. Genise. 2002. Bird-like fossil footprints from the Late Triassic. *Nature* 417: 936–938.

———. 2004. Middle Jurassic mammalian and dinosaur footprints and petrified forests from the volcaniclastic La Matilde Formation. In E. Bellosi and R. Melchor (Eds.), *Ichnia 2004, First International Congress on Ichnology, Fieldtrip guidebook*, 47–63. Trelew: Museo Paleontológico Egidio Feruglio.

Meyer, C. A., D. Hippler, and M. G. Lockley. 2001. The Late Cretaceous vertebrate ichnofacies of Bolivia—facts and implications. VII International Symposium on Mesozoic Terrestrial Ecosystems. *Asociación Paleontológica Argentina, Publicación Especial N°* 7: 133–138.

Mezzalira, S. 1980. Aspectos paleoecológicos da Formação Bauru. *Mesa Redonda: a Formação Bauru no Estado de São Paulo e regiões adjacentes. Sociedade Brasileira de Geologia, Publicação Especial* 7: 1–9.

Mikhailov, K. E. 1997. Eggs, eggshells, and nests. In P. J. Currie and K. Padian (Eds.), *Encyclopedia of dinosaurs*, 205–209. San Diego: Academic Press.

Milner, A. C. 1997. Spinosauridae and Baryonychidae. In P. J. Currie and K. Padian (Eds.), *Encyclopedia of dinosaurs*, 699–700. San Diego: Academic Press.

———. 2001. Fish-eating theropods: a short review of the systematic, biology and palaeobiogeography of spinosaurs. In *Actas de las II Jornadas de Paleontologia de dinosaurios y su entorno*, 129–138. Burgos, Spain: Colectivo Arqueológico-Paleontológico de Salas.

Mojica, J., and J. Dorado. 1987. El Jurásico anterior a los movimientos intermálmicos en los Andes colombianos. In W. Volkheimer (ed.), *Biostratigrafía de los sistemas regionales del Jurásico y Cretácico de América del Sur, Mendoza*, 49–110. Buenos Aires: Comité Sudamericano del Jurásico y Cretácico.

Molnar, R. E. 1980a. An ankylosaur (Ornithischia: Reptilia) from the Lower Cretaceous of southern Queensland. *Memoirs of the Queensland Museum* 20: 77–87.

———. 1980b. Australian Late Mesozoic tetrapods: Some implications. *Mémoires de la Société Géologique de France* 139: 131–143.

———. 1990. Problematic Theropoda: "Carnosaurs." In D. W. Weishampel, P. Dodson, and H. Osmólska (Eds.), *The Dinosauria*, 306–317. Berkeley: University of California Press.

———. 1996a. Observations on the Australian ornithopod dinosaur, *Muttaburrasaurus*. In F. E. Novas and R. E. Molnar (Eds.), *Proceedings of the Gondwanan dinosaur symposium*, 639–652. Brisbane: Queensland Museum.

———. 1996b. Preliminary report on a new ankylosaur from the Early Cretaceous of Queensland, Australia. In F. E. Novas and R. E. Molnar (Eds.), *Proceedings of the Gondwanan dinosaur symposium*, 653–668. Brisbane: Queensland Museum.

———. 2001. A reassessment of the phylogenetic position of Cretaceous sauropod dinosaurs from Queensland, Australia. *Asociación Paleontológica Argentina, Publicación Especial N°* 7: 139–144.

Mones, A. 1980. Nuevos elementos de la paleoherpetofauna del Uruguay (Crocodilia y Dinosauria). In *Actas II Congreso Argentino de Paleontologia y Bioestratigrafia*, 265–277. Buenos Aires.

Moody, J. M. 1997. Theropod teeth from the Jurassic of Venezuela. *Boletín de la Sociedad Venezolana de Geología* 22(2):37–42.

Moratalla, J. J., and J. E. Powell. 1994. Dinosaur nesting patterns. In K. Carpenter, K. F. Hirsch, and J. R. Horner (Eds.), *Dinosaur eggs and babies*, 37–46. Cambridge: Cambridge University Press.

Moreno, K., and M. J. Benton. 2005. Occurrence of sauropod dinosaur tracks in the Upper Jurassic of Chile (redescription of *Iguanodonichnus frenki*). *Journal of South American Earth Sciences* 20: 253–257.

Moreno, K., N. Blanco, and A. Tomlinson. 2004. Nuevas huellas de dinosaurios del Jurásico Superior en el norte de Chile. *Ameghiniana* 41(4): 535–543.

Moreno, K., and D. Rubilar. 1997. Presencia de nuevas pistas de dinosaurios (Theropoda-Ornithopoda) en la Formación Baños del Flaco, Provincia de Colchagua, VI Región, Chile. In *Actas VIII Congreso Iberoamericano de Biodiversidad y Zoología de Vertebrados*, 95. Concepción.

Moreno, K., D. Rubilar, and N. Blanco. 2000. Icnitas de dinosaurios de la Formación Chacarilla, I y II Región, norte de Chile. *Ameghiniana* 37(Suppl. 4): 30R.

Mourier, T., P. Bengtson, M. Bonhomme, E. Buge, H. Cappetta, J. Crochet, M. Feist, K. Hirsch, E. Jaillard, G. Laubacher, J. Lefranc, M. Moullade, C. Noblet, D. Pons, and J. Rey. 1988. The Upper Cretaceous–Lower Tertiary marine to continental transition in the Bagua basin, northern Peru: Paleontology, biostratigraphy, radiometry, correlations. *Newsletter in Stratigraphy* 19(3): 143–177.

Mourier, T., E. Jaillard, G. Laubacher, C. Noblet, A. Pardo, B. Sigé, and P. Taquet. 1986. Découverte de restes dinosauriens et mammaliens d'âge Cretacé supérieur à la base des couches rouges du synclinal de Bagua (Andes nord-péruviennes): Aspects stratigraphiques, sédimentologiques et paléogeographiques concernant la régression fin-Crétacée. *Bulletin de la Société Géologique de France* 8: 171–175.

Murry, P. A., and R. A. Long. 1989. Geology and paleontology of the Chinle Formation, Petrified Forest National Park and vicinity, Arizona and a discussion of vertebrate fossils of the southwestern Upper Triassic. In S. G. Lucas and A. P. Hunts (Eds.), *Dawn of the age of dinosaurs in the American southwest*, 29–64. Albuquerque: New Mexico Museum of Natural History.

Naish, D. 2000. A small, unusual theropod (Dinosauria) femur from the Wealden Group (Lower Cretaceous) of the Isle of Wight, England. *Neues Jahrbuch für Geologie und Paläontologie, Monatschafte*, 217–234.

Naish, D., D. M. Martill, and E. Frey. 2004. Ecology, systematics and biogeographical relationships of dinosaurs, including a new theropod, from the Santana Formation (?Albian, Early Cretaceous) of Brazil. *Historical Biology* 16(2): 57–70.

Nesbitt, S. J., and M. A. Norell. 2005. Extreme convergence in the body plans of an early suchian (Archosauria) and ornithomimid dinosaurs (Theropoda). *Proceedings of the Royal Society of London B* 273(1590): 1045–1048.

Nesbitt, S. J., R. B. Irmis, and W. G. Parker. 2007 A critical re-evaluation of the Late Triassic dinosaur taxa of North America. *Journal of Systematic Palaeontology* 5(2): 209–243.

Noblet, C., G. Leonardi, P. Taquet, R. Marocco, and E. Córdova. 1995. Novelle découverte d'empreintes laissées par des dinosaures dans la Formation des Couches Rouges (bassin de Cuzco-Sicuani, sud du Perou): Conséquences stratigraphiques et tectoniques. *Comptes Rendus de l'Académie des Sciences de Paris* (Série II) 320: 785–791.

Norell, M. A., Q. Ji, K. Gao, C. Yuan, Y. Zhao, and L. Wang. 2002. "Modern" feathers on a non-avian dinosaur. *Nature* 416: 36–37.

Norell, M. A., and P. J. Makovicky. 1999. *Important features of the dromaeosaurid skeleton. II: Information from newly collected specimens of* Velociraptor mongoliensis. American Museum Novitates no. 3282. New York: American Museum of Natural History.

———. 2004. Dromaeosauridae. In D. B. Weishampel, P. Dodson, and H. Osmólska (Eds.), *The Dinosauria*, 196–209. Berkeley: University of California Press.

Noriega, J., and C. P. Tambussi. 1995. A Late Cretaceous Presbyornithidae (Aves: Anseriformes) from Vega Island, Antarctic Peninsula: paleobiogeographic implications. *Ameghiniana* 32(1): 57–61.

Norman, D. B. 2004. Basal Iguanodontia. In D. B. Weishampel, P. Dodson, and H. Osmólska (Eds.), *The Dinosauria*, 413–437. Berkeley: University of California Press.

Norman, D. B., H. D. Sues, L. M. Witmer, and R. A. Coria. 2004. Basal Ornithopoda. In D. B. Weishampel, P. Dodson, and H. Osmólska (Eds.), *The Dinosauria*, 393–412. Berkeley: University of California Press.

Norman, D. B., L. Witmer, and D. B. Weishampel. 2004. Basal Ornithischia. In D. B. Weishampel, P. Dodson, and H. Osmólska (Eds.), *The Dinosauria*, 325–334. Berkeley: University of California Press.

Novas, F. E. 1986. Un probable terópodo (Saurischia) de la Formación Ischigualasto (Triásico Superior), San Juan, Argentina. In *IV Congreso Argentino de Paleontología y Bioestratigrafía*, 2:1–6. Mendoza, Argentina.

———. 1989a. The tibia and tarsus in Herrerasauridae (Dinosauria, incertae sedis) and the origin and evolution of the dinosaurian tarsus. *Journal of Paleontology* 63(5): 677–690.

———. 1989b. Los dinosaurios carnívoros de la Argentina. Doctoral Dissertation, Universidad Nacional de La Plata, Facultad de Ciencias Naturales.

———. 1991. Relaciones filogenéticas de los dinosaurios terópodos ceratosaurios. *Ameghiniana* 28(3–4): 410.

———. 1992a. La evolución de los dinosaurios carnivoros. In J. L. Sanz, B. Meléndez, F. E. Novas, J. E. Powell, J. J. Moratalla, M. G. Lockley, A. D. Buscalioni, J. Morales, and M. C. Diéguez (Eds.), *Los Dinosaurios y su entorno biótico*, 125–163. Cuenca, Spain: Instituto Juan de Valdés.

———. 1992b. Phylogenetic relationships of basal dinosaurs, the Herrerasauridae. *Palaeontology* 35(1): 51–62.

———. 1994. New Information on the Systematics and Postcranial Skeleton of *Herrerasaurus ischigualastensis* (Theropoda: Herrerasauridae) from the Ischigualasto Formation (Upper Triassic) of Argentina. *Journal of Vertebrate Paleontology* 13(4): 400–423.

———. 1996a. Dinosaur monophyly. *Journal of Vertebrate Paleontology* 16(4):723–741.

———. 1996b. Alvarezsauridae, Late Cretaceous maniraptorans from Patagonia and Mongolia. In F. E. Novas and R. E. Molnar (Eds.), *Proceedings of the Gondwanan dinosaur symposium*, 675–702. Brisbane: Queensland Museum.

———. 1997a. Abelisauridae. In P. J. Currie and K. Padian (Eds.), *Encyclopedia of dinosaurs*, 1–2. San Diego, Calif.: Academic Press.

———. 1997b. Herrerasauridae. In P. J. Currie and K. Padian (Eds.), *Encyclopedia of dinosaurs*, 303–311. San Diego, Calif.: Academic Press.

———. 1997c. South American dinosaurs. In P. J. Currie and K. Padian (Eds.), *Encyclopedia of dinosaurs*, 678–689. San Diego, Calif.: Academic Press.

———. 1997d. Anatomy of *Patagonykus puertai* (Theropoda, Avialae, Alvarezsauridae), from the Late Cretaceous of Patagonia. *Journal of Vertebrate Paleontology* 17(1): 137–166.

———. 1998. *Megaraptor namunhuaiquii* gen. et. sp. nov., a large-clawed, Late Cretaceous Theropod from Argentina. *Journal of Vertebrate Paleontology* 18(1): 4–9.

———. 1999. Dinosaur. In *McGraw-Hill yearbook of science and technology*, 1999, 133–135. New York: McGraw-Hill Book Company.

———. 2004. Avian traits in the ilium of *Unenlagia comahuensis* (Maniraptora, Avialae). In P. Currie, E. Koppelhus, and E. Martin (Eds.), *Feathered dragons: Studies on the transition from dinosaurs to birds*, 150–166. Bloomington: Indiana University Press.

Novas, F. E., and F. L. Agnolín. 2004. *Unquillosaurus ceibali* Powell, a giant maniraptoran (Dinosauria, Theropoda) from the Late Cretaceous of Argentina. *Revista del Museo Argentino de Ciencias Naturales Bernardino Rivadavia* 6(1): 61–66.

Novas, F. E., F. L. Agnolín, and S. Bandyopadhyay. 2004. Cretaceous Theropods from India: a review of specimens described by Huene and Matley (1933). *Revista del Museo Argentino de Ciencias Naturales Bernardino Rivadavia* 6(1): 67–103.

Novas, F. E., and S. Bandyopadhyay. 1999. New approaches on the Cretaceous theropods from India. VII International Symposium on Mesozoic Terrestrial Ecosystems (Buenos Aires). *Asociación Paleontológica Argentina, Publicación Especial* N° 7: 46–47.

———. 2001. Abelisaurid pedal unguals from the Late Cretaceous of India. *Asociación Paleontologica Argentina, Publicación Especial* N° 7: 145–149.

Novas, F. E., E. Bellosi, and A. Ambrosio. 2002. Los "Estratos con Dinosaurios" del Lago Viedma y Río La Leona (Cretácico, Santa Cruz): sedimentología y contenido fosilífero. *Actas XV Congreso Geológico Argentino*, article 315. CD-ROM. El Calafate.

Novas, F. E., L. C. Borges Ribeiro, and I. S. Carvalho. 2005. Maniraptoran theropod ungual from the Marilia Formation (Upper Cretaceous), Brazil. *Revista del Museo Argentino de Ciencias Naturales Bernardino Rivadavia* 7(1): 31–36.

Novas, F. E., I. S. Carvalho, L. C. Borges Ribeiro, and A. H. Méndez. 2008. First abelisaurid bone remains from the Maastrichtian Marília Formation, Bauru Basin, Brazil. *Cretaceous Research* 29: 625–635.

Novas, F. E., A. V. Cambiaso, and A. Ambrosio. 2004. A new basal iguanodontian (Dinosauria, Ornithischia) from the Upper Cretaceous of Patagonia. *Ameghiniana* 41(1): 75–82.

Novas, F. E., A. V. Cambiaso, J. Lirio, and H. Núñez. 2002. Paleobiogeografía de los dinosaurios cretácicos polares de Gondwana. *Ameghiniana* (Resúmenes) 39(4): 15R.

Novas, F. E., J. I. Canale, and M. P. Isasi. 2004. Giant deinonychosaurian theropod from the Late Cretaceous of Patagonia. *Journal of Vertebrate Paleontology* 24(Suppl. 3): 98A.

Novas, F. E., S. Chatterjee, D. K. Rudra, and P. M. Datta. In press. New abelisaurid theropod from the Late Cretaceous of India. Plenum Press.

Novas, F. E., F. Dalla Vecchia, and D. Pais. 2005. Theropod pedal unguals from the Late Cretaceous (Cenomanian) of Morocco, Africa. *Revista del Museo Argentino de Ciencias Naturales Bernardino Rivadavia* 7(2):167–175.

Novas, F. E., S. de Valais, P. Vickers-Rich, and T. H. Rich. 2005. A large Cretaceous theropod from Patagonia, Argentina, and the evolution of carcharodontosaurids. *Naturwissenschaften* 92: 226–230.

Novas, F. E., and M. D. Ezcurra. 2005. The evolutionary radiation of Triassic dinosauriforms. *Ameghiniana* 42(Suppl. 4): 36–37R.

———. 2006. Reinterpretation of the dorsal vertebrae of *Argentinosaurus huinculensis* (Sauropoda, Titanosauridae). *Ameghiniana* 43(Suppl. 4):48–49R.

Novas, F. E., M. D. Ezcurra, and F. L. Agnolín. 2006. Humerus of a basal abelisauroid theropod from the Late Cretaceous of Patagonia. *Revista del Museo Argentino de Ciencias Naturales Bernardino Rivadavia* 8(1): 63–68.

Novas, F. E, M. D. Ezcurra, and A. Lecuona. 2008. *Orkoraptor burkei* nov. gen. et sp., a large coelurosaurian theropod from the Maastrichtian Pari Aike Formation, Southern Patagonia, Argentina. *Cretaceous Research* 29(2008): 468–480.

Novas, F. E., M. Fernández, Z. B. Gasparini, J. M. Lirio, H. J. Núñez, and P. F. Puerta. 2002. *Lakumasaurus antarcticus*, n. gen. et sp., a new mosasaur (Reptilia, Squamata) from the Upper Cretaceous of Antarctica. *Ameghiniana* 39(2): 245–249.

Novas F. E., R. D. Martínez, S. de Valais, and A. Ambrosio. 1999. Nuevos registros de Carcharodontosauridae (Dinosauria, Theropoda) en el Cretácico de Patagonia. *Ameghiniana* 36(Suppl. 4): 17A.

Novas, F. E., and D. Pol. 2002. Alvarezsaurid relationships reconsidered. In L. M. Chiappe, and L. Witmer (Eds.), *Mesozoic birds: Above the heads of dinosaurs*, 121–125. Berkeley: University of California Press.

———. 2005. New evidence on deinonychosaurian dinosaurs from the Late Cretaceous of Patagonia. *Nature* 433: 858–861.

Novas, F. E., D. Pol, J. I. Canale, J. D. Porfiri, and J. O. Calvo. 2009. A bizarre Cretaceous theropod dinosaur from Patagonia and the evolution of Gondwanan dromaeosaurids. *Proceedings of the Royal Society of London* 276: 1101–1107.

Novas, F. E., and P. F. Puerta. 1997. New evidence concerning avian origins from the Late Cretaceous of Patagonia. *Nature* 387: 390–392.

Novas, F. E., L. Salgado, J. O. Calvo, and F. L. Agnolín. 2005. Giant titanosaur (Dinosauria, Sauropoda) from the Late Cretaceous of Patagonia. *Revista del Museo Argentino de Ciencias Naturales Bernardino Rivadavia* 7(1): 37–41.

Nowinski, A. 1971. *Nemegtosaurus mongoliensis* n. gen., n. sp. (Sauropoda) from the uppermost Cretaceous of Mongolia. *Palaeontologia Polonica* 25: 57–81.

O'Connor, P. M. 2007. The postcranial axial skeleton of *Majungasaurus crenatissimus* (Theropoda: Abelisauridae) from the Late Cretaceous of Madagascar. In S. D. Sampson and D. W. Krause (Eds.), *Majungasaurus crenatissimus from the Late Cretaceous of Madagascar*, 127–162. [Northbrook, Ill.]: Society of Vertebrate Paleontology.

Olsen, P. E., and H. D. Sues. 1986. Correlation of continental Late Triassic and Early Jurassic sediments, and patterns of the Triassic–Jurassic tetrapod transition. In K. Padian (ed.), *The beginning of the age of dinosaurs*, 321–351. Cambridge: Cambridge University Press.

Olson, S., and D. Parris. 1987. The Late Cretaceous birds of New Jersey. *Smithsonian Contributions to Paleobiology* 63: 1–22.

Osmólska, H. 1980. The Late Cretaceous vertebrate assemblages of the Gobi Desert, Mongolia. *Memoirs de la Société Géologique de France* 139: 145–150.

Osmólska, H., and R. Barsbold. 1990. Troodontidae. In D. B. Weishampel, P. Dodson, and H. Osmólska (Eds.), *The Dinosauria*, 259–268. Berkeley: University of California Press.

Osmólska, H., P. J. Currie, and R. Barsbold. 2004. Oviraptorosauria. In D. B. Weishampel, P. Dodson, and H. Osmólska (Eds.), *The Dinosauria*, 165–183. Berkeley: University of California Press.

Ostrom, J. H. 1969. Osteology of *Deinonychus antirrhopus*, an unusual theropod from the Lower Cretaceous of Montana. *Bulletin of the Peabody Museum of Natural History* 30: 1–165.

———. 1978. The osteology of *Compsogathus longipes* Wagner. *Zitteliana* 4: 73–118.

Ostrom, J. H., and J. S. McIntosh. 1966. *Marsh's dinosaurs: The collections from Como Bluff*. New Haven, Conn.: Yale University Press.

Padian, K., and C. L. May. 1993. The earliest dinosaurs. In S. G. Lucas and M. Morales (Eds.), *The nonmarine Triassic: Transactions of the international symposium and field trip on the nonmarine*

Triassic, 3:379–381. Albuquerque: New Mexico Museum of Natural History and Science.

Page, R., A. Ardolino, R. Barrio, M. Franchi, A. Lizuain, S. Page, and D. Nieto. 1999. Estratigrafía del Jurásico y Cretácico del Macizo de Somún Curá, provincias de Río Negro y Chubut. In R. Caminos (ed.), *Geología Argentina*, 460–488. Buenos Aires: Instituto de Geología y Recursos Minerales.

Parker, W. G., R. B. Irmis, S. J. Nesbitt, J. W. Martz, and L. S. Browne. 2005. The Late Triassic pseudosuchian *Revueltosaurus callenderi* and its implications for the diversity of early ornithischian dinosaurs. *Proceedings of the Royal Society of London B* 272(1566): 963–969.

Parrish, J. M. 1997. Evolution of archosaurs. In J. O. Farlow and M. K. Brett-Surman (Eds.), *The complete dinosaur*, 191–203. Bloomington: Indiana University Press.

Paul, G. S. 1988. *Predatory dinosaurs of the world: A complete illustrated guide*. New York: Simon and Schuster.

Paulina Carabajal, A., R. A. Coria, and P. J. Currie. 2003. Primer hallazgo de Abelisauria en la Formación Lisandro (Cretácico Superior), Neuquén. *Ameghiniana* 40(Suppl. 4): 65R.

Pereda-Suberbiola, X., F. Torcida Fernández Baldor, L. A. Izquierdo, P. Huerta, D. Montero, and G. Pérez. 2003. First rebbachisaurid dinosaur (Sauropoda, Diplodocoidea) from the early Cretaceous of Spain: palaeobiogeographical implications. *Bulletin de la Société Géologique de France* 174(5): 471–479.

Perle, A., L. M. Chiappe, R. Barsbold, J. M. Clark, and M. A. Norell. 1994. *Skeletal morphology of* Mononykus olecranus *(Theropoda: Avialae) from the Late Cretaceous of Mongolia*. American Museum Novitates no. 3105. New York: American Museum of Natural History.

Perle, A., M. A. Norell, L. M. Chiappe, and J. M. Clark. 1993. Flightless bird from the Cretaceous of Mongolia. *Nature* 362: 623–626.

Perle, A., M. A. Norell, and J. M. Clark. 1999. A new maniraptoran theropod—*Achillobator giganticus* (Dromaeosauridae)—from the Upper Cretaceous of Burkhant, Mongolia. *Contributions of the Department of Geology of the National University of Mongolia* 101: 1–106.

Pitman, W. C., III, S. Cande, J. LaBrecque, and J. Pindell. 1993. Fragmentation of Gondwana: The separation of Africa from South America. In P. Goldblatt (ed.), *Biological relationships between Africa and South America*, 15–34. New Haven, Conn.: Yale University Press.

Pol, D., and J. E. Powell. 2005. Anatomy and phylogenetic relationships of *Mussaurus patagonicus* (Dinosauria, Sauropodomorpha) from the Late Triassic of Patagonia. In *Boletim de Resumos: II Congresso Latino-Americano de Paleontologia de Vertebrados*, 208. Rio de Janeiro: Museo Nacional.

———. 2007a. Skull anatomy of *Mussaurus patagonicus* (Dinosauria: Sauropodomorpha) from the Late Triassic of Patagonia. *Historical Biology* 19(1): 125–144.

———. 2007b. New information on *Lessemsaurus sauropoides* (Dinosauria: Sauropodomorpha) from the Upper Triassic of Argentina. In P. M. Barrett and D. J. Batten (Eds.), *Evolution and Palaeobiology of Early Sauropodomorph Dinosaurs*, 223–243. Special Papers in Palaeontology No. 77. London: Palaeontological Association.

Porfiri, J. D., J. O. Calvo, and F. E. Novas. 2005. Hallazgo de un nuevo Theropoda del Cretácico Tardío en Lago Barreales, Neuquén, Patagonia, Argentina. In *Boletim de resumos: II Congresso Latino-Americano de Paleontología de Vertebrados*, 209–210. Rio de Janeiro: Museo Nacional.

Powell, J. E. 1979. Sobre una asociación de dinosaurios y otras evidencias de vertebrados del Cretácico Superior de la región de La Candelaria, Prov. de Salta, Argentina. *Ameghiniana* 16(1–2): 191–204.

———. 1980. Sobre la presencia de una armadura dérmica en algunos dinosaurios titanosáuridos. *Acta Geologica Lilloana* 15: 41–47. Tucumán: Instituto Miguel Lillo.

———. 1985. Hallazgo de nidadas de huevos de Dinosaurios (Sauropoda-Titanosauridae) del Cretácico Superior del Salitral Ojo de Agua, provincia de Río Negro. In *II Jornadas Argentinas de Paleontología de Vertebrados*, 15. Tucumán: Universidad Nacional de Tucumán.

———. 1986. Revisión de los titanosáuridos de América del Sur. Doctoral thesis, Universidad Nacional de Tucumán, Tucumán, Argentina.

———. 1987a. Morfología del esqueleto axial del los dinosaurios titanosáuridos (Saurischia, Sauropoda) del estado de Minas Gerais, Brasil. In *Anais do X Congreso Brasileiro de Paleontología, Río de Janeiro*, 155–171. Rio de Janeiro.

———. 1987b. The Late Cretaceous fauna of Los Alamitos, Patagonia, Argentina. Part VI. The Titanosaurids. *Revista del Museo Argentino de Ciencias Naturales Bernardino Rivadavia, Paleontología* 3(3): 147–153.

———. 1987c. Hallazgo de un dinosaurio hadrosáurido (Ornithischia: Ornithopoda) en la Formación Allen (Cretacico Superior) de Salitral Moreno, Provincia de Río Negro, Argentina. In *X Congreso Geológico Argentino*, San Miguel de Tucumán, Actas III, 149–152.

———. 1990. *Epachtosaurus sciuttoi* (gen. et sp. nov.) un dinosaurio sauropodo del Cretácico de Patagonia. In *V Congreso Argentino de Paleontología y Bioestratigrafía*, Tucumán, Actas I, 123–128.

———. 1992. Osteología de *Saltasaurus loricatus* (Sauropoda-Titanosauridae) del Cretácico Superior del Noroeste argentino. In J. L. Sanz and A. D. Buscalioni (Eds.), *Los Dinosaurios y su entorno biótico*, 4:165–230. Cuenca, Spain: Instituto Juan de Valdés.

———. 2003. Revision of South American titanosaurid dinosaurs: Palaeobiological, palaeobiogeographical and phylogenetic aspects. *Records of the Queen Victoria Museum* 111: 1–173.

Price, L. I. 1947. Sedimentos mesozoicos da baía de São Marcos, Estado do Maranhão. *Divisão de Geologia e Mineralogia, Notas Preliminares e Estudos* 40: 1–7.

———. 1951. Um ovo de dinossaurio na Formação Bauru do Cretáceo do estado de Minas Gerais. *Divisão de Geologia e Mineria, Notas Preliminares e Estudos* 53: 1–9.

———. 1959. Sobre um crocodilídeo notossúquio do Cretácico brasileiro. *Divisão de Geologia e Mineralogia Boletim* 188: 1–55.

———. 1961. Sobre os dinossaurios do Brasil. *Anais da Academia Brasileira de Ciências* 33(3–4): 28–29.

Proserpio, C. A. 1987. Descripción geológica de la hoja 44e. Valle General Racedo, provincia del Chubut. *Dirección Nacional de Minería y Geología* 201: 1–102.

Raath, M. 1969. A new coelurosaurian dinosaur from the Forest Sandstone of Rhodesia. *Arnoldia* 4: 1–25.

Rapela, C., and E. Llambías. 1999. El magmatismo gondwánico y los ciclos fanerozoicos. In R. Caminos (ed.), *Geología Argentina*, 373–376. Buenos Aires: Instituto de Geología y Recursos Minerales.

Rapela C., and R. J. Pankhurst. 1996. Monzonite suites: The innermost Cordilleran plutonism of Patagonia. *Transactions of the Royal Society of Edinburgh, Earth Sciences* 87: 193–203.

Rauhut, O. W. M. 1998. *Elaphrosaurus bambergi* and the early evolution of theropod dinosaurs. *Journal of Vertebrate Paleontology* 18(Suppl. 3): 71A.

———. 2002. Dinosaur evolution in the Jurassic: a South American perspective. *Journal of Vertebrate Paleontology* 22(Suppl. 3): 98A.

———. 2003a. Interrelationships and evolution of basal theropod dinosaurs. *Special Papers in Palaeontology* No. 69: 1–215. London: The Palaeontological Association.

———. 2003b. Revision of *Amygdalodon patagonicus* Cabrera, 1947 (Dinosauria, Sauropoda). *Mitteilungen aus dem Museum fuer Naturkunde in Berlin: Geowissenschaftliche Reihe*, 6:173–181.

———. 2003c. A dentary of *Patagosaurus* (Sauropoda) from the Middle Jurassic of Patagonia. *Ameghiniana* 40(3): 425–432.

———. 2004. Provenance and anatomy of *Genyodectes serus*, a large-toothed ceratosaur (Dinosauria: Theropoda) from Patagonia. *Journal of Vertebrate Paleontology* 24(4): 894–902.

———. 2004. Braincase structure of the Middle Jurassic theropod dinosaur *Piatnitzkysaurus*. *Canadian Journal of Earth Sciences* 41: 1109–1122.

———. 2005a. Osteology and relationships of a new theropod dinosaur from the Middle Jurassic of Patagonia. *Palaeontology* 48(1): 87–110.

———. 2005b. Post-cranial remains of 'coelurosaurs' (Dinosauria, Theropoda) from the Late Jurassic of Tanzania. *Geological Magazine* 142(1): 97–107.

Rauhut, O. W. M., G. Cladera, P. Vickers-Rich, and T. H. Rich. 2003. Dinosaur remains from the Lower Cretaceous of the Chubut Group, Argentina. *Cretaceous Research* 24(5): 487–497.

Rauhut O. W. M., A. López Arbarello, and P. Puerta. 2001 Jurassic vertebrates from Patagonia. *Journal of Vertebrate Paleontology* 21(Suppl. 3): 91A.

Rauhut, O. W. M., T. Martin, E. Ortiz-Jaureguizar, and P. Puerta. 2002. A Jurassic mammal from South America. *Nature* 416: 165–168.

Rauhut, O. W. M., K. Remes, R. Fechner, G. Cladera, and P. Puerta. 2005. Discovery of a short-necked sauropod dinosaur from the Late Jurassic period of Patagonia. *Nature* 435: 670–672.

Rauhut, O. W. M., and C. Werner. 1995. First record of the family Dromaeosauridae (Dinosauria: Theropoda) in the Cretaceous of Gondwana (Wadi Milk Formation, northern Sudan). *Paläontologische Zeitschrift* 69: 475–489.

Reig, O. A. 1963. La presencia de dinosaurios saurisquios en los "Estratos de Ischigualasto" (Mesotriásico superior) de las Provincias de San Juan y La Rioja (República Argentina). *Ameghiniana* 3(1): 3–20.

Reisz, R. R., D. Scott, H. D. Sues, D. C. Evans, and M. A. Raath. 2005. Embryos of an early Jurassic prosauropod dinosaur and their evolutionary significance. *Science* 309(5735): 761–764.

Rich, T. H., and P. Vickers-Rich. 1989. Polar dinosaurs and biotas of the early Cretaceous of southeastern Australia. *National Geographic Research* 5(1): 15–53.

———. 1994. Neoceratopsians and ornithomimosaurs: Dinosaurs of Gondwanan origin? *Research & Exploration: A Scholarly Publication of the National Geographic Society* 10: 129–131.

———. 2003. Protoceratopsian? ulnae from Australia. *Records of the Queen Victoria Museum* 113: 1–12.

Rich, T. H., P. Vickers-Rich, M. Fernández, and S. Santillana. 1999. A probable hadrosaur from Seymour Island, Antarctic Peninsula. In Y. Tomida, T. H. Rich, and P. Vickers-Rich (Eds.), *Proceedings of the Second Gondwanan Dinosaur Symposium*, 219–222. Tokyo: National Science Museum.

Rich, T. H., P. Vickers-Rich, O. Giménez, R. Cúneo, P. Puerta, and R. Vacca. 1999. A new sauropod dinosaur from Chubut province, Argentina. In Y. Tomida, T. Rich, and P. Vickers-Rich (Eds.), *Proceedings of the Second Gondwanan Dinosaur Symposium*, 61–84. Tokyo: National Science Museum.

Rich, T. H., P. Vickers-Rich, F. E. Novas, R. Cúneo, P. Puerta, and R. Vacca. 2000. Theropods from the "Middle" Cretaceous Chubut Group of the San Jorge sedimentary basin, central Patagonia. A preliminary note. *Gaia* 15: 111–115.

Rogers, R. R., C. C. Swisher, III, P. C. Sereno, A. M. Monetta, C. A. Forster, and R. N. Martínez. 1993. The Ischigualasto tetrapod assemblage, Late Triassic, Argentina, and 40Ar/39Ar dating of dinosaurs origins. *Science* 260(5109): 794–797.

Romer, A. S. 1956. *Osteology of the reptiles*. Chicago: University of Chicago Press.

———. 1971. The Chañares (Argentina) Triassic reptile fauna. X. Two new but incompletely known long-limbed pseudosuchians. *Breviora* 378: 1–10.

———. 1972. The Chañares (Argentina) Triassic reptile fauna. XV. Further remains of the thecodonts *Lagerpeton* and *Lagosuchus*. *Breviora* 394: 1–7.

Rougier, G., M. de la Fuente, and A. B. Arcucci. 1995. Late Triassic turtles from South America. *Science* 268(5212): 855–858.

Rowe, T. 1989. A new species of the theropod dinosaur *Syntarsus* from the Early Jurassic Kayenta Formation of Arizona. *Journal of Vertebrate Paleontology* 9(2): 125–136.

Rowe, T., and J. A. Gauthier. 1990. Ceratosauria. In D. B. Weishampel, P. Dodson, and H. Osmólska (Eds.), *The Dinosauria*, 151–168. Berkeley: University of California Press.

Rowe T., R. S. Tykoski, and J. Hutchinson. 1997. Ceratosauria. In P. J. Currie and K. Padian (Eds.), *Encyclopedia of dinosaurs*, 106–110. San Diego: Academic Press.

Rubert, R. R., and C. L. Schultz. 2004. Um novo horizonte de correlacao para o Triásico Superior do Rio Grande so Sul. *Pesquisas em Geociencias* 31(1): 71–88.

Rubilar, D. 2003. Registro de dinosaurios en Chile. *Boletín del Museo Nacional de Historia Natural* 52: 137–150.

Rubilar, D., K. Moreno, and N. Blanco. 2000. Huellas de dinosaurios ornitópodos en la Formación Chacarilla (Jurásico Superior-Cretácico Inferior), I Región de Tarapaca, Chile. In *Actas: IX Congreso Geológico Chileno*, 550–554. Puerto Varas.

Rubilar, D., K. Moreno, and A. Vargas. 1998. A revision of dinosaur trackways (Theropoda–Sauropoda–Ornithopoda) at Baños del Flaco Formation, VI Región, Chile. *Ameghiniana* 36(1): 80.

Rusconi, C. 1933. Sobre reptiles cretáceos del Uruguay (*Uruguaysuchus aznarezi*) y sus relaciones con los notosúquidos de Patagonia. *Instituto de Geología y Perforaciones Boletín* 19: 1–64.

Russell, D. A. 1967. A census of dinosaur specimens collected in Western Canada. *Natural History Papers of the National Museum of Canada* 36: 1–13.

———. 1993. The role of Central Asia in dinosaurian biogeography. *Canadian Journal of Earth Sciences* 30(10–11): 2002–2012.

———. 1996. Isolated dinosaur bones from the Middle Cretaceous of the Tafilalt, Morocco. *Bulletin du Muséum National d'Histoire Naturelle* (Série 4) 18: 349–402.

Russell, D. A., P. Béland, and J. S. McIntosh. 1980. Paleoecology of the dinosaurs of Tendaguru (Tanzania). *Mémoires de la Société Géologique de France* 139: 169–175. Paris: F.-G. Levrault.

Russell, D. A., and J. M. Dong. 1993. The affinities of a new theropod from the Alxa Desert, Inner Mongolia, People's Republic of China. *Canadian Journal of Earth Sciences* 30(10–11): 2107–2127.

Russell, D. E., O. Odreman Rivas, B. Battail, and D. A. Russell. 1992. Discovery of fossil vertebrates in the La Quinta Formation, Jurassic of Western Venezuela. *Comptes Rendus de la Academie de Sciences de París* 314: 1247–1252. New York: Elsevier Science.

Sabino, I. F., J. A. Salfity, and R. A. Marquillas. 1998. La Formación La Yesera (Cretácico) en el depocentro de Alemanía, Cuenca del Grupo Salta, Noroeste de la Argentina. In *Actas VII Reunión Argentina de Sedimentologia*, 128–135. Salta.

Salfity, J. A. 1982. Evolución paleogeográfica del Grupo Salta (Cretácico-Eogénico), Argentina. In *Actas 5th Congreso Latinoamericano de Geologia*, 1:11–26. Buenos Aires.

Salfity, J. A., and R. A. Marquillas. 1981. Las unidades estratigráficas cretácicas del Norte de la Argentina. In W. Wolkheimer and E. Musacchio (Eds.), *Cuencas sedimentarias del Jurásico y Cretácico de América del Sur*, 1:303–307. Buenos Aires: Comité Sudamericano del Jurásico y Cretácico.

———. 1999. La Cuenca Cretácico-Terciaria del Norte Argentino. In R. Caminos (Ed.), *Geología Argentina*, 613–626. Buenos Aires: Institut de Geología y Recursos Minerales.

Salgado, L. 1993. Comments on *Chubutisaurus insignis* Del Corro (Saurischia, Sauropoda). *Ameghiniana* 30(3): 265–270.

———. 1996. *Pellegrinisaurus powelli* nov. gen. et sp. (Sauropoda, Titanosauridae) from the Upper Cretaceous of lago Pellegrini, northwestern Patagonia, Argentina. *Ameghiniana* 33(4): 355–365.

———. 1999. The macroevolution of Diplodocimorpha (Dinosauria; Sauropoda): a developmental model. *Ameghiniana* 36(2): 203–216.

———. 2003a. Los saurópodos de Patagonia: sistemática, evolución y paleobiología. In *Actas de las II Jornadas Internacionales sobre Paleontología de Dinosaurios y su Entorno*, 139–168. Burgos, Spain: Colectivo Arqueológico-Paleontológico de Salas.

———. 2003b. Considerations on the bony plates assigned to titanosaur (Dinosauria, Sauropoda). *Ameghiniana* 40(3): 441–456.

———. 2003c. Should we abandon the name Titanosauridae? Some considerations on the taxonomy of titanosaurian sauropods (Dinosauria). *Revista Española de Paleontología* 18(1): 15–21.

Salgado, L., S. Apesteguía, and S. Heredia. 2005. A new specimen of *Neuquensaurus australis*, a Late Cretaceous saltasaurine titanosaur from North Patagonia. *Journal of Vertebrate Paleontology* 25(3): 623–634.

Salgado, L., and C. Azpilicueta. 2000. Un nuevo saltasaurino (Sauropoda, Titanosauridae) de la provincia de Río Negro (Formación Allen, Cretácico Superior), Patagonia, Argentina. *Ameghiniana* 37(3): 259–264.

Salgado, L., and J. F. Bonaparte. 1991. Un nuevo saurópodo Dicraeosauridae, *Amargasaurus cazaui*, gen. et sp. nov., de la Formación La Amarga, Neocomiano de la Provincia del Neuquén, Argentina. *Ameghiniana* 28(3–4): 333–346.

———. 2007. Sauropodomorpha. In Z. Gasparini, L. Salgado, and R. Coria (Eds.), *Patagonian Mesozoic reptiles*, 188–229. Bloomington: Indiana University Press.

Salgado, L., and J. O. Calvo. 1992. Cranial osteology of *Amargasaurus cazaui* Salgado and Bonaparte (Sauropoda, Dicraeosauridae) from the Neocomian of Patagonia. *Ameghiniana* 29(4): 337–346.

———. 1993. Report of a sauropod with amphiplatyan mid-caudal vertebrae from the Late Cretaceous of Neuquén Province (Argentina). *Ameghiniana* 30(2): 215–218.

———. 1997. Evolution of titanosaurid sauropods. II: The cranial evidence. *Ameghiniana* 34(1): 33–48.

Salgado, L., J. O. Calvo, and R. A. Coria. 1991. Un dinosaurio saurópodo de caudales anfiplaticas en el Cretácico Superior de la Provincia del Neuquén. *Ameghiniana* 28(3–4): 412.

Salgado, L., I. S. Carvalho, and A. Garrido. 2006. *Zapalasaurus bonapartei*, un nuevo dinosaurio saurópodo de La Formación La Amarga (Cretácico Inferior), noroeste de Patagonia, Provincia de Neuquén, Argentina. *Geobios* 39: 695–707.

Salgado, L., and R. A. Coria. 1993. El género *Aeolosaurus* (Sauropoda, Titanosauridae) en la Formación Allen (Campaniano-Maastrichtiano) de la Provincia de Río Negro, Argentina. *Ameghiniana* 30(2): 119–128.

———. 1996. First evidence of an armoured ornithischian dinosaur in the Late Cretaceous of North Patagonia, Argentina. *Ameghiniana* 33(4): 367–371.

———. 2005. Sauropods of Patagonia: systematic update and notes on global sauropod evolution. In V. Tidwell and K. Carpenter (Eds.), *Thunderlizards: The sauropodomorph dinosaurs*, 430–452. Bloomington: Indiana University Press.

Salgado, L., R. A. Coria, and J. O. Calvo. 1997. Evolution of titanosaurid sauropods. I: Phylogenetic analysis based on the postcranial evidence. *Ameghiniana* 34(1): 3–32.

Salgado, L., R. A. Coria, and L. M. Chiappe. 2001. Cráneos de embriones de titanosaurios (Sauropoda) del Cretácico Superior de Patagonia (Auca Mahuevo, Neuquén, Argentina). *Ameghiniana* 38(Suppl. 4): 19R.

Salgado, L., R. A. Coria, and S. Heredia. 1997. New materials of *Gasparinisaura cincosaltensis* (Ornithischia, Ornithopoda) from the Upper Cretaceous of Argentina. *Journal of Paleontology* 71(5): 933–940.

Salgado, L., and R. García. 2002. Variación morfológica en la secuencia de vértebras caudales de algunos saurópodos titanosaurios. *Revista Española de Paleontología* 17(2): 211–216.

Salgado, L., A. Garrido, S. E. Cocca, and J. R. Cocca. 2004. Lower Cretaceous rebbachisaurid sauropods from Cerro Aguada del León (Lohán Cura Formation), Neuquén Province, Northwestern Patagonia, Argentina. *Journal of Vertebrate Paleontology* 24(4): 903–912.

Salgado, L., and Z. B. de Gasparini. 2004. El registro más antiguo de Dinosauria en la Cuenca Neuquina (Aaleniano, Jurásico Medio). *Ameghiniana* 41(3): 505–508.

———. 2006. Reappraisal of an ankylosaurian dinosaur from the Upper Cretaceous of James Ross Island (Antarctica). *Geodiversitas* 28(1): 119–135.

Salgado, L., and R. D. Martínez. 1993. Relaciones filogenéticas de los titanosáuridos basales. *Andesaurus delgadoi* y *Epachthosaurus sp. Ameghiniana* 30(3): 339.

Salinas, G. C., R. D. Juárez Valieri, and L. E. Fiorelli. 2005. Los dinosaurios ornitisquios de América del Sur y su importancia paleobiogeográfica. In *Boletim de Resumos: II Congresso Latino-Americano de Paleontologia de Vertebrados*, 240–241. Rio de Janeiro: Museo Nacional.

Salinas, P., P. Sepúlveda, and L. G. Marshall. 1991. Hallazgo de restos óseos de dinosaurios (saurópodos), en la Formación Pajonales (Cretácico Superior) Sierra de Almeyda, II Región de Antofagasta, Chile: implicancia cronológica. In *Actas VI Congreso Geológico Chileno*, 534–537. Viña del Mar.

Sampson, S. D., M. T. Carrano, and C. A. Forster. 2001. A theropod dinosaur with bizarre dentition from the late Cretaceous of Madagascar. *Nature* 409: 504–506.

Sampson, S. D., and D. W. Krause. 2007. *Majungasaurus crenatissimus* (Theropoda: Abelisauridae) from the Late Cretaceous of Madagascar. In S. D. Sampson and D. W. Krause (Eds.), *Majungasaurus crenatissimus from the Late Cretaceous of Madagascar*, 1–184. [Northbrook, Ill.]: Society of Vertebrate Paleontology.

Sampson, S. D., D. W. Krause, P. Dodson, and C. A. Forster. 1996. The premaxilla of *Majungasaurus* (Dinosauria: Theropoda), with implications for Gondwanan paleobiogeography. *Journal of Vertebrate Paleontology* 16(4): 601–605.

Sampson, S. D., and L. M. Witmer. 2007. Craniofacial Anatomy of *Majungasaurus crenatissimus* (Theropoda: Abelisauridae) from the Late Cretaceous of Madagascar. In S. D. Sampson and D. W. Krause (Eds.), *Majungasaurus crenatissimus from the Late Cretaceous of Madagascar*, 32–102. [Northbrook, Ill.]: Society of Vertebrate Paleontology.

Sampson, S. D., L. M. Witmer, C. A. Forster, D. W. Krause, P. M. O'Connor, P. Dodson, and F. Ravoavy. 1998. Predatory dinosaur remains from Madagascar: Implications for the Cretaceous biogeography of Gondwana. *Science* 280(5366): 1048–1051.

Sánchez-Villagra, M. R. 1994. An ornithischian from the Jurassic of the Venezuelan Andes. *Journal of Vertebrate Paleontology* 14(3): 44A.

Sánchez-Villagra, M. R., and J. M. Clark. 1994. An ornithischian from the Jurassic of the Venezuelan Andes. *Journal of Vertebrate Paleontology* 14(3): 44A.

Santucci, R. M., and R. J. Bertini. 2006. A new titanosaur from western São Paulo State, Upper Cretaceous Bauru Group, South-East Brazil. *Palaeontology* 49(1): 59–66.

Sanz, J. L., J. E. Powell, J. le Loeuff, R. Martínez, and X. Pereda-Suberbiola. 1999. Sauropod remains from the Upper Cretaceous of Laño (northcentral Spain). Titanosaur phylogenetic relationships. *Estudios del Museo de Ciencias Naturales de Álava* 14: 235–255.

Sarjeant, W. A. S., J. B. Delair, and M. G. Lockley. 1998. The footprints of *Iguanodon*: A history and taxonomic study. *Ichnos* 6(3): 183–202.

Scherer, C. M. S., U. F. Faccini, and E. L. Lavina. 2000. Arcabouço estratigráfico do Mesozóico da Bacia do Paraná. In M. Holz and L. F. De Ros (Eds.), *Geologia do Rio Grande do Sul*, 335–354. Porto Alegre: Editora da Universidade/UFRGS.

Schubert, C. 1986. Stratigraphy of the Jurassic La Quinta Formation, Mérida Andes, Venezuela: Type section. *Zeitschrift Deutsche Geologie* 137: 391–411.

Schultz, C. L., and M. C. Langer. 2007. Tetrápodes Triássicos do Rio Grande do Sul, Brasil. In I. S. Carvalho et al. (Eds.), *Paleontologia: Cenarios de vida*, 1:184–188. Río de Janeiro: Editora Interciencia.

Schweitzer, M. H., F. D. Jackson, L. M. Chiappe, G. Schmitt, G. James, J. O. Calvo, and D. E. Rubilar. 2002. Late Cretaceous avian eggs with Embryos from Argentina. *Journal of Vertebrate Paleontology* 22(1): 191–195.

Sciutto, J. C., and R. D. Martínez. 1994. Un nuevo yacimiento fosilífero de la Formación Bajo Barreal (Cretácico Tardío) y su fauna de saurópodos. *Naturalia Patagónica (Ciencias de la Tierra)* 2: 27–47.

Scotese, C. R., and A. J. Golonka. 1992. PALEOMAP Paleogeographic Atlas. In *PALEOMAP Progress Report 20*. Arlington, Texas: Department of Geology, University of Texas at Arlington.

Seeley, H. G. 1887. On the classification of the fossil animals commonly named Dinosauria. *Proceedings of the Royal Society of London B* 43(206): 165–171.

———. 1888. The classification of the Dinosauria. *Report of the British Association for the Advancement of Science 1887*, 698–699. London: J. Murray.

Senra, M. C. E., and L. H. Silva e Silva. 1999. Moluscos dulçaqüícolas e microfósseis vegetais associados da Formação Marília, Bacia Bauru (Cretáceo Superior), Minas Gerais, Brasil. In *Simpósio sobre o Cretáceo do Brasil*, 5, *Serra Negra*: 497–500. UNESP.

Senter, P. 2005. Function in the stunted forelimbs of *Mononykus olecranus* (Theropoda), a dinosaurian anteater. *Paleobiology* 31(3): 373–381.

Sereno, P. C. 1986. Phylogeny of the bird-hipped dinosaurs (Order Ornithischia). *National Geographic Research* 2(2): 234–256.

———. 1991a. Basal archosaurs: phylogenetic relationships and functional implications. *Journal of Vertebrate Paleontology* 11(Suppl. 3): 1–53.

———. 1991b. *Lesothosaurus*, "fabrosaurids," and the early evolution of Ornithischia. *Journal of Vertebrate Paleontology* 11(2): 168–197.

———. 1994. The pectoral girdle and forelimb of the basal theropod *Herrerasaurus ischigualastensis*. *Journal of Vertebrate Paleontology* 13(4): 425–450.

———. 1997. The origin and evolution of dinosaurs. *Annual Review of Earth Planetary Sciences* 25: 435–489. Palo Alto, Calif.: Annual Reviews Inc.

———. 1998. A rationale for phylogenetic definitions, with application to the higher-level taxonomy of Dinosauria. *Neues Jahrbuch für Geologie und Paleontologie Abhandlungen* 210: 41–83.

———. 1999. The evolution of dinosaurs. *Science* 284(5423): 137–147.

———. 2001. Alvarezsaurids: Birds or ornithomimosaurs? In J. Gauthier and L. F. Gall (Eds.), *New perspectives on the origin and early evolution of birds: Proceedings of the International Symposium in Honor of John H. Ostrom*, 69–98. New Haven, Conn.: Peabody Museum of Natural History, Yale University.

———. 2007. The phylogenetic relationships of early dinosaurs: a comparative report. *Historical Biology* 19(1): 145–155.

Sereno, P. C., and A. B. Arcucci. 1993. Dinosaurian precursors from the Middle Triassic of Argentina: *Lagerpeton chanarensis*. *Journal of Vertebrate Paleontology* 13(4): 358–399.

———. 1994. Dinosaurian precursors from the Middle Triassic of Argentina: *Marasuchus lilloensis*, gen. nov. *Journal of Vertebrate Paleontology* 14(1): 53–73.

Sereno, P. C., A. L. Beck, D. B. Dutheil, H. C. E. Larsson, G. H. Lyon, B. Moussa, R. W. Sadleir, C. A. Sidor, D. J. Varricchio, G. P. Wilson, and J. A. Wilson. 1999. Cretaceous sauropods from the Sahara and the uneven rate of skeletal evolution among dinosaurs. *Science* 286(5443): 1342–1347.

Sereno, P. C., D. B. Dutheil, M. Iarochene, H. C. E. Larsson, G. Lyon, P. M. Magwene, C. A. Sidor, D. J. Varricchio, and J. A. Wilson. 1996. Predatory dinosaurs from the Sahara and Late Cretaceous faunal differentiation. *Science* 272(5264): 986–991.

Sereno, P. C., C. A. Forster, R. R. Rogers, and A. M. Monetta. 1993. Primitive dinosaur skeleton from Argentina and the early evolution of Dinosauria. *Nature* 361: 64–66.

Sereno, P. C., H. C. E. Larsson, C. A. Sidor, and B. Gado. 2001. The giant crocodyliform *Sarcosuchus* from the Cretaceous of Africa. *Science* 294(5546): 1516–1519.

Sereno, P. C., R. N. Martinez, J. A. Wilson, D. J. Varricchio, O. A. Alcober, and H. C. E. Larsson. 2008. Evidence for avian intrathoracic air sacs in a new predatory dinosaur from Argentina. *PLoS ONE* 3(9): 1–20.

Sereno, P. C., and F. E. Novas. 1990. Dinosaur origins and the phylogenetic position of pterosaurs. *Journal of Vertebrate Paleontology* 10(Suppl. 3): 42A.

———. 1992. The complete skull and skeleton of an early dinosaur. *Science* 258(5085): 1137–1140.

———. 1994. The skull and neck of the basal theropod *Herrerasaurus ischigualastensis*. *Journal of Vertebrate Paleontology* 13(4): 451–476.

Sereno, P. C., and J. A. Wilson. 2005. Structure and evolution of a sauropod tooth battery. In K. A. Curry Rogers and J. A. Wilson (Eds.), *The sauropods: Evolution and paleobiology*, 157–177. Berkeley: University of California Press.

Sereno, P. C., J. A. Wilson, and J. L. Conrad. 2004. New dinosaurs link southern landmasses in the Mid-Cretaceous. *Proceedings of the Royal Society of London B* 271(1546): 1325–1330.

Shubin, N. H., and H. D. Sues. 1991. Biogeography of early Mesozoic continental tetrapods: patterns and implications. *Paleobiology* 17(3): 214–230.

Sigé, B. 1968. Dents de micromammiferes et fragments de coquilles d'oeufs de dinosauriens dans la faune de vertibres du Cretace superieur de Laguna Umayo (Andes peruviennes). *Comptes Rendus de l'Académie de Sciences de Paris* 267: 1495–1498.

Simbras, F., G. Oliveira, D. A. Campos, and A. W. A. Kellner. 2007. On a large sauropod cervical vertebra from the Presidente Prudente Formation, Bauru Group (Late Cretaceous), São Paulo, Brazil. In *Anais do XX Congresso Brasileiro de Paleontologia*, 274. Buzios, Brasil.

Simón, M. E., 2001. A giant sauropod from the Upper Cretaceous of El Chocón, Neuquén, Argentina. *Ameghiniana* 38 (Suppl. 4): 19R.

Skelton, P. W. (ed.). 2003. *The Cretaceous world*. Cambridge: Cambridge University Press.

Smith, J. B., M. C. Lamanna, K. J. Lacovara, P. Dodson, J. R. Smith, J. C. Poole, R. Giegengack, and Y. Attia. 2001. A giant sauropod dinosaur from an Upper Cretaceous mangrove deposit in Egypt. *Science* 292(5522): 1704–1706.

Smith, N., P. J. Makovicky, W. R. Hammer, and P. J. Currie. 2007. Osteology of *Cryolophosaurus ellioti*. *Zoological Journal of the Linnean Society* 151(2): 377–421.

Soares, P. 1981. Estratigrafía das formaçoes Jurassico-Cretáceas na Bacia do Paraná, Brasil. In W. Volkheimer and E. A. Musacchio (Eds.), *Cuencas sedimentarias del Jurásico y Cretácico de América del Sur*, 1:271–304. Buenos Aires: Comité Sudamericano del Jurásico y Cretácico.

Sprechmann, P., J. Bossi, and J. da Silva. 1981. Cuencas del Jurásico y Cretácico del Uruguay. In W. Volkheimer and E. A. Musacchio (Eds.), *Cuencas sedimentarias del Jurásico y Cretácico de*

América del Sur, 1:239–270. Buenos Aires: Comité Sudamericano del Jurásico y Cretácico.

Stevens, K. A., and J. M. Parrish. 1999. Neck posture and feeding habits of two Jurassic sauropod dinosaurs. *Science* 284(5415): 798–800.

Stipanicic, P. N., and J. F. Bonaparte. 1979. Cuenca triásica de Ischigualasto-Villa Unión (provincias de La Rioja y San Juan). In J. C. Turner (ed.), *II Simposio Geología Regional Argentina*, 1:523–575. Córdoba, Argentina: Academia Nacional de Ciencias de Córdoba.

Stipanicic, P., and C. Marsicano. 2002. Léxico Estratigráfico de la Argentina. Volumen VIII. Triásico. *Asociación Geológica Argentina* (Série B) 26:1–370.

Stipanicic, P., and A. O. Reig. 1957. El "Complejo Porfírico de la Patagonia extraandina" y su fauna de anuros. *Acta Geologica Lilloana* 1: 185–297. Tucumán: El Instituto Miguel Lillo.

Stromer, E. 1915. Ergebnisse der Forschungsreisen Prof. E. Stromers in den Wüsten Ägyptens. II. Wirbeltierreste der Baharije-Stufe (unterstes Cenoman). 3. Das Original des Theropoden *Spinosaurus aegyptiacus* nov. gen., nov. spec. *Abhandlungen der Königlich Bayerischen Akademie der Wissenschaften, Mathematisch-Physikalische Klasse* 28: 1–32.

———. 1931. Ergebnisse der Forschungsreisen Prof. E. Stromers in den Wüsten Ägyptens. II. Wilbertierreste der Baharîje-Stufe (unterstes Cenoman). 10. Ein Skelett-rest von *Carcharodontosaurus* nov. gen. *Abhandlungen der Bayerischen Akademie der Wissenschaften Mathematisch-naturwissenschaftliche Abteilung, Neue Folge* 9: 1–23.

———. 1934. Ergebnisse der Forschungsreisen Prof. E. Stromers in den Wüsten Ägyptens. II. Wirbeltierreste der Baharîje-Stufe. 13 Dinosauria. *Abhandlungen der Bayerischen Akademie der Wissenschaften Mathematisch-naturwissenschaftliche Abteilung, Neue Folge* 22: 1–79.

———. 1936. Ergebnisse der Forschungsreisen Prof. E. Stromers in den Wüsten Ägyptens. VII. Baharîje-Kessel und Stufe mit deren Fauna und Flora. Eine ergänzenden Zusammenfassung. *Abhandlungen der Bayerischen Akademie der Wissenschaften Mathematisch-naturwissenschaftliche Abteilung, Neue Folge* 33: 1–102.

Suárez Riglos, M. 1995. Huellas de dinosaurios en Sucre. *Asociación Sucrence de Ecología. Anuario* 95: 44–48.

Sues, H. D., E. Frey, D. M. Martill, and D. M. Scott. 2002. *Irritator challengeri*, a spinosaurid (Dinosauria: Theropoda) from the Lower Cretaceous of Brazil. *Journal of Vertebrate Paleontology* 22(3): 535–547.

Sues, H. D., and D. B. Norman. 1990. Hypsilophodontidae, *Tenontosaurus*, Dryosauridae. In D. B. Weishampel, P. Dodson, and H. Osmólska (Eds.), *The Dinosauria*, 498–509. Berkeley: University of California Press.

Sues, H. D., and P. Taquet. 1979. A pachycephalosaurid dinosaur from Madagascar and a Laurasia-Gondwanaland connection in the Cretaceous. *Nature* 279: 633–635.

Sullivan, R. M., and S. G. Lucas. 1999. *Eucoelophysis baldwini*, a new theropod dinosaur from the Upper Triassic of New Mexico, and the status of the original types of *Coelophysis*. *Journal of Vertebrate Paleontology* 19(1): 81–90.

Tapia, A. 1918. Una mandíbula de dinosaurio procedente de Patagonia. *Physis* 4: 369–370.

Taquet, P. 1976. Géologie et paléontologie du gisement de Gadoufaoua (Aptien du Niger). *Cahiers de Paléontologie*, 1–191. Paris: Centre National de la Recherche Scientifique.

Taquet, P., and D. Russell. 1999. A massively-constructed iguanodont from Gadoufaoua, Lower Cretaceous of Niger. *Annales de Paléontologie* 85(1): 85–96.

Thulborn, R. A. 1975. Dinosaur polyphyly and the classification of Archosaurs and birds. *Australian Journal of Zoology* 23(2): 249–270.

Thulborn, T. 1990. *Dinosaur tracks*. London: Chapman and Hall.

Torcida Fernández Baldor, F., X. Pereda Suberbiola, P. Huerta Hurtado, L. A. Izquierdo, D. Montero, and G. Pérez. 2003. Descripción preliminar de un dinosaurio rebaquisáurido (Sauropoda Diplodocoidea) del Cretácico Inferior de Burgos (España). In *Actas de las II Jornadas Internacionales sobre Paleontología de Dinosaurios y su Entorno*, 203–211. Burgos, Spain: Colectivo Arqueológico-Paleontológico de Salas.

Turner, A., D. Pol, J. Clarke, G. Erickson, and M. A. Norell. 2007. A basal dromaeosaurid and size evolution preceding avian flight. *Science* 317(5843): 1378–1381.

Uliana, M. A., and K. T. Biddle. 1988. Mesozoic-Cenozoic paleogeographic and geodynamic evolution of southern South America. *Revista Brasileira de Geociências* 18(2): 172–190.

Uliana, M. A., and L. Legarreta. 1999. Jurásico y Cretácico de la cuenca del Golfo de San Jorge. In R. Caminos (ed.), *Geologia Argentina*, 496–510. Buenos Aries: Instituto de Geología y Recursos Minerales.

Upchurch, P. 1998. The phylogenetic relationships of sauropod dinosaurs. *Zoological Journal of the Linnean Society* 124(1): 43–103.

———. 1999. The phylogenetic relationships of the Nemegtosauridae (Saurischia, Sauropoda). *Journal of Vertebrate Paleontology* 19(1): 106–125.

Upchurch, P., P. M. Barrett, and P. Dodson. 2004. Sauropoda. In D. B. Weishampel, P. Dodson, and H. Osmólska (Eds.), *The Dinosauria*, 259–322. 2nd ed. Berkeley: University of California Press.

Urien, C. M., J. J. Zambrano, and L. R. Martins. 1981. The basins of southeastern South America (southern Brazil, Uruguay and eastern Argentina) including the Malvinas Plateau and Southern South Atlantic paleogeographic evolution. In W. Volkheimer and E. Musacchio (Eds.), *Cuencas sedimentarias del Jurásico y Cretácico de América del Sur*, 1: 45–125. Buenos Aires: Comité Sudamericano del Jurásico y Cretácico.

Van Valen, L. 1969. What was the largest dinosaur? *Copeia* 1969(3): 624–626.

Vargas, A., M. Suárez, D. Rubilar, and K. Moreno. 2000. A titanosaurid vertebra from Pichasca, Formación La Vinita (Late Cretaceous), IV Region, northern Chile. *Ameghiniana* 37(Suppl. 4): 35R.

Veralli, C., and J. O. Calvo. 2004. Dientes de terópodos carcharodontosáuridos del Turoniano superior-Coniaciano inferior del Neuquén, Patagonia, Argentina. *Ameghiniana* 41(4): 587–590.

Vianey-Liaud M., K. Hirsch, A. Sahni, and B. Sigé. 1997. Late cretaceous Peruvian eggshells and their relationships with Laurasian and eastern Gondwanian material. *Geobios* 30(1): 75–90.

Vickaryous, M. K., T. Maryańska, and D. B. Weishampel. 2004. Ankylosauria. In D. B. Weishampel, P. Dodson, and H. Osmólska (Eds.), *The Dinosauria*, 363–392. 2nd ed. Berkeley: University of California Press.

Vickers-Rich, P., T. H. Rich, D. R. Lanus, L. S. V. Rich, and R. Vacca. 1999. "Big tooth" from the early Cretaceous of Chubut Province, Patagonia: A possible carcharodontosaurid. In Y. Tomida, T. H. Rich, and P. Vickers-Rich (Eds.), *Proceedings of the Second Gondwanan Dinosaur Symposium*, 13–45. Tokyo: National Science Museum.

Vilas Boas, I., I. S. Carvalho, M. A. Medeiros, and H. Pontes. 1999. Dentes de *Carcharodontosaurus* (Dinosauria, Tyrannosauridae) do Cenomaniano, Bacia de São Luís (Norte do Brasil). *Anais da Academia Brasileira de Ciências* 71(4): 846–847.

Vildoso Morales, C. A. 1991. Tetrápodos del miembro inferior de la Formación Bagua (Cretácico tardío-Paleoceno) del norte peruano. In *VII Jornadas Argentinas de Paleontología de Vertebrados*, *Ameghiniana* 26(3–4): 251–252.

Vildoso Morales, C. A., and M. P. Sciammaro. 2005. El género *Titanosaurus* (Sauropoda. Titanosauridae) en el Cretácico tardío del norte peruano. In *XIX Congresso Brasileiro de Paleontologia*, 224R. Aracaju.

Volkheimer, W. 1969. Esporas y granos de pólen del Jurásico de Neuquén (República Argentina). II. Asociaciones microflorísticas, aspectos paleoecológicos y paleoclima. *Ameghiniana* 6: 127–145.

von Huene, F. 1914. Saurischia and Ornithischia. *Geological Magazine* 6: 444–445.

———. 1927a. Short review of the present knowledge of the Sauropoda. *Memoirs of the Queensland Museum* 9(1): 121–126.

———. 1927b. Sichtung der Grundlagen der jetzigen Kenntnis der Sauropoden. *Eclogae Geologicae Helvetiae* 20(3): 444–470.

———. 1927c. Contribución a la paleobiogeografía de Sud América. *Boletín de la Academia de Ciencias de Córdoba* 30: 231–294.

———. 1929a. Los saurisquios y ornitisquios del Cretáceo Argentino. *Anales del Museo de La Plata* (2)3: 1–196.

———. 1929b. Terrestriche Oberkreide in Uruguay. *Centralblatt für Mineralogie, Geologie und Paläontologie* 8: 107–112.

———. 1931a. Die fossilen Färhten im Rhät von Ischigualasto in Nordwest-Argentinien. *Palaeobiologica* 4: 99–112.

———. 1931b. Verschiedene mesosoische Wirbeltierreste aus Sudamerika. *Neues Jahrbuch für Mineralogie, Geologie und Paläontologie* 66 (B): 181–198.

———. 1932. Die fossile Reptil-Ordnung Saurischia, ihre Entwicklung und Geschichte. *Monographien zur Geologie und Palaeontologie* (Serie 1) 4: 1–361.

———. 1934a. Nuevos dientes de saurios del Cretáceo del Uruguay. *Boletín del Instituto Geológico de Uruguay* 21: 13–20.

———. 1934b. Neue Saurier-Zahne aus der Kreide von Uruguay. *Zentralblatt für Mineralogie und Palaeontologie*, 182–189.

———. 1942. *Die fossilen Reptilien des sudamerikanischen Gondwanalandes*. Munich: C. H. Beck.

von Huene, F., and C. A. Matley. 1933. The Cretaceous Saurischia and Ornithischia of the central provinces of India. *Memoirs of the Geological Survey of India (Palaeontologia Indica)* 21(1): 1–74.

Walker, C. A. 1981. New subclass of birds from the Cretaceous of South America. *Nature* 292: 51–53.

Wedel, M. J., R. L. Cifelli, and R. K. Sanders. 2000. Osteology, paleobiology, and relationships of the sauropod dinosaur *Sauroposeidon*. *Acta Palaeontologica Polonica* 45(4): 343–388.

Weishampel, D. B. 1990. Dinosaurian distribution. In D. B. Weishampel, P. Dodson, and H. Osmólska (Eds.), *The Dinosauria*, 63–139. Berkeley: University of California Press.

———. 2004. Ornithischia. In D. B. Weishampel, P. Dodson, and H. Osmólska (Eds.), *The Dinosauria*, 323–324. 2nd ed. Berkeley: University of California Press.

Weishampel, D. B., P. M. Barrett, R. A. Coria, J. Le Loeuff, X. Xu, X. Zhao, A. Sahni, E. M. Gomani, and C. R. Noto. 2004. Dinosaur distribution. In D. B. Weishampel, P. Dodson and H. Osmólska (Eds.), *The Dinosauria*, 517–606. 2nd ed. Berkeley: University of California Press.

Weishampel, D. B., C. M. Jianu, Z. Csiki, and D. B. Norman. 2003. Osteology and phylogeny of *Zalmoxes* (n.g.), an unusual euornithopod dinosaur from the latest Cretaceous of Romania. *Journal of Systematic Palaeontology* 1(2): 1–56.

Weishampel, D. B., and L. Witmer. 1990. *Lesothosaurus*, *Pisanosaurus* and *Technosaurus*. In D. B. Weishampel, P. Dodson, and H. Osmólska (Eds.), *The Dinosauria*, 417–425. Berkeley: University of California Press.

Welles, S. 1984. *Dilophosaurus wetherilli* (Dinosauria, Theropoda): Osteology and comparisons. *Palaeontographica A* 185: 85–180.

Wellnhofer, P. 1988. Ein neues Exemplar von *Archaeopteryx*. *Archaeopteryx* 6: 1–30.

———. 1993. Das siebte Exemplar von *Archaeopteryx* aus den Solnhofener Schichten. *Archaeopteryx* 11: 1–48.

Wichmann, R. 1916. Las capas con dinosaurios en la costa sur del Río Negro, frente a General Roca. *Revista de la Sociedad Argentina de Ciencias Naturales* 2: 258–262.

———. 1924. Nuevas observaciones geológicas en la parte oriental del Neuquén y en el territorio de Río Negro. *Ministerio de Agricultura. Sección Geología* 2: 1–22.

Wiffen, J. 1996. Dinosaurian palaeobiology: A New Zealand perspective. In F. E. Novas and R. E. Molnar (Eds.), *Proceedings of the Gondwanan dinosaur symposium*, 725–731. Brisbane: Queensland Museum.

Wilson, J. A. 2002. Sauropod dinosaur phylogeny: Critique and cladistic analysis. *Zoological Journal of the Linnean Society* 136: 217–276.

———. 2005. Redescription of the Mongolian sauropod *Nemegtosaurus mongoliensis* Nowinski (Dinosauria: Saurischia) and comments on Late Cretaceous sauropod diversity. *Journal of Systematic Palaeontology* 3(3): 283–318.

Wilson, J. A., and M. T. Carrano. 1999. Titanosaurs and the origin of "wide-gauge" trackways: a biomechanical and systematic perspective on sauropod locomotion. *Paleobiology* 25(2): 252–267.

Wilson, J. A., R. Martínez, and O. Alcober. 1999. Distal tail segment of a titanosaur (Dinosauria: Sauropoda) from the Upper Cretaceous of Mendoza, Argentina. *Journal of Vertebrate Paleontology* 19(3): 591–594.

Wilson, J. A., and P. C. Sereno. 1998. Early evolution and higher-level phylogeny of sauropod dinosaurs. *Society of Vertebrate Paleontology Memoir* 5: 1–68.

Wilson, J. A., P. C. Sereno, S. Srivastava, D. K. Bhatt, A. Khosla, and A. Sahni. 2003. A new abelisaurid (Dinosauria, Theropoda) from the Lameta Formation (Cretaceous, Maastrichtian) of India. *Contributions from the Museum of Paleontology, University of Michigan* 31(1): 1–42.

Wilson, J. A., and P. Upchurch. 2003. A revision of *Titanosaurus* Lydekker (Dinosauria-Sauropoda), the first dinosaur genus with a "Gondwanan" distribution. *Journal of Systematic Palaeontology* 1(3): 125–160.

Wing, S. L., and H. D. Sues. 1992. Mesozoic and early Cenozoic terrestrial ecosystems. In A. K. Behrensmeyer, J. D. Damunth, W. A. DiMichele, R. Potts, H. D. Sues, and S. L. Wing (Eds.), *Terrestrial ecosystems through time: Evolutionary paleoecology of terrestrial plants and animals*. Chicago: University of Chicago Press.

Woodward, A. S. 1901. On some extinct reptiles from Patagonia, of the genera *Miolania*, *Dinilysia*, and *Genyodectes*. *Proceedings of the Zoological Society of London* 1901: 169–184.

Xu, X. 2002. Deinonychosaurian fossils from the Jehol Group of Western Lioning and the coelurosaurian evolution. Doctoral Dissertation, Chinese Academy of Sciences, Beijing.

Xu, X., X. Wang, and X. Wu. 1999. A dromaeosaurid dinosaur with a filamentous integument from the Yixian Formation of China. *Nature* 401: 262–266.

Xu, X., Z. Zhou, and R. O. Prum. 2001. Branched integumentary structures in *Sinornithosaurus* and the origin of feathers. *Nature* 410: 200–204.

Xu, X., Z. Zhou, and X. Wang. 2000. The smallest known non-avian theropod dinosaur. *Nature* 408: 705–708.

Yates, A. M. 2003. A new species of the primitive dinosaur, *Thecodontosaurus* (Saurischia: Sauropodomorpha) and its implications for the systematics of early dinosaurs. *Journal of Systematic Palaeontology* 1(1): 1–42.

Yates, A. M., and J. W. Kitching. 2003. The earliest known sauropod dinosaur and the first steps towards sauropod locomotion. *Proceedings of Royal Society of London B* 270(1525): 1753–1758.

Zambrano, J. J. 1981. Distribución y evolución de las cuencas sedimentarias en el continente sudamericano durante el Jurásico y el Cretácico. In W. Volkheimer and E. Musacchio (Eds.), *Cuencas Sedimentarias del Jurásico y Cretácico de América del Sur*, 9–44. Buenos Aires: Comité Sudamericano del Jurásico y Cretácico.

Zerfass, H., E. L. Lavina, C. L. Schultz, A. J. Vasconcellos Garcia, U. F. Faccini, and F. Chemale. 2003. Sequence stratigraphy of continental Triassic strata of Southernmost Brazil: A contribution to Southwestern Gondwana palaeogeography and palaeoclimate. *Sedimentary Geology* 161(1–2): 85–105.

Zinsmeister, W. J. 1982. Late Cretaceous-Early Tertiary Molluscan Biogeography of the Southern Circum-Pacific. *Journal of Paleontology* 56(1): 84–102.

INDEX

I

Page numbers in italics indicate illustrations.

Abdala, F., 59
Abelichnus, 153, 338, 339
Abelichnus astigarrae, 338, 339
abelisaurid maxilla from Bajo Barreal Formation, 286, 287
Abelisauridae, 244, 245, 250, 252, 253, 255, 259, 261, 272–286, 291, 294, 306; phylogenetic relationships, 246, 273, 286
Abelisauroidea, 149, 154, 244, 246, 249–272, 250, 319, 390; abelisauroids documented in South America, 266–272; caudal vertebrae, 259; convergent resemblances in skull between abelisaurids and carcharodontosaurids, 255; functional morphology and behaviour, 259; geographic and stratigraphic distribution, 250–252; ossified tendons, 259; pelvic girdle and hindlimbs, 262–264; phylogenetic relationships, 246, 252, 253; shoulder girdles and forelimbs, 261, 262; skin impressions, 264–266; skulls, jaws, and teeth, 253–256; vertebral column, 256–261
Abelisaurus, 157, 244, 249, 251, 252, 255, 255, 268, 272–275, 274, 277, 281, 286, 392
Abelisaurus comahuensis, 157, 244, 249, 251, 255, 273–275, 274, 281
Abrantes, E., 368
Achillesaurus, 310, 317
Achillesaurus manazzonei, 310, 317
Achillobator, 322, 333, 334
Acrocanthosaurus, 255, 295, 298, 302, 388
Adamantina Formation (Brazil), 142, 143, 144, 220, 227, 336
Adamantisaurus, 144, 219, 224

Adamantisaurus mezzalirai, 144, 224
Los Adobes Formation (Argentina), 159
Aegyptosaurus baharijensis, 183
Aeolosaurinae, 219–224
Aeolosaurini, 219
Aeolosaurus, 158, 162, 168, 187, 193, 196, 209, 210, 215, 219, 220–224, 220, 226, 228, 232, 397
Aeolosaurus rionegrinus, 158, 219, 220, 221, 223
Aetosauroides scagliai, 30
aetosaurs, 3, 30, 32
Agnolín, Federico, 270, 271, 336
Agnosphitys, 80
Agua de la Peña Group (Argentina), 27
Agua del Choique (Argentina), 240
Aguada del Caño (Argentina), 155, 215
Agustinia, 152, 169, 171, 172, 172, 199, 389, 391
Agustinia ligabuei, 152, 169, 171, 172, 172, 199
Los Alamitos farm (Argentina), 158, 372, 397, 398
Los Alamitos Formation (Argentina), 158, 219, 372, 398
Alamosaurus sanjuanensis, 182
Alcântara Formation (Brazil), 141, 142, 177, 288, 368, 369
Alcober, Oscar, xix, 40
Algoasaurus bauri, 176
Allen Formation (Argentina), 151, 157, 158, 209, 219, 233, 235, 236, 285, 332, 336, 340, 348, 351, 352, 372, 377, 378, 380, 396, 397, 398, 399
Allosaurus, 7, 8, 41, 46, 117, 118, 119, 291, 293, 295, 298, 302, 303, 306, 312
Alonso, R., 378
Alvarenga, Herculano, 336

Alvarezsauridae, 244, 264, 305, 306, 310; behaviour, 312; environmental conditions and geographical distribution, 314; forelimbs, 311, 312; phylogenetic relationships, 314; vertebral column, 314
Alvarezsaurus, 156, 157, 244, 304, 310, 314, 315, 316, 317, 395, 396
Alvarezsaurus calvoi, 156, 157, 244, 304, 310, 314, 315, 316, 317
Alwalkeria, 80
Alxasaurus, 320
La Amarga Formation (Argentina), 112, 151, 152, 153, 169, 172, 266, 346, 347, 387, 388
Amargasaurus, 111, 112, 114, 152, 153, 168, 170, 171–174, 174, 387, 391, 392
Amargasaurus cazaui, 112, 152, 153, 168, 170, 172–174, 174
Amazonsaurus, 142, 168, 177, 179–181, 181, 199, 389
Amazonsaurus maranhensis, 142, 168, 177, 179–181, 181
Amboró National Park (Bolivia), 146
Ameghinichnus patagonicus, 99
Ameghino, Carlos, 161
Ameghino, Florentino, 243, 366
amniotes, 3
Ampelosaurus atacis, 182
Amygdalodon, 88, 95, 100–102, 102, 128
Amygdalodon patagonicus, 88, 95, 100–102, 102
Anabisetia, 153, 345, 352, 353, 356–360, 357, 359, 363, 365, 366, 394, 395
Anabisetia saldiviai, 153, 345, 356–359, 357, 359
Anacleto Formation (Argentina), xx, 155, 157, 209, 213, 227, 231, 236, 237, 273, 279, 294, 338, 354
Andean Orogeny, 136

439

Andesaurus, 152, 187, 199, 200, 201, 202, 207, 217, 228, 389
Andesaurus delgadoi, 152, 187, 199, 201
angiosperms, 181, 386, 388, 394
Angostura Colorada Formation (Argentina), 158, 219, 398
Aniksosaurus, 162, 245, 304–307, 308
Aniksosaurus darwini, 162, 245, 304–307, 308
La Anita farm (Argentina), 164
Ankylosauria, 343, 344, 346; faunal interchange, 399; fossil record of southern ankylosaurs, 347; phylogenetic relationships, 348; record in General Roca, 349; record in Puerto Yeruá, 349; record in Rancho Avila, 349; record in Salitral Moreno, 348; trackways from Cal Orko, 351, 382; trackways from Piranhas Formation, 351, 351
Ankylosauridae, 347, 348, 385
Anseriformes, 337
Antarctica, xx, 87, 135, 137, 138, 181, 183, 252, 337, 338, 347, 348, 351, 371, 395, 399, 400, 402
Antarctopelta oliveroi, 347
Antarctosaurus, 144, 155, 157, 167, 176, 185, 187, 189, 190, 205, 211–215, 214, 215, 217, 219, 220, 231, 235, 349, 351, 395
Antarctosaurus wichmannianus, 144, 157, 167, 176, 185, 189, 211, 213–215, 214, 349
Antarctosaurus?, 205, 215–217, 217, 394, 398
Antarctosaurus? giganteus, 205, 215–217, 217, 394, 398
Antenor Navarro Formation (Brazil), 141, 238, 369, 389
Antetonitrus, 70
Apatosaurus, 110, 111, 177, 382
Apesteguía, Sebastián, xi, 152
Aquatilavipes, 339
Araraquara (Brazil), 93, 123, 128
Araripe Basin (Brazil), 136, 138, 140, 246, 288, 369, 389
Araripesuchus gomesii, 140
Archaeopteryx, 314, 315, 320, 325, 327, 328, 329, 332, 396
Archaeornithomimus, 315
Archosauria, 3, 6, 8, 12, 16, 20, 28; phylogenetic relationships, 2

arctometatarsalian foot, 264, 308, 330
Arcucci, Andrea, xix, 20, 28, 55, 57
Argana Formation (Morocco), 81
Argentinosaurus, 4, 153, 169, 187, 193, 199, 202, 204, 205, 207, 207, 215, 218, 301, 390, 392, 394, 398
Argentinosaurus huinculensis, 153, 202, 204, 205, 207, 207, 301, 398
Argyrosaurus, 144, 149, 161, 162, 167, 193, 209, 210, 215, 217, 218, 218, 227, 231, 392, 394
Argyrosaurus superbus, 144, 149, 161, 162, 167, 217, 218, 218
aridity: in the Early Cretaceous, 130; in the Jurassic, 87, 130; regulation of body temperature or storing energy in arid environments, 392; zone of major aridity of South America, 131
Aristonectes, 400
Aristosuchus pusillus, 309
Aroifilla Formation (Perú), 145
Atacama (Chile), 146, 147
Atlantic Ocean, 136; incursions onto the eastern side of South America, 372, 401, 402; marine transgressions, 137, 138
Atlascopcosaurus, 353
Auca Mahuida fossil site (Argentina), 157, 237
Aucasaurus, 157, 251, 252, 254, 260, 261, 262, 262, 264, 267, 269, 270, 272, 273, 276, 277, 279–281, 281, 283–286
Aucasaurus garridoi, 157, 251, 279–281, 281
Austral Basin (Argentina), 97, 163, 402, 139. *See also* Magallanes Basin
Australia, xx, 168, 183, 246, 252, 304, 319, 324, 347, 379, 385, 390, 395, 399, 400, 402
Austroraptor, 158, 245, 304, 322, 324, 325, 332–334, 335, 388, 397
Austroraptor cabazai, 158, 245, 304, 322, 332–334, 335, 397
Aves, 13, 245, 305, 325, 332, 334–338. *See also* birds
Avila, L., 368
Azendohsaurus laarousii, 81
Azevedo, S., 59, 61, 220
Azpelicueta, C., 233

Bactrosaurus, 375
Báez, A. M., 74, 75
Bagua Basin (Perú), 144
Bahariasaurus, 293, 392
Bahía Laura Group (Argentina), 97
Bailey, J., 174
Bajada Moreno (Argentina), 246
Bajo Barreal Formation (Argentina), 176, 190, 245, 286, 287, 306; Lower Member of, 162, 179, 181, 202, 211, 211, 217, 285, 306, 365, 389, 395; Upper Member of, 162, 219, 374, 380
Bajo de la Carpa Formation (Argentina), xx, 155, 156, 157, 212, 271, 314, 317, 336, 396
Bajo de Santa Rosa (Argentina), 156, 158, 332
Bakker, Robert, 12, 13, 387
Baños del Flaco Formation (Chile), 92, 114, 121, 127, 369
Barapasaurus tagorei, 100
Barberena, Mario Costa, 36
Los Barreales Lake (Argentina), 154, 155, 201, 203, 323, 329
Barredo, S., 77, 78, 79
Barrosopus slobodai, 339
Barsboldia, 378
Barun Goyot Formation (Mongolia), 314
Baryonyx, 290, 291, 292, 293, 294
Bauru Basin (Brazil), 138, 140, 142–144
Bauru Group (Brazil), 142, 143, 144, 167, 183, 209, 227, 251, 283, 398
Baurutitan, 143, 183, 187, 218, 223, 225, 226, 226
Baurutitan britoi, 143, 225, 226, 226
Bayo Mesa Hill (Argentina), 153, 356
Benton, M., 59, 114,
Berberosaurus liassicus, 246
Bertini, R., 144, 221, 224
Bigarella, J., 92
bipedalism, 4, 77, 196
birds: 3, 4, 6, 10, 11, 14, 44, 79, 80, 115, 140, 147, 149, 153, 156, 157, 157, 158, 212, 245, 253, 259, 286, 287, 303, 305, 307, 314, 316, 319, 320, 321, 322, 325, 327, 328, 329, 330, 332, 334, 335, 336, 336, 337, 338, 339, 396, 402. *See also* Aves
Bittencourt, J., 290

Blanco, N., 119
Los Blanquitos Formation (Argentina), 147, 322, 397
Bonaparte, José, xix, 12, 13, 27, 29, 31, 36, 37, 38, 39, 39, 50, 51, 53, 54, 57, 59, 61, 63, 65, 66, 67, 68, 70, 73, 95, 100, 103, 104, 105, 108, 112, 117, 127, 129, 130, 147, 155, 158, 161, 162, 164, 167, 168, 171, 174, 181, 202, 204, 205, 209, 217, 231, 232, 234, 244, 248, 266, 267, 268, 269, 271, 272, 276, 279, 314, 315, 324, 335, 336, 343, 344, 346, 372, 374, 375, 380, 385, 396
Bonatitan, 158, 187, 230, 229, 234–236, 237, 397
Bonatitan reigii, 158, 229, 234–236, 397
Bonitasaura, 190, 212, 213, 214
Bonitasaura salgadoi, 212, 213
Borgomanero, G., 140
Botucatu Formation (Brazil), 33, 88, 92, 97, 100, 119, 121, 123, 127, 128, 130, 132, 387
Brachiosaurus, 4, 8, 181, 185, 187, 190, 191, 199, 207, 208, 211
Brachytrachelopan, 89, 96, 100, 101, 111–114, 112, 128, 129
Brachytrachelopan mesai, 89, 96,100, 112–114, 112
Brasilichnium elusivum, 93, 127
Brasilitherium riograndensis, 35
Brasilodon quadrangularis, 35
Bressanichnus, 153, 338, 339
Bressanichnus patagonicus, 338, 339
El Brete (Argentina), 147, 149, 232, 268, 270, 271, 272, 335, 336
El Brete abelisauroid putatively described as an oviraptorosaur, 270, 271, 272
Brett-Surman, Michael, 4, 343, 372, 374, 375
Brontopodus birdi, 240
Browne, L., 74
Bueno dos Reis, F., 128
La Buitrera fossil site (Argentina), 152, 179, 393
Buitreraptor, 152, 245, 304, 322, 324–328, 330–332, 333, 395
Buitreraptor gonzalezorum, 152, 245, 304, 322, 331, 332, 333

Cabrera, Ángel, 100, 101
Cajones Formation (Bolivia), 146

Cajual Island (Brazil), 141, 368
Cal Orcko (Bolivia), 146, 240, 240, 351
Calama (Chile), 147
Calquenque Formation (Argentina), 99, 107
Calvo, Jorge, xix, 108, 110, 112, 155, 168, 177, 188, 201, 201, 212, 221, 222, 230, 240, 281, 282, 293, 323, 327, 329, 330, 338, 353, 356, 359, 361, 363, 369
Camarasauromorpha, 108–110, 185
Camarasaurus, 63, 65, 68, 103, 108, 185, 211
Cambiaso, Andrea, 359
Campos, Diógenes de Almeida, 168, 225, 283, 285, 288, 290, 398
Camposaurus arizonensis, 54, 80
Camptosaurus, 363
Campylodon ameghinoi, 162, 167
Campylodoniscus ameghinoi, 189, 211, 212
Canale, Juan, 283
Candeiro, R., 144, 368
Candelaria town (Brazil), 33, 34, 35, 50
Candeleros Formation (Argentina), 151, 152, 153, 154, 176, 177, 199, 240, 281, 290, 298, 331, 338, 338, 370, 389, 393
Canudo, J., 154, 290
Cañadón Asfalto Basin (Argentina), 138, 159–162
Cañadón Asfalto Formation (Argentina), 95, 97, 101, 103, 119, 160
Cañadón Calcáreo Formation (Argentina), 95, 96, 101, 109, 112, 129
Cañadón Largo Formation (Argentina), 33
Cañón del Colorado Formation (San Juan Province), 93
Carcharodontosauridae, 154, 245, 274, 291, 293, 294–303, 306, 390, 393; cervical vertebrae, 297; dentary, 297; geographical distribution, 297; pelvic girdle, 297; skull, 295, 297
Carcharodontosaurus, 255, 293, 295, 297, 298, 301, 306
Carcharodontosaurus saharicus, 255, 295
Caririchnium, 369
Caririchnium magnificum, 369

Carnotaurus, 161, 244, 249, 250, 251, 252–257, 255, 256, 257, 259–264, 259, 261–263, 266, 267–279, 277, 279, 281–286, 298, 392
Carnotaurus sastrei, 161, 244, 249, 250, 251, 255, 257, 264, 276–279, 277, 279
Carrano, Matthew, 251
Carvalho, Ismar de Souza, xix, 79, 93, 128, 179, 382
Casamiquela, Rodolfo, xix, 36, 37, 38, 39, 61, 99, 100, 101, 114, 121, 127, 338, 343, 372
Casamiquelichnus navesorum, 99, 121, 123
Case, Judd, 330
Cassuarius cassuarius, 253
Castillo Formation (Argentina), 162, 211
Cathartesaura anaerobica, 152, 179
Caturrita Formation (Brazil), 33, 35, 36, 50, 59, 75, 79
Caudipteryx, 319
Cerapoda, 71, 73, 74, 345, 352
Ceraptopsia, 379–382; record in Cinco Saltos (Argentina), 379; trackways from Itapecurú Formation (Brazil), 382
Ceratosauria, 54, 57, 115, 119, 245, 246–248, 249, 250, 252, 257, 266, 271, 286; phylogenetic relationships, 246
Cerritosaurus binsfeldi, 34
Cerro Barcino Formation (Argentina), 159, 161, 198, 248, 251, 298
Cerro Carnerero Formation (Argentina), 95, 101
Cerro Cóndor (Argentina), 95, 96, 103, 105, 107, 112, 119
Cerro Guillermo (Argentina), 155, 200
Cerro Lisandro Formation (Argentina), 152, 153, 154, 290, 336, 337, 356, 368
Cerro Los Leones (Argentina), 152, 171, 199
Cerro Policía town (Argentina), 212, 349
Cetiosaurus, 71, 101, 104, 105, 107
Chacarilla Formation (Chile), 92, 114, 119, 121, 127, 128, 129, 146
Chaco Basin (Argentina), 88, 132, 388
Chaliminia musteloides, 31, 32

chanaresuchids, 34
Chanaresuchus bonapartei, 28
Chanaria platyceps, 28
Los Chañares Formation (Argentina), 27, 28, 29, 34, 83
Los Chañares, fossil site (Argentina), 38
Charadriiformes, 337
Charruodon tetracuspidatus, 34
Chaunaca Formation (Perú), 145
chelonian, 31
Chiappe, Luis, xix, 149, 213, 237, 270, 312, 319, 334, 336
Chico River (Argentina), 99, 162, 217, 374, 380
chigutisaurids, 30
China Muerta Hill (Argentina), 152
Chindesaurus briansmalli, 42, 80
Chiniquá town (Brazil), 33, 34
Chiniquodon sanjuanensis, 30
Chinle Formation (United States), 42, 81
Chirostenotes, 320
chirotheroid footprints, 31
El Chocón Lake (Argentina), 153, 154, 238, 240, 240, 283, 299, 370
Chon Aike Formation (Argentina), 97
Chong Díaz, G., 147
Chubut Group (Argentina), 159, 160, 161, 162, 211, 394
Chubut Province, xx, 95, 97, 101, 103, 105, 112, 158, 159, 161, 162, 167, 198, 202, 211, 217, 242, 248, 251, 276, 286, 291, 298, 306, 342, 343, 345, 365, 374, 380
Chubutisaurus, 161, 168, 185, 187, 198, 199, 200, 389
Chubutisaurus insignis, 161, 168, 187, 198, 199, 200
Chure, D., 132
Cinco Saltos town (Argentina), 149, 157, 167, 209, 231, 243, 343, 354, 380, 397, 398, 399
Clarens Formation (Zimbabwe), 93
Clarke, Julia, 336
Clasmodosaurus spatula, 164, 243
Clevosaurus brasiliensis, 35
Cochabamba (Bolivia), 145
Coelophysis, 44, 46, 47, 51, 54, 55, 56, 80, 252
Coelophysoidea, 54–57, 252
coelophysoids, 54, 57, 82, 119, 252
Coelurosauria, 243, 244, 266, 287, 303–310, 307, 314, 327; phylogenetic relationships, 305, 307; record in the southern continents, 304
coelurosaurians, 11, 44, 117, 118, 119, 140, 161, 164, 244, 245, 246, 251, 264, 291, 293, 294, 304, 305, 307, 309, 310, 311, 319, 321, 322, 323, 334, 354, 388, 390, 393, 394
Colalura Sandstone (Australia), 246
Colbert, Edwin, 36
Colhue Huapi Lake (Argentina), 161, 162, 217, 380
Coli Toro Formation (Argentina), 158, 343
La Colonia Formation (Argentina), 158, 160, 161, 276, 372
Coloradisaurus, 31, 32, 38, 40, 57, 59, 66, 67, 67, 83
Coloradisaurus brevis, 31, 32, 38, 57, 66, 67, 83
Los Colorados Formation (Argentina), 27, 30–32, 31, 35, 40, 54, 55, 57, 66, 68, 83
competition and opportunism in Triassic ecosystems, 82–84
Compsognathidae, 305
compsognathids, 245, 309, 310
Compsognathus longipes, 309
Condorraptor, 88, 95, 119, 121, 129
Condorraptor currumilli, 88, 95, 119, 121, 129
Confuciusornis, 327
Confusão Creek (Brazil), 144, 183
Cooper Canyon Formation (United States), 81
Corda Formation (Brazil), 142, 238
Coria, Rodolfo, xix, 55, 57, 99, 107, 108, 109, 123, 155, 168, 204, 205, 213, 230, 243, 262, 275, 281, 285, 294, 298, 299, 301, 302, 338, 339, 348, 349, 354, 356, 359, 366, 367, 368
Corythosaurus, 375, 377
Costa Franco-Rosas, Aldirene, 219
Cowen, R., 132
Cretaceous dinosaurs, xix; Aptian-Albian faunal turnover in North America, 393; Aptian-Cenomanian largest dinosaurs, 390; Cenomanian-Turonian faunal turnover in Gondwana, 393; desert conditions in the Early Cretaceous, 387; dinosaur assemblage of the Santana Formation, 389; dinosaur taxa shared by South America and Africa, 390; dinosaurs from Argentina, 147–164; dinosaurs from Bolivia, 145, 146; dinosaurs from Brazil, 138–144; dinosaurs from Chile, 146, 147; dinosaurs from Patagonia and Antarctica, 400; dinosaurs from Perú, 144, 145; dinosaurs from Uruguay, 144; distinctions in faunal composition between northern and southern Gondwanan regions, 391; distinctions between fossil record of vertebrates from Bajo de la Carpa and Allen formations, 396; effects of the expansion of lacustrine and marine-influenced environments in faunal composition, 397; evolution of gigantic sauropods in relation to land areas, 398; evolutionary stages, 386; faunal composition at the beginning of the Cretaceous, 387; faunal interchange between the Americas during Campanian-Maastrichtian, 399; faunal structure for the Early Cretaceous in South and North America, 387; faunal succession 384; fossil record in South America, 138, 143; isolation of Gondwanan faunas, 388; non-homologous spiny projections in Early Cretaceous dinosaurs and environmental conditions, 391
Cretaceous: about, xx, 3, 4, 33, 57, 79, 88, 88, 89, 92, 99, 108, 111, 112, 114, 115, 121, 127, 129, 130, 131, 132, 133, 134, 135–138, 137, 138, 141, 143, 149, 151, 159, 160, 163; volcanic events, 132
Crurotarsi, 3
Cúneo, R., 109
Currie, Philip, xx, 294, 299, 301, 302, 338
Curry Rogers, K., 214
Las Curtiembres Formation (Argentina), 147

cynodonts, 28, 30, 32, 34, 35, 38, 59, 82

da Rosa, A., 59, 61
da Silva, R., 79
Dalla Vecchia, F., 367, 391
de Carlés, Enrique, 149
de Valais, Silvina, 79, 99, 123
Deccan Traps (India), 132, 133
Deferrariischnium, 153, 338, 339
Deferrariischnium mapuchensis, 338, 339
Deinonychosauria, 321, 322, 323; record in Gondwana, 291, 294, 304, 305, 306, 320, 325, 330
Deinonychus, 305, 311, 315,316, 320, 321, 322, 325–327, 329, 427
Del Corro, Guillermo, 198
Deltadromeus agilis, 250, 251
Depéret, Charles, 196
dermal plates, 171, 219, 343, 349
Deseado Massif (Argentina), 88, 97–99, 99, 130, 159
Dicraeosaurinae, 111, 112, 172; dicraeosaurid extinction, 393
Dicraeosaurus hansemanni, 111, 112, 172
Dicraeosaurus sattleri, 111, 112
Dicroidium flora, 29, 33, 87
dicynodonts, 28, 30, 34, 35, 38
Dilophosaurus, 56, 118, 252, 436
Dinodontosaurus brevirostris, 28
Dinodontosaurus platygnathus, 28
Dinodontosaurus Zone, 34
Dinosaur Park Formation (Canada), xx
Dinosauria: acetabulum, 11, 12, 14, 18; astragalar anatomy, 16; brevis shelf, 18, 22; diagnostic features, 17, 21, 74, 75, 108, 231, 250, 253, 371; forelimbs, 4, 8, 10, 21, 22; hindlimbs, 4, 8, 10, 11, 12, 14, 21, 22; history of classification, 11–13; ischium, 14, 11, 16, 18, 19, 20, 21, 22; locomotor apparatus, 4, 20; monophyletism, 13; ossified tendons, 10; pectoral girdle, 4; pelvic girdle, 4, 7, 8, 11, 21, 22; phylogenetic relationships among Dinosauria, 2, 3, 27, 40, 42; polyphyletism, 12; sacral count, 17; skull, 5, 6, 7, 17, 21, 22; systematic position within amniotes, 3, 2; tarsal bones, 20; tibial astragalar articulation, 19; vertebral column, 4, 7, 7, 8, 10, 11, 18, 22
Dinosauriformes, 13, 14, 16, 19, 20, 22, 34, 41, 47, 77, 80, 81
dinosaur-mimic archosaurs, 81
Dinosauromorpha, 20, 22, 406
dinosauromorphs, 3, 19, 20, 22, 28, 73
Diplodocimorpha, 110, 112, 168, 169,170 172, 175, 382, 387, 389, 392, 393; phylogenetic relationships among diplodocimorphs, 109
Diversification of dinosaurs: Cretaceous, 20, 249, 304, 337, 343, 385, 386, 388, 393, 394, 396, 402; Jurassic, 101, 128, 245; Triassic, 8, 80, 82, 83, 84
Djadokhta Formation (Mongolia), 314
Dravidosaurus blanfordi, 346
Dromaeosaurus, 321
Dryomorpha, 359
Dryosaurus, 14, 17, 129, 307, 357, 359, 361, 363, 368
Durham, J., 108

Ecteninion lunensis, 30
Edmontosaurus, 351, 378
eggs, xx, 33, 63, 138, 142, 144, 145, 156, 158, 185, 236, 237, 238, 239, 279, 334, 338, 340
Ekrixinatosaurus, 152, 251, 272, 273, 276, 281, 282, 282, 283, 285, 286, 390
Ekrixinatosaurus novasi, 152, 251, 281, 282
Elaphrosaurus bambergi, 246, 261
Elasmaria, 353
Elliot Formation (Zimbabwe), 70, 93, 100
elmisaurid-like ungual from Brazil, 320
Elmisaurus, 320
Elongaloolithidae, 340
Enantiornis leali, 149
Enantiornithes, 245, 335
Engelmann, G., 132
Entre Ríos Province (Argentina), 144, 149, 342, 349, 397
Eoraptor, 30, 30, 39, 41, 49, 50, 50, 53, 80, 83
Eoraptor lunensis, 30, 39, 49, 50
Epachthosaurus, 162, 168, 187–189, 193, 193, 194, 194, 197, 202–205, 204, 209, 217, 218, 222, 224, 228, 232, 351
Epachthosaurus sciuttoi, 162, 189, 193, 194, 202, 204
Eucoelophysis baldwini, 81
Euhelopus, 185, 207, 208
Euoplocephalus, 351
eurymetatarsal foot, 363
Eusauropoda, 100, 101, 102, 103, 107, 108
Exaeretodon frenguellii, 30
Exaeretodon riograndensis, 34
Ezcurra, Martín, 55, 56, 57
Ezequiel Ramos Mexía Lake (Argentina), 153, 177, 240, 240, 338, 339, 370
La Esquina fossil site (Argentina), 66, 68

Faccini, U., 92
Faccio, G., 237
Fasolasuchus tenax, 31, 32
faunal interchange, 344, 399
Faxinal do Soturno (Brazil), 35, 36, 50
Federal University of Río Grande do Sul (Porto Alegre, Brazil), 36
Ferigolo, Jorge, 51, 53, 75, 76
Fernandes, Marcelo Adorna, 93, 128
Fiorillo, A., 132
flapping flight, acquisition of, 328
footprints: avian-like footprints from Santo Domingo Formation, 78, 79, 80; bird-like, 32, 78, 79, 80, 424; ceratopsian, 351, 382; chiroteroid, 31; Cretaceous ornithopods, 141, 142, 146, 153, 369, 370, 370, 398; hadrosaurids, 149, 378; Jurassic ornithischians, 127; Jurassic sauropods, 92, 114; Jurassic theropods, 89, 92, 119, 121, 122; tridactyl, 29, 77, 79, 119, 121, 127, 128. *See also* ichnological evidence
Forasiepi, Analía, 234
Forster, Cathy, 40, 42, 50, 185, 214, 251, 374, 375, 377
Frankfurt, N., 270, 319
Frenguellisaurus ischigualastensis, 47
Frey, E., 319, 389
Fundo El Triunfo Formation (Perú), 144, 145

Futalognkosaurus, 155, 169, 188, 201, 202, 202
Futalognkosaurus dukei, 155, 201, 202, 202

Galliformes, 335, 336
Galton, Peter, 12, 13, 57, 59, 68, 70
García, R., 107, 221
Gasparini, Zulma, 107, 155, 167, 168, 209
Gasparinisaura, 157, 157, 345, 352–354, 354, 356, 357, 357, 359, 363, 365, 394
Gasparinisaura cincosaltensis, 157, 157, 345, 354, 354, 357
Gauthier, Jacques, 13, 115
Gaviidae, 335
Genasauria: phylogenetic relationships, 346
General Roca city (Argentina), 213, 219, 233, 285, 290, 336, 348, 349, 377
Genise, J., 79
Genusaurus sisteronis, 249, 251
Genyodectes, 161, 243, 245, 246, 248, 248
Genyodectes serus, 161, 243, 246, 248, 248
Geologic time scale: Cretaceous, 134; Jurassic, 86; Triassic, 24
Gichón-Yeruá Basin (Uruguay and NE Argentina), 144, 149
Giganotosaurus carolinii, 152, 177, 255, 295, 297, 298, 301, 302
Gilmoreosaurus, 375
Giménez, O., 109
Girón Formation (Colombia), 90, 108
Gobi Desert (Mongolia), 304
Godoy, L., 93
Gomphodontosuchus brasiliensis, 34
Gondwanatitan, 144, 183, 187, 219–221, 221, 223, 224, 226
Gondwanatitan faustoi, 144, 220, 221, 223
González Riga, Bernardo, xix, 221, 222, 240
Gracilisuchus stipanicicuorum, 28
Grallator, 99
Gryposaurus incurvimanus, 372, 374
Guaibasauridae, 53
Guaibasaurus, 17, 18, 35, 39, 41, 44, 49–51, 53, 53, 54, 59, 80

Guaibasaurus candelariensis, 35, 39, 50, 53
Gualosuchus reigi, 28
Guichón Formation (Uruguay), 143, 349
Gutierrez, F., 146

Hadrosaurichnus australis, 378
Hadrosaurichnus titicacaensis, 378
Hadrosauridae, 352, 370–383; biotic interchange between the Americas, 344; footprints and eggshells, 378–383; fossil record in Argentina in contrast with other regions of South America, 352, 370, 371, 375, 378; fossil record in North America, 371; hadrosaurids and littoral paleoenvironments, 372; hadrosaurids from Patagonia, 343, 344, 345, 377, 97, 398; phylogenetic relationships, 371
Haplocanthosaurus, 170, 177, 196, 197
Haubold, H., 351
Hay, W., 138
Hemiprotosuchus leali, 31, 32
herbivory among basal dinosauriforms, 81
Herrerasauridae: as large predators, 43; dorsal vertebrae, 44, 53; evidence of diet in, 43; evolutionary radiation, 82; pectoral girdle and forelimbs, 44, 49, 53; pelvic girdle and hindlimbs, 46, 50; phylogenetic relationships, 46; sacrum, 44, 49, 51; skull, 46
Herrerasaurus, 14, 14, 16, 18, 18, 19, 20, 30, 30, 37, 39–44, 42, 44, 46–49, 51, 58, 80, 82
Herrerasaurus ischigualastensis, 30, 37, 39, 41, 42, 44, 46, 47, 82
cf. *Heterodontosaurus*, 71, 74, 75
Heterodontosaurus, 73, 75, 359
Holtz, T., 4
Hoplitosaurus, 349
Hoplitosuchus raui, 34, 41
Horner, J., 374, 375, 377
Hornitos Formation (Chile), 147
Los Hornos Hill (Argentina), 164, 205
Hoyada del Cerro Las Lajas fossil site (Argentina), 71

von Huene, Frederich, xx, 20, 33, 35, 37, 41, 76, 100, 142, 144, 149, 161, 167, 168, 211, 212, 213, 214, 215, 231, 232, 343, 349, 351, 366, 369, 380
Huincul Formation (Argentina), 152, 153, 179, 204, 275, 283, 301, 389, 393
Humaca town (Bolivia), 145, 239, 240
Hutchinson, J., 309, 310
Hypacrosaurus, 375, 377, 378
Hyperodapedon huenei, 34
Hyperodapedon mariensis, 34
Hyperodapedon sanjuanensis, 30, 34
Hypsilophodon, 18, 356, 357, 361, 363
Hypsilophodontidae, 352, 365

ichnofaunas from La Matilde and Botucatu formations, 130
ichnological evidence, 76, 77, 79, 83, 114, 119, 338, 344, 352, 382. *See also* footprints
Ignotornis, 339
Iguanodon, 71, 363, 366
Iguanodonichnus frenkii, 92, 114, 127, 369
Iguanodontia: clades endemic from Gondwana, 114, 127, 352; diversity in Cenomanian through Maastrichtian times, 352; documentation in the Cretaceous of South America, 352, 356, 359, 395
Ilha do Medo (Brazil), 369, 382
Ilokelesia, 153, 250, 251, 253, 268, 269, 271, 272, 275, 276, 276, 277, 283
Ilokelesia aguadagrandensis, 153, 250, 251, 275, 276, 276
India, xx, 100, 129, 168, 183, 197, 213, 249, 250, 251, 259, 263, 264, 264, 346, 393, 397, 400, 402
Indosaurus matleyii, 252
Indosuchus raptorius, 252
Ingenia, 319
Ingeniero Jacobacci town (Argentina), 158, 338, 397, 398
Instituto Miguel Lillo (Tucumán, Argentina), 38, 66
Irajatherium hernandezi, 35
Iriarte, J., 147
Irmis, R., 74

Irritator, 140, 246, 288–290, 290, 389
Irritator challengeri, 140, 246, 288–290, 290, 389
Ischignathus sudamericanus, 30
Ischigualastia jenseni, 30
Ischigualasto Formation (Argentina), xix, xx, 29, 30, 34, 42, 47, 49, 71, 82
Ischigualasto-Villa Unión Basin (Argentina), 27, 32, 76, 93
Ischisaurus cattoi, 37, 47
Isisaurus colberti, 188
Islas Malvinas town (Argentina), 158
Itapecuru Formation (Brazil), 142, 172, 176, 179, 382, 389

Jachaleria candelariensis, 35
Jagüel Formation (Argentina), 157
Jehol Group (China), 321, 436
La Juanita farm (Argentina), 298
Juárez Valieri, R., xi, 285, 398
Jujuy Province (Argentina), 149
Jurassic, 87, 88; dinosaur record from South America, 19, 57, 88, 89, 90, 90, 92, 93, 95, 99, 100, 102, 107, 108, 114, 115, 119, 121, 123, 127, 127, 128, 129, 130, 135, 338; dinosaur-bearing beds in South America, 26, 88, 89, 158, 163, 402; dinosaurs from Argentina, 88, 89, 93, 95, 99, 100, 102, 107, 115, 19, 121, 129, 135; dinosaurs from Brazil, 57, 89, 90, 92, 93, 100, 119, 123, 128, 130; dinosaurs from Chile, 89, 90, 92, 114, 119, 121, 127, 128, 129; dinosaurs from Colombia, 88, 90, 107, 108, 128, 135, 338; dinosaurs from Venezuela, 88–90, 90, 100, 123, 127, 128; effects of intense volcanic activity on mass extinctions, 97, 131, 132, 133; lava flows, 88, 92, 131, 132, 386; regionalization in dinosaur distribution, 129; volcanic events, 88, 97, 131, 133

Kangnasaurus, 352, 353, 363
Kellner, Alexander, xix, 59, 61, 144, 168, 255, 227, 283, 285, 288, 290, 307, 309, 327, 329, 330, 398
Kentrosaurus aethiopicus, 346
Khöler, G., 378

Kielan-Jaworowska, Sofia, 385
Kirkwood Formation (South Africa), 346, 387
Kischlat, E., 41
Kitching, J., 59
Krause, Dave, 251
"*Kritosaurus*" *australis*, 158, 372, 373, 374, 375, 377, 377, 378, 397

Lacovara, Kenneth, xx, 164
Laevisuchus, 250, 252, 267
Laevisuchus indicus, 250, 252
Lagerpeton, 4, 12, 13, 14, 15, 16, 16, 18, 19, 20, 22, 28, 38, 40, 80
Lagerpeton chanarensis, 12, 28, 38
Lagosuchus, 12, 20, 28, 38
Lagosuchus talampayensis, 12, 28, 38. See also *Marasuchus lilloensis*
Laguna Colorada Formation (Argentina), 33, 74
Laguna Manantiales farm (Argentina), 99, 130
Laguna Palacios Formation (Argentina), 162, 374
Lakumasaurus, 400
lambeosaurines, 345; record in Salitral Moreno, 375, 377, 378
Lambeosaurus, 375, 377
Lameta Formation (India), 251, 252, 259, 263, 264
Lametasaurus, 249, 252, 259, 263, 272, 273, 284, 285
Lametasaurus indicus, 249, 252, 263
Lamanna, Matthew, xx, 274, 286, 393
Langer, Max Cardoso, xix, 39, 41, 42, 46, 53, 59, 75, 76
Langston, W., 107
Laplatasaurus, 144, 157, 167, 188, 196, 209, 210, 210, 214, 214, 217, 231, 235, 397, 399
Laplatasaurus araukanicus, 144, 157, 167, 188, 209, 210, 210, 214, 214, 397
large theropods possibly related to *Megaraptor*, 294
Lavina, E., 92
Lavocat, René, 176
Le Loeuff, J., 197
Leal, L., 59, 61
Lecho Formation (Argentina), 145, 147, 149, 149, 232, 268, 270, 335
Lectavis bretincola, 149

La Leona River (Argentina), 164, 205
Leonardi, Giuseppe, xix, 93, 127, 130, 140, 142, 238, 239, 344, 351, 369
Leptoceratops, 382
Lesothosaurus, 8, 18, 58, 71, 73, 125
Lessemsaurus, 31, 38, 40, 57, 68, 70, 70, 83
Lessemsaurus sauropoides, 31, 38, 57, 68–71, 70
Lewisuchus admixtus, 28. See also *Pseudolagosuchus major*
Liaoning (China), 304, 321, 332
Ligabueichnium, 351, 351
Ligabueichnium bolivianum, 351, 351
Ligabueino, 152, 244, 250, 251, 257, 262, 266, 267, 267, 268, 270, 271
Ligabueino andesi, 152, 244, 250, 251, 266–268, 267
Ligabuesaurus, 152, 169, 187, 198, 199, 200, 200, 207, 389, 390
Ligabuesaurus leanzai, 152, 187, 199, 200
Liliensternus, 54, 55, 57
Liliensternus liliensterni, 54
Limayichnus, 153, 154, 352, 370, 370, 390, 394
Limayichnus major, 370, 370
Limaysaurus, 152, 168, 170, 171, 174, 176, 177, 178, 179, 179, 180, 181, 182, 199, 389, 390
Limaysaurus tessonei, 152, 168, 170, 177–179, 179, 180, 181, 182
Limenavis, 158, 335, 336
Limenavis patagonica, 158, 336
Lío, Gabriel, 336
Lirainosaurus, 182, 196, 235
Lirainosaurus astibiae, 182
Lockley, Martin, 239
Lognkosauria, 188, 201
Lohan Cura Formation (Argentina), 151, 152, 171, 176, 177, 199, 389
Lonco Trapial Group (Argentina), 95
Loncoche Formation (Argentina), 157, 158, 240, 372
Loncosaurus, 164, 243, 366, 367, 367
Loncosaurus argentinus, 164, 243, 366, 367, 367
Loricosaurus scutatus, 231, 343, 344
Luperosuchus fractus, 28

Lurdusaurus, 352, 393
Lydekker, Richard, xx, 161, 167, 168, 188, 231

Machado, E., 290
Macrogryphosaurus, 155, 201, 345, 352, 353, 354, 357, 361, 363–365, 365, 386, 395
Macrogryphosaurus gondwanicus, 155, 345, 361, 363–365, 365
Madagascar, xx, 81, 103, 168, 183, 197, 249, 250, 251, 304, 332, 345, 393, 397, 398, 400, 402
Maevarano Formation, 251
Magallanes Basin (Argentina and Chile), 138, 151, 163, 164
Magdalena, Department of (Colombia), 90, 107
Mahler, L., 251
Majungasaurus, 250, 251, 252, 253, 255, 255, 256, 256, 259, 260, 261, 262, 264, 267, 271, 272, 273, 274, 275, 276, 277, 279, 282, 283, 286, 345
Majungasaurus athopus, 250, 251, 255, 345
Makay Formation (Madagascar), 81
Makovicky, Peter, 330, 331, 332
Malargüe Group (Argentina), 157
Malawisaurus, 183, 187, 189, 196, 200, 214
Malawisaurus dixeyi, 183
Malvinas (Falkland) Plateau, 136
Manera de Bianco, T., 378
Maniraptora: phylogenetic relationships, 305, 307, 308, 314
maniraptorans, 44, 117, 143, 147, 155, 244, 259, 304, 305, 314, 315, 319, 320, 322, 325, 327, 329, 331, 334, 368, 395, 396, 397
maps of fossil localities of South America that yielded dinosaurs: Cretaceous, 138, 143, 149, 150, 159, 160, 163, 167, 242, 342; Jurassic, 90, 93, 95, 96, 98; Triassic, 26, 28, 34
Mapusaurus, 293, 297, 297, 301, 302, 302
Mapusaurus roseae, 297, 301, 302, 302
Maranhão State (Brazil), 141, 177, 179, 288, 368, 382

Marasuchus, 4, 12, 13, 13,14, 14, 15, 16, 16, 18, 18, 19, 20, 20, 22, 28, 29, 40, 44, 46, 47, 49, 51, 58, 65, 80, 82. See also *Lagosuchus*
Marasuchus lilloensis, 12, 13, 14, 28, 82. See also *Lagosuchus talampayensis*
Marifil Formation (Argentina), 96
Marília Formation (Brazil), 142, 143, 144, 222, 225, 236, 320, 321
Marquillas, R., 378
Marshall, L., 146
Marsicano, C., 74, 75, 77, 79
Martill, D., 309, 319, 389
Martinelli, Agustín, 51, 234, 270, 271, 317
Martínez, Ricardo N., 93
Martínez, Rubén D., xix, 161, 162, 191, 202, 205, 211, 306, 365, 374
Martz, J., 74
Masiakasaurus, 250, 251, 255, 259, 261, 264, 267, 268, 270, 271, 272, 279, 286
Masiakasaurus knopfleri, 250, 251
Massetognathus, 28, 83
Massetognathus major, 28
Massetognathus pascuali, 28
Massetognathus terugii, 28
Massospondylus, 63, 65, 67, 70, 88, 93, 95, 129
Massospondylus carinatus, 63
mastodonsaurids, 30
Mata Amarilla Formation (Argentina), 163, 191, 366
Mata Formation (Brazil), 33
Matasiete Formation (Argentina), 162
La Matilde Formation (Argentina), 97, 99, 99, 114, 119, 121, 125, 130, 132
Maxakalisaurus, 144, 183, 209, 225, 227, 228
Maxakalisaurus topai, 144, 225, 227, 228
Medeiros, Manuel, 142, 367, 391
Megagomphodon oligodens, 28
Megaloolithus, 237
Megaraptor, 155, 162, 201, 245, 288, 291, 292, 292, 293, 294, 306, 323, 388, 393, 395
Megaraptor namunhuaiquii, 155, 245, 288, 291–294, 292, 323
Megatherium, 196
Melanorosauridae, 66, 68, 83

Melchor, R., 79, 99
Mendoza Province (Argentina), 26, 107, 149, 155, 157, 167, 200, 240, 345, 398
Mendozasaurus, 155, 188, 189, 192, 196, 200, 201, 202, 202, 207, 227
Mendozasaurus neguyelap, 155, 200, 201, 202
Mercedes Formation (Uruguay), 144, 167, 236
Microceratops, 382
Microraptor, 4, 115, 321, 330
Microraptor zhaoianus, 321
Microvenator, 320
Microvenator celer, 320
migratory route between Antarctica and South America, 138
Minas Gerais State (Brazil), 142, 143, 144, 167, 183, 222, 227, 320
Minmi, 347, 348, 351, 390
Minmi paravertebra, 347, 390
Mirischia, 140, 245, 246, 304, 305, 309, 310, 311
Mirischia asymmetrica, 140, 245, 246, 304, 305, 309, 310, 311
Moanasaurus, 400
El Molino Formation (Bolivia), 145, 146, 239, 351, 351
Molnar, Ralph, 243, 248, 366, 380
Mones, Alvaro, 237
Monetta, A., 42, 50
Mongolia, 183, 304, 310, 396
Mononykus, 244, 310, 311, 312, 313, 314, 315, 316, 317
Mononykus olecranus, 244, 310, 313
Moody, John, 123
Moratalla, J., 237
Moreno, K., 114, 119, 282
Morrison Formation (United States), 95, 129, 132, 246, 290
Morturneria, 400
Museo Argentino de Ciencias Naturales (Buenos Aires, Argentina), 39, 40, 257, 373
Museo de La Plata (La Plata, Argentina), 37, 149, 161, 217, 349
Museum of Comparative Zoology (Harvard, United States), 35
Mussaurus, 33, 38, 40, 57, 61, 62, 63, 64, 65, 65, 66, 83
Mussaurus patagonicus, 33, 38, 57, 61–66, 65, 83

Musters Lake (Argentina), 161, 162, 202, 211
Muttaburrasaurus, 352, 353, 386

Naashoibitosaurus, 372, 374
Naashoibitosaurus ostromi, 374
Naish, D., 310, 389
Nesbitt, S., 74
Nemegtosaurus, 182, 190, 190, 211, 212, 214, 215
Nemegtosaurus mongoliensis, 182, 190
Neoaetosauroides engaeus, 31
Neogaeornis wetzeli, 147, 336
Neornithes: diversification of modern groups of birds, 337; fossil remains in South America, 245; southern origin, 338; Turonian record in Patagonia, 336
Neosauropoda, 108; phylogenetic relationships, 109, 169
Neotheropoda: phylogenetic relationships, 245
nests, 33, 138, 156, 185, 236–238, 279, 312
Neuquén Group (Argentina), 136, 151, 152–158, 314, 349
Neuquén Province (Argentina), 93, 107, 149, 153, 154, 154, 155, 156, 157, 158, 167, 169, 171, 172, 177, 179, 181, 190, 191, 199, 204, 221, 236, 237, 239, 242, 251, 266, 275, 281, 283, 291, 294, 301, 315, 325, 329, 330, 336, 336, 337, 338, 342, 346, 356, 361, 368, 370
Neuquenornis, 156, 157, 335, 336, 396
Neuquenornis volans, 156, 157, 335
Neuquenraptor, 155, 245, 304, 322, 324, 329, 330, 331, 331, 332, 334
Neuquenraptor argentinus, 155, 245, 304, 322, 329–331, 331
Neuquensaurus, 157, 183, 187, 188, 194, 195, 196, 207, 209, 210, 219, 220, 222, 227, 228, 229, 230, 231, 232, 233, 233, 234, 235, 236, 398
Neuquensaurus australis, 157, 183, 188, 195, 210, 228, 229, 231, 232, 233
Neuquina Basin (Argentina), 99, 136, 138, 149–158, 150, 384, 389
Nigersaurus, 176, 182, 215
Nigersaurus taqueti, 176, 182

Noasaurus, 149, 244, 249, 250, 251, 257, 266, 267, 268, 269, 269, 270, 271, 272, 274, 276, 283, 286
Noasaurus leali, 149, 244, 249, 250, 251, 268–270, 269, 271
Noblet, D., 145
Nodosauridae, 347, 348, 349, 385
Nomingia, 319
Nopsca, Franz, 329
Norman, D., 72, 73
Nothofagus flora, 400
Notobatrachus degiustoi, 99
Notohypsilophodon, 162, 345, 353, 357, 363, 365, 366, 366, 394
Notohypsilophodon comodoroensis, 162, 345, 365, 366, 366
Nqwebasaurus thwasi, 304

Ocho Hermanos farm (Argentina), 162, 211, 286
Oliveira, F., 93, 127
Opisthocoelicaudia, 182, 187, 189, 191, 194, 207, 215, 218, 219, 220, 222, 227, 231, 232, 396
Opisthocoelicaudia skarzynskii, 182, 187
Ornithischia: abundance and diversity of southern ornithischians during the Cretaceous, 346, 400; definition, 71; footprints of Jurassic ornithischians, 127; meaning of the name, 3; ornithischians in the Gondwanan realm during the Cretaceous, 400; ornithischians in the Jurassic of Venezuela, 123; pelvic girdle and hindlimb morphology, 12, 73; phylogenetic relationships 13, 17, 46, 71, 72, 74, 75, 125, 346; Triassic record in South America, 71, 80
Ornithodira: monophyly, 3, 13; locomotor capabilities, 3, 19, 20, 22, 77
Ornitholestes, 315, 320
Ornithopoda, 128, 352–379; cladistic analysis, 359; Cretaceous footprints, 369, 370, 370; Cretaceous record in Gondwana, 352; ornithosuchids, 30, 83; phylogenetic relationships, 352, 353

ossifications, 171, 172, 196, 197, 198, 346, 351, 363, 399
Ouranosaurus-like caudal vertebrae from NE Brazil, 367–368, 368
Oviraptor, 319
oviraptorosaur from the Santana Formation, 319
Oviraptorosauria, 149, 270, 271, 304, 305, 306, 319
Owen, Richard, 11, 12

pachycephalosaurians, 345, 399
Pais, D., 251, 367, 391
Pajcha Pata town (Bolivia), 145
Paleochersis talampayensis, 31
Paleodesert of Botucatu (Brazil), 88, 130, 131, 135, 387
paleogeographic maps: Cretaceous, 137, 400; Early Cretaceous 88; Late Jurassic, 131; Triassic, 26
Pampa de Agnia (Argentina), 101
La Pampa Province (Argentina), 149, 157, 158, 236, 345, 398
Paralititan, 183, 217, 392
Paralititan stromeri, 183
Paramylodon, 196, 197
Paraná Basin (in Brazil), 25, 26, 33, 88, 92, 128, 130, 132, 135, 144, 387, 388
Paranthodon africanus, 346, 377
parasagital crest in *Zupaysaurus* and *Syntarsus*, 57
Parasaurolophus, 370, 375, 377
Paraves: phylogenetic relationships, 323
Pari Aike Formation (Argentina), 163, 164, 205, 305, 352, 361, 398
Parker, W., 74,
Parnaíba Basin (Brazil), 88, 92, 132, 135, 136, 138, 140, 142, 172, 179
Parvicursor remotus, 310
Paso Córdova (Argentina), 157, 213, 317
Paso del Sapo Formation (Argentina), 157, 158, 372
Patagonichnornis venetiorum, 338
Patagonykus, 155, 244, 304, 310, 312, 315, 316, 317, 319, 368, 396
Patagonykus puertai, 155, 244, 304, 310, 312, 315–317, 319, 368
Patagopteryx, 156, 157, 245, 304, 307, 327, 335, 336, 396

Index 447

Patagopteryx deferrariisi, 156, 157, 245, 304, 336
Patagosaurus, 68, 88, 95, 97, 100, 101, 103, 104, 105, 105, 106, 107, 107, 108, 109, 128, 129
Patagosaurus fariasi, 88, 95, 97, 100, 103–107, 107, 109,
Patterson, Bryan, 38
Paul, G., 248
Paulina Carabajal, A., 99, 123
Paulo Creek (Brazil), 283
Pellegrini Lake (Argentina), 157, 227, 273, 397, 399
Pellegrinisaurus, 157, 193, 215, 219, 220, 225, 226, 227, 228, 229, 397
Pellegrinisaurus powelli, 157, 225, 227, 228, 229, 397
Pelorocephalus ischigualastensis, 30
Perle, A., 312
Petrified Forest National Monument at Madre e Hija Hill (Argentina), 99
Phuwiangosaurus sirindhornae, 183
phytosaur, 35
Piatnitzkysaurus, 88, 95, 97, 115, 117, 117, 118, 118, 119, 129, 293, 315
Piatnitzkysaurus floresi, 88, 95, 97, 115–119 117, 118
Pichasca (Chile), 146
Picún Leufú village (Argentina), 152, 171, 199
Picunichnus, 153, 338, 339
Picunichnus benedettoi, 338, 339
Piranhas Formation (Brazil), 141, 351
Pisanosaurus, 12, 17, 19, 30, 30, 38, 39, 71, 71, 72, 73, 73, 74, 80, 81, 82, 83
Pisanosaurus mertii, 30, 38, 39, 71, 74, 73
Planicie de Renteria (Argentina), 154, 155
Plaza Huincul town (Argentina), 153, 204, 301, 338, 356
Pleurocoelus nanus, 182
Plottier Formation (Argentina), 154, 155, 215
Poblete, F., 293
Pol, Diego, 61, 63, 64, 65, 66, 70, 329
Polacanthidae, 347, 348
Polarornis, 338
Porfiri, Juan, 203, 293, 323, 327, 329, 330

Portezuelo Formation (in La Rioja Province, Argentina), 77, 78, 79, 80, 83, 155
Portezuelo Formation (in Neuquén Province, Argentina), 154, 155, 191, 201, 294, 314, 315, 323, 325, 329, 336, 361, 363, 368
Porto Alegre city (Brazil), 33, 36
Postosuchus, 43
Powell, Jaime, xix, 61, 63, 64, 65, 66, 70, 149, 168, 196, 202, 205, 209, 215, 217, 225, 227, 229, 231, 232, 237, 244, 268, 269, 322, 343, 344, 377, 378, 380
Presidente Prudente town (Brazil), 183, 209, 336
Prestosuchus, 43
Price, Llewellyn Ivor, xix, 35, 140, 143, 167, 168, 225
Probainognathus jenseni, 28
Probelesodon lewisi, 28
Probelesodon minor, 28
procolophonids, 34
Promastodonsaurus bellmani, 30
Prosauropoda, 57, 59, 70
Proterochampsa barrionuevoi, 30
Proterochampsa nodosa, 34
proterochampsids, 30
Prozostrodon brasiliansis, 34
Pseudohesperosuchus jachaleri, 31, 32
Pseudolagosuchus, 13, 14, 14, 15, 16, 19, 20, 20, 21, 22, 28, 40, 46, 74, 80
Pseudolagosuchus major, 28
pteridosperms, 25, 87
Puerta, P., 109
Puertasaurus, 164, 169, 192, 192, 202, 204, 205, 206, 207, 208, 208, 209, 215, 398
Puertasaurus reuili, 164, 192, 202, 204, 205–209, 208
Puerto Yeruá Formation (Argentina), 144, 149, 349
Pumares, J., 67, 68
Pycnonemosaurus, 246, 251, 272, 283, 284, 284, 285
Pycnonemosaurus nevesi, 246, 251, 283, 284, 284

Quaesitosaurus, 182, 190, 212, 214
Quaesitosaurus orientalis, 182
Quantasaurus, 353
Quebrada Monardes Formation (Chile), 146

Quebrada Pajonales Formation (Chile), 146, 147
Quilmesaurus, 158, 251, 263, 284, 285, 285
Quilmesaurus curriei, 158, 251, 285, 285
La Quinta Formation (Venezuela), 89, 90, 123, 125
Quiriquina Formation (Chile), 147, 336

Rahiolisaurus, 252, 264, 272, 273
Rahiolisaurus grujatensis, 252
Rahonavis, 304, 322, 324, 325, 327, 328, 332
Rahonavis ostromi, 304, 332
Rajasaurus, 252, 259, 259, 272, 273
Rajasaurus narmadensis, 252,
Rapetosaurus, 183, 187, 190, 191, 212, 214, 215, 227, 235
Rapetosaurus krausei, 183
Los Rastros Formation (Argentina), 27, 28–30, 76, 77, 79, 80
Rauhut, Oliver, xx, 100, 101, 102, 110, 112, 117, 119, 246, 248
rauisuchids, 3, 83
Rauisuchus tiradentes, 34
Rayoso Formation (Argentina), 152, 176, 179, 181
Rayososaurus, 152, 177, 179, 182, 199, 389, 391
Rayososaurus agrioensis, 152, 177, 179, 182
Rebbachisauridae, 111, 168, 169, 175–177; in Cretaceous dinosaur communities, 389
rebbachisaurids, 152, 171, 175, 176, 177, 179, 385, 386, 389, 391, 392, 393
Rebbachisaurus, 176, 177, 179, 182
Rebbachisaurus garasbae, 177, 179, 182
"*Rebbachisaurus*" *tessonei*, 177. See also *Limaysaurus tessonei*
Reig, Osvaldo, xix, 29, 36, 37, 38, 39, 47, 234
Reptilia, 3
Revueltosaurus, 74, 81, 82
Revueltosaurus callenderi, 81
Rhadinosuchus gracilis, 34
Rhynchosaur Zone, 34, 47, 58
rhynchosaurs, 30, 34, 35, 43, 82, 83, 84
Rich, Thomas, xx, 109

Richter, M., 59
Riggs, Elmer, 161, 217, 343
Rinchenia, 319
Rincón de los Sauces (Argentina), 155, 190, 221
Rinconsaurus, 155, 187, 212, 219, 221, 222, 222, 227
Rinconsaurus caudamirus, 155, 221, 222, 222
Río Colorado Subgroup (Argentina), 152, 155, 213, 279, 335, 397, 398, 399
Rio do Peixe Basin (Brazil), 138, 140, 141
Rio Grande do Sul State (Brazil), 33, 34, 36, 36, 50, 58, 75, 93, 130
Río Limay Subgroup (Argentina), 152, 153, 154, 199, 204
Río Negro Province (Argentina), 149, 152, 154–158, 166, 167, 179, 198, 209, 212, 213, 219, 227, 231, 233, 236, 237, 240, 240, 242, 243, 251, 273, 285, 290, 317, 332, 336, 338, 340, 342, 343–345, 348, 349, 354, 372, 377, 378, 397
Río Neuquén Subgroup (Argentina), 152, 154, 155, 200, 215, 221, 394
Riograndia guaibensis, 35
La Rioja Province (Argentina), 26, 27, 28, 29, 30, 32, 38, 55, 66, 68, 71, 77, 79, 155, 167, 236
Riojasaurus, 18, 31, 32, 38, 40, 57, 59, 62, 65, 66, 67, 67, 68, 68, 70, 83
Riojasaurus incertus, 31, 32, 38, 57, 66–68, 67, 68, 83
Riojasuchus tennuiceps, 31
Roca Formation (Argentina), 157
Rocasaurus, 158, 187, 228, 229, 230, 233, 234, 234, 235, 397
Rocasaurus muniozi, 158, 228, 229, 233, 234, 234, 397
Rogers, R., 50, 214
Romer, Alfred, xx, 4, 27, 29, 36, 37, 38, 38, 39
Roth, Santiago, 149, 243
Rougier, Guillermo, xix, 153, 372
Rubilar, D., 114, 119, 121, 127
Rubilar-Rogers, D., 282
Rugops, 250, 251, 253, 255, 255, 272, 275, 283, 286
Rugops primus, 250, 251, 255
Russell, Dale, 112, 251, 367

Sacisaurus, 35, 71, 76, 76, 81, 83
Sacisaurus agudoensis, 35, 71, 76, 76
Salamanca Formation (Argentina), 374
Salamuni, R., 92
Salgado, Leonardo, xix, 107–110, 112, 168, 170, 174, 177, 181, 185, 189, 198, 205, 212, 213, 217, 221, 228, 230, 232, 233, 243, 275, 298, 348, 349, 354, 356, 357, 366, 367, 389, 393, 394
Salinas, P., 147
Salitral Moreno (Argentina), 149, 158, 198, 219, 233, 237, 336, 344, 348, 349, 351, 375, 377, 397, 398, 399
Salta Group, 147, 149, 149
Salta Province (Argentina), 149, 167, 229, 232, 242, 268, 270, 322, 336, 342, 378, 398
Saltasaurinae, 187, 210, 224, 229–236; opening of the Atlantic Ocean and diversification of saltasaurines, 230; titanosaurids probably related to, 225–228
Saltasaurus, 100, 149, 168, 187, 187, 190, 193, 194, 195, 196, 198, 200, 207, 207, 209, 210, 214, 215, 218–220, 222, 224, 227–236, 230, 232–233, 343, 351, 398
Saltasaurus loricatus, 149, 193, 195, 196, 198, 200, 228, 229, 230, 232–233, 343
Saltenia ibanezi, 147
Sampson, Scott, 251
San Jorge Basin (Argentina), 97, 138, 159–163, 402
San Jorge Formation (Argentina), 374
San Salvador Formation (Chile), 119
Sanga do Cabral Formation (Brazil), 33
Santa Cruz Province (Argentina), 28, 33, 61, 74, 95, 97, 99, 99, 121, 159, 162, 163, 163, 191, 205, 242, 251, 342, 361, 366
Santa María Formation (Brazil), 33, 34, 35, 41, 47, 58, 81
Santa María Supergroup (Brazil), 25, 33
Santa María town (Brazil), 33, 34, 36, 40, 59

Santana do Cariri (Brazil), 140, 289
Santana Formation (Brazil), 136, 140, 245, 246, 288, 291, 294, 307, 309, 334, 344, 389, 390
Santanaraptor, 140, 245, 246, 304, 305, 307–309, 309, 310
Santanaraptor placidus, 140, 245, 246, 304, 305, 307–309, 309
Santo Domingo Formation (Argentina), 32, 78, 79, 80
Santucci, R., 144, 223, 224
Sanz, J. L., 205
São Luis Basin (Brazil), 138, 140, 141–142, 177, 183
São Paulo State (Brazil), 92, 93, 123, 128, 130, 142, 143, 144, 167, 183, 209, 220, 224, 336
Sarmientichnus, 88, 99, 121–123, 125, 129
Sarmientichnus scagliai, 88, 99, 121–123, 125
Saturnalia, 17, 18, 34, 35, 39, 44, 46, 47, 49, 51, 53, 54, 57–59, 61, 61, 62, 80, 83, 84
Saturnalia tupiniquim, 34, 39, 57, 58–59, 61, 80
Saurischia: definition, 41; meaning of the name, 12; phylogenetic relationships, 42
saurischians, 11, 12, 13, 22, 30, 35, 40, 41, 44, 46, 49–53, 80, 83
"The saurischians and ornithischians from the Cretaceous of Argentina" (von Huene), 167
Saurolophus, 372, 374
Sauropoda, 57, 65, 78, 80, 83, 87, 100–108, 127, 212; absence of sauropods in the Botucatu paleodesert, 128, 132; abundance and diversity of titanosauriforms and diplodocimorphs in the Aptian-Cenomanian interval, 389, 390; Cretaceous faunas dominated by sauropods, 181, 385, 398; Cretaceous record in South America, 238–242; eggs and nests, 236–238; feeding habits in titanosauriforms and diplodocimorphs, 390; Jurassic footprints, 114; Jurassic record in South America, 107; phylogenetic relationships, 66, 70, 100, 101, 103, 172

Sauropodomorpha, 17, 41, 42, 53, 57–71, 78, 80, 230; phylogenetic relationships among basal Sauropodomorpha, 58; Triassic record from South America, 35–40, 57–71
Saurosuchus, 30, 43, 82
Saurosuchus galilei, 30, 82
Scherer, C., 92
Schiller, Walter, 149
Schultz, César, xi, 367
Schwanke, C., 79
Secernosaurus, 161, 162, 343, 374–375, 377, 377, 378, 380, 397
Secernosaurus koerneri, 161, 162, 343, 374–375, 377, 377, 378
Seeley, Harry, 11, 12
segnosaurian, 47
Senoniano Lacustre (Argentina), 147, 159
Senter, Phil, 312
separation of South America from Africa, 136, 138, 390
Serendipaceratops arthurclarkei, 380
Sereno, Paul, xx, 13, 20, 22, 39, 40, 42, 50, 185, 246, 251, 279, 286
Serra do Veadinho (Brazil), 142, 222, 224, 225, 320
Serra Geral Formation (Brazil), 33, 92, 135, 142
Shuvosaurus inexpectatus, 81
Shuvuuia deserti, 310
Sierra Barrosa (Argentina), 154, 155, 157, 294, 338, 339
Sierra de San Bernardo (Argentina), 161, 162, 202, 211, 217, 286
Sierra del Portezuelo (Argentina), 153, 154, 155, 156, 315, 325, 330, 368
Sigilmassasaurus brevicollis, 367
Silesaurus, 14, 15, 51, 74, 75, 76, 81, 82, 83, 125
Sillosuchus longicervix, 30
Sinornithoides, 315
Sinornithosaurus, 321, 330, 333
Sinosauropteryx prima, 309
Sinraptor, 295, 301, 302, 305
skin impressions, xx, 110, 185, 251, 264, 265, 266
Skorpiovenator, 252, 272, 283
Skorpiovenator bustingorry, 252, 283
Smith, N., 293

Smith-Woodward, Arthur, 243
Snow Hill Island Formation (Antarctica), 330
Somuncura Massif (Argentina), 95, 96, 158, 159, 159, 160, 161, 402
Sonco Formation (Perú), 145
Soroavisaurus australis, 149
Soto, Matías, xi, 369
Soturnia caliodon, 35
Sousa Basin (Brazil), 138, 140, 140–141, 369
Sousa Formation (Brazil), 141, 369
Sousaichnium, 369, 370, 370
Sousaichnium monettae, 370, 370
Sousaichnium pricei, 369
South American dinosaur faunas: during the Early Cretaceous (Berriasian through Barremian), 387–388; during the Mid- to Early Late Cretaceous interval (Aptian through Cenomanian), 388–394; during the Late Senonian Interval (Turonian through Campanian), 394–397; during the Maastrichtian Interval, 397–402
Souto, P., 128
Sphaerouvum erbeni, 237
sphenodontian, 35, 393
Sphaeroolithidae, 379
spinosaurid tooth from the Candeleros Formation, 290–291
spinosaurid remains from the Santana Formation, 289–290
Spinosauridae, 154, 245, 287–294; documentation in the fossil record, 288
Spinosaurus aegyptiacus, 117, 288
Spinostropheus gautieri, 246
Spondylosoma, 34, 35
Spondylosoma absconditum, 34, 35
Staurikosaurus, 17, 18, 34, 36, 39, 41, 42, 44, 46–49, 49, 51, 80
Staurikosaurus pricei, 34, 36, 39, 41, 46–49, 49
Stegosauria, 345–346, 347, 349; extinction before Aptian times, 389; record in the Early Cretaceous of South America and Africa, 345; record in Patagonia, 345; spiny projections, 391
Stegosaurus, 41, 346

Stratigraphy of dinosaur-bearing formations in South America: Cretaceous, 141, 148, 151; Jurassic, 91, 94; Triassic, 24
Strenusaurus procerus, 66
Stromer, Ernst, 295, 367
subaerial connections between South and North America, 137
Sub-Andean Basin (Argentina, Bolivia, and Perú), 138, 146, 147–149
Suchomimus, 174, 292
Sucre town (Bolivia), 146, 238, 239, 351
survival of non-dinosaurian dinosauriforms into Carnian and Norian times, 81, 82
Synapsida, 3
Syntarsus, 12, 14, 54, 55, 56, 57, 118, 252
Syntarsus kayentakatae, 54, 57

Talampaya Formation (Argentina), 27
Talenkauen, 164, 345, 352, 353, 356, 357, 359, 361–363, 363, 365, 366, 386, 394
Talenkauen santacrucensis, 164, 345, 361–363, 363
Tanius, 375
Taniwhasaurus, 400
Tapia, Augusto, 343, 380
Taponichnus, 378
Taponichnus donottoi, 378
Tarapaca (Chile), 146
Tarjados Formation (Argentina), 27
Tehuelchesaurus, 88, 96, 101, 108–110, 111, 128
Tehuelchesaurus benitezii, 88, 96, 108, 109–110, 111
Telmatosaurus, 370, 375
Telosichnus saltensis, 378
temnospondyls, 29
Tendaguru Formation (Tanzania), 112, 346
Tenontosaurus, 352, 359, 363
Tetanurae, 54, 55, 115–119, 245, 252, 286–338; phylogenetic relationships, 288
Tethys Sea, 25, 87, 129, 135
Tetrapodosaurus, 382
Teyuwasu, 41
Teyuwasu barberenai, 41
Thecodontosaurus, 59, 61, 63, 67

Therioherpeton cargnini, 34
Therizinosauroidea, 305, 319
Theropoda: Cenomanian-Turonian faunal turnorver, 393; Cretaceous record in South America, 242, 243–343; evolution on the southern continents during the Cretaceous, 388, 390–393, 395–397; geographic distribution of carcharodontosaurids and spinosaurids, 297–298, 385, 391, 392; Jurassic record in South America, 88, 114–123; phylogenetic relationships among basal Theropoda, 55, 115; spiny projections, 392; Triassic, 54–57; vicariance model, 396
Thescelosaurus, 352, 363, 365
Thulborn, T., 127, 351
Thyreophora, 71, 345, 346, 348, 349
Timimus, 324
Tiourarén Formation (Niger), 246
Titanosauria, 101, 154, 168, 181–236; abundance in the Upper Cretaceous, 167, 181, 182, 394; Cretaceous paleobiogeography, 182; dermal ossifications, 196–197, 198; disparity in neck anatomy, 208; diversity in the Santonian-Maastrichtian transition, 397; ecological relation with angiosperms, 181, 394; facultative bipedalism in titanosaurs, 196; eggs, nests, and embyos, 236–238; girdles and limbs, 195–196; hyposphene-hypanthrum articulations in titanosaurs, 193, 197, 202–204, 207, 218, 220, 222; long-necked and short-necked titanosaurs, 192; morphotypes of titanosaurs in the Late Cretaceous, 399; phylogenetic relationships, 185, 185–189; skull, 189–191; tooth crowns, 191; vertebral column and thorax, 191–194; wide-gauge trackmakers, 181, 195, 240
Titanosaurus nanus, 167, 182
Tolar Formation (Chile), 146, 147
Tomlinson, A., 119
Tordillo Formation (Argentina), 99, 107
Toro Toro Formation (Bolivia), 145, 239, 240, 351, 351

Toro Toro town (Bolivia), 145, 239, 351
Torvosaurus, 16, 290, 291, 293
El Tranquilo Basin (Argentina), 26, 33, 37, 38, 40, 61, 83
Traversondontid Zone, 34
Trialestes romeri, 30, 37
Triangulo Mineiro (Brazil), 142
Triassic, xix, xx, 3, 12, 15, 24, 25–84; dinosaur-bearing beds in South America, 26–35; dinosaurs from Argentina, 27–33; dinosaurs from Brazil, 33–35; paleogeographic map of South America, 26; sedimentary accumulations in South America, 25, 26
Triceratops, 352, 382
Trigonosaurus, 143, 183, 205, 207, 209, 219, 222–224, 223, 224, 226, 232
Trigonosaurus pricei, 143, 222–224, 223, 224, 226
Troodon, 333
Troodontidae, 264, 305, 320, 321, 325, 330
Tropidosuchus romerii, 28
Tsintaosaurus, 375, 377
tyrannosaurids, 10, 244, 255, 297, 304, 305, 309, 311, 325, 330, 354, 388, 393, 396
Tyrannosaurus, 4, 260, 295, 327
Tyrannotitan, 161, 255, 293, 295, 297, 297, 298, 298, 301, 302
Tyrannotitan chubutensis, 161, 255, 295, 297, 298, 298

Unaysaurus, 35, 39, 57, 59–62, 63, 79, 80
Unaysaurus tolentinoi, 35, 39, 57, 59–61, 63, 79
uncinate processes, 361, 363, 364
Unenlagia, 155, 201, 245, 304, 322–330, 327, 332, 334
Unenlagia comahuensis, 155, 245, 304, 322, 325–329, 327
Unenlagia paynemili, 155, 245, 304, 322, 323, 329–331, 332
Unenlagiinae, 324–334
Universidad Nacional de San Juan (San Juan, Argentina), 39, 40
University of Chicago (Chicago, United States), 39, 40
University of Tübingen (Tübingen, Germany), 35

Unnamed basal titanosaur from Bajo Barreal Formation, 211, 211
Unnamed dromaeosaurid from Portezuelo Formation, 323–324
Unquillosaurus, 147, 244, 304, 322–323, 324, 334, 388, 397
Unquillosaurus ceibalii, 147, 244, 304, 322–323, 324, 397
Upchurch, Paul, 57, 59, 70, 81, 103, 106, 185, 188, 189
urolite, 128
Uruguaysuchus, 144, 369
Utahraptor ostrommaysi, 333

Vacca, R., 109
Valdosaurus, 352
Valley of the Moon, 29. *See also* Ischigualasto Formation
Vegavis, 338
Velociraptor mongoliensis, 321
Velocisaurus, 156, 244, 249, 250, 251, 264, 266, 271–272, 273, 286, 395
Velocisaurus unicus, 156, 249, 250, 251, 271–272, 273
Venaticosuchus rusconii, 30, 82
Vera, E., 317
Veralli, C., 293
Vickers-Rich, Patricia, 109,
Vilquechico Formation (Perú), 145, 146, 236, 378
Vince, Martín, 63, 65, 105
Viñita Formation (Chile), 146
Volkheimer, W., 130
Volkheimeria, 88, 95, 101, 103, 104, 128, 129
Volkheimeria chubutensis, 88, 95, 103, 104

Wadi Milk Formation (Sudan), 322
Walker, Cyril, 335
Weishampel, D., 72, 73, 374, 375, 377
Wichmann, Ricardo, 157, 167, 213
Wildeichnus, 88, 99, 121, 125, 129
Wildeichnus navesi, 88, 99, 121, 125
Wilson, J., 108, 185, 188, 222
Witmer, L., 72, 73

Xenotarsosaurus, 162, 251, 264, 272, 284–286
Xenotarsosaurus bonapartei, 162, 251, 264, 285–286

Yates, A., 59
Yixian Formation (China), 304
Yungavolucris brevipedalis, 149
Yunnanosaurus, 67

Zapalasaurus, 152, 169–171
Zapalasaurus bonapartei, 152, 169–171

Zupaysaurus, 31, 49, 54, 55–57, 57
Zupaysaurus rougieri, 31, 54, 55–57, 57

FERNANDO E. NOVAS is a researcher of the National Council of Scientific and Technical Research (CONICET) at Museo Argentino de Ciencias Naturales (Buenos Aires). He is the author of many scientific papers on dinosaurs and he was also a contributor to the *Encyclopedia of Dinosaurs*.

The Age of Dinosaurs in South America was designed by Jamison Cockerham at Indiana University Press, set in type by Jamie McKee at MacKey Composition, and printed by Sheridan Books, Inc. June Silay was the project editor and Robert Sloan was the sponsoring editor.

The text type is Electra, designed by William A. Dwiggins (circa 1935) and the headings and captions are set in Frutiger, designed by Adrian Frutiger (circa 1975), both issued by Adobe Systems. The display type is Museo, designed by Jos Buivenga (in 2008), issued by exljbris.